Solid Mechanics and Its Applications

Volume 201

Series Editor

G. M. L. Gladwell
Department of Civil Engineering, University of Waterloo, Waterloo, Canada

For further volumes:
http://www.springer.com/series/6557

Aims and Scope of the Series

The fundamental questions arising in mechanics are: *Why?*, *How?*, and *How much?* The aim of this series is to provide lucid accounts written by authoritative researchers giving vision and insight in answering these questions on the subject of mechanics as it relates to solids.

The scope of the series covers the entire spectrum of solid mechanics. Thus it includes the foundation of mechanics; variational formulations; computational mechanics; statics, kinematics and dynamics of rigid and elastic bodies: vibrations of solids and structures; dynamical systems and chaos; the theories of elasticity, plasticity and viscoelasticity; composite materials; rods, beams, shells and membranes; structural control and stability; soils, rocks and geomechanics; fracture; tribology; experimental mechanics; biomechanics and machine design.

The median level of presentation is the first year graduate student. Some texts are monographs defining the current state of the field; others are accessible to final year undergraduates; but essentially the emphasis is on readability and clarity.

Meinhard Kuna

Finite Elements in Fracture Mechanics

Theory—Numerics—Applications

 Springer

Meinhard Kuna
Institute für Mechanik und Fluiddynamik
TU Bergakademie Freiberg
Freiberg
Germany

ISSN 0925-0042
ISBN 978-94-007-9873-1   ISBN 978-94-007-6680-8 (eBook)
DOI 10.1007/978-94-007-6680-8
Springer Dordrecht Heidelberg New York London

Originally published in the German language by Vieweg+Teubner, 65189 Wiesbaden, Germany, as "Kuna, Meinhard; Numerische Beanspruchungsanalyse von Rissen: Finite Elemente in der Bruchmechanik"
© Viegweg+Teubner
Springer Fachmedien Wiesbaden GmbH 2010
Springer Fachmedien is part of Springer Science+Business Media
Partly translated by Aaron Kuchle

Printed on acid-free paper

Springer is part of Springer Science+Business Media (www.springer.com)

Preface

Prevention and assessment of fracture and damage processes play an essential role in the development and dimensioning of engineering constructions, components, and facilities in order to ensure their technical safety, durability, and reliability. In the case of failure, mistakes made by engineers in this respect can have catastrophic consequences for the lives of people, the environment, and even for the economy. In many engineering components and materials, defects may exist resulting from manufacturing or operation, which cannot always be avoided. Therefore, the fracture mechanical assessment of crack-like defects is of great importance. In the context of technical surveillance and studies of causes of failure cases, besides materials characterization the analysis of the mechanical loading situation at cracks, notches, and similar defects under in-service conditions is of particular interest.

In the past 50 years, *fracture mechanics* has been developed into an independent interdisciplinary scientific field, which resides between engineering mechanics, materials science, and solid-state physics. Fracture mechanics defines load parameters and criteria in order to quantitatively assess crack behavior in materials and components under static, dynamic, or cyclic loading.

Additionally, numerical methods of applied mechanics are used nowadays for fracture-mechanical stress analysis. The finite element method (FEM) has been established in many areas of engineering as a universal and efficient tool of modern engineering design and stress analysis. Numerous software packages are available, which offer not only standard methods of structural mechanics but also fracture-mechanical options of more recent invention. However, the treatment of crack issues requires particular theoretical precognition and numerical algorithms, much of which has not yet been integrated into the engineer's education and practice to the necessary extent, but has been available mostly to »fracture-mechanical experts« only.

The intention of the present monograph consists in closing this gap. In the introduction, we present the essential theoretical basics of fracture mechanics, whose parameters are to be determined using the FEM. The main part of the book is focused on specific numerical techniques to analyze plane and spatial crack problems in elastic and plastic materials under all technically relevant loads. Finally, worked samples for the solutions of practical problems will be provided for each area.

This textbook is addressed to graduate students of engineering study courses, especially those in mechanical engineering, civil engineering, vehicle design, materials science, aerospace industry, or computational engineering. It shall provide an introduction into this area of expertise to graduates and postgraduates of these fields and support in their own research activities. Moreover, I consider as a target audience engineers in design and computation departments of many industrial branches and officials in technical controlling institutions, who are confronted with issues of dimensioning, assessment, and supervision of strength and durability of engineering constructions. Furthermore, this textbook should build a bridge for materials scientists and materials engineers to theoretical fracture mechanics in order to use numerical techniques for materials modeling or to analyze materials and components tests using computations. This textbook requires a basic knowledge of continuum mechanics, strength of materials, material theory, and the finite-element method. The essential basics of mechanics of materials are reviewed for convenience in the Appendix.

I was gratified at the positive response by which the scientific community in Germany has appreciated the first edition of this book entitled "Numerische Beanspruchungsanalyse von Rissen—Finite-Elemente in der Bruchmechanik", edited by Vieweg-Teubner publisher in 2008. Many readers confirmed to me personally that the book was a useful help to understand fracture mechanics concepts and a real assistance in performing their own numerical computations. Meanwhile, the second improved edition appeared in 2010. Therefore, I feel encouraged to offer this book to a wider audience in the English language.

Many persons contributed to the preparation of the book. First of all, my very sincere thanks go to Ms. M. Beer for making all the excellent drawings. Numerous numerical examples were elaborated during the pleasant joint work with my former and current Ph.D.-students or co-workers at the institute. In particular, I would like to express my thanks to Dr. M. Abendroth, Dr. M. Enderlein, Dr. E. Kullig, Th. Leibelt, C. Ludwig, Dr. U. Mühlich, Dr. F. Rabold, Dr. B. N. Rao, Prof. Dr. A. Ricoeur, Dr. A. Rusakov, L. Sommer, and L. Zybell.

I appreciate the fruitful cooperation lasting for many years with my colleagues Prof. Dr. G. Pusch (TU Freiberg) and Prof. Dr. P. Hübner (University Mittweida), from where the reported engineering applications of fracture assessment have emerged. Also, I am indebted to Prof. Dr. M. Fulland (University Zittau) and Dr. I. Scheider (GKSS Geesthacht), who provided me kindly graphical material for additional examples. My thanks go to Prof. Dr. W. Brock for reviewing the German manuscript and giving constructive comments on the scientific presentation of the subject.

In the course of translating and revising the English manuscript, I appreciate the great assistance by A. Kuchle (Chaps. 4–7) E. Beschler, and J. Bergemann.

Finally, special thanks should be expressed to Springer Science Media for the favor to publish this book. Sincere gratitude is due to Ms. N. Jacobs, Publishing editor, and Ms. C. Feenstra for their cooperation and assistance in printing the book in an excellent form.

Last but not least, I cordially want to thank my wife, Christine Kuna, for her great understanding and infinite patience.

Freiberg, December 2012   Meinhard Kuna

Contents

Glossary

α	Hardening coefficient
α_{cf}	Plastic constraint factor
α_d	Dilatation wave ratio
α_{ij}	Anisotropic elastic constants
α_t	Coefficient of thermal expansion
$\boldsymbol{\alpha}^t$, α_{ij}^t	Coefficients of thermal expansion tensor
α_s	Shear wave ratio
β_T	Biaxial parameter
β_{ij}	Thermal stress coefficients
$\boldsymbol{\beta}$, $\boldsymbol{\beta}_B$	Internal hybrid Ansatz coefficients
Γ	Integration path
Γ^+, Γ^-	Upper, lower crack faces
Γ_ε	Crack tip integration path
γ	Sliding
γ	Material constant
γ	Specific surface energy
γ_I	Principal shear strains
γ_{II}	Principal shear strains
γ_{III}	Principal shear strains
γ_d	Aspect ratio dilatation waves
γ_D	Dynamic surface energy
γ_s	Aspect ratio shear waves
γ_t	Crack opening angle
$\Delta\sigma$	Cyclic stress range
ΔK	Cyclic stress intensity factor
ΔK_{eff}	Effective cyclic stress intensity
ΔK_{th}	Threshold value fatigue
δ	Variational symbol
δ	Separation (cohesive zone model)
δ_c	Decohesion length
δ_n	Separation (normal)

δ_s	Separation (transversal)
δ_t	Separation (tangential)
δ_t	Crack opening displacement CTOD
δ_T	Total shear separation
$\boldsymbol{\delta} = [\boldsymbol{u}]$	Separation vector (cohesive model)
ϵ	Dielectric constant
ϵ	Bimaterial constant
ϵ_{ijk}	Permutation tensor LEVI − CEVITA
ε_0	Reference strain ($\approx \sigma_{F/E}$)
ε_I	Principal strain
ε_{II}	Principal strain
ε_{III}	Principal strain
ε^H	Dilatational strain tensor
ε_v^p	Equivalent plastic strain
ε_M^p	Equivalent plastic matrix strain
$\boldsymbol{\varepsilon}, \varepsilon_{ij}$	Strain tensor
$\boldsymbol{\varepsilon}^D, \varepsilon_{ij}^D$	Strain deviator
$\boldsymbol{\varepsilon}^e, \varepsilon_{ij}^e$	Elastic strains
$\boldsymbol{\varepsilon}^p, \varepsilon_{ij}^p$	Plastic strains
$\boldsymbol{\varepsilon}^t, \varepsilon_{ij}^t$	Thermal strains
$\boldsymbol{\varepsilon}$	Strain matrix
$\boldsymbol{\varepsilon}^e$	Elastic strain matrix
$\boldsymbol{\varepsilon}^p$	Plastic strain matrix
$\boldsymbol{\varepsilon}^*$	Initial strain matrix
ζ	Complex variable
η	Shear-tension ratio (cohesive model)
η	Error indicator FEM global
$\eta(x_1)$	Gradient function
$\eta(a/w)$	Geometry function J_p-integral
η_e	Error indicator element e
$\boldsymbol{\eta}, \eta_{mn}$	EULER-ALMANSI strain tensor
θ	Polar coordinate, angle
θ_c	Crack propagation angle
θ_d	Angle for dilatation waves
θ_s	Angle for shear waves
ϑ	Heat transition coefficient
κ	Elastic constant
$\varkappa$	Node distortion parameter
$\varkappa$	Crack tip position
$\varkappa$	Dynamic exaggeration factor
$\dot{\Lambda}$	Plastic LAGRANGE multiplier
λ	Exponent of complex stress function
λ	LAME's elasticity constant

μ	Shear modulus
μ	Shear shape factor
ν	POISSON's number
ξ, ξ_i	Natural element coordinates
ξ^g, ξ_i^g	Coordinates of integration points
Π_C	Principle of complementary energy
Π_{CH}	Hybrid stress principle
Π_{GH}	Hybrid mixed principle
Π_{MH*}	Simplified hybrid mixed principle
Π_P	Principle of potential energy
Π_{PH}	Hybrid displacement principle
Π_R	HELLINGER-REISSNER principle
Π_{ext}	Potential of external loads
Π_{int}	Internal mechanical potential
$\widehat{\Pi}_{ext}$	Complementary external potential
$\widehat{\Pi}_{int}$	Complementary internal potential
ρ	Notch radius
ρ	Density (current configuration)
ρ_0	Density (reference configuration)
ϱ	Mixed-mode-ratio
σ	Normal stress (cohesive zone model)
σ_c	Cohesive strength tension
σ_0	Reference stress ($\approx \sigma_F$)
σ_F	Yield stress
σ_{F0}	Initial yield stress
σ^H	Dilatational stress tensor
σ_I	Principal normal stress
σ_{II}	Principal normal stress
σ_{III}	Principal normal stress
σ_c	Critical stress
σ_M	Matrix yield stress
σ_{max}	Maximum stress
σ_{min}	Minimum stress
σ_n	Nominal tensile stress
σ_v	V. MISES equivalent stress
$\boldsymbol{\sigma}$, σ_{ij}	CAUCHY's stress tensor
$\boldsymbol{\sigma}^D$, σ_{ij}^D	Stress deviator
$\boldsymbol{\sigma}$	CAUCHY-stress matrix
τ	Shear stress
τ_c	Cohesive strength shear
τ_t	Shear stress tangential
τ_s	Shear stress transversal
τ_F	Shear yield stress
τ_{F0}	Initial shear yield stress

τ_{I}	Principal shear stress
τ_{II}	Principal shear stress
τ_{III}	Principal shear stress
τ_{ij}	Shear stress components
τ_{n}	Nominal shear stress
Φ	Yield condition, dissipation function
φ	Electric potential
φ	Angular coordinate for elliptical cracks
φ	Scalar wave potential
$\phi(z)$	Complex stress function
$\chi(z)$	Complex stress function
χ	Crack opening function
ψ	Phase angle
ψ_{e}	Elastic potential
$\boldsymbol{\psi}, \psi_i$	Vectorial wave potential
$\Omega(z)$	Complex stress function
Ω	Integration domain J-integral
$\overline{\Omega}$	Integration domain J-integral
ω	Damage variable
$\bar{\omega}$	Surface charge density
A	Complex stress coefficient
A	Crack face
A	Surface (reference configuration)
A_{I}	Factors energy release rate
A_{II}	Factors energy release rate
A_{III}	Factors energy release rate
A_{σ}	Stress coefficient
A_{B}	Fracture process zone
A_i	Coefficients of eigenfunctions
$\mathbf{A}^{(e)}$	Correlation matrix
a	Surface (current configuration)
a	Crack length
a	Semi-axis of elliptical cracks
$\dot{a}$	Crack velocity
$\ddot{a}$	Crack acceleration
a_0	Initial crack length
a_c	Critical crack length
a_{eff}	Effective crack length
a_i	Coefficients of eigenfunctions
a_{th}	Crack length from threshold value
$\boldsymbol{a}, a_i$	Acceleration vector
B	Specimen thickness
B	Complex stress coefficient
B_{I}	BUECKNER-singularity

$\mathbf{B}$	Strain-displacement-matrix
$\overline{\mathbf{B}}$	Non-linear strain-displacement matrix
$\widetilde{\mathbf{B}}$	Hybrid element matrix
b	Ligament length
b_i	Coefficient of eigenfunctions
b_T	Biaxial loading parameter
$\boldsymbol{b}, b_{mn}$	Left CAUCHY-GREEN deformation-tensor
$\bar{\boldsymbol{b}}, \bar{b}_i$	Body force vector
C	Closed Integration path
C	Complex stress coefficient
C	PARIS-coefficient
$\boldsymbol{C}, C_{MN}$	Right CAUCHY-GREEN deformation tensor
$\mathbf{C}$	Material matrix
$\mathbf{C}^{e}$	Elasticity matrix
$\mathbf{C}^{ep}$	Elastic-plastic material matrix
$C_{\alpha\beta}$	Elasticity matrix
$\mathbb{C}, C_{ijkl}$	Elasticity tensor 4th order
c	Larger semi-axis of elliptical cracks
c_d	Dilatational wave velocity
c_i	Coefficients eigenfunctions
c_R	RAYLEIGH's wave velocity
c_s	Shear wave velocity
c_v	Specific heat capacity
$\mathcal{D}$	Dissipation energy
D	Plate stiffness
$D(\dot{a})$	RAYLEIGH's function
$\boldsymbol{D}, D_i$	Electric flux density
$\mathbf{D}$	Differentiation matrix
d_p	Extension of plastic zone
dA	Area element
dS	Surface element
dV	Volume element
ds	Line element
$\boldsymbol{d}, d_{ij}$	Deformation velocity tensor
E	Elasticity modulus
$E(k)$	Elliptical integral 2. kind
$\boldsymbol{E}, E_{MN}$	GREEN-LANGRANGE strain tensor
$\boldsymbol{E}, E_i$	Electric field strength
$\mathbf{E}$	GREEN-LAGRANGE strain matrix
$\boldsymbol{e}_i$	Basis vectors
e	EULER number e ≈ 2.718
F	Single force
$F(\boldsymbol{x})$	AIRY's stress function
F_L	Plastic limit load (collapse load)

$\tilde{F}_i^{(n)}$	Eigenfunctions mode I
$\boldsymbol{F}, F_{mM}$	Deformation gradient
$\mathbf{F}$	System load vector
$\mathcal{F}$	Flux integral
f	Void volume fraction
f^*	Modified void volume fraction
f_0	Initial void volume fraction
f_c	Critical void volume fraction
f_f	Void volume fraction at failure
f_N	Void density for nucleation
f_{ij}^L	Angular functions crack tip fields $(L = \mathrm{I}, \mathrm{II}, \mathrm{III})$
$\mathbf{f}$	Element load vector
G	Energy release rate
G_{I}	Energy release rate für crack mode I, II, III
G_{II}	Energy release rate für crack mode I, II, III
G_{III}	Energy release rate für crack mode I, II, III
G_{c}	Critical energy release rate
G^{dyn}	Dynamic energy release rate
$G_i^{\mathrm{I}}, G_i^{\mathrm{II}}$	Fracture mechanical weight functions
$\tilde{G}_i^{(n)}$	Eigenfunctions mode II
$\mathbf{G}$	Hybrid element matrix
$g(a, w)$	Geometry function for K-factors
g_i^L	Angular functions crack tip fields $(L = \mathrm{I}, \mathrm{II}, \mathrm{III})$
$\boldsymbol{g}, g_i$	Temperature gradient
H	Height crack element
$H(\tau)$	HEAVISIDE's jump function
H_α	Hardening function
$H_i^{\mathrm{I}}, H_i^{\mathrm{II}}$	Fracture mechanical weight functions
H_{ij}	IRWIN-matrix anisotropy
$\tilde{H}_3^{(n)}$	Eigenfunctions Modus III
$\mathbf{H}$	Hybrid element matrix
$\mathbf{H}$	Matrix hardening function
$\overline{\mathbf{H}}$	Displacement gradient matrix
h	Thickness (plates, sheets)
$\hbar$	Stress triaxiality number
h_α	Hardening variable
$\boldsymbol{h}, h_i$	Heat flux vector
$\mathbf{h}$	Matrix hardening variables
I_1^A, I_2^A, I_3^A	Invariants of tensors A
$\boldsymbol{I}, \delta_{ij}$	KRONECKER's symbol, unity tensor
I_p, I_{pi}	Momentum vector
$\mathrm{i} = \sqrt{-1}$	Imaginary unit number
J	J-integral

J	Determinant of deformation gradient det $|F|$
J^*	Dynamic J-integral (stationary crack)
J^{dyn}	Dynamic J-integral (moving crack)
$\hat{J}$	3D disk-shaped integral
J_{Ic}	Critical material parameter
$J_{\text{R}}(\Delta a)$	Crack resistance curve (EPFM)
J_{e}	Elastic part of J
J_{p}	Plastic part of J
J, J_k	J-integral vector
$\tilde{J}, \tilde{J}_k$	Elastic-plastic J-integral
$J^{\text{te}}, J_k^{\text{te}}$	Thermoelastic J-integral
$\mathbf{J}$	JACOBI's functional matrix
$\mathcal{K}$	Kinetic energy
K	Compression modulus
K_D	Dielectric displacement intensity factor
K_{I}	Stress intensity factors
K_{II}	Stress intensity factors
K_{III}	Stress intensity factors
K_{I}^{d}	Dynamic stress intensity factor
$K_{\text{Ic}}, K_{\text{IIc}}$	Static fracture toughness
K_{ID}	Dynamic fracture toughness (moving crack)
K_{Id}	Dynamic fracture toughness (stationary crack)
K_{Ia}	Crack arrest toughness
K_{max}	Maximum stress intensity
K_{min}	Minimum stress intensity
K_{op}	Crack opening intensity factor
K_{v}	Equivalent stress intensity factor
K_1, K_2	Stress intensity factors for interface crack
$\widetilde{K}$	Complex stress intensity factor
$\mathbf{K}$	System stiffness matrix
k	Specimen stiffness
k	Heat conduction coefficient
k_1, k_2	Stress intensity factors for plates
$\mathbf{k}$	Element stiffness matrix
L	Length crack element
$\mathcal{L}(u, \dot{u}, t)$	LAGRANGE's function
$\tilde{L}_{ij}^{(n)}$	Eigenfunctions Mode III
$\mathbf{L}$	Hybrid displacement matrix
l, l_{ij}	Velocity gradient
dL	Line element length (reference configuration)
dl	Line element length (current configuration)
Δl_k	Virtual displacement of crack front
$\tilde{M}_{ij}^{(n)}$	Eigenfunctions Mode I

$\mathbf{M}$	System mass matrix
m	PARIS-exponent
m_i	Bending moments (plate theory)
$\mathbf{m}$	Element mass matrix
N	Number of load cycles
N_K	Number of all nodes FEM-system
$N_a(\xi_i)$	Shape functions
N_B	Load cycles until fracture
$\tilde{N}_{ij}^{(n)}$	Eigenfunctions Mode II
$\widehat{\mathbf{N}}, \widehat{N}_{ij}$	Normal direction in stress space
$\mathbf{N}$	Matrix of shape functions
n_D	Number of dimensions
n_E	Number of finite elements
n_G	Number of GAUSS-points
n_H	Number of hardening variables
n_K	Number of nodes per element
n_L	Number of individual forces
n_f	Number of rigid body degrees of freedom
$\boldsymbol{n}, n_i$	Normal vector
P	Global load parameter
P	Crack face forces
$\boldsymbol{P}, P_k$	Generalized configurational force
$\boldsymbol{P}, P_{Mn}$	1st PIOLA-KIRCHHOFF stress tensor
$\mathbf{P}$	Hybrid stress matrix
$p(x_1)$	Crack face loads
$p(\boldsymbol{x})$	Surface loads (plate theory)
$\boldsymbol{p}, p_i$	Material body force vector
$\mathcal{Q}$	Thermal energy
Q	Constraint factor (EPFM)
Q	Crack face forces
$\boldsymbol{Q}, Q_{ij}$	Energy-momentum-tensor
q	Displacement of load point
q	Weighting function 2D
$q(x_1)$	Crack face loads
q_1, q_2, q_3	Parameters GURSON-Model
q_i	Transversal forces (plate theory)
q_k	Weighting function 3D
R	Stress ratio $K_{\min}/K_{\max}$
$R(\varepsilon^{\mathrm{p}})$	Isotropic hardening variable
R_m	Ultimate tensile strength
$R_{\mathrm{p0,2}}$	Yield strength
$R(\Delta a)$	Crack growth resistance curve (LEFM)
$\boldsymbol{R}, R_{nM}$	Rotation tensor
$\mathbf{R}$	Hybrid boundary stress matrix

$\mathbf{R}$	Residual vector
r	Polar coordinate, radius
r_B	Size of fracture process zone
r_F	Radius plastic Zone
r_J	Dominance radius J-field
r_K	Dominance radius K-field
r_p	Size of plastic zone
r_d	Radius of dilatation waves
r_s	Radius of shear waves
$\mathbf{r}$, r_{ij}	Transformation matrix for rotation
S	Surface
$\widetilde{S}$	Interelement boundary
$S(\theta)$	Energy density factor
S^+, S^-	Upper, lower crack surface
S_ε	Surface crack tube
S_{end}	End faces
S_t	Boundary with given $\bar{t}$
S_u	Boundary with given $\bar{u}$
$\boldsymbol{S}$, S_k	Sectional force (reference configuration)
$\mathbb{S}$, $\mathbb{S}_{ijkl}$	Elastic compliance tensor
$\mathbf{S}$	Elastic compliance matrix
s	Arc length
$\boldsymbol{s}$, s_k	Sectional force (current configuration)
T	Temperature field
T_{ij}	Stress components 2nd order
T_k^*	Generalized energy integral
$\boldsymbol{T}$, T_{MN}	2nd PIOLA-KIRCHHOFF stress tensor
$\mathbf{T}$	2nd PIOLA-KIRCHHOFF stress matrix
$\boldsymbol{t}$, t_i	Sectional traction vector
$\bar{\boldsymbol{t}}$, $\bar{t}_i$	Traction vector
$\bar{\mathbf{t}}$	Boundary stress matrix
$\boldsymbol{t}$	Traction vector (cohesive zone model)
t^c, t_i^c	Crack face tractions
U	Strain energy density
U	Crack opening factor
$\hat{U}$	Complementary strain energy density
$\check{U}$	Specific stress work density
U^e	Elastic strain energy density
U^p	Plastic stress work density
$\check{U}^{te}$	Thermoelastic strain energy density
$\boldsymbol{U}$, U_{MN}	Right stretch tensor
$\boldsymbol{u}$, u_i	Displacement vector
$\tilde{\boldsymbol{u}}$, $\tilde{u}_i$	Element boundary displacements
$\bar{\boldsymbol{u}}$, $\bar{u}_i$	Boundary displacement vector

u	Displacement matrix
V	Notch opening displacement COD
V	Volume (reference configuration)
$\boldsymbol{V}$, V_{mn}	Left stretch tensor
V	System nodal displacements
v	Volume (current configuration)
$\boldsymbol{v}$, v_i	Velocity vector
v	Nodal displacements vector
$\mathcal{W}_{\text{ext}}$	External mechanical work
$\mathcal{W}_{\text{int}}$	Internal mechanical work
$\widehat{\mathcal{W}}_{\text{ext}}$	Complementary external work
$\widehat{\mathcal{W}}_{\text{int}}$	Complementary internal work
$\mathcal{W}_{\text{c}}$	Work for crack opening
$\mathcal{W}_B$	Work in fracture process zone
w	Specimen width
$w(\boldsymbol{x})$	Deflection (plate theory)
w_g	Weighted integration rules
$\boldsymbol{w}$, w_{ij}	Spin tensor
X, X_{M}	Coordinates (material)
$\boldsymbol{X}$, X_{ij}	Kinematic hardening variable
$\boldsymbol{x}$, x_{m}	Coordinates (spatial)
x	Element coordinate matrix
$\hat{\mathbf{x}}$	Nodal coordinate matrix
z	Complex variable
z	Arbitrary FEM output quantity
1D	One-dimensional
2D	Two-dimensional
3D	Three-dimensional
ASTM	American Society for Testing of Materials
BVP	Boundary Value Problem
CT	Compact Tension Specimen
CTE	Crack Tip Element
CTOA	Crack Tip Opening Angle
CTOD	Crack Tip Opening Displacement
DIM	Displacement Interpretation Method
EDI	Equivalent Domain Integral
EPFM	Elastic-Plastic fracture mechanics
ESIS	European Structural Integrity Society
FAD	Failure Assessment Diagram
FEM	Finite Element Method
IBVP	Initial Boundary Value Problem
LEFM	Linear-Elastic Fracture Mechanics
LSY	Large-Scale Yielding
MCCI	Modified Crack Closure Integral

ODE	Ordinary Differential Equation
PC	Plastic Collapse
PDE	Partial Differential Equation
QPE	Quarter-Point Elements
RSE	Regular Standard Elements
SENB	Single Edge Notched Bending Specimen
SINTAP	Structural Integrity Assessment procedure
SSY	Small-Scale Yielding
SZH	Stretched Zone Height
VCE	Virtual Crack Extension

Chapter 1
Introduction

1.1 Fracture Phenomena in Nature and Engineering

The term »fracture« describes the local detachment of material cohesion in a solid body. It concerns a process that either partially disrupts the body which leads to the development of incipient cracks or entirely destroys it. The actual fracture process occurs locally by means of elementary failure mechanisms on a microscopic level of the materials and is determined by its physical and micro-structural properties as the example on Fig. 1.1 shows. The global form of appearance of the fracture on a macroscopic level consists in the formation and propagation of one or multiple cracks in the body, whereby complete mechanical failure is finally induced. On this level, fracture processes can be effectively described using methods of solid mechanics and mechanics of materials. Fracture processes in nature and engineering are sufficiently known to everyone. Very impressive are cracks and fractures of natural materials such as stone and ice, especially if they appear in great geological formations as rock failures, crevasses and earthquakes, see Fig. 1.2.

Engineering products and developments of mankind were and still are especially confronted with issues of safety and durability and have always posed a challenge to engineering. Spontaneous fracture is the most dangerous type of failure of a mechanically stressed construction! Nowadays boldly conceived buildings made of concrete and steel, reliable airplanes and high speed trains, crash-tested cars and strength-optimized high-tech materials document the technical progress in those areas. On the contrary, a considerable number of engineering failure cases testifies to the painful experiences on the way there. Examples are cracking in buildings and engine parts, the entire collapse of bridges, the burst of vessels and the breakup of vehicle components (see Figs. 1.3 and 1.4).

In most cases the reasons are undetected defects in material or components, insufficient dimensioning of the construction compared to the actual load, or the application of materials with deficient strength. In the modern Industrial Age the guarantee of safety, durability and reliability of technical constructions, components and facilities holds great importance. Engineering mistakes in this area can have catastrophic

M. Kuna, *Finite Elements in Fracture Mechanics*, Solid Mechanics and Its
Applications 201, DOI: 10.1007/978-94-007-6680-8_1,
© Springer Science+Business Media Dordrecht 2013

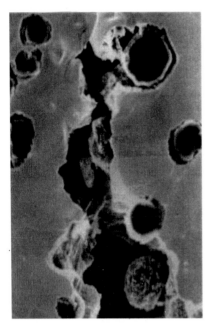

Fig. 1.1 Micro-crack in the structure of ductile cast iron

Fig. 1.2 Macro-crack (crevasse) in the Fründel glacier, Switzerland

Fig. 1.3 ICE railway accident near Eschede in 1998 as a result of a broken wheel rim

consequences for the life of people and the environment as well as for the economy and availability of certain products in case of failure. Therefore, scientific concepts for the assessment and prevention of fracture and damage processes play a decisive role.

Fig. 1.4 Bridge collapse during an earthquake at Northridge in 1994, USA

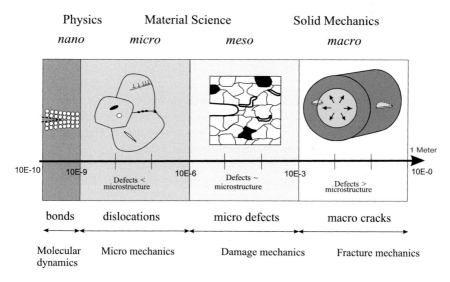

Fig. 1.5 Fracture processes at different scales and levels

Fracture and failure processes appear on all length scales. While the engineer mainly prefers the mentioned macroscopic scale, the materials scientist is interested in mesoscopic processes that take place in the material or the underlying microscopic phenomena. Solid-state physicists are preferentially interested in the nanoscopic structures of atomic bonds. All drafted scopes in Fig. 1.5 contribute to understanding of the strength properties of materials and their fracture behavior. They can be easily classified by the ratio of the defects size to the structures dimension. Nowadays models are developed on each scale using methods such as molecular dynamics, micromechanics, damage mechanics and fracture mechanics. Moreover they are linked by means of scale-bridging techniques. The numerical simulation of cracks and defects is an essential tool on all levels of modeling.

Many different areas in engineering are concerned with evaluating the fracture strength and durability of constructions. In order to contribute to a better understanding and a clarification of the terms, an introductory classification will be given in the following:

The classical *theory of strength* acts on the assumption of deformable bodies of given geometry (G), which are free of any defects and pose an ideal continuum. Using the computational methods of applied mechanics, the strains and stresses inside the component are determined as a result of the external load (L) by assuming a specific deformation law (elasticity, plasticity, etc.) of the material (M). Based on these results, failure hypotheses are formulated and strength parameters are calculated mostly in terms of effective stress σ_v, which characterizes the stress state in each material point. Using tests on simple samples with elementary loading conditions (e. g. tension test), critical values σ_c of the material's strength are determined measuring the onset of failure (e. g. fracture). In order to guarantee the safety of components, the maximum occurring loads need to stay below the critical strength parameters, which is commonly expressed as strength criterion:

$$\sigma_v(G, L, M) \leq \sigma_{tol}(M) = \frac{\sigma_c}{S}.$$

The admissible load σ_{tol} is defined by the material parameter σ_c divided by a safety factor $S > 1$. It is assumed that the parameters determined by laboratory samples actually represent true (geometry independent) material properties and can therefore be transferred to the components' geometry (*transfer principle*).

The mentioned relation describes a *local* strength hypothesis applied to each material point. In contrast to this, also *global* failure criteria are known, such as e. g. the plastic limit load F_L, which quantifies the loss of loading capacity of the entire component. A local loss does not immediately have to lead to global failure. Depending on loading and geometry, the construction can withstand the propagation of damage. This behavior is described with the terms *safety reserve* and *damage tolerance*.

Depending on the temporal process, one can distinguish between static, dynamic and cyclic loading. The *service strength theory* has been established as a sub-category in the case of regular cyclic or stochastic random loading occurring frequently in practice.

However, the traditional strength theories and the therein used material parameters (yield strength, tensile strength, endurance limit, ultimate strain, impact energy) often fail in predicting and avoiding fracture processes, as the reoccurring failure cases in practice show. The reason for this is that fracture processes primarily originate from points of concentrated stresses on crack-like defects. In such cases, the classical strength criteria provide no usable quantitative correlation between loading situation, geometry and material property.

A rather modern discipline is *continuum damage mechanics — CDM*. At this point the same methods as in classical strength theory are used with the difference that when expressing the law of materials, it is assumed that the material possesses small continuously distributed defects, e. g. microcracks or micropores. But these defects are not discretely and individually treated but enter only implicitly as averaged defect density per volume in homogenized form into the material law. The defect density expresses a measure for the damage D of the component and is used as an internal variable in the material law. It may change in the course of stress until a critical threshold D_c of the damage is reached, which correlates to the creation of an incipient crack. According to that, a damage-mechanical material law describes both the deformation and the failure properties of the material in local form on every point inside the material of the structure and hence implicitly contains a *local* failure criterion in the form of:

$$D(G, L, M) \leq D_c(M).$$

Damage mechanics is therefore suitable for the modeling of micro-mechanical failure processes in a component, before a macrocrack is formed or for the modeling of the fracture process zone at the tip of a macrocrack.

1.2 Fracture Mechanics

The specific field which deals with fracture and failure processes in engineering materials and constructions is called *fracture mechanics*. In contrast to the two above-mentioned theories, in fracture mechanics it is assumed that every component and every real material inevitably possesses flaws or other defects. The reason for this is that due to manufacturing (initial cracks, pores, inhomogeneity in materials, delamination, flaws or similar) defects are present in many technical materials or that flaws can form in the course of mechanical, thermal or corrosive service loading. It is well known, that the real strength of a material is orders of magnitudes lower than the theoretically possible strength of defect-free, ideal atomic bonds. Moreover, defects (casting defects, quenching cracks, incomplete fusions in welded joints and others) can originate during the manufacturing by means of technological processes leading to cracking. Often geometric notches or abrupt material discontinuities cannot be avoided due to the constructive requirements of a component, which cause high localized stresses. Additionally, it is important to note that the methods of non-destructive testing have physical resolution limits, so that defects due to manufacturing or operating cannot always be excluded without doubt. That means, the at least hypothetical existence of flaws of this size has to be expected! Unavoidable defects of this kind can escalate to macroscopic cracks and generate the decisive cause for the initiation of fracture.

For this reason, the existence of such defects is *explicitly* assumed in fracture mechanics and modeled *as cracks* of the size a. Such a discrete crack is surrounded by defect-free material which is described by the established material laws of continuum mechanics. Using the computational methods of applied mechanics, the stress and deformation states at the crack are determined. It is clear that very high inhomogeneous states of stress and deformation develop at the tip of the crack. Such concentrations of stress however, cannot be treated by the classical strength concepts of mechanics. Therefore, appropriate fracture-mechanical parameters B need to be found, which identify the loading condition at the crack. These will then be compared to fracture-mechanical material parameters B_c, characterizing the specific material's resistance against crack propagation. For this purpose, specific fracture-mechanical material test methods have been developed, whereby simple specimens with a crack are loaded until failure. Based on this, quantitative statements can be gained about the crack behavior, e. g. under which conditions the crack propagates further or what needs to be done to avoid it. Analogously to the above-mentioned theories, a fracture-mechanical strength criterion has the form:

$$B(G, L, M, a) \leq B_c(M).$$

This conceptual approach of fracture mechanics is presented in Fig. 1.6. Compared to established strength hypotheses, the essential generalization is the introduction of an additional geometric variable, the crack length a. From this fact one can suppose that size effects will play an important role. Therefore, fracture mechanics provides a relation between the component's geometry (G), the position and size (a) of the crack-like defect, the external loading (L), the local crack load (B) and the material resistance against crack propagation B_c. Depending on which of these parameters are known and which are sought, fracture mechanics offers correlations to assess the strength, durability and reliability of components. Therewith the following questions can be answered in the subsequently mentioned phases of a technical construction or structural component.

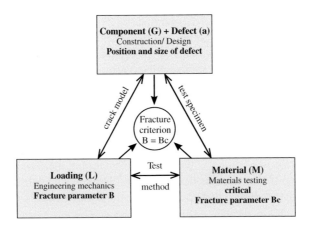

Fig. 1.6 Schematic sketch of the fracture-mechanical assessment concept

(a) During the design phase:

- How is the construction to be dimensioned and the maximal load to be specified in order to prevent inevitable defects in the material or component from growing and eventually causing a fracture?
- Which material (fracture toughness B_c) needs to be chosen so that cracks of given size do not become critical under service loading?
- How high, according to statistics, is the remaining risk of a total failure?

(b) Throughout the production process:

- How can cracks and material damages be technologically avoided?
- How can defects of unacceptable size be discovered by methods of non-destructive testing during quality control?

(c) During the operational phase:

- By how much does the loading capacity of a component decrease, if a crack of length a is discovered?
- What is the critical size of a crack a_c, which causes fracture under the given circumstances?
- How long does it take for a crack to grow from its initial size a_0 to the critical length a_c?
- How are the inspection intervals to be chosen, during which a verification of cracking and crack propagation is necessary?

(d) After a technical case of failure:

- What were the reasons of failure? Cracks overlooked during the inspection? A missing fracture-mechanical safety analysis? Non admissibly high operational loads? Wrong use of materials or negative material changes?
- Which remedial actions are necessary and possible in the future?
- How great is the percentage of loss in the product line due to fracture reliability?

In many fields of industry and technology the established classical criteria of strength suffice. In certain areas of application however, fracture-mechanical safety assessments are a necessary addition and legally required:

- Constructions and facilities with extremely high safety-related requirements for the protection of people and environment such as power plants, buildings and bridges as well as components in nuclear engineering and aerospace.
- Structural components that require great reliability and durability such as rail wheels, machine parts, automobiles, turbine blades, glow filaments or microelectronic systems.

But the scientific comprehension and mastery of fracture processes can also be beneficial as in the following fields:

- At processes and technologies in which the fracture process is volitional and performed on purpose as in mining and geotechnical engineering (blasting, tunnel
  construction, mining for raw materials) or in materials preparation and crushing
  technology (crushers, mills, recycling) in order to optimize the machines, tools
  and processes as well as minimize the energy consumption.
- During the development of new materials with outstanding strength and fracture
  toughness properties, fracture-mechanical simulations can contribute to optimiza-
  tion of the microstructural design. On the contrary, new materials require the development of specific fracture-mechanical strength hypotheses for the dimensioning
  of the construction according to the materials. Examples are high-performance
  ceramics (improvement of toughness), fiber-reinforced materials (delamination
  cracks, anisotropy), turbine blades made of monocrystalline superalloys and
  others.

Fracture mechanics has been established in the last 50 years as an independent
scientific discipline. According to the nature of fracture processes, fracture mechanics combines the knowledge and model approaches of engineering mechanics,
materials research and solid-state physics. Hence it comprises an interdisciplinary
field in whose further development mechanical engineers, continuum mechanics
researchers, materials scientists and physicists are involved. Excellent fracture-
mechanical knowledge is part of the state of the art in many fields of industry by
now. Numerous technical design rules, test specifications and national controlling
authorities ensure that this professional knowledge is implemented into practice on
behalf of engineering safety.

Fracture mechanics is divided into the following subtasks:

- Analysis of the mechanical loading condition at cracks on the basis of continuum-
  mechanical models using analytical or numerical methods of structural mechanics.
- Derivation of material specific fracture parameters and fracture-mechanical failure
  criteria for the initiation and progress of crack propagation.
- Development of test methods for the experimental determination of suitable material parameters, which characterize the resistance of a material against crack
  propagation.
- Application of the fracture-mechanical failure criteria and concepts to constructions with cracks in order to gain quantitative statements about their fracture
  resistance and remaining service life.

1.3 Computational Methods for Cracks

For all above-mentioned subtasks of fracture mechanics, computational methods
to analyze crack models are of vital importance. From the historical perspective,
developments and applications in fracture mechanics have always been closely linked
to progress in the analytical and numerical techniques of engineering mechanics and

continuum mechanics. The wish of this textbook is to make the reader acquainted with the modern numerical methods of computation currently used in fracture mechanics to analyze components with cracks. But first, a historical review will be given.

Near the turn of the 20th century, the methods of elasticity theory had mathematically matured enough so that for the first time plane problems in homogeneous, linear-elastic plates with holes or notches could be solved (Kirsch, Inglis). Groundbreaking was the development of complex stress functions by Kolosov in 1909 which were extended in the thirties by Muskhelishvili, Savin, Westergaard, Föppl and others to a powerful tool for plate analyses. The first solution of a crack in a plane originates from Inglis, 1913. It presented the foundation for the first fracture-mechanical concept of energy release rate by Griffith in the year 1921. Westergaard inserted his method of complex stress functions for crack-related problems into plates. In 1946 Sneddon managed to find the solution for circular and elliptic cracks in space with the method of integral transformation. In 1957 Williams calculated the proper series of eigenfunctions for the stress distribution around crack tips on a plane. Irwin recognized in 1957 that stress fields at all sharp crack tips feature a singularity of the same type. Hereupon he established the concept of stress intensity factors, which is being used very successfully in fracture mechanics to this day. Other semi-analytic solution techniques for plane crack problems are singular integral equations (Muskhelishvili, Erdogan and others). The methods of calculation available until that point were limited to two and three-dimensional isotropic-elastic boundary value problems for simple crack configurations, mostly in infinite domains.

Numerical solution methods of engineering mechanics (such as the finite difference method, collocation methods, Fourier-transformations) could not be effectively realized until the rapid development of electronic computer science in the 1960s. All these methods were initially used for crack problems but soon replaced by the essentially more universal and efficient *finite elements method (FEM)* . Pioneers in the field of FEM development for structural analyses were Zienkiewicz, Argyris, Wilson, Bathe and others. Soon after, the *boundary element method (BEM)* appeared which was especially well elaborated for crack problems by Cruse and Brebbia. The first international conference on the application of numerical methods in fracture mechanics took place in Swansea/GB in 1978. Thanks to these methods, great progress was made in fracture mechanics of ductile materials. Today mainly the FEM and the BEM (for special tasks) are used for continuum-mechanical stress analyses, materials-mechanical models and numerical simulations as indispensable tools for computation in fracture research. With these methods it is by now possible to analyze complicated crack configurations in real technical structures under complex loads with non-linear material behavior. The amount of publications dealing with the further development and application of these numerical methods in fracture research that was released in the past decades is almost unmanageable. Since that time, new numerical methods are already developing such as the »mesh-free« FEM/BEM, the discrete element method, particle methods and the extended X-FEM, conquering the application field of fracture mechanics.

1.4 Basic Literature on Fracture Mechanics

In the following, some important classical books are listed that contain fundamental representations of theoretical, experimental and engineering aspects of fracture mechanics. These literature is recommended for those readers, who want to get a broader survey about this field or who wish to learn more about a specific topic. Please note that this selection does not claim to be exhaustive.

Theoretical fundamentals

- Sih, G.: Mechanics of Fracture, Noordhoff, Leyden, Netherlands, 1975
- Broberg, K.: Cracks and Fracture. Academic Press, London, 1999
- Karihaloo, B., Knauss, in: Comprehensive Structural Integrity, W. G. Milne, I., Ritchie, R. O. and Karihaloo, B. (Eds.), Vol. 2: Fundamental theories and mechanisms of failure. Elsevier Pergamon, Amsterdam, 2003
- Gross, D. and Seelig, T.: Fracture mechanics with an introduction to micromechanics. Springer Berlin, 2011

Dynamic fracture

- Freund, L.: Dynamic fracture mechanics, Cambridge University Press, 1993
- Ravi-Chandar, K.: Dynamic fracture, Elsevier Science, Amsterdam, 2004

Nonlinear fracture mechanics

- Hutchinson, J.: Nonlinear Fracture Mechanics. Technical university of Denmark. Department of solid mechanics and DTH., 1979
- Kanninen, M. F. and Popelar, C. H.: Advanced Fracture Mechanics. Clarendon Press Oxford, 1985

Fracture of brittle materials

- Bazant, Z., Planas, J.: Fracture and Size Effects in Concrete and other Quasibrittle Materials. CRC Press, Boca Raton, 1997
- Cotterell, B., Mai, Y.: Fracture Mechanics of Cementitious Materials. Blackie Academic and Professional, 1996
- Lawn, B.: Fracture of Brittle Solids. Cambridge University Press, Cambridge, 1993

Creep fracture

- Riedel, H.: Fracture at High Temperature. Springer Berlin, 1987
- Miannay, D.: Fracture Mechanics. Springer New York, 1998

Fatigue fracture

- Suresh, S.: Fatigue of Materials. Cambridge University Press, Cambridge, 1998
- Ravichandran, K. S., Ritchie, R. O. and Murakami, Y.: Small fatigue cracks. mechanics, mechanisms and applications. Elsevier Amsterdam, 1999

- Schijve, J.: Fatigue of Structures and Materials. Kluwer Academic Publishers, Dordrecht, 2001

Engineering applications

- Atluri, S. N.: Structural integrity and durability. Tech Science Press, 1997
- Saxena, A.: Nonlinear fracture mechanics for engineers. CRC Press Boca Raton, 1998
- Anderson, T.: Fracture Mechanics: Fundamentals and Application. CRC Press, Boca Raton, 2005
- R6-Revision 4: Assessments of the integrity of structures containing defects, British Energy Generation Ltd, Barnwood, Gloucester, 2009
- Forschungskuratorium Maschinenbau: FKM Richtlinie: Bruchmechanischer Festigkeitsnachweis für Maschinenbauteile (3. Auflage) VDMA Verlag Frankfurt, 2009
- Zerbst, U., Schödel, M., Webster, S., Ainsworth, R.: Fitness for Service—Fracture Assessment of Structures Containing Cracks. A workbook based on the European SINTAP/FITNET procedure, Elsevier, 2007

Computational methods and solutions

- Atluri, S. N.: Computational methods in the mechanics of fracture. Elsevier Science Publ., 1986
- Aliabadi, M. H., Rooke D. P.: Numerical fracture mechanics. Kluwer Academic Publishers, Dordrecht, 1991
- Munz, D. and Fett, T.: Stress intensity factors and weight functions. Computational Mechanics Publications, Southampton, 1997
- Murakami, Y.: Stress Intensity Factors Handbook. Pergamon Press, Vol.: 1–5, New York, 1987
- de Borst, R. and Mang, H. A.: in: Comprehensive Structural Integrity, Milne, I., Ritchie, R. O. and Karihaloo, B. (Eds.), Vol. 3: Numerical and computational methods, Elsevier Pergamon, Amsterdam, 2003
- Ingraffea, A R.: Computational Fracture Mechanics. in: Encyclopedia of Computational Mechanics, E. Stein, R. de Borst, T. Hughes (eds.) Volume 2, Chapter 11, John Wiley and Sons, 2004

Fracture of piezoelectric materials

- Qin, Q.: Fracture Mechanics of Piezoelectric Materials. WIT Press Southampton, 2001

Chapter 2
Classification of Fracture Processes

Fracture processes are classified based on quite different individual aspects. The reason for that is the tremendous variety in which fracture processes appear and the diverse reasons leading to failure. First and foremost, a fracture depends on the properties of the considered material because the damage processes happening on a micro-structural level in the material determine its characteristic behavior. These microscopic structures and failure mechanisms vary diversely in the lineup of engineering materials. Just as important for fracture behavior is the type of external loading of the component. In this category one can differentiate between e.g. fractures due to static, dynamic or cyclic loading. Further important factors are the temperature, the multiaxiality of the loading, the rate of deformation and the chemical or environmental conditions.

2.1 Macroscopic Manifestations of Fracture

The macroscopic classification of fracture processes corresponds to the view of the designer and computation engineer. Fracture of a structure is inevitably connected to the propagation of one or more cracks which can eventually lead to entire rupture and loss of its load carrying capacity.

That is why particular emphasis is placed on the temporal and spatial progress of crack propagation. In fracture mechanics it is assumed that a macroscopic crack exists. This crack may be present from the very beginning due to a material defect or due to the component manufacturing. Often cracks originate in consequence of operational loading and material fatigue, which is the subject matter of the field of service strength of materials. After all, hypothetical cracks, which have to be assumed for purpose of safety assessment, are part of it as well. The macroscopic mechanical aspects of fracture can be categorized with respect to the load and fracture progression as follows:

M. Kuna, *Finite Elements in Fracture Mechanics*, Solid Mechanics and Its Applications 201, DOI: 10.1007/978-94-007-6680-8_2,
© Springer Science+Business Media Dordrecht 2013

(a) Type of loading

According to their temporal progress, mechanical loads are divided into *static*, *dynamic* and (periodically-cyclic or random) *variable* loads, the respective types of fracture to which they can be assigned. Fracture processes under static load are typical for load-bearing constructions e.g. in civil engineering. Impact, drop or crash processes are associated with highly dynamically accelerated deformations and inertia forces. In mechanical engineering and vehicle construction, much attention needs to be paid to variable loads which can, in contrast to static loading, lead to cracks and crack propagation at considerably lower amplitudes. About 60 % of all technical failures happen because of material fatigue or propagation of fatigue cracks.

(b) Orientation of a crack in relation to its principal stresses

As it is known from the classical theory of strength of materials, failure is in most cases controlled by the local stress which is clearly determined by the principal stresses σ_I, σ_{II} and σ_{III} and their axes. Depending on the material, either hypotheses of the maximum principal stress (Rankine), the maximum shear stress (Coulomb) or extended mixed criteria (Mohr) are used. The macroscopic image of fracture is therefore often affected by the principle stress trajectories. A distinction is being made between:

- The normal-planar crack or *cleavage fracture* exists, when the fracture faces are located perpendicularly to the direction of the highest principal stress $\sigma_{max} = \sigma_I$.
- The shear-planar crack or *shear fracture* exists, when the fracture faces coincide with the intersection planes of the maximum shear stress $\tau_{max} = (\sigma_I - \sigma_{III})/2$.

The situation is outlined for a simple tension rod in Fig. 2.1. However, it can be assigned to the local stress state at any point of the body. On a torsion rod (shaft) the fracture faces would run either vertically or inclined by 45° to the axis, depending on whether a shear or a cleavage fracture is assumed.

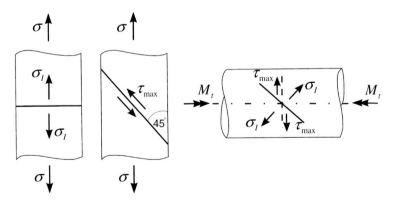

Fig. 2.1 Orientation of crack surfaces with respect to principal stress directions

(c) Stability of crack propagation

In the initial situation, a crack has a specific size and shape. As long as it does not change, the crack is regarded as a static or *stationary* crack. The moment in which the crack propagation starts due to critical loading, is called *crack initiation*. The crack size now increases and the crack is called *unsteady*.

An important feature of fracture is the stability of the crack propagation. The fracture process is then marked as *unstable* if the crack grows abruptly without the need to increase external loading. The critical condition is exceeded for the first time and persists without any additional energy supply. A typical example is the crack in the American Liberty Bell (Fig. 2.2), which developed spontaneously (allegedly on G. Washington's birthday), supposedly due to a casting defect. In contrast to this, if an additional increase of the external load is necessary in order to let the crack grow further, it is called *stable crack growth*. This means the critical condition needs to be induced by supplying additional energy again and again. Decisive for the stability of crack growth is the issue of how the stress situation changes in the body and at the crack itself due to the growth of the crack. Stable crack growth is often connected with plastic, energy consuming deformations in the component, which the failure case of a tube (Fig. 2.5) shows. Yet, this connection is by no means sufficient, which the example of a slowly growing crack in a car's windshield made of brittle glass teaches us.

If the crack propagation in the body comes to a standstill, it is called *crack arrest*.

(d) Magnitude of inelastic deformations

Depending on the amount of inelastic deformations or accumulated plastic work in the body that precede or accompany crack growth, distinctions are made between:

- **deformation-poor, or macroscopically brittle fracture** The nominal stresses are far below the plastic yield limit, the plastic or viscoplastic zones are very small and the load-deformation diagram runs linearly until crack initiation.
- **deformation-rich, or macroscopically ductile fracture** appears when the fracture process is connected with large inelastic deformations. The load-deformation diagram displays a distinctive non-linearity and the inelastic domains spread out over the entire cross-section (plastic limit load exceeded).

(e) Subcritical crack growth

In contrast to the above-mentioned types of crack propagation, there are fracture processes that happen far below the critical load and develop in a stable manner with a very low rate of growth . To describe them, the term *subcritical crack growth* was introduced. The most important form of appearance is *fatigue crack growth*, whereby the crack gradually grows under alternating loads. A characteristic failure case on a shaft loaded cyclically by rotation-bending is shown in Fig. 2.4. Subcritical constant loading in connection with viscoplastic deformations can lead to the so-called *creep fracture*. If a corrosive medium acts on the crack slit, a crack growth due to *stress corrosion cracking* is observed in spite of subcritical loading.

Fig. 2.2 Macroscopic brittle fracture of the Liberty Bell, Philadelphia 1752

Fig. 2.3 Failure case of a gas pipeline with dynamic crack growth

Fig. 2.4 Failure case due to fatigue crack growth on a shaft

Fig. 2.5 Macroscopic ductile fracture of a tube made of steel

(f) Crack growth rate

In contrast to the dynamic, impulsive load of a stationary crack, the dynamics of the fracture process itself will be considered. In most cases the crack propagation happens so slowly, that all dynamic effects in the structure may be neglected. In that case a *quasi-static analysis* is sufficient. If the crack growth rate reaches the level of acoustic wave speeds in the solid, velocity terms, inertia forces as well as interactions between the crack and the sound waves need to be taken into account. Additionally to that, failure mechanisms in the material depend on the deformation rate, which mostly leads to an embrittlement on fast running cracks. In this way, dynamic crack growth processes have already caused catastrophic failures, as the example of a gas pipeline in Fig. 2.3 shows.

Fig. 2.6 Transcrystalline brittle fracture of steel at room temperature

Fig. 2.7 Intercrystalline cleavage fracture of steel St52 at −196 °C

2.2 Microscopic Appearances of Fracture

For a better understanding of the material-specific failure mechanisms during fracture, it is necessary and especially useful to visit the »scene of the crime«—the fracture surface. To do that, the best choice is to use a scanning electron microscope, because of its depth of sharpness, chemical element analysis and material contrast. It is also possible to infer the reasons of damage from the characteristic patterns of the fracture surface (fractography). The different failure mechanisms lead to characteristic patterns of the fracture faces. These typical »faces« of all the various fracture types are catalogued in fractographic atlases, see e.g. [1]. Hereby, the view point of material scientists and failure case studies is put into focus. The most important microscopic appearances of a crack are:

- The *cleavage fracture* is characterized by plane fracture faces and minor deformations. The reason for this is brittle cracking along preferred crystallographic orientations due to high normal stresses. Body-centered cubic metals at low temperatures (Fig. 2.7) and ceramic materials (Fig. 2.8) tend towards cleavage fracture.
- In polycrystalline ceramic and metallic materials, characteristic differences of the fracture surfaces can be observed. Depending on whether cracking occurs along the boundaries between the individual grains or separates the grains by cleavage, it is called *intercrystalline* or *transcrystalline*, respectively. The difference can be seen very clearly by comparing Figs. 2.6 and 2.7.
- During a *dimple fracture*, the failure mechanism is associated with large plastic deformations in the process zone. Due to this, microscopic voids form, grow and eventually coalesce, which leads to a distinctive dimple structure of the fracture surface. Figure 2.9 shows the typical fracture pattern of high-alloyed steel.

- The *fatigue fracture* is quite smooth and mixed with fatigue striations due to very minor plastic deformations as the overall picture in Fig. 2.4 shows. It usually proceeds in a transcrystalline manner. Distinctive traces of cyclic plastic straining can be identified in the detailed microscopic view of Fig. 2.10.
- *Creep fracture* occurs often in metals due to damage of the grain boundaries, where creep pores are forming due to diffusion processes. This eventually leads to intercrystalline failure. Figure 2.11 shows a fracture surface of an aluminium alloy at high temperatures.

The diversity of failure mechanisms is further illustrated by the fractographic views of a crack resulting from stress corrosion cracking (Fig. 2.12) and a modern fiber-reinforced glass composite (Fig. 2.13).

Understanding of micro-structural failure mechanisms during fracture is not only important for material scientists and failure analysts. It also provides continuum mechanics engineers with useful information about which stress or deformation states control these mechanisms in order to describe them properly by means of macroscopic parameters and criteria.

2.3 Classification of Fracture Processes

In summary a classification of the fracture processes shall be given in the way it is used today. The overview in Fig. 2.14 is geared to the deformation properties of the materials, which are explained in detail in Appendix A. The following chapters on fracture mechanics are structured according to this classification.

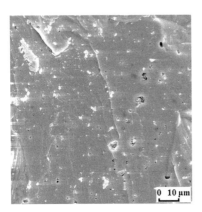

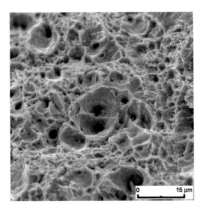

Fig. 2.8 Fracture surface of a brittle sinter ceramic (transcrystalline)

Fig. 2.9 Ductile dimple fracture of the high-alloyed steel 27MnSiVS6

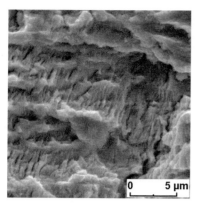

Fig. 2.10 Fractographic view of a fatigue fracture of steel C15

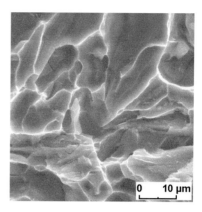

Fig. 2.11 Fracture surface of a creep fracture in aluminum AlSi10Mg at 300 °C

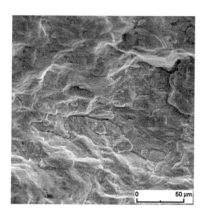

Fig. 2.12 Fracture surface of a stress corrosion crack in a CuZn37 alloy

Fig. 2.13 Fracture surface of the fiber composite Fortadur (SiC fibers in Duran glass)

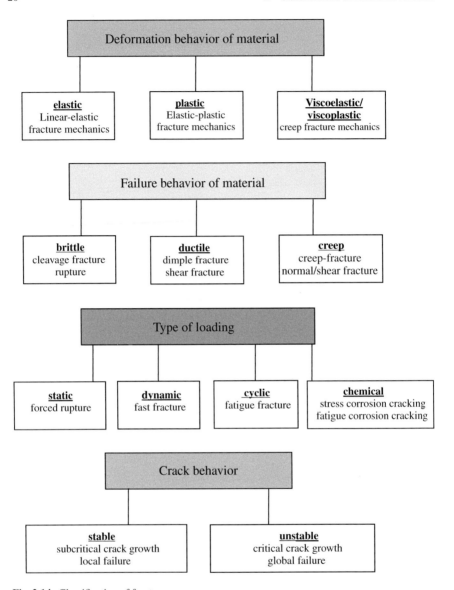

Fig. 2.14 Classification of fracture processes

Reference

1. Engel L, Klingele H (1982) Rasterelektronenmikroskopische Untersuchungen von Metallschä-
den. Gerling-Institut für Schadensforschung und Schadensverhütung, Köln

Chapter 3
Basics of Fracture Mechanics

The theoretical foundations of fracture mechanics will be presented in this chapter. The main focus lies on the description of the available continuum-mechanical solutions for cracks. On the basis of stress and deformation situations determined this way, suitable parameters, which clearly describe the loading states during fractures, are then selected. These *loading and fracture parameters* shape the foundation for the formulation of *fracture criteria*. With their help, the behavior of cracks can be quantitatively evaluated. These usually closed mathematical solutions are the preconditions to being able to calculate the sizes of cracks with numerical methods later on. Naturally, the experimental test methods of fracture mechanics used to evaluate the material parameters are based on the understanding of the loading situation as well.

3.1 Model Assumptions

In fracture mechanics the behavior of cracks in bodies is described from a macroscopic point of view in the context of continuum mechanics. The term »body« is supposed to include technical constructions, components and facilities as well as material structures on all scales. In the following, the crack is considered in a geometrically idealized form as a mathematical cut or slit in the body. First of all, this means that a purely plane separation of the body is assumed, which leads to either two *crack faces* (2D) or to two *crack surfaces* (3D). At the *crack tip* (2D) or the *crack front* (3D) the crack faces or crack surfaces converge respectively. Secondly, an ideal sharp crack tip with a notch radius $\rho = 0$ is assumed. Actually, the tips of physical cracks always have of course a finite radius of curvature. However, in comparison to the crack length and the body dimensions, it can be regarded as infinitely small. Therefore, the shape of the crack tip is clearly defined. In this point fracture mechanics differs from the theory of notch stresses. Regarding the deformation of a crack, a distinction is drawn between three independent movements of the two crack

M. Kuna, *Finite Elements in Fracture Mechanics*, Solid Mechanics and Its
Applications 201, DOI: 10.1007/978-94-007-6680-8_3,
© Springer Science+Business Media Dordrecht 2013

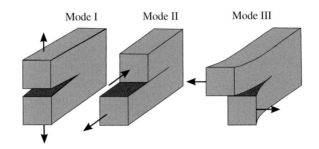

Fig. 3.1 Definition of the three crack opening modes

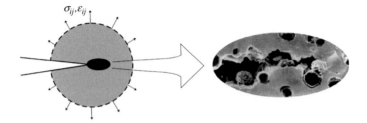

Fig. 3.2 Zone of micromechanical fracture processes at the crack tip

faces relative to each other. The so-called *crack opening modes* are schematically described in Fig. 3.1 and are defined as follows:

Mode I: Opening Mode: The crack opens perpendicular to the crack plane. This can be caused by tensile loading.

Mode II: In-plane sliding mode: The crack faces are displaced on their plane, normal to the crack front, which correlates to a transversal shearing load.

Mode III: Out-of-plane tearing mode: The crack faces are displaced on their plane, parallel to the crack front, which is related to anti-plane longitudinal shearing load.

Every type of crack deformation can be regarded as a superposition of these three basic kinematic modes. In the three-dimensional case, Fig. 3.1 is to be understood as a local section around a segment of the crack front, whereas the size of the modes will change along the crack front.

In general, the continuum-mechanical modeling of cracks *cannot* describe the physical fracture process of the material separation. Rather, the stresses and deformations in the body are—just like in the classical theory of strength—calculated using *material laws of deformation* (see Sect. A. 4). Based on these results the strength hypotheses are finally formulated. The stress and strain state practically describes the » boundary conditions « under which the micromechanical processes in the *fracture process zone* proceed at the crack tip (Fig. 3.2). If the failure phenomena were to be included in the continuum-mechanical simulation, the material law would have to be expanded by models of material *failure*. This approach is taken in the context of continuum damage mechanics and cohesive zone models. But for now we will not pursue this way further. Of course it is possible to switch to the meso or micro scale in order to simulate discrete cracks, their nucleation and interaction in a fracture process zone, using continuum mechanics or fracture mechanics as well.

3.2 Linear-Elastic Fracture Mechanics

In linear-elastic fracture mechanics, crack problems are analyzed in bodies whose deformation behavior can be assumed to be linear-elastic according to the generalized Hooke's law (Sect. A.4.1). Apart from very brittle materials, in truth there are physical or geometrical non-linearities in almost all structures, particularly at notches and crack tips. In many cases, the non-linear effects are limited to small areas which may be neglected in comparison to crack size or the component dimensions. The elastic material can basically be anisotropic. For now we will restrict ourselves to the simpler case of isotropy. The term of linearity implies small displacements and infinitesimal deformations (A.30).

3.2.1 Two-Dimensional Crack Problems

Crack Under Mode I Loading

We investigate a slit-like straight crack of the length $2a$ in an infinitely large sheet of isotropic linear-elastic material. The load is assumed to act vertically to the crack with constant tension σ, see Fig. 3.3. A Cartesian coordinate system is positioned with its origin at the middle of the crack so that the positions of the crack faces Γ^+ and Γ^- are determined by:

$$-a \leq x_1 \leq +a, \quad x_2 = \pm 0 . \tag{3.1}$$

In order to find a solution to this boundary value problem of elasticity theory, we use the method of complex functions explained in the Appendix A.5.2 with the complex variables $z = x_1 + ix_2$. The boundary conditions of this problem are given by zero tractions $\bar{t}_i = 0$ at the crack faces Γ^+ and Γ^-, whose normal vectors n_j only point in direction $\mp x_2$ ($n_1 = 0, n_2 = \mp 1$). According to the Cauchy formula,

$$t_i = \sigma_{ij}n_j = \mp \sigma_{i2} = \bar{t}_i = 0 \quad \Rightarrow \tau_{12} = 0, \sigma_{22} = 0 \text{ on } \Gamma^+ \text{ and } \Gamma^- \tag{3.2}$$

As an additional boundary condition, the undisturbed uniaxial homogeneous tension state needs to be reached in infinite distance from the crack:

$$\sigma_{22} = \sigma, \quad \sigma_{11} = \tau_{12} = 0 \quad \text{for } |z| = \sqrt{x_1^2 + x_2^2} \to \infty . \tag{3.3}$$

Due to lack of space it is not possible to specify the approach here (see [1]), but only the result is given in the form of complex stress functions $\phi(z)$ and $\chi(z)$:

$$\phi(z) = \frac{\sigma}{4} z + \frac{\sigma}{2}\left[\sqrt{z^2 - a^2} - z\right], \quad \chi'(z) = \frac{\sigma}{2} z - \frac{\sigma}{2}\frac{a^2}{\sqrt{z^2 - a^2}}. \tag{3.4}$$

With the help of Kolosov's formulas (A.158), the stress and deformation fields in the entire plate can be derived. The first two terms in (3.4) represent correctly the decay behavior according to (3.3), which can be easily shown by inserting in (A.158). The second terms in (3.4) vanish for $|z| \to \infty$ and hence describe the actual effect of the crack on the stress distribution in the sheet ($\Re( ) \; \hat{=}$ real part, $\Im( ) \; \hat{=}$ imaginary part).

$$S_1 := \sigma_{11} + \sigma_{22} = 4\Re\phi' = \sigma\Re\left[\frac{2z}{\sqrt{z^2 - a^2}} - 1\right]$$

$$S_2 := \sigma_{22} - \sigma_{11} + 2i\tau_{12} = 2\left[\bar{z}\phi'' + \chi''\right] = \sigma\left[1 + a^2\frac{z - \bar{z}}{(z^2 - a^2)^{3/2}}\right] \tag{3.5}$$

$$\sigma_{11} = \frac{1}{2}\Re(S_1 - S_2), \quad \sigma_{22} = \frac{1}{2}\Re(S_1 + S_2), \quad \tau_{12} = \frac{1}{2}\Im(S_2)$$

Firstly, the stress distribution on the ligament ($|x_1| \geq a$, $x_2 = 0$) is to be analyzed. Because of symmetry regarding x_2, here the shear stress is $\tau_{12} \equiv 0$. The normal stresses are obtained from (3.5):

$$\sigma_{11} = \sigma\left[\frac{x_1}{\sqrt{x_1^2 - a^2}} - 1\right], \quad \sigma_{22} = \sigma\frac{x_1}{\sqrt{x_1^2 - a^2}}. \tag{3.6}$$

The result indicates that the normal stresses at the crack tips ($x_1 \to \pm a$) grow to infinity (Fig. 3.3). The evaluation of (3.5) at the crack faces $|x_1| < a$ confirms that the boundary conditions (3.2) are fulfilled.

The deformation state is determined by using the 3rd Kolosov's formula(A.158):

$$2\mu(u_1 + iu_2) = \frac{\sigma}{2}\left[\kappa\sqrt{z^2 - a^2} + \frac{a^2 - z\bar{z}}{\sqrt{z^2 - a^2}} - \frac{1}{2}(\kappa - 1)z - \bar{z}\right]. \tag{3.7}$$

The calculation of the displacements of the upper and lower crack face ($|x_1| < a$, $x_2 = \pm 0$) show that the opened crack has the shape of an ellipse (Fig. 3.3).

$$u_1 = \mp\frac{1 + \kappa}{8\mu}\sigma x_1, \quad u_2 = \pm\frac{1 + \kappa}{4\mu}\sigma\sqrt{a^2 - x_1^2} . \tag{3.8}$$

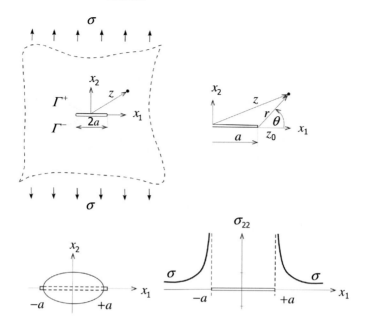

Fig. 3.3 Crack in infinite sheet under tension *top*) coordinate system, *bottom*) crack opening and distribution of stress

Here μ denotes the shear modulus. The elastic constant κ amounts to $\kappa = 3 - 4\nu$ in the *state of plane strain* and to $\kappa = (3 - \nu)/(1 + \nu)$ in the *state of plane stress*, see Appendix A.5.2.

Of particular interest is the local stress distribution in the immediate proximity of the crack tips. Therefore, the coordinate system (r, θ) drafted in Fig. 3.3 is introduced directly at the crack tip (in this example at $z = z_0 = +a$):

$$z = a + re^{i\theta}, \quad z - z_0 = re^{i\theta} = a\zeta e^{i\theta} \quad \text{with } \zeta = \frac{r}{a}. \tag{3.9}$$

By inserting z and $\bar{z}$ into the Eq. (3.5), the following terms can be obtained:

$$S_1 = \sigma\Re\left[\frac{2\left(1 + \zeta e^{i\theta}\right)}{\sqrt{2\zeta e^{i\theta} + \zeta^2 e^{2i\theta}}} - 1\right] \text{ and } S_2 = \sigma\left[1 + \frac{\zeta\left(e^{i\theta} - e^{-i\theta}\right)}{\left(2\zeta e^{i\theta} + \zeta^2 e^{2i\theta}\right)^{3/2}}\right]. \tag{3.10}$$

They can be approximated for $\zeta = \frac{r}{a} \ll 1$ as follows:

$$S_1 = \sigma_{11} + \sigma_{22} \approx \sigma\Re\left[\frac{2}{\sqrt{2\zeta e^{i\theta}}}\right] = \sigma\sqrt{\frac{a}{2r}}2\cos\frac{\theta}{2}$$

$$S_2 = \sigma_{22} - \sigma_{11} + 2i\tau_{12} \approx \sigma\frac{\zeta 2i\sin\theta}{\left(2\zeta e^{i\theta}\right)^{3/2}} = \sigma\sqrt{\frac{a}{2r}}i\sin\theta\, e^{-i3\theta/2}. \tag{3.11}$$

The result is the stress state at the crack tip in polar coordinates (r, θ). The factor $\sigma\sqrt{\pi a} = K_{\mathrm{I}}$ can be split off.

$$
\begin{Bmatrix} \sigma_{11} \\ \sigma_{22} \\ \tau_{12} \end{Bmatrix} = \frac{K_{\mathrm{I}}}{\sqrt{2\pi r}} \begin{Bmatrix} \cos\frac{\theta}{2}\left[1 - \sin\frac{\theta}{2}\sin\frac{3\theta}{2}\right] \\ \cos\frac{\theta}{2}\left[1 + \sin\frac{\theta}{2}\sin\frac{3\theta}{2}\right] \\ \sin\frac{\theta}{2}\cos\frac{\theta}{2}\cos\frac{3\theta}{2} \end{Bmatrix} = \frac{K_{\mathrm{I}}}{\sqrt{2\pi r}} \begin{Bmatrix} f_{11}^{\mathrm{I}}(\theta) \\ f_{22}^{\mathrm{I}}(\theta) \\ f_{12}^{\mathrm{I}}(\theta) \end{Bmatrix}
$$

$$(3.12)$$

The stresses σ_{33} in thickness direction are, assuming a state of plane stress, zero and assuming a state of plane strain:

$$
\sigma_{33} = \nu(\sigma_{11} + \sigma_{22}) = \frac{K_{\mathrm{I}}}{\sqrt{2\pi r}} 2\nu \cos\frac{\theta}{2} . \tag{3.13}
$$

Using the two-dimensional Hooke's law (A.145) or (A.150), the resultant strain components are calculated to:

$$
\begin{Bmatrix} \varepsilon_{11} \\ \varepsilon_{22} \\ \gamma_{12} \end{Bmatrix} = \frac{K_{\mathrm{I}}}{2\mu\sqrt{2\pi r}} \begin{Bmatrix} \cos\frac{\theta}{2}\left[\frac{\kappa-1}{2} - \sin\frac{\theta}{2}\sin\frac{3\theta}{2}\right] \\ \cos\frac{\theta}{2}\left[\frac{\kappa-1}{2} + \sin\frac{\theta}{2}\sin\frac{3\theta}{2}\right] \\ 2\sin\frac{\theta}{2}\cos\frac{\theta}{2}\cos\frac{3\theta}{2} \end{Bmatrix} . \tag{3.14}
$$

The strains in thickness direction ε_{33} are zero for the plane strain state according to definition and for the plane stress state:

$$
\varepsilon_{33} = -\frac{K_{\mathrm{I}}}{\mu\sqrt{2\pi r}} \frac{\nu}{1+\nu} \cos\frac{\theta}{2} . \tag{3.15}
$$

In a similar way, it is possible to expand the displacement field (3.7) using the approach (3.9) near the crack tip for $\zeta = \frac{r}{a} \ll 1$ which yields:

$$
\begin{Bmatrix} u_1 \\ u_2 \end{Bmatrix} = \frac{K_{\mathrm{I}}}{2\mu}\sqrt{\frac{r}{2\pi}} \begin{Bmatrix} \cos\frac{\theta}{2}[\kappa - \cos\theta] \\ \sin\frac{\theta}{2}[\kappa - \cos\theta] \end{Bmatrix} = \frac{K_{\mathrm{I}}}{2\mu}\sqrt{\frac{r}{2\pi}} \begin{Bmatrix} g_1^{\mathrm{I}}(\theta) \\ g_2^{\mathrm{I}}(\theta) \end{Bmatrix} . \tag{3.16}
$$

With the help of the performed asymptotic analysis, the displacement, strain and stress fields at the crack tip have been successfully calculated. This solution is called a *crack tip field* or an *asymptotic near field*. Studying the results of (3.12), (3.14) and (3.16), the following characteristics can be identified:

- The stresses and deformations behave singularly with $1/\sqrt{r}$ if the crack tip $r \to 0$ is approached.

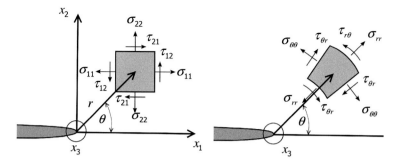

Fig. 3.4 Stresses at the crack tip in Cartesian and cylindrical coordinates

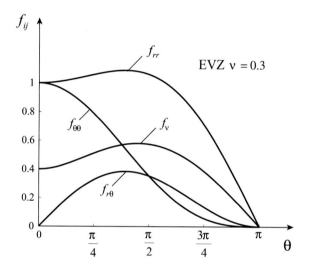

Fig. 3.5 Angular distribution of stresses around the crack tip (Mode I)

- The displacement fields are proportional to the square root of the distance to the crack tip $\sqrt{r}$. The crack opens parabolically.
- All field variables at the crack tip are proportional to $K_I = \sigma\sqrt{\pi a}$. This factor K_I is called *stress intensity factor*. The Index I represents the crack opening mode I.
- The intensity of the near field solution does not only rise linearly (as expected) with the tensile load σ but it also depends on the length a of the crack!

Occasionally, the description of the near field solution is advantageous in cylindrical coordinates, see Fig. 3.4. Applying the common transformation rules (see (A.54)) to expressions (3.12), (3.14) and (3.16), the displacement, deformation and stress components are gained in the (r, θ, x_3)–system:

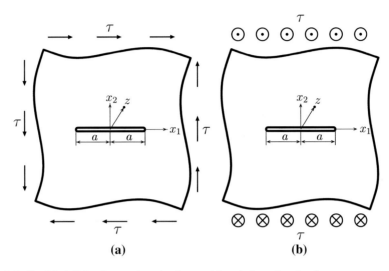

Fig. 3.6 Crack in infinite sheet under **a** in-plane and **b** anti-plane shear loading

$$
\left\{
\begin{array}{c}
\sigma_{rr} \\
\sigma_{\theta\theta} \\
\tau_{r\theta}
\end{array}
\right\}
=
\frac{K_{\mathrm{I}}}{4\sqrt{2\pi r}}
\left\{
\begin{array}{c}
5\cos\frac{\theta}{2} - \cos\frac{3\theta}{2} \\
3\cos\frac{\theta}{2} + \cos\frac{3\theta}{2} \\
\sin\frac{\theta}{2} + \sin\frac{3\theta}{2}
\end{array}
\right\}
=
\frac{K_{\mathrm{I}}}{\sqrt{2\pi r}}
\left\{
\begin{array}{c}
f_{rr}(\theta) \\
f_{\theta\theta}(\theta) \\
f_{r\theta}(\theta)
\end{array}
\right\}
\tag{3.17}
$$

$$
\left\{
\begin{array}{c}
\varepsilon_{rr} \\
\varepsilon_{\theta\theta} \\
\gamma_{r\theta}
\end{array}
\right\}
=
\frac{K_{\mathrm{I}}}{8\mu\sqrt{2\pi r}}
\left\{
\begin{array}{c}
[2\kappa - 1]\cos\frac{\theta}{2} - \cos\frac{3\theta}{2} \\
[2\kappa - 3]\cos\frac{\theta}{2} + \cos\frac{3\theta}{2} \\
2\sin\frac{\theta}{2} + 2\sin\frac{3\theta}{2}
\end{array}
\right\}
\tag{3.18}
$$

$$
\left\{
\begin{array}{c}
u_r \\
u_\theta
\end{array}
\right\}
=
\frac{K_{\mathrm{I}}}{4\mu}\sqrt{\frac{r}{2\pi}}
\left\{
\begin{array}{c}
[2\kappa - 1]\cos\frac{\theta}{2} - \cos\frac{3\theta}{2} \\
-[2\kappa + 1]\sin\frac{\theta}{2} + \sin\frac{3\theta}{2}
\end{array}
\right\}
\tag{3.19}
$$

The strains ε_{33} and stresses σ_{33} in the thickness direction stay unchanged during this transformation. The angular dependence $f_{ij}(\theta)$ of the stresses (3.17) as well as the v. Mises–effective stress $f_{\mathrm{v}}(\theta)$ are described in Fig. 3.5.

Crack Under Mode II Loading

In a similar way as in the previous section, it is possible to analyze the crack in an infinite sheet under in-plane shear loading $\tau \cong \tau_{21}$. Only the boundary conditions (3.3) have a different form compared to the tension problem

$$
\sigma_{11} = \sigma_{22} = 0, \quad \tau_{12} = \tau \quad \text{for } |z| = \sqrt{x_1^2 + x_2^2} \to \infty .
\tag{3.20}
$$

With the help of Fig. 3.6 (left), it is immediately clear that the solution has to be antisymmetric with respect to the x_2–coordinate. The complex stress functions for this case read

$$\phi(z) = -\mathrm{i}\frac{\tau}{4}z - \mathrm{i}\frac{\tau}{2}\left(\sqrt{z^2 - a^2} - z\right), \quad \chi'(z) = \mathrm{i}\tau z + \mathrm{i}\frac{\tau}{2}\left(\frac{2z^2 - a^2}{\sqrt{z^2 - a^2}} - 2z\right).$$

$$(3.21)$$

The first terms in both stress functions respectively represent the pure, undisturbed shear stress situation in a plane free of crack. The second terms describe the part of the solution due to the crack. It is left as an exercise for the reader to calculate the corresponding solutions for the field variable $\sigma_{ij}(x_1, x_2)$, $u_i(x_1, x_2)$, as well as to carry out the asymptotic expansion for $r \to \pm a$. Only the result for the crack tip near field will be given here. In addition it is necessary to introduce the *stress intensity factor* K_{II} for mode II loading, which takes the following value for this boundary value problem:

$$K_{\mathrm{II}} = \tau\sqrt{\pi a},$$

$$(3.22)$$

$$\begin{Bmatrix} \sigma_{11} \\ \sigma_{22} \\ \tau_{12} \end{Bmatrix} = \frac{K_{\mathrm{II}}}{\sqrt{2\pi r}} \begin{Bmatrix} -\sin\frac{\theta}{2}\left[2 + \cos\frac{\theta}{2}\cos\frac{3\theta}{2}\right] \\ \sin\frac{\theta}{2}\cos\frac{\theta}{2}\cos\frac{3\theta}{2} \\ \cos\frac{\theta}{2}\left[1 - \sin\frac{\theta}{2}\sin\frac{3\theta}{2}\right] \end{Bmatrix} = \frac{K_{\mathrm{II}}}{\sqrt{2\pi r}} \begin{Bmatrix} f_{11}^{\mathrm{II}}(\theta) \\ f_{22}^{\mathrm{II}}(\theta) \\ f_{12}^{\mathrm{II}}(\theta) \end{Bmatrix}.$$

$$(3.23)$$

and in cylindrical coordinates

$$\begin{Bmatrix} \sigma_{rr} \\ \sigma_{\theta\theta} \\ \tau_{r\theta} \end{Bmatrix} = \frac{K_{\mathrm{II}}}{4\sqrt{2\pi r}} \begin{Bmatrix} -5\sin\frac{\theta}{2} + 3\sin\frac{3\theta}{2} \\ -3\sin\frac{\theta}{2} - 3\sin\frac{3\theta}{2} \\ \cos\frac{\theta}{2} + 3\cos\frac{3\theta}{2} \end{Bmatrix}.$$

$$(3.24)$$

The displacements near the crack tips are

$$\begin{Bmatrix} u_1 \\ u_2 \end{Bmatrix} = \frac{K_{\mathrm{II}}}{2\mu}\sqrt{\frac{r}{2\pi}} \begin{Bmatrix} \sin\frac{\theta}{2}[\kappa + 2 + \cos\theta] \\ -\cos\frac{\theta}{2}[\kappa - 2 + \cos\theta] \end{Bmatrix} = \frac{K_{\mathrm{II}}}{2\mu}\sqrt{\frac{r}{2\pi}} \begin{Bmatrix} g_1^{\mathrm{II}}(\theta) \\ g_2^{\mathrm{II}}(\theta) \end{Bmatrix}.$$

$$(3.25)$$

$$\begin{Bmatrix} u_r \\ u_\theta \end{Bmatrix} = \frac{K_{\mathrm{II}}}{4\mu}\sqrt{\frac{r}{2\pi}} \begin{Bmatrix} -[2\kappa - 1]\sin\frac{\theta}{2} + 3\sin\frac{3\theta}{2} \\ -[2\kappa + 1]\cos\frac{\theta}{2} + 3\cos\frac{3\theta}{2} \end{Bmatrix}.$$

$$(3.26)$$

Crack Under Mode III Loading

At last, the plane crack under anti-plane shear loading $\tau \hat{=} \tau_{23}$ is to be considered, see Fig. 3.6 (right). This time, the boundary conditions for the stress state on the crack and at infinity are

$$\tau_{23} = 0 \quad \text{for } |x_1| \leq a \quad \text{and} \quad \tau_{13} = 0, \tau_{23} = \tau \quad \text{for } |z| \to \infty . \tag{3.27}$$

In order to solve this type of boundary value problems, a complex stress function $\Omega(z)$ has been introduced in Appendix A.5.4. For the mode III crack problem the solution is found by the setting

$$\Omega(z) = -i\tau\sqrt{z^2 - a^2} . \tag{3.28}$$

The antisymmetric displacement u_3 of the crack faces against each other has an elliptic form once again

$$\mu u_3 = \Re\Omega(z) = \pm\tau\sqrt{a^2 - x_1^2} \quad \text{for } x_2 = \pm 0 . \tag{3.29}$$

The calculation of the shear stresses according to (A.165) on the ligament in front of the crack tips reveals a singularity here as well:

$$\tau_{23} = -\Im\Omega'(z) = \frac{\tau x_1}{\sqrt{x_1^2 - a^2}}, \quad \tau_{13} = \Re\Omega'(z) = 0 \text{ for } |x_1| > 0, \quad x_2 = 0. \tag{3.30}$$

If the solution is expanded in the same way as in mode I around the crack tip $z = a + re^{i\theta}$, the result reads

$$u_3 = \tau\sqrt{\pi a}\frac{2}{\mu}\sqrt{\frac{r}{2\pi}}\sin\frac{\theta}{2} = \frac{2K_{III}}{\mu}\sqrt{\frac{r}{2\pi}}\sin\frac{\theta}{2} = \frac{K_{III}}{2\mu}\sqrt{\frac{r}{2\pi}}g_3^{III}(\theta) \tag{3.31}$$

$$\begin{Bmatrix} \tau_{13} \\ \tau_{23} \end{Bmatrix} = \frac{K_{III}}{\sqrt{2\pi r}}\begin{Bmatrix} -\sin\frac{\theta}{2} \\ +\cos\frac{\theta}{2} \end{Bmatrix} = \frac{K_{III}}{\sqrt{2\pi r}}\begin{Bmatrix} f_{13}^{III}(\theta) \\ f_{23}^{III}(\theta) \end{Bmatrix} . \tag{3.32}$$

Here $K_{III} = \tau\sqrt{\pi a}$ denotes the stress intensity factor for mode III. Qualitatively, the same $1/\sqrt{r}$–singularity of the stresses and the same $\sqrt{r}$–behavior of the displacements occur at the crack tip just as in mode I and mode II loading.

3.2.2 Eigenfunctions of the Crack Problem

In the previous section we have the stress singularity at the crack tip extracted from the complete solution of the boundary value problem for cracks in two-dimensional domains. Obviously, the singular behavior is causally associated with an »infinitely sharp« crack tip. For that reason, the elastic solution at an isolated crack tip will be further investigated in an infinite plane, see Fig. 3.7. Expediently, we place a polar coordinate system (r, θ) in the crack tip. For the solution of this particular boundary value problem, the two complex potentials are set as simple power series

$$\phi(z) = Az^\lambda, \quad \chi(z) = Bz^{\lambda+1}, \quad z = re^{i\theta}, \tag{3.33}$$

whereas the coefficients A and B are complex numbers. The exponent λ needs to be a positive real number in order to prevent infinite displacements at the crack tip. Therefor, the stresses in polar coordinates are calculated according to (A.161), with particular interest in the circumferential stress $\sigma_{\theta\theta}$ and the shear stress $\tau_{r\theta}$. They are obtained by adding the two first Kolosov equations (A.158).

$$\sigma_{\theta\theta} + i\tau_{r\theta} = \phi'(z) + \overline{\phi'(z)} + \left(\bar{z}\phi''(z) + \chi''(z)\right)e^{2i\theta}$$

$$= \lambda Az^{\lambda-1} + \lambda\overline{A}\bar{z}^{\lambda-1} + \left[\lambda(\lambda-1)A\bar{z}z^{\lambda-2} + \lambda(\lambda+1)Bz^{\lambda-1}\right]e^{2i\theta}$$

$$= \lambda r^{\lambda-1}\left[Ae^{i(\lambda-1)\theta} + \overline{A}e^{-i(\lambda-1)\theta}\right.$$

$$\left. +A(\lambda-1)e^{i(\lambda-1)\theta} + (\lambda+1)Be^{i(\lambda+1)\theta}\right] \tag{3.34}$$

As boundary conditions the normal and shear stresses must be zero for all r on the traction-free crack faces $\theta = \pm\pi$. This means $\sigma_{\theta\theta} + i\tau_{r\theta} = 0$. Because of that, the term in [ ]–brackets (with $e^{\pm i\pi} = 1$) has to vanish:

$$\theta = +\pi : \quad A\lambda e^{i\lambda\pi} + \overline{A}e^{-i\lambda\pi} + (\lambda+1)Be^{i\lambda\pi} = 0 \tag{3.35}$$

$$\theta = -\pi : \quad A\lambda e^{-i\lambda\pi} + \overline{A}e^{i\lambda\pi} + (\lambda+1)Be^{-i\lambda\pi} = 0 . \tag{3.36}$$

These relations form a homogeneous system of 2 complex (4 real) equations for the 2 complex (4 real) coefficients A and B, which need to be determined. As necessary condition for the solution, the coefficient determinant has to be set to zero, which results in a transcendent equation for the exponent (eigenvalue) λ. An easier approach would be to multiply Eqs. (3.35) and (3.36) by $e^{-i\lambda\pi}$ or by $e^{+i\lambda\pi}$ respectively, and subtract them from each other, getting

$$\overline{A}\left(e^{2i\lambda\pi} - e^{-2i\lambda\pi}\right) = 0 . \tag{3.37}$$

Setting the term in parentheses to zero results in $\sin(2\lambda\pi) = 0$, which leads to the real value λ:

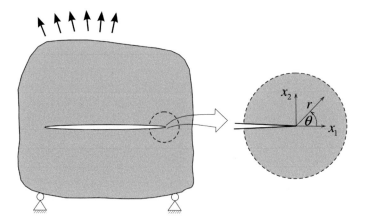

Fig. 3.7 Analysis of the near field at the crack tip

$$\lambda = \frac{n}{2} \quad \text{mit } n = 1, 2, 3 \ldots \tag{3.38}$$

Thus it was shown that an infinite number of eigenvalues $\lambda = \frac{n}{2}$ exists. The corresponding eigenfunctions can be obtained by using the Ansatz (3.33). The entire solution of the boundary value problem consists of the superposition of these eigenfunctions with undetermined coefficients A_n and B_n.

$$\phi = \sum_{n=1}^{\infty} A_n z^{\frac{n}{2}}, \quad \chi = \sum_{n=1}^{\infty} B_n z^{\frac{n}{2}+1} \tag{3.39}$$

The expressions (3.35) or (3.36) now provide the correlation

$$\frac{n}{2} A_n + (-1)^n \overline{A}_n + \left(\frac{n}{2} + 1\right) B_n = 0, \tag{3.40}$$

so that B_n can be replaced by the coefficient $A_n = a_n + i b_n$.

By inserting (3.39) and (3.40) into Kolosov's Eq. (A.158), the radial and angular functions are determined for the nth eigenfunction in real notation, which was discovered for the first time by Williams [2] in 1957:

$$\sigma_{11}^{(n)}(r, \theta) = r^{\frac{n}{2}-1} \{a_n \tilde{M}_{11}^{(n)}(\theta) + b_n \tilde{N}_{11}^{(n)}(\theta)\}$$
$$\sigma_{22}^{(n)}(r, \theta) = r^{\frac{n}{2}-1} \{a_n \tilde{M}_{22}^{(n)}(\theta) + b_n \tilde{N}_{22}^{(n)}(\theta)\} \tag{3.41}$$
$$\tau_{12}^{(n)}(r, \theta) = r^{\frac{n}{2}-1} \{a_n \tilde{M}_{12}^{(n)}(\theta) + b_n \tilde{N}_{12}^{(n)}(\theta)\}$$

with

$$\tilde{M}_{11}^{(n)} = \frac{n}{2} \left\{ \left[2 + (-1)^n + \frac{n}{2} \right] \cos \left(\frac{n}{2} - 1 \right) \theta - \left(\frac{n}{2} - 1 \right) \cos \left(\frac{n}{2} - 3 \right) \theta \right\}$$

$$\tilde{N}_{11}^{(n)} = \frac{n}{2} \left\{ \left[-2 + (-1)^n - \frac{n}{2} \right] \sin \left(\frac{n}{2} - 1 \right) \theta + \left(\frac{n}{2} - 1 \right) \sin \left(\frac{n}{2} - 3 \right) \theta \right\}$$

$$\tilde{M}_{22}^{(n)} = \frac{n}{2} \left\{ \left[2 - (-1)^n - \frac{n}{2} \right] \cos \left(\frac{n}{2} - 1 \right) \theta + \left(\frac{n}{2} - 1 \right) \cos \left(\frac{n}{2} - 3 \right) \theta \right\}$$

$$\tilde{N}_{22}^{(n)} = \frac{n}{2} \left\{ \left[-2 - (-1)^n + \frac{n}{2} \right] \sin \left(\frac{n}{2} - 1 \right) \theta - \left(\frac{n}{2} - 1 \right) \sin \left(\frac{n}{2} - 3 \right) \theta \right\}$$

$$\tilde{M}_{12}^{(n)} = \frac{n}{2} \left\{ \left(\frac{n}{2} - 1 \right) \sin \left(\frac{n}{2} - 3 \right) \theta - \left[\frac{n}{2} + (-1)^n \right] \sin \left(\frac{n}{2} - 1 \right) \theta \right\}$$

$$\tilde{N}_{12}^{(n)} = \frac{n}{2} \left\{ \left(\frac{n}{2} - 1 \right) \cos \left(\frac{n}{2} - 3 \right) \theta - \left[\frac{n}{2} - (-1)^n \right] \cos \left(\frac{n}{2} - 1 \right) \theta \right\} \quad (3.42)$$

and

$$u_1^{(n)}(r, \theta) = \frac{1}{2\mu} r^{\frac{n}{2}} \left\{ a_n \tilde{F}_1^{(n)}(\theta) + b_n \tilde{G}_1^{(n)}(\theta) \right\}$$

$$u_2^{(n)}(r, \theta) = \frac{1}{2\mu} r^{\frac{n}{2}} \left\{ a_n \tilde{F}_2^{(n)}(\theta) + b_n \tilde{G}_2^{(n)}(\theta) \right\} \quad (3.43)$$

with

$$\tilde{F}_1^{(n)} = \left[\kappa + (-1)^n + \frac{n}{2} \right] \cos \frac{n}{2} \theta - \frac{n}{2} \cos \left(\frac{n}{2} - 2 \right) \theta$$

$$\tilde{G}_1^{(n)} = \left[-\kappa + (-1)^n - \frac{n}{2} \right] \sin \frac{n}{2} \theta + \frac{n}{2} \sin \left(\frac{n}{2} - 2 \right) \theta$$

$$\tilde{F}_2^{(n)} = \left[\kappa - (-1)^n - \frac{n}{2} \right] \sin \frac{n}{2} \theta + \frac{n}{2} \sin \left(\frac{n}{2} - 2 \right) \theta \quad (3.44)$$

$$\tilde{G}_2^{(n)} = \left[\kappa + (-1)^n - \frac{n}{2} \right] \cos \frac{n}{2} \theta + \frac{n}{2} \cos \left(\frac{n}{2} - 2 \right) \theta .$$

The terms with a_n correspond to the crack opening mode I, while mode II is associated with the coefficient b_n. For $n = 1$, the well- known singular solutions (3.12) and (3.23) are obtained showing $r^{-1/2}$ (use addition theorems !). The *stress intensity factors* K_I and K_{II} are related to the coefficients a_1, b_1 of the first eigenfunction in the following way

$$K_I - i K_{II} = \sqrt{2\pi} (a_1 + i b_1). \quad (3.45)$$

Of particular importance is the 2nd eigenfunction $n = 2$, which describes only a constant stress state parallel to the crack faces, the so-called T-*stress*. (The functions belonging to b_2 only accomplish a stress-free rigid body rotation.)

$$\sigma_{11}^{(2)} = r^0 4a_2 = T_{11} = \text{const.}, \quad \sigma_{22}^{(2)} \equiv \tau_{12}^{(2)} \equiv 0$$

$$2\mu u_1^{(2)} = a_2(\kappa + 1)x_1, \quad 2\mu u_2^{(2)} = a_2(\kappa - 3)x_2 \quad (3.46)$$

It is possible to develop eigenfunctions at the crack tip for mode III as well, by using again a power series with the exponent $\lambda > 0$ and a complex coefficient C for the stress function Ω.

$$\Omega(z) = Cz^\lambda = Cr^\lambda e^{i\lambda\theta} . \tag{3.47}$$

Concerning the boundary conditions, the shear stress τ_{23} at the crack faces $\theta = \pm\pi$ has to be zero:

$$\tau_{23} = -\Im\Omega'(z) = \left(\overline{\Omega'(z)} - \Omega'(z)\right) \tag{3.48}$$

$$\begin{aligned} \theta = +\pi : \quad & \lambda r^{\lambda-1}\left[\overline{C}e^{-i(\lambda-1)\pi} - Ce^{i(\lambda-1)\pi}\right] = 0 \\ \theta = -\pi : \quad & \lambda r^{\lambda-1}\left[\overline{C}e^{i(\lambda-1)\pi} - Ce^{-i(\lambda-1)\pi}\right] = 0. \end{aligned} \tag{3.49}$$

In order to find the solution of this homogeneous system of equations for C and $\overline{C}$, it is necessary to set the coefficient determinant to zero, which leads to the eigenvalue equation λ:

$$\sin(2\lambda\pi) = 0 \quad\Rightarrow\quad \lambda = \frac{n}{2} \quad n = 1, 2, 3, \ldots \tag{3.50}$$

This reveals the same eigenvalues as in mode I and mode II. The entire solution can now be composed by combining all eigenfunctions with the coefficient C_n. The relation $\overline{C}_n = (-1)^n C_n$ follows from (3.49), which means the coefficients are alternating either purely real or imaginary:

$$\Omega(z) = \sum_{n=1}^{\infty} C_n z^{\frac{n}{2}}, \quad C_n = -i^n c_n . \tag{3.51}$$

Finally, the corresponding eigenfunctions can be calculated from (3.47) using the relations (A.165):

$$u_3^{(n)}(r,\theta) = \frac{c_n}{2\mu} r^{\frac{n}{2}} \tilde{H}_3^{(n)}(\theta), \quad \tilde{H}_3^{(n)} = \begin{cases} 2\sin\frac{n}{2}\theta & \text{for } n = 1, 3, \cdots \\ 2\cos\frac{n}{2}\theta & \text{for } n = 2, 4, \cdots \end{cases} \tag{3.52}$$

$$\tau_{13}^{(n)}(r,\theta) = c_n r^{\frac{n}{2}-1} \tilde{L}_{13}^{(n)}(\theta), \quad \tilde{L}_{13}^{(n)} = \begin{cases} \frac{n}{2}\sin(\frac{n}{2}-1)\theta & \text{for } n = 1, 3, \cdots \\ \frac{n}{2}\cos(\frac{n}{2}-1)\theta & \text{for } n = 2, 4, \cdots \end{cases}$$

$$\tau_{23}^{(n)}(r,\theta) = c_n r^{\frac{n}{2}-1} \tilde{L}_{23}^{(n)}(\theta), \quad \tilde{L}_{23}^{(n)} = \begin{cases} \frac{n}{2}\cos(\frac{n}{2}-1)\theta & \text{for } n = 1, 3, \cdots \\ -\frac{n}{2}\sin(\frac{n}{2}-1)\theta & \text{for } n = 2, 4, \cdots \end{cases} \tag{3.53}$$

Considering $n = 1$, the asymptotic singularity according to (3.31) and (3.32) is exactly reproduced, whereby the relation $K_{III} = c_1\sqrt{\pi/2}$ applies. The eigenfunction for $n = 2$ corresponds to a constant shear stress $\tau_{13} = T_{13} = c_2$.

The derived eigenfunctions apply to all plane elastic crack problems. Therefore, the solution of the boundary value problem can be formulated as complete series expansion of the terms (3.41), (3.43), (3.52) and (3.53):

$$\sigma_{ij}(r,\theta) = \sum_{n=1}^{\infty} r^{\frac{n}{2}-1} \left[a_n \tilde{M}_{ij}^{(n)}(\theta) + b_n \tilde{N}_{ij}^{(n)}(\theta) + c_n \tilde{L}_{ij}^{(n)}(\theta) \right] \qquad (3.54)$$

$$u_i(r,\theta) = \frac{1}{2\mu} \sum_{n=1}^{\infty} \left[a_n \tilde{F}_i^{(n)}(\theta) + b_n \tilde{G}_i^{(n)}(\theta) + c_n \tilde{H}_i^{(n)}(\theta) \right]. \qquad (3.55)$$

The unknown coefficients a_n, b_n and c_n have to be determined from the boundary conditions of the actual crack problem and represent the modes I, II and III respectively.

These eigenfunctions form an indispensable basis for many numerical methods used to treat crack problems in finite bodies. Their usage started at the first calculations by means of a boundary collocation method in the 1960s [3] and lasts until the recent development of special elements for crack tips [4].

3.2.3 Three-Dimensional Crack Problems

In many cases in practice, the crack configuration is of three-dimensional character. If e.g. a planar flaw is embedded in a spatial structure, usually curvilinear crack faces and fronts occur, see Fig. 3.8a. Even for a plane crack geometry, a three-dimensional crack problem is present when the stress state changes along the crack. This often appears at through cracks in specimens of finite thickness, see Fig. 3.8b. Of practical importance are surface cracks where the crack front intersects the body's external surface as Fig. 3.8c shows. Closed-form analytical solutions for spatial crack configurations only exist for a limited number of simple cases, mostly in infinite domains.

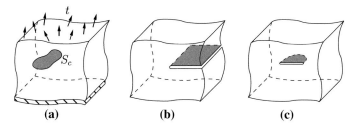

Fig. 3.8 Spatial crack configurations

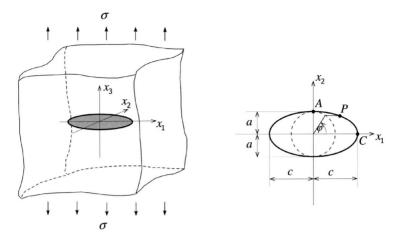

Fig. 3.9 Elliptic internal crack in infinite domain with coordinate system

Internal Elliptical Crack Under Tension

An important example is the elliptic planar internal crack in an infinite domain, see Fig. 3.9. For this type of crack configuration, many solutions under different types of loading have been found by Sneddon [5] using integral transformations and by Fabrikant [6] using spatial stress potential functions. As an example, the result for tensile loading σ perpendicular to the crack plane is given here. The elliptic crack has two semi-axes a and c. A point P of the crack front is defined by the angle φ over $x_1 = c \cos \varphi$ and $x_2 = a \sin \varphi$. This is a pure mode I loading with a variable K_I–factor along the crack front.

$$K_I = \frac{\sigma \sqrt{\pi a}}{E(k)} \left(\sin^2 \varphi + \frac{a^2}{c^2} \cos^2 \varphi \right)^{\frac{1}{4}} \tag{3.56}$$

$E(k)$ is the complete elliptic integral of second kind $k = \sqrt{1 - a^2/c^2}$

$$E(k) = \int_0^{\frac{\pi}{2}} \sqrt{1 - k^2 \sin^2 \alpha} \, d\alpha \approx \sqrt{1 + 1.464 \left(\frac{a}{c} \right)^{1.65}} \quad \text{for } (a \leq c). \tag{3.57}$$

The maximum value of K_I is reached at the apex A of the minor semi-axis ($a < c$) and the minimum value at C.

$$K_{Imax} = K_{IA} = \frac{\sigma \sqrt{\pi a}}{E(k)}, \quad K_{Imin} = K_{IC} = K_{IA} \sqrt{\frac{a}{c}} \tag{3.58}$$

In the special case of a circular crack, the K_I–factor has everywhere the same value

$$K_I = \frac{2}{\pi}\sigma\sqrt{\pi a} \quad \text{for } c = a.$$ (3.59)

Spatial Crack Tip Field

By means of studies on elliptic internal cracks [5] and through cracks in plates with straight and curved crack fronts [7, 8], it has been shown that, in principle, the same near fields exist at three-dimensional crack tips, as we discovered in the two-dimensional case. However, the asymptotic solutions now only apply locally with respect to a point on the crack front. That is why according to Fig. 3.10, an associated Cartesian coordinate system (n, v, t) is introduced along the crack front (coordinate s), where $n \triangleq x_1$ lies normally toward the crack front. $t \triangleq x_3$ runs tangentially along it and $v \triangleq x_2$ lies vertically upon the crack plane. Performing a limit process $r \to 0$ within the normal plane (n, v) toward the crack front, the near tip solutions of the two-dimensional crack problem under the plane strain state can be found: In the general case, these are composed of the mode I, II and III components.

Therefore, the following applies to the crack tip field:

$$\sigma_{ij}(r, \theta, s) = \frac{1}{\sqrt{2\pi r}}\left[K_I(s)f_{ij}^I(\theta) + K_{II}(s)f_{ij}^{II}(\theta) + K_{III}(s)f_{ij}^{III}(\theta)\right] + T_{ij}(s),$$ (3.60)

whereby the angular functions $f_{ij}^{I,II,III}$ describe the terms (3.12), (3.23) and (3.32). The term T_{ij} includes all stress components of the $n = 2$nd order, representing constant finite values in the (x_1, x_3)–crack plane.

$$[T_{ij}] = \begin{bmatrix} T_{11} & 0 & T_{13} \\ 0 & 0 & 0 \\ T_{31} & 0 & T_{33} \end{bmatrix}$$ (3.61)

Fig. 3.10 Coordinate system along a crack front in space

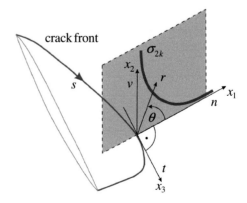

In the same way it is possible to summarize the asymptotic displacement fields of the three modes (3.16), (3.25) and (3.31) using the angular functions $g_i^{\mathrm{I,II,III}}$ for the plane strain:

$$u_i(r, \theta, s) = \frac{1}{2\mu}\sqrt{\frac{r}{2\pi}}\left[K_{\mathrm{I}}(s)g_i^{\mathrm{I}}(\theta) + K_{\mathrm{II}}(s)g_i^{\mathrm{II}}(\theta) + K_{\mathrm{III}}(s)g_i^{\mathrm{III}}(\theta)\right] . \quad (3.62)$$

By means of local superposition of the three modes with the respective stress intensity factors as coefficients, the near field in the vicinity of the crack front is uniquely defined by these relations. Here, the three K–factors and the T_{ij}–stresses are functions of the crack front position s.

It is possible to show that a state of plane strain asymptotically prevails at any point of the crack front in the body. Special considerations are necessary for such points at the crack front, which cross the surface, since here a situation similar to the plane stress state exists, where in most cases a different type of singularity occurs.

3.2.4 Stress Intensity Factors: K-Concept

For isotropic linear-elastic material behavior the asymptotic near field solutions are always of the same mathematical form (3.60) and (3.62). The strength of this crack tip field is entirely determined by the *stress intensity factors* K_{I}, K_{II} and K_{III}, which quasi represent» still free « coefficients. The magnitudes of the three stress intensity factors have to be determined by the solution of the specific boundary value problem defined by a body with crack. Thus, the K–factors depend on the geometry of the body, the size and position of the crack as well as on the load and the bearing conditions. To determine them, it is generally necessary to find first the complete solution of the boundary value problem using analytical or numerical calculation methods and then to analyze the crack tip field. A closer look at the stress fields on the ligament in front of the crack ($\theta = 0$) of the three crack opening modes I (3.12), II (3.23) and III (3.32) shows that only those stress components are different from zero, which corresponds to the respective crack opening mode or the far field loading, i. e., σ_{22} in mode I, τ_{21} in mode II and τ_{23} in mode III. The related angular function approaches just the value one. By solving the relations for the stress intensity factors and taking the limit towards the crack tip, the conditional equations are obtained:

$$\left\{\begin{array}{c} K_{\mathrm{I}} \\ K_{\mathrm{II}} \\ K_{\mathrm{III}} \end{array}\right\} = \lim_{r\to 0} \sqrt{2\pi r}\left\{\begin{array}{c} \sigma_{22}(r, \theta = 0) \\ \tau_{21}(r, \theta = 0) \\ \tau_{23}(r, \theta = 0) \end{array}\right\} . \quad (3.63)$$

In the following chapters of this book we will elaborate in detail on applications of the finite element method for determining stress intensity factors. It is beyond our scope to thoroughly address analytic calculation methods for elastic crack problems. Only the most important methods are mentioned below, as well as the respective literature ([9] that provides an overview):

- Complex function theory (complex analyses, conform mappings, series expansion): Muskhelishvili [1], Tamusz [10]
- Integral transforms (Laplace, Hankel): Sneddon and Lowengrub [5]
- Singular integral equations: Erdogan [11], Muskhelishvili [1]
- Three-dimensional potential approaches: Kassir and Sih [12], Fabrikant [6].

Generally, the stress intensity factors for all crack problems can be written in the following form (example only for K_I):

$$
\begin{aligned}
K_I &= K_I \text{ (geometry, fracture, load, material)} \\
&= \sigma_n \sqrt{\pi a}\; g \text{ (geometry, material)} ,
\end{aligned}
\tag{3.64}
$$

whereby a represents the crack length, σ_n denotes a representative nominal stress and the function g describes the influence of the body and crack geometry as well as, in certain circumstances, the elastic material properties.

Inspecting this formula makes clear that in fracture mechanics the common similarity relations do not apply anymore. That is, if a body with a crack is geometrically enlarged by the factor β, while the nominal load σ_n stays unchanged, the K_I–value is increased by the factor $\sqrt{\beta}$ because of the extended crack length! The stress intensity factors have the quite peculiar dimension [stress]$\times$ [length]$^{1/2}$ and are given in the units $\mathrm{Nmm}^{-3/2}$ or $\mathrm{MPa}\sqrt{\mathrm{m}}$.

In previous decades, numerous K–factor solutions for various crack configurations and types of loading have been calculated in many different ways. They are compiled in hand books, see Murakami [13], Rooke and Cartwright [14], Tada, Paris and Irwin [15] and Theilig and Nickel [16].

The stress intensity factors form an excellent basis for formulating fracture criteria. The idea arose from Irwin [17], who suggested the *concept of stress intensity factors* in 1957. It is based on the following ideas, which will at first be described for the case of mode I:

- The singular crack tip solution with coefficient K_I describes the loading state in a finite region around the crack tip with radius r_K. At a greater distance $r > r_K$, its dominance fades because other terms of the series (3.39) gain influence, see Fig. 3.11.
- In reality (even in brittle materials), the K_I–singularity does of course not reach up to the very crack tip $r \to 0$, because here a fracture process zone develops, inside of which the limits of elasticity theory are exceeded, compare Fig. 3.2. If,

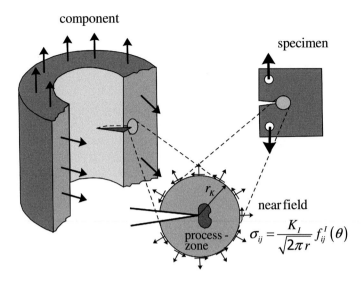

Fig. 3.11 Dominance of the near field solution at the crack tip in all test specimen and components

however, it is assumed that the size r_B of the process zone is much smaller than the domain r_K of validity of K_I–solution, all fracture processes are controlled by the near field solution, acting as »boundary condition« . Given the same K_I, the same process happens, independently of which type of crack configuration is present, see Fig. 3.11. Conversely, this assumption confirms that the process zone has negligible reactions on the near field solution.

- By means of this »autonomy principle of crack tip singularity« we have reduced the entire body geometry and loading via (3.64) to the K_I–factor.
- The crack propagation will initiate just when a critical material state in the process zone is reached. This material-specific limit value of load carrying capacity is called *fracture toughness*.

Thus, the stress intensity concept provides the fracture criterion

$$K_I = K_{Ic} . \qquad (3.65)$$

The fracture-mechanical loading quantity K_I stands on the left side of the equation. On the right-hand side, the fracture toughness K_{Ic} represents the material resistance against crack initiation.

Introduction of the *stress intensity concept* marks a milestone in the development of fracture mechanics. To this day the concept has proven itself many times.

The same considerations can be transferred to other types of crack opening if they solely occur as pure mode II or pure mode III loading, respectively

$$K_{\mathrm{II}} = K_{\mathrm{IIc}} \quad \text{and} \quad K_{\mathrm{III}} = K_{\mathrm{IIIc}}. \tag{3.66}$$

K_{IIc} and K_{IIIc} represent the corresponding fracture toughness values. Unfortunately, these modes rarely appear isolated. In the general case, a combined loading of the crack consists of all three modes so that the fracture process is controlled by K_{I}, K_{II} and K_{III}. Then, the fracture criterion has to be formulated using a generalized stress parameter B and an assigned material parameter B_{c}.

$$B(K_{\mathrm{I}}, \ K_{\mathrm{II}}, \ K_{\mathrm{III}}) = B_{\mathrm{c}} \tag{3.67}$$

The respective criteria are introduced in Sect. 3.4.4

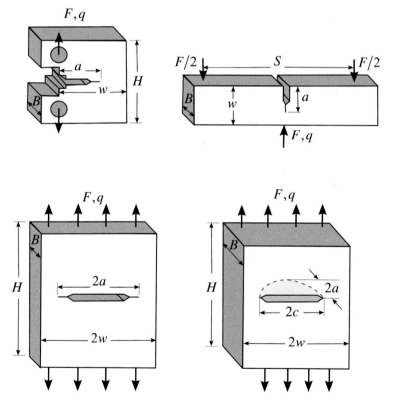

Fig. 3.12 Common forms of specimen used to determine fracture-mechanical parameters compact tension (CT) test, Singe-edged notched bending (SENB) test, Central-crack tension (CCT) test and tensile test with semi-elliptical surface crack M(T)

The *fracture toughness* K_{Ic} of a material has to be experimentally determined on a specimen having an initial crack by means of a standardized fracture experiment. For this, suitable specimen geometries are developed. The well-known ones are displayed in Fig. 3.12. A fatigue crack is created at the notch tip in order to gain reproducible starting conditions. Afterwards, the specimen is stressed in a testing machine with a force F and the deformation path q is recorded. During elastic material behavior the force-deformation diagram takes a linear course until the initiation of brittle fracture at the force F_c. Taking this force and the crack length a, the K-factor at fracture is calculated with the help of the known geometry functions $g(a, w)$ of the specimen. Subject to certain validity criteria, this value represents the fracture toughness K_{Ic} .

$$K_{Ic} = F_c \, g(a, w) \sqrt{\pi a} \tag{3.68}$$

The specimen preparation, test procedure and evaluation of test results to determine the material parameters of linear-elastic fracture mechanics have all been standardized. Mandatory documents are the international norm ISO 12135 [18], the ESIS-P2 regulations [19] in Europe and the ASTM 1820 standard [20] in the USA. Detailed information on fracture-mechanical material testing can be found e.g. in Blumenauer & Pusch[21].

3.2.5 Energy Balance During Crack Propagation

Global Energy Release Rate

The energy balance during crack propagation is investigated in a body with a crack. For this purpose we consider the boundary value problem shown in Fig. 3.13. Surface tractions $\bar{t}$ are imposed on the part S_t of the external boundary, Eventually body forces $\bar{b}$ act in the volume V, and displacements $\bar{u}$ are prescribed on the boundary part S_u. Applying the 1st law of thermodynamics to a deformable body provides the change in energy per time:

$$\dot{W}_{ext} + \dot{Q} = \dot{W}_{int} + \dot{\mathcal{K}} + \dot{\mathcal{D}}. \tag{3.69}$$

On the left-hand side of the equation stands the energy input into the body per time as a result of the power of the external mechanical loading

$$\dot{W}_{ext} = \int_{S_t} \bar{t}_i \dot{u}_i \, dS + \int_V \bar{b}_i \dot{u}_i \, dV \tag{3.70}$$

and the exchange of thermal energy $\dot{Q}$ by way of thermal flux or internal heat sources. On the right-hand side of the balance equation stand those types of energy absorbed by the body per time, i. e. the internal energy, which for the purely mechanical case correlates to

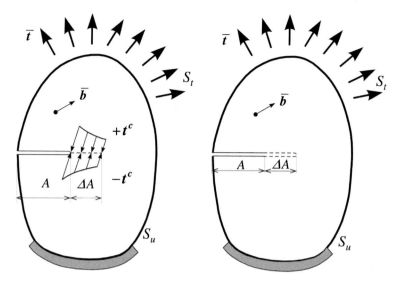

Fig. 3.13 Energy balance during crack propagation by ΔA

$$\mathcal{W}_{\text{int}} = \int_V U \, \mathrm{d}V \,, \quad U(\varepsilon_{kl}) = \int_0^{\varepsilon_{kl}} \sigma_{ij}(\varepsilon_{mn}) \, \mathrm{d}\varepsilon_{ij} \,, \tag{3.71}$$

and the kinetic energy $\mathcal{K}$ (with density ρ)

$$\mathcal{K} = \frac{1}{2} \int_V \rho \, \dot{u}_i \dot{u}_i \, \mathrm{d}V \,. \tag{3.72}$$

In addition, there is the mainly dissipative energy $\mathcal{D}$, consumed during the crack propagation in the process zone. Since it is directly related to the creation of new surfaces, this term is set proportionally to the crack area A with the material constant γ. The factor 2 accounts for the fact that a fracture leads to two new surfaces.

$$\mathcal{D} = 2\gamma A \tag{3.73}$$

In this section we will concentrate on static problems, so that $\mathcal{K} = 0$. For purely elastic deformations $U = U^e$, the internal energy possesses the character of an internal potential $\Pi_{\text{int}} = \mathcal{W}_{\text{int}}$. Furthermore, we consider the body to be an adiabatically closed system without any internal heat sources, so that $\dot{\mathcal{Q}} = 0$ as well. Finally, it is assumed that external loads are conservative forces (gravity, springs), resulting from a potential Π_{ext}, which decreases with the performed external work $\dot{\Pi}_{\text{ext}} = -\dot{\mathcal{W}}_{\text{ext}}$. Thereby, the energy balance (3.69) simplifies to

$$\dot{\mathcal{W}}_{\text{ext}} = \dot{\mathcal{W}}_{\text{int}} + \dot{\mathcal{D}} \,, \quad -\dot{\Pi}_{\text{ext}} = \dot{\Pi}_{\text{int}} + \dot{\mathcal{D}} \,. \tag{3.74}$$

With these assumptions we will now analyze the propagation of a crack, having the initial size $A^{(1)} = A$ at the time $t^{(1)}$ and enlarging in a quasi-static process to the area $A^{(2)} = A + \Delta A$ at the time $t^{(2)} = t^{(1)} + \Delta t$, see Fig. 3.13. Therefore, the following energy difference related to the time increment Δt or equally to the change of the crack area ΔA, exists between the final and starting state:

$$W_{\text{ext}}^{(2)} - W_{\text{ext}}^{(1)} = W_{\text{int}}^{(2)} - W_{\text{int}}^{(1)} + 2\gamma\,\Delta A$$
$$\Delta W_{\text{ext}} = \Delta W_{\text{int}} + \Delta\mathcal{D} \implies \frac{\Delta W_{\text{ext}}}{\Delta A} = \frac{\Delta W_{\text{int}}}{\Delta A} + \frac{\Delta\mathcal{D}}{\Delta A} . \tag{3.75}$$

Introducing the internal and external potentials, which can be combined to the total potential $\Pi = \Pi_{\text{int}} + \Pi_{\text{ext}}$, we obtain

$$\frac{\Delta(W_{\text{ext}} - W_{\text{int}})}{\Delta A} = -\frac{\Delta\Pi}{\Delta A} \overset{!}{=} \frac{\Delta\mathcal{D}}{\Delta A} = 2\gamma . \tag{3.76}$$

Physically, this result can be interpreted as follows: the left-hand side describes the available amount of potential energy $-\Delta\Pi$, which is supplied by the external load and the elastically stored internal energy during crack propagation by ΔA (the minus sign indicates the decrease of the potential energy). This quantity is therefore called *energy release rate* and is defined for finite or infinitesimal crack propagation as follows

$$\overline{G} = -\frac{\Delta\Pi}{\Delta A}, \quad G = -\lim_{\Delta A \to 0}\frac{\Delta\Pi}{\Delta A} = -\frac{d\Pi}{dA} . \tag{3.77}$$

The right-hand side of (3.76) displays the fracture energy $2\Delta A$ required for material separation and formation of new surfaces. Its quantity depends on the material behavior and represents the critical material parameter $G_{\text{c}} = 2\gamma$. This energy balance during crack propagation has been compiled by A. A. Griffith [22] and named after him.

Energetic fracture criterion by Griffith:

$$-\frac{d\Pi}{dA} = G = G_{\text{c}} = 2\gamma, \tag{3.78}$$

It states the following: In order to initiate and maintain quasi-static crack propagation in a conservative system, the provided energy release rate has to be equal to the dissipative energy needed for fracture per crack area. The Dimension of G is [force $\times$ length/length^2] and is mostly specified in J/m^2 or N/m.

Griffith determined the energy release rate $G = 2\pi\sigma^2 a/E' = 4\gamma$ for a crack of length $2a$ in a sheet under tension (Fig. 3.3). For a given crack length, the critical fracture stress can be found as

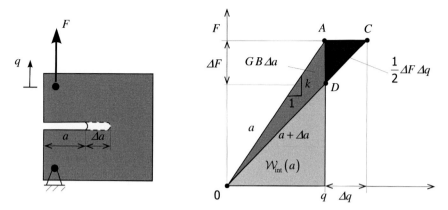

Fig. 3.14 Correlation between force–deformation–curve and energy release rate

$$\sigma_c = \sqrt{\frac{2E'\gamma}{\pi a}} = \frac{K_{Ic}}{\sqrt{\pi a}} \,. \tag{3.79}$$

Solved for a, the criterion provides the critical crack length a_c that is necessary for fracture under the given load

$$a_c = \frac{2E'\gamma}{\pi\sigma^2} \quad \text{with} \quad E' = E \text{ (plane stress)} \quad \text{and} \quad E' = E/(1-\nu^2) \text{ (plane strain)}. \tag{3.80}$$

The Griffith–criterion also applies to any finite crack propagation ΔA and in an extreme case even to the complete generation of a crack from zero state $A^{(1)} = 0$ to its final state $A^{(2)} = A$.

Finally, the relation between the energy release rate and the force-deformation diagram of a specimen with crack will be established, see Fig. 3.14. In this case, the imposed external loads consist only of the concentrated force F, which causes a displacement $q = F/k(a)$ of the load point. Hereby, $k(a)$ represents the stiffness of the specimen depending on the crack length. The strain energy in this elastic body equals the area below the linear force–deformation–curve

$$\mathcal{W}_{int} = \frac{1}{2}Fq = \frac{1}{2}kq^2 = \frac{1}{2}\frac{F^2}{k} \,. \tag{3.81}$$

If the crack length is increased by Δa, then the stiffness of the specimen will decrease as shown in Fig. 3.14. Depending on how the external load behaves during crack propagation, two extreme cases can be distinguished:

(a) Fixed grips ($q = $ const)
This corresponds to a very stiff loading device. The external work is zero $d\mathcal{W}_{ext} = F\, dq = 0$, and the potential energy for crack propagation arises only from the stored strain energy so that (3.76) gives

$$G = -\frac{d\Pi}{dA} = -\frac{dW_{int}}{dA} = -\frac{d}{Bda}\left[\frac{1}{2}k(a)q^2\right] = -\frac{q^2}{2B}\frac{dk(a)}{da}$$

$$= -\frac{1}{2B}\frac{F^2}{k^2}\frac{dk(a)}{da} = -\frac{1}{2Bda}q\,dF$$

$$(3.82)$$

(b) Dead load ($F = $ const)

This case is realized by a weight or a very soft loading device that performs the external work $dW_{ext} = F\,dq(a) > 0$ during crack propagation. If the last term of (3.81) is used for W_{int}, the energy release rate reads

$$G = -\frac{d\Pi}{dA} = \frac{d}{Bda}[W_{ext} - W_{int}] = \frac{1}{B}\left[F\frac{dq}{da} - \frac{1}{2}F^2\frac{d}{da}\left(\frac{1}{k(a)}\right)\right]$$

$$= -\frac{1}{2B}\frac{F^2}{k^2}\frac{dk(a)}{da} = -\frac{1}{2Bda}F\,dq. \qquad (3.83)$$

Comparing the results (3.82) and (3.83) leads to the surprising conclusion that in both cases the energy release rate G is identical and naturally positive (since $dk/da < 0$). While in case (a) the crack-driving energy is due to the decrease of W_{int}, it is supplied in case (b) by the external work, which in addition increases the strain energy W_{int} by the same value, compare [ ]–brackets in (3.83). This statement is generally true for elastic systems with cracks, since according to the law of Clapeyron $W_{ext} = 2W_{int}$ holds. From Fig. 3.14 it is apparent that the released energy $-d\Pi = GBda$ matches the triangular area between the two force–deformation–curves of a and $a + \Delta a$. The area OAD is to be assigned to case (a) and the area OAC to case (b), respectively. Both areas only differ in the triangle $ACD \equiv \frac{1}{2}\Delta F\Delta q$, which vanishes for $\Delta a \to 0$ of higher order.

Local Energy Release Rate

Continuing our considerations with the help of Fig. 3.13, all mechanical field variables experience a change from the initial state (1) to the final state (2) during crack propagation. That transition is carried out theoretically in the following way: In (1), sectional tractions $t^c = t^{c(1)}$ apply to the ligament ΔA in front of the crack, which are by definition equal and opposite on both faces $t_i^c \equiv t_i^{c+} = -t_i^{c-}$. We open the crack by cutting along ΔA and replace the tractions by external boundary tractions of equal value, so that the crack stays closed as in (1). Subsequently, all these boundary tractions are being quasi-statically reduced to zero, wherewith the crack expands to the final state (2) and attains a stress-free surface ΔA. The relative displacements $\Delta u_i = u_i^+ - u_i^-$ of the crack faces change from the closed state ($\Delta u_i^{(1)} = 0$) to the opened state $\Delta u_i^{(2)}$. During this process the tractions perform work at the crack face displacements $t_i^{c+}du_i^+ + t_i^{c-}du_i^- = t_i^c(du_i^+ - du_i^-) = t_i^c d\Delta u_i$, so that the total work to extend the crack by ΔA can be expressed by the following integral

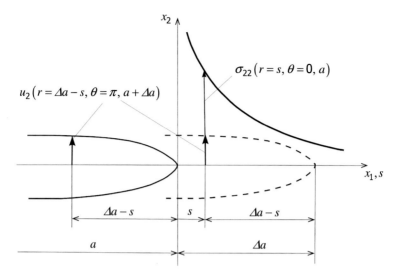

Fig. 3.15 Load releasing work during crack propagation

$$\Delta W_c = \int_{\Delta A} \int_{(1)}^{(2)} t_i^c \, \mathrm{d}\Delta u_i \mathrm{d}A \implies \int_{\Delta A} \frac{1}{2} t_i^c \Delta u_i^{(2)} \mathrm{d}A < 0 . \tag{3.84}$$

Since hereby the system gives off energy, it has to be negative. For linear–elastic material behavior the inner integral can be evaluated to $t_i^c \Delta u_i^{(2)}/2$ and we get the formula on the right.

This work term ΔW_c must be added to the total balance of energy per crack propagation.

$$\frac{\Delta W_{\mathrm{ext}}}{\Delta A} - \frac{\Delta W_{\mathrm{int}}}{\Delta A} + \frac{\Delta W_c}{\Delta A} = 0 \tag{3.85}$$

Using the definition (3.76) of the potential energy $\Pi = \Pi_{\mathrm{ext}} + \Pi_{\mathrm{int}} = -W_{\mathrm{ext}} + W_{\mathrm{int}}$, the energy release rate finally follows.

$$G = -\frac{\Delta \Pi}{\Delta A} = \frac{\Delta W_{\mathrm{ext}}}{\Delta A} - \frac{\Delta W_{\mathrm{int}}}{\Delta A} = -\frac{\Delta W_c}{\Delta A} . \tag{3.86}$$

Hereby it has been shown that the change of potential energy of the system (body plus load) equals the work that is needed locally to »set free« the new crack surfaces ΔA. Note that this is a virtual elastic unloading process, which is not concerned with the energy dissipation $\mathcal{D}$ during a real fracture process! The result (3.86) therefore represents an alternative calculation method for the energy release rate—the crack driving force. In principle, this method is also applicable to finite crack extension ΔA.

The Crack Closure Integral

In the following section this calculation method will be applied to the asymptotic crack tip solution for isotropic–elastic material (Sect. 3.2.2). This will first be
explained using crack opening type I. The tip of a crack of initial length a, as shown in
Fig. 3.15, is displaced by Δa during crack propagation. We assume that this process
takes place in the domain of validity of the K_I–dominated crack tip solution (3.12),
(3.16). This can always be achieved if the crack tip is approached closely enough
taking the limits $r \to 0$ and $\Delta a \to 0$. In mode I, a symmetric stress and deformation
state exists with respect to the crack plane. The tractions on the ligament Δa correspond to the σ_{22}–stresses of the near field solution (3.12) for the initial crack length
a

$$- t_2^c(s) = \sigma_{22}(r = s, \theta = 0, a) = \frac{K_I(a)}{\sqrt{2\pi s}}, \quad t_1^c = t_3^c = 0. \tag{3.87}$$

After crack propagation by Δa, the crack opening displacements result from (3.16)
for the final crack length $a + \Delta a$ to

$$u_2^\pm(r = \Delta a - s, \theta = \pm\pi, a + \Delta a) = \pm \frac{\kappa + 1}{2\mu} K_I(a + \Delta a)\sqrt{\frac{\Delta a - s}{2\pi}}. \tag{3.88}$$

During a virtual load releasing process, the work of the stresses $t_2^c(s)$ performed with
the displacements $u_2(s)$ of the crack faces, results according to (3.84) for linear–
elastic behavior in

$$\Delta W_c = -\int_0^{\Delta a} \frac{1}{2}\sigma_{22}(u_2^+ - u_2^-)ds = -\int_0^{\Delta a} \frac{K_I(a)}{\sqrt{2\pi s}} \frac{\kappa + 1}{2\mu} K_I(a + \Delta a)\sqrt{\frac{\Delta a - s}{2\pi}} ds$$

$$= -\frac{\kappa + 1}{4\pi\mu} K_I(a) K_I(a + \Delta a) \int_0^{\Delta a} \sqrt{\frac{\Delta a - s}{s}} ds. \tag{3.89}$$

Hence the energy release rate G (which we now regard per uniform thickness
$B = 1$ m in the plane problem) can be calculated by means of (3.86). Additionally,
we carry out the limiting process to a differential crack propagation $\Delta a \to 0$.

$$G = \lim_{\Delta a \to 0}\left(-\frac{\Delta W_c}{B\Delta a}\right) = \frac{\kappa + 1}{4\pi\mu} K_I^2(a) \lim_{\Delta a \to 0} \underbrace{\frac{1}{\Delta a}\int_0^{\Delta a}\sqrt{\frac{\Delta a - s}{s}} ds}_{\pi/2} \tag{3.90}$$

Thus the following correlation exists between the infinitesimal energy release rate G and the stress intensity factor K_I

$$G \mathrel{\hat{=}} G_I = \frac{\kappa + 1}{8\mu} K_I^2 = K_I^2 / E'. \tag{3.91}$$

Physically, this means that within linear–elastic fracture mechanics the stress intensity concept by Irwin and the energy criterion by Griffith are equal and can be converted into each other.

The correlation derived here goes back to Irwin [23], who calculated the work at a virtual crack closure of $a + \Delta a$ to a and named it *crack closure integral*. But within the theory of elasticity, crack closure and crack expansion are equivalent and reversible processes, which lead to the same result. For non-linear material behavior however, this is not valid anymore so that, in general, one should denote it as a *crack opening integral*.

Finally, the generalization for mode II and mode III will be elaborated. During mode II, the tractions $t_1^c \mathrel{\hat{=}} \tau_{12}$ and crack face displacements $u_1^{\pm} \neq 0$ have components solely in the direction of x_1 (note (3.23) and (3.25)), which are proportional to K_{II}. Similarly, in mode III only components in direction of x_3 appear according to (3.31) and (3.32) with the factor K_{III}, so that the result for the energy release rate is

$$\text{Modus II: } G_{II} = K_{II}^2 / E', \quad \text{Modus III: } G_{III} = K_{III}^2 / 2\mu = K_{III}^2 (1+\nu)/E. \tag{3.92}$$

If mode I, II and III loading are combined, the energy release rate during infinitesimal crack propagation In the direction of x_1 is calculated using the sum

$$G = G_I + G_{II} + G_{III} = \frac{1}{E'} \left(K_I^2 + K_{II}^2 \right) + \frac{1+\nu}{E} K_{III}^2. \tag{3.93}$$

Stability of Crack Propagation

The fracture criterion by Griffith (3.78) sets the necessary energetic conditions for a crack to be able to propagate at all. In order to assess the further course of crack propagation–especially the issue of stability–it is crucial to see how the fracture condition itself changes. The energy release rate G is both a function of crack length a and of loading, which, depending on the types of boundary conditions, can be controlled either by forces (F) or displacements (q). On the other hand, in many brittle materials such as ceramic or concrete, one can observe that the crack growth

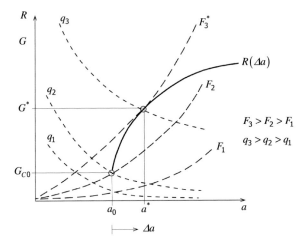

Fig. 3.16 Regarding stability of crack propagation

resistance G_c increases during crack propagation Δa from an initial value G_{c0} to a saturation value. The reason for this is the formation of the process zone until its final shape. This material-specific behavior is described by the *crack growth resistance curve* $R(\Delta a)$ (see Fig. 3.16)

$$G_c = R(\Delta a) \quad (R - curve),\tag{3.94}$$

which is measured during fracture experiments with a steadily expanding crack.

During crack propagation, the thus modified fracture criterion must be fulfilled

$$G(F, q, a) = R(\Delta a),\tag{3.95}$$

which represents in a way the state of balance between the crack driving force and the crack resistance. In order to assess the stability of the fracture process, we need to compare the changes of both values during crack propagation as functions of the crack length, i.e.

$$\left.\frac{\partial G}{\partial a}\right|_{F,q} \lesseqgtr \frac{\partial R}{\partial a} \quad \begin{cases} \text{stable} \\ \text{indifferent} \\ \text{unstable} \end{cases}\tag{3.96}$$

whereby the load (F or q) is fixed. The crack behavior is called *stable*, when the crack resistance R increases faster than the driving energy release rate G. In this case, it is necessary to enhance the load to let the crack expand further. In Fig. 3.16 the crack resistance curve as well as the energy supply for crack propagation G are displayed as functions of the crack length for a set of external loads (dashed lines). At given force F, these curves show a monotonic increasing course with a. On the contrary, at fixed displacements q, the $G(a)$–curves have a falling tendency, because the crack relaxes itself by its growth. The crack behavior is regarded as *unstable* as soon as the energy supply increases faster than the crack resistance does. Then, the slope of the G–curve in Fig. 3.16 is equal to or greater than the one of the R–curve (values marked with *). The excessive energy causes an accelerated, dynamic crack propagation. Considering however the G–curves for fixed displacements in Fig. 3.16, it becomes apparent that after crack initiation at q_2 the crack can never become unstable despite an increasing load. Everybody knows this kind of behavior from splitting wood with a wedge. It can also happen under strain-controlled thermal stresses in a cracked component.

Finally, it is worth noting that all considerations presented with the help of G and R can similarly be assigned to the stress intensity factor K_I and to a crack length-dependent fracture toughness $K_c(\Delta a)$, since both criteria are equivalent in linear-elastic fracture mechanics.

3.2.6 The J-Integral

Independently, Cherepanov [24] (1967) and Rice [25] (1968) introduced another fracture-mechanical load parameter – the J-integral. This parameter has proven extremely valuable not only in linear-elastic fracture mechanics, but it could also be applied very successfully in fracture mechanics at inelastic material behavior. Subsequently, diverse extensions of the classic J integral have been made with regard to types of loading, material laws and field problems, which will be elab-

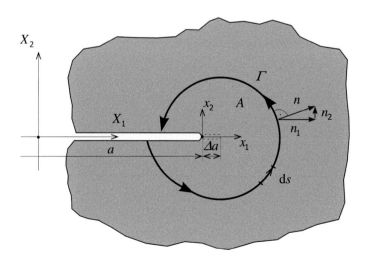

Fig. 3.17 Definition of J as line integral around the crack tip

orated in Sect. 6. Particularly in connection with numerical stress analysis of cracks the J-integral holds special meaning.

In this section we will concentrate on application of the J-integral to elastic materials, which may also be non-linear. For an easier understanding the following elaborations are limited to infinitesimal deformations.

Derivation of the J-Integral

In the following section it will be proven that the change of potential energy during infinitesimal crack propagation–the energy release rate $G = -\mathrm{d}\Pi/\mathrm{d}A$–can be expressed with the help of a path-independent line integral. Figure 3.17 shows a linear crack problem (uniform thickness B). We choose an arbitrary domain A around the crack tip surrounded by the curve Γ, which runs from the lower to the upper crack face in a mathematically positive sense. The normal unit vector n_j points toward the outside. In order to calculate the potential energy of the system, we would have to consider the entire body. But we will find out that the result does not depend on the chosen domain. From outside the sectional stresses $t_i = \sigma_{ij}n_j$ act on Γ. They are supposed to keep constant during crack growth $\mathrm{d}a$. The body forces are zero. The crack expands along its initial direction by $\mathrm{d}a$ and the domain A is displaced along with it. During this, all field variables change directly and implicitly with the crack length. Therefore, besides the fixed coordinates (X_1, X_2), a moving system $(x_1 = X_1 - a, x_2 = X_2)$ at the crack tip is introduced (see Fig. 3.17), so that the total derivative is read

$$\frac{\mathrm{d}(\cdot)}{\mathrm{d}a} = \frac{\partial(\cdot)}{\partial a} + \frac{\partial x_1}{\partial a}\frac{\partial(\cdot)}{\partial x_1} = \frac{\partial(\cdot)}{\partial a} - \frac{\partial(\cdot)}{\partial x_1}. \tag{3.97}$$

Thereby, we differentiate the potential energy, which is a function of the displacement field u_i, with respect to the crack length

$$-\frac{\mathrm{d}\Pi(u_i)}{\mathrm{d}a} = \frac{\mathrm{d}}{\mathrm{d}a}\{W_{\text{ext}}(u_i) - W_{\text{int}}(u_i)\} = \frac{\mathrm{d}}{\mathrm{d}a}\left\{\int_\Gamma t_i u_i\,\mathrm{d}s - \int_A U\,\mathrm{d}A\right\}$$

$$= \int_A \frac{\partial U(u_i)}{\partial x_1}\,\mathrm{d}A - \int_\Gamma t_i\frac{\partial u_i}{\partial x_1}\,\mathrm{d}s + \left[-\int_A \frac{\partial U}{\partial a}\,\mathrm{d}A + \int_\Gamma t_i\frac{\partial u_i}{\partial a}\,\mathrm{d}s\right]. \tag{3.98}$$

Using $\frac{\partial U}{\partial a} = \frac{\partial U}{\partial \varepsilon_{ij}}\frac{\partial \varepsilon_{ij}}{\partial a} = \sigma_{ij}\frac{\partial u_{i,j}}{\partial a}$, converting the line integral into an area integral by means of Gauss's divergence theorem and applying the equilibrium equations $\sigma_{ij,j} = 0$, respectively, the term in [ ]–brackets of (3.98) disappears. The 1st integral of (3.98) can also be converted by Gauss's theorem, and transformed using the arc length $\mathrm{d}s$ along Γ.

$$\int_A U_{,j}\,\delta_{1j}\,\mathrm{d}A = \int_\Gamma U n_1\,\mathrm{d}s = \int_\Gamma U\,\mathrm{d}x_2 \tag{3.99}$$

Thereby the energy release rate G can be calculated along the curve Γ using a line integral, which is denoted as J integral:

$$G = -\frac{d\Pi}{da} = J \equiv \int_{\Gamma} \left[U \, dx_2 - t_i \frac{\partial u_i}{\partial x_1} \, ds \right]. \tag{3.100}$$

Path Independence of the J-Integral

In order to prove that J is independent of the choice of area A and integration path Γ, we compare two paths Γ_1 and Γ_2 around the crack tip (Fig. 3.18). The paths are completed by the segments Γ^+ and Γ^- along the upper and lower crack faces respectively, which results in a closed curve $C = \Gamma_2 + \Gamma^+ - \Gamma_1 + \Gamma^-$. C entirely encloses the differential area $\bar{A} = A_2 - A_1$. The crack tip was circumvented here! The integrals over Γ^+ and Γ^- are zero, because on the crack faces $t_i = 0$ is valid and $dx_2 = 0$. Evaluating J along C, the line integral is now converted back into an area integral over $\bar{A}$ and the Cauchy theorem $t_i = \sigma_{ij} n_j$ is applied

$$\int_C \left[U n_1 - \sigma_{ij} n_j \frac{\partial u_i}{\partial x_1} \right] ds = \int_C \left[U \delta_{1j} - \sigma_{ij} \frac{\partial u_i}{\partial x_1} \right] n_j \, ds$$
$$= \int_{\bar{A}} \frac{\partial}{\partial x_j} \left[U \delta_{1j} - \sigma_{ij} \frac{\partial u_i}{\partial x_1} \right] d\bar{A}. \tag{3.101}$$

The strain energy U is a function of the strains $\varepsilon_{ij} = \left(u_{i,j} + u_{j,i} \right)/2$. According to (A.74), its differentiation gives the stresses σ_{ij} in case of a (non-)linear–elastic material. Hence, differentiating the integrand (3.101) by the chain rule we obtain

$$\frac{\partial U}{\partial \varepsilon_{ij}} \frac{\partial \varepsilon_{ij}}{\partial x_1} - \frac{\partial \sigma_{ij}}{\partial x_j} \frac{\partial u_i}{\partial x_1} - \sigma_{ij} \frac{\partial}{\partial x_1} \left(\frac{\partial u_i}{\partial x_j} \right)$$
$$= \sigma_{ij} \frac{\partial \varepsilon_{ij}}{\partial x_1} - 0 - \sigma_{ij} \frac{\partial}{\partial x_1} \frac{1}{2} \left(\frac{\partial u_i}{\partial x_j} + \frac{\partial u_j}{\partial x_i} \right) = 0 \tag{3.102}$$

Because of the equilibrium conditions, (A.71) the 2nd term disappears, since the inertia and volume forces are not considered. Moreover, in the 3rd term the symmetry $\sigma_{ji} = \sigma_{ij}$ was used, whereby it directly compensates the 1st term. The result is the following:

$$\int_C (\cdot) \, ds = \int_{\Gamma_1} (\cdot) \, ds - \int_{\Gamma_2} (\cdot) \, ds + \int_{\Gamma^+ + \Gamma^-} (\cdot) \, ds = 0. \tag{3.103}$$

Since the crack face integrals over $\Gamma^+ + \Gamma^-$ are zero, the path independency has been proven

Fig. 3.18 Towards path independence of the J-integral

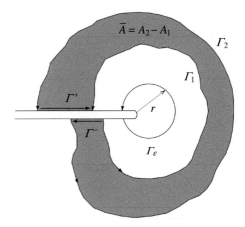

$$J_{\Gamma_1} = \int_{\Gamma_1} (\cdot)\, \mathrm{d}s = J_{\Gamma_2} = \int_{\Gamma_2} (\cdot)\, \mathrm{d}s\,. \qquad (3.104)$$

Looking back, we summarize the preconditions that were necessary

- There are no volume loads $\bar{b}_i = 0$ and no inertia forces $\rho\ddot{u}_i = 0$.
- The crack faces Γ^+ and Γ^- are stress-free $\bar{t}_i = 0$.
- The elastic strain energy density U is a potential function: $\partial U/\partial \varepsilon_{ij} = \sigma_{ij}$.
- U does not explicitly depend on x_1, only implicitly via $\varepsilon_{ij}(x_1)$ (homogeneous material).

3.2.7 Cracks in Anisotropic Elastic Bodies

Many modern materials in light weight constructions have anisotropic elastic properties. These include the big group of fiber-reinforced plastics and metals, in which
high-strength fibers (glass, carbon, whisker) are embedded in a matrix material in
order to increase the strength and stiffness. The reinforcement can either be in unidirectional orientation or as fiber mats in various layers of different orientation.
Besides these composites and laminates, especially metallic and ceramic materials
with anisotropic crystalline structure are of interest. Examples are nickel-base mono-
crystals for turbine blades or silicon components in micro-system-technology. After
all, reinforced concrete in civil engineering, finned steel constructions, stringers in
airplanes and others belong to the category of anisotropic structures, too. Therefore,
anisotropy has to be considered when dealing with fracture–mechanical problems.

We analyze a crack in the infinite plane under mode I, II and III loading, see
Fig. 3.19. The principal axes of anisotropy are fixed by the material coordinate system
$(\bar{x}_1, \bar{x}_2, \bar{x}_3)$, whereby we restrict the considerations to orthogonality and the $\bar{x}_3$ axis
running parallel to the x_3 coordinate. In the case of plane stress, (A.82) results in the
anisotropic compliance matrix formulated in engineering constants in the material

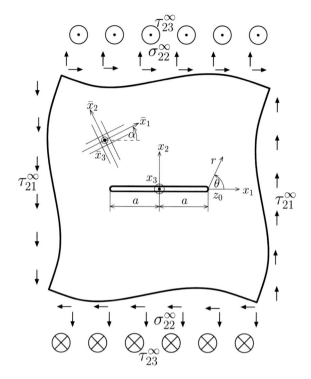

Fig. 3.19 Crack in a sheet of anisotropic elastic material

coordinate system

$$
\begin{bmatrix} \bar{\varepsilon}_{11} \\ \bar{\varepsilon}_{22} \\ \bar{\gamma}_{12} \end{bmatrix} = \begin{bmatrix} \dfrac{1}{E_1} & -\dfrac{\nu_{21}}{E_2} & 0 \\ -\dfrac{\nu_{12}}{E_1} & \dfrac{1}{E_2} & 0 \\ 0 & 0 & \dfrac{1}{G_{12}} \end{bmatrix} \begin{bmatrix} \bar{\sigma}_{11} \\ \bar{\sigma}_{22} \\ \bar{\tau}_{12} \end{bmatrix} . \tag{3.105}
$$

Using the rules of tensor algebra, this relation is transformed into the global coordinate system by means of the rotation matrix **r** depending on the angle α, which provides a full compliance matrix $[a_{\alpha\beta}] \widehat{=} [S_{\alpha\beta}]$ with $\alpha, \beta = \{1, 2, 6\}$

$$
\begin{bmatrix} \varepsilon_{11} \\ \varepsilon_{22} \\ \gamma_{12} \end{bmatrix} = \begin{bmatrix} a_{11} & a_{12} & a_{16} \\ a_{12} & a_{22} & a_{26} \\ a_{16} & a_{26} & a_{66} \end{bmatrix} \begin{bmatrix} \sigma_{11} \\ \sigma_{22} \\ \tau_{12} \end{bmatrix} . \tag{3.106}
$$

For plane strain the stress–strain–relation reads with modified constants $b_{\alpha\beta}$ as

$$
\begin{bmatrix} \varepsilon_{11} \\ \varepsilon_{22} \\ \gamma_{12} \end{bmatrix} = \begin{bmatrix} b_{11} \; b_{12} \; b_{16} \\ b_{12} \; b_{22} \; b_{26} \\ b_{16} \; b_{26} \; b_{66} \end{bmatrix} \begin{bmatrix} \sigma_{11} \\ \sigma_{22} \\ \tau_{12} \end{bmatrix}, \quad b_{\alpha\beta} = a_{\alpha\beta} - \frac{a_{\alpha3}a_{\beta3}}{a_{33}}. \tag{3.107}
$$

The stress in thickness direction is then $\sigma_{33} = -(a_{13}\sigma_{11} + a_{23}\sigma_{22} + a_{26}\tau_{12})/a_{33}$.
Detailed instructions to material modeling of laminates can be found in [26].

For the solution of boundary value problems in plane anisotropic elasticity, generalized complex stress functions have been developed by Lekhnitskii [27] and Stroh [28]. On this basis, Sih, Paris and Irwin [29] have found such a crack solution for the first time, which will be presented here. The stresses and displacements are derived from two complex holomorphic functions $\phi_k(z_k)$ ($k = \{1, 2\}$) of the complex variables $z_k = x_1 + s_k x_2$. The complex constants s_k can be found from the roots of the characteristic equation subject to the elastic constants $a_{\alpha\beta}$

$$
a_{11}s^4 - 2a_{16}s^3 + (2a_{12} + a_{66})s^2 - 2a_{26}s + a_{22} = 0, \tag{3.108}
$$

which provides two conjugated complex solutions

$$
s_1 = \gamma_1 + i\delta_1, \quad s_2 = \gamma_2 + i\delta_2, \quad s_3 = \bar{s}_1, \quad s_4 = \bar{s}_2. \tag{3.109}
$$

Applying the same approach as used for isotropic material behavior in Sect. 3.2.7, it is now possible to calculate the crack tip fields [29]. In polar coordinates (r, θ) the result around the crack tip is

Mode I

$$
\sigma_{11} = \frac{K_I}{\sqrt{2\pi r}}\Re\left[\frac{s_1 s_2}{s_1 - s_2}\left(\frac{s_2}{\sqrt{\cos\theta + s_2\sin\theta}} - \frac{s_1}{\sqrt{\cos\theta + s_1\sin\theta}}\right)\right]
$$

$$
\sigma_{22} = \frac{K_I}{\sqrt{2\pi r}}\Re\left[\frac{1}{s_1 - s_2}\left(\frac{s_1}{\sqrt{\cos\theta + s_2\sin\theta}} - \frac{s_2}{\sqrt{\cos\theta + s_1\sin\theta}}\right)\right] \tag{3.110}
$$

$$
\tau_{12} = \frac{K_I}{\sqrt{2\pi r}}\Re\left[\frac{s_1 s_2}{s_1 - s_2}\left(\frac{1}{\sqrt{\cos\theta + s_1\sin\theta}} - \frac{1}{\sqrt{\cos\theta + s_2\sin\theta}}\right)\right]
$$

$$
u_1 = K_I\sqrt{\frac{2r}{\pi}}\Re\left[\frac{1}{s_1 - s_2}\left(s_1 p_2\sqrt{\cos\theta + s_2\sin\theta} - s_2 p_1\sqrt{\cos\theta + s_1\sin\theta}\right)\right]
$$

$$
u_2 = K_I\sqrt{\frac{2r}{\pi}}\Re\left[\frac{1}{s_1 - s_2}\left(s_1 q_2\sqrt{\cos\theta + s_2\sin\theta} - s_2 q_1\sqrt{\cos\theta + s_1\sin\theta}\right)\right]
$$

$$
\tag{3.111}
$$

Mode II

$$\sigma_{11} = \frac{K_{II}}{\sqrt{2\pi r}}\Re\left[\frac{1}{s_1 - s_2}\left(\frac{s_2^2}{\sqrt{\cos\theta + s_2\sin\theta}} - \frac{s_1^2}{\sqrt{\cos\theta + s_1\sin\theta}}\right)\right]$$

$$\sigma_{22} = \frac{K_{II}}{\sqrt{2\pi r}}\Re\left[\frac{1}{s_1 - s_2}\left(\frac{1}{\sqrt{\cos\theta + s_2\sin\theta}} - \frac{1}{\sqrt{\cos\theta + s_1\sin\theta}}\right)\right] \quad (3.112)$$

$$\tau_{12} = \frac{K_{II}}{\sqrt{2\pi r}}\Re\left[\frac{1}{s_1 - s_2}\left(\frac{s_1}{\sqrt{\cos\theta + s_1\sin\theta}} - \frac{s_2}{\sqrt{\cos\theta + s_2\sin\theta}}\right)\right]$$

$$u_1 = K_{II}\sqrt{\frac{2r}{\pi}}\Re\left[\frac{1}{s_1 - s_2}\left(p_2\sqrt{\cos\theta + s_2\sin\theta} - p_1\sqrt{\cos\theta + s_1\sin\theta}\right)\right]$$

$$u_2 = K_{II}\sqrt{\frac{2r}{\pi}}\Re\left[\frac{1}{s_1 - s_2}\left(q_2\sqrt{\cos\theta + s_2\sin\theta} - q_1\sqrt{\cos\theta + s_1\sin\theta}\right)\right] \quad (3.113)$$

with the material-dependent constants ($k = \{1, 2\}$):

$$p_k = a_{11}s_k^2 + a_{12} - a_{16}s_k \quad \text{and} \quad q_k = a_{12}s_k + \frac{a_{22}}{s_k} - a_{26} \quad (3.114)$$

The mode III problem during anti-planar shear stress τ_{23}^∞ remains decoupled from the in-plane problem in case of orthotropy, i. e. Hooke's law has the form

$$\begin{bmatrix}\gamma_{23}\\\gamma_{13}\end{bmatrix} = \begin{bmatrix}a_{44} & a_{45}\\a_{45} & a_{55}\end{bmatrix}\begin{bmatrix}\tau_{23}\\\tau_{13}\end{bmatrix}. \quad (3.115)$$

The near fields at the crack consist merely of the shear stresses τ_{13} and τ_{23} as well as of the u_3–displacement, each as a function of (x_1, x_2), respectively

$$\tau_{13} = -\frac{K_{III}}{\sqrt{2\pi r}}\Re\left[\frac{s_3}{\sqrt{\cos\theta + s_3\sin\theta}}\right], \quad \tau_{23} = \frac{K_{III}}{\sqrt{2\pi r}}\Re\left[\frac{1}{\sqrt{\cos\theta + s_3\sin\theta}}\right] \quad (3.116)$$

$$u_3 = K_{III}\sqrt{\frac{2r}{\pi}}\Re\left[\sqrt{\cos\theta + s_3\sin\theta}/(c_{45} + s_3 c_{44})\right]. \quad (3.117)$$

Thereby, $s_3, \bar{s}_3$ are the complex roots of the equation $c_{44}s^2 + 2c_{45}s + c_{55} = 0$, which depend on the elastic stiffness constants $c_{\alpha\beta}$, ($\alpha, \beta = \{4, 5\}$) of the material

$$c_{44} = \frac{a_{55}}{c}, \quad c_{55} = \frac{a_{44}}{c}, \quad c_{45} = -\frac{a_{45}}{c}, \quad c = a_{44}a_{55} - a_{45}^2. \quad (3.118)$$

The definition of the stress intensity factors K_I, K_{II} and K_{III} doesn't change compared to the isotropic case. They are determined by extrapolation of the singular stress behavior on the ligament in front of the crack using (3.63). For the infinite sheet with given far field loads, the same relationship exists as in the isotropic cases

$$K_{\mathrm{I}} = \sigma_{22}^{\infty}\sqrt{\pi a}, \quad K_{\mathrm{II}} = \tau_{21}^{\infty}\sqrt{\pi a}, \quad K_{\mathrm{III}} = \tau_{23}^{\infty}\sqrt{\pi a}. \tag{3.119}$$

Thus, the radial behavior of the anisotropic crack tip field is characterized by the same $1/\sqrt{r}$–singularity of stresses and strains as well as by the $\sqrt{r}$–dependency of the displacements as in the isotropic–elastic case. The only differences are the angular functions of (3.110)–(3.117). Evaluating the real parts leads to fairly complicated mathematical expressions, but in principle it is possible.

Finally, the relation between the energy release rate (during self-similar crack expansion) and the stress intensity factors will be stated. It can be gained with the help of Irwin's crack closure integral from the asymptotic solution.

$$G_{\mathrm{I}} = -\frac{a_{22}}{2}K_{\mathrm{I}}\Im\left[\frac{K_{\mathrm{I}}(s_1 + s_2) + K_{\mathrm{II}}}{s_1 s_2}\right], \quad G_{\mathrm{III}} = K_{\mathrm{III}}^2\Im\left[\frac{c_{45} + s_3 c_{44}}{2c_{44}c_{55}}\right]$$

$$G_{\mathrm{II}} = \frac{a_{11}}{2}K_{\mathrm{II}}\Im\left[K_{\mathrm{II}}(s_1 + s_2) + K_{\mathrm{I}}s_1 s_2\right]. \tag{3.120}$$

In the special case if the crack coincides with a symmetry plane of the orthotropy ($\alpha = 0$ in Fig. 3.19), the following real expressions are obtained

$$G = G_{\mathrm{I}} + G_{\mathrm{II}} + G_{\mathrm{III}}, \quad G_{\mathrm{I}} = K_{\mathrm{I}}^2\sqrt{\frac{a_{11}a_{22}}{2}}\left[\sqrt{\frac{a_{22}}{a_{11}}} + \frac{2a_{12} + a_{66}}{2a_{11}}\right]^{1/2}$$

$$G_{\mathrm{II}} = K_{\mathrm{II}}^2\frac{a_{11}}{\sqrt{2}}\left[\sqrt{\frac{a_{22}}{a_{11}}} + \frac{2a_{12} + a_{66}}{2a_{11}}\right]^{1/2}, \quad G_{\mathrm{III}} = K_{\mathrm{III}}^2\frac{1}{2\sqrt{c_{44}c_{55}}}. \tag{3.121}$$

3.2.8 Interface Cracks

Often cracks appear in the interface between two materials with different mechanical properties. Such cracks are referred to as interface cracks, see [30]. Interface cracks can be found especially in joint connections (gluing, bonding, welding, soldering), because the strength of the bonding materials is often less than the strength of both join partners. Interface cracks are also very important for sandwich materials (e. g. fiber–reinforced laminate materials, coating systems and coatings of all kinds where crack propagation leads to delamination of layers. After all, even the strength of many construction materials is essentially influenced by failure mechanisms at inner interfaces (boundaries between grains, phases and the like).

Furthermore, we will focus on the stress situation at an interface crack between two isotropic–elastic materials and the relevant fracture–mechanical parameters. The near tip field solution is studied for a crack in an infinite sheet located in an interface between materials (1) and (2) with the elastic constants E_1, ν_1 and E_2, ν_2, see

Fig. 3.20 Crack in the interface between two dissimilar
materials

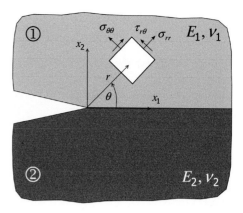

Fig. 3.20. The difference between the elastic properties of both materials is specified
by the parameter ϵ (Bi–material–constant), calculated from the individual constants
$\mu_m = E_m/2(1 + \nu_m)$ and $\kappa_m = 3 - 4\nu_m$ for $m = 1,\ 2$.

$$\epsilon = \frac{1}{2\pi} \ln \frac{\mu_2 \kappa_1 + \mu_1}{\mu_1 \kappa_2 + \mu_2}, \quad 0 \le |\epsilon| \le 0,175, \tag{3.122}$$

In the limit case of identical materials (1)=(2), ϵ becomes zero. Similarly to a crack
in homogeneous material (Sect. 3.3.2), we use the method of complex functions. To
evaluate the eigenfunctions for this crack, we use a series expansion for $\phi^{(m)}(z)$ and
$\chi^{(m)}(z)$, just as in (3.33), which has to differ for both materials $m = 1,\ 2$ in the upper
and lower half–plane, however.

$$\phi^{(m)}(z) = A^{(m)} z^\lambda, \quad \chi^{(m)}(z) = B^{(m)} z^{\lambda+1}, \quad z = re^{i\theta} \tag{3.123}$$

At the crack faces the tractions are zero

$$\theta = \pi : \quad \sigma_{\theta\theta}^{(1)} + i\tau_{r\theta}^{(1)} = 0, \quad \theta = -\pi : \quad \sigma_{\theta\theta}^{(2)} + i\tau_{r\theta}^{(2)} = 0 \tag{3.124}$$

and at the interface ahead of the crack the displacements and stresses of both regions
need to be continuous

$$\theta = 0 : \quad u_1^{(1)} + iu_2^{(1)} = u_1^{(2)} + iu_2^{(2)}, \quad \sigma_{\theta\theta}^{(1)} + i\tau_{r\theta}^{(1)} = \sigma_{\theta\theta}^{(2)} + i\tau_{r\theta}^{(2)}. \tag{3.125}$$

If these four boundary and transition conditions are expressed by the complex functions of (3.34) and (A.161), we obtain a homogeneous system of equations for the 4
complex (8 real) coefficients $A^{(m)}$, $B^{(m)}$. The characteristic polynomial of this eigenvalue system provides a conditional equation for the exponent λ that can assume the
following eigenvalues

$$\lambda = -\frac{1}{2} + n + i\epsilon \quad \text{mit} \quad n = 1, 2, 3, \ldots \tag{3.126}$$

In contrast to the homogeneous case, the eigenvalues $\lambda(n)$ are complex numbers with positive real part. To describe the near field $r \to 0$, only the dominating term of the solution is of interest associated with the smallest eigenvalue

$$\lambda(n = 1) = \frac{1}{2} + i\epsilon . \tag{3.127}$$

Inserting (3.123) with (3.127) into Kolosov's formulas (A.161), we gain the complete displacement and stress fields in polar coordinates at the crack tip (see [31]). The intensity of these fields is calculated by the coefficient of the $n = 1$th eigenfunction that has been introduced by Rice [32] as complex stress intensity factor

$$\widetilde{K} = K_1 + iK_2 = |\widetilde{K}|e^{i\psi}/l^{i\epsilon}, \quad |\widetilde{K}| = \sqrt{K_1^2 + K_2^2}, \quad \psi = \arctan(K_2/K_1) . \tag{3.128}$$

Here, l denotes a reference length (e. g. crack length a) and the phase angle ψ describes the ratio of the crack opening modes. This makes it possible to calculate the stresses ahead of the crack and the displacement jumps $\Delta u_i = u_i^{(1)}(r, \pi) - u_i^{(2)}(r, -\pi)$ across the crack faces in the following way (angular functions are normalized to 1 at $\theta = 0$)

$$\sigma_{22}(r, 0) + i\tau_{12}(r, 0) = \frac{\widetilde{K}}{\sqrt{2\pi r}} r^{i\epsilon}$$

$$\Delta u_2(r) + i\Delta u_1(r) = \frac{8}{1 + 2i\epsilon} \frac{\widetilde{K}}{E^* \cosh(\pi\epsilon)} \sqrt{\frac{r}{2\pi}} r^{i\epsilon} . \tag{3.129}$$

E^* is the averaged elastic modulus $1/E^* = (1/E_1' + 1/E_2')/2$ in plane strain.

As in homogeneous materials we get radial functions of the type $1/\sqrt{r}$ or $\sqrt{r}$ respectively. However, due to the imaginary part of (3.127), they are extended by

$$r^{i\epsilon} = e^{i\epsilon \ln r} = \cos(\epsilon \ln r) + i \sin(\epsilon \ln r) \tag{3.130}$$

which leads to two consequences:

Firstly, the crack opening modes I and II always occur coupled. Because of the complex product $\widetilde{K}r^{i\epsilon}$, the crack tip fields cannot be split into separate functions with their own coefficients K_I and K_{II} as in the homogeneous case of (3.45). That is why the new terms K_1 and K_2 are used here. Therefore, it is not possible to e. g. relate the crack face displacements u_2 and u_1 or the stresses σ_{22} and τ_{12} ahead of the crack in a unique way to the modes I or II anymore. Their ratio even changes with the distance r! That is easily recognized if the crack face displacements of (3.129) are written in real representation

$$
\begin{Bmatrix} \Delta u_1(r) \\ \Delta u_2(r) \end{Bmatrix} = \sqrt{\frac{r}{2\pi}} \frac{8}{(1+2\mathrm{i}\epsilon)E^*\cosh(\pi\epsilon)} \begin{Bmatrix} K_1\sin(\epsilon\ln r) + K_2\cos(\epsilon\ln r) \\ K_1\cos(\epsilon\ln r) + K_2\sin(\epsilon\ln r) \end{Bmatrix}.
$$
(3.131)

In the homogeneous case of ($\epsilon = 0$) these terms merge into their corresponding Eqs. (3.16) and (3.25) using the identity $K_1 = K_\mathrm{I}$ and $K_2 = K_\mathrm{II}$.

Secondly, Eqs. (3.129) and (3.130) reveal that the stresses and displacements oscillate due to the interval $[-1, +1]$ of angular functions. The oscillations become faster the closer we come to the crack tip, since $\ln r \to -\infty$. This leads to the physically absurd result that the crack faces interpenetrate each other. In order to avoid this oscillating singularity, contact zone models have been proposed in [30]. In engineering practice however, the approach of Rice has prevailed and been proved. The reason for this is that for practically relevant material combinations, the bi–material constants ($\epsilon < 0, 05$) are rather small. The largest radius r_c, where crack face contact occurs for the first time, can be estimated to $r_\mathrm{c}/l = \exp(-(\psi + \pi/2)/\epsilon)$ [32]. For a mixed–mode–ratio of $K_2/K_1 = 1$ ($\psi = -\pi/4$) and $\epsilon = 0, 05$ one finds $r_\mathrm{c} \approx 2 \cdot 10^{-9}l$. If the crack length is chosen as the reference length $l = 2a$, the oscillation region is therefore negligibly small and is quantified by $\widetilde{K}$ dominating outside of the contact zone.

The complex stress intensity factor $\widetilde{K}$ has to be determined as a function of geometry, crack length and load. The solution for the interface crack of the length $l = 2a$ in an infinite plane under combined tensile and shear load by the normal stress σ and shear stress τ is known

$$
\widetilde{K} = K_1 + \mathrm{i}K_2 = (\sigma + \mathrm{i}\tau)\sqrt{\pi a}(2a)^{-\mathrm{i}\epsilon}(1 + 2\mathrm{i}\epsilon).
$$
(3.132)

Generally for interface cracks, $\widetilde{K}$ has the generic form

$$
\widetilde{K} = (\sigma_\mathrm{n} + \mathrm{i}\tau_\mathrm{n})\sqrt{\pi a}(2a)^{-\mathrm{i}\epsilon}g(a, w, \epsilon),
$$
(3.133)

whereby σ_n and τ_n are nominal stresses and the function g exemplifies the dependency on geometry and material combination. From relation (3.132) it is obvious that a global tensile stress σ also causes a local shear load $K_2 \approx 2\epsilon K_1$ and vice versa! Furthermore, the mixed–mode–ratio $K_2/K_1 = \tan\psi$ changes along with the crack length because of the complex term $(2a)^{-\mathrm{i}\epsilon}$.

The formulation of a fracture criterion for interface cracks based on $\widetilde{K}$—similar to the K_I concept for homogeneous crack configurations—encounters several fundamental difficulties, however. First of all, the definition (3.128) of $\widetilde{K}$ gives a complex dimension MPa $\cdot$ m$^{-\mathrm{i}\epsilon}$, which is difficult to understand. Secondly, the critical size of $\widetilde{K}$ does not only depend on the absolute value $|\widetilde{K}|$ but also on the phase angle ψ occurring in the specimen. This means a two-parameter criterion $\widetilde{K}_c = K_c e^{\mathrm{i}\psi_c}$ or $K_{\mathrm{I}c}(\psi)$ is required. Thirdly, during the transfer from the specimen (case 1) to a component (case 2) not only the absolute values of the intensity factors should agree, but also their phase angles have to match the relation

$$\widetilde{K}_1 = |\widetilde{K}_1|e^{i\psi_1}(2a_1)^{-i\epsilon} = |\widetilde{K}_2|e^{i\psi_2}(2a_2)^{-i\epsilon} = \widetilde{K}_2$$
$$|\widetilde{K}_1| = |\widetilde{K}_2| \quad \text{and} \quad \psi_2 = \psi_1 - \epsilon \ln(a_1/a_2),$$

$$(3.134)$$

to ensure the same crack tip loading.

Rice suggested a pragmatic way out of this complication: the relations (3.129) can be converted into the classical (homogeneous form) if the stress is coupled to a certain distance $\hat{r}$

$$K_I + iK_{II} = \widetilde{K}\hat{r}^{i\epsilon} = (K_1 + iK_2)\hat{r}^{i\epsilon},$$

$$(3.135)$$

This means, exactly that mode ratio is taken over which exists in the interface solution at $\hat{r}$ with the phase angle $\hat{\psi} = \psi + \epsilon \ln(\hat{r}/a)$. Due to physical reasons it is rational to choose for $\hat{r}$ just the material–specific size of the crack process zone which should be smaller than the crack length a and bigger than the oscillation region r_c: $r_c < \hat{r} \ll a$.

Alternatively, it is possible to use the energy release rate G for the interface crack as fracture–mechanical parameter. With the help of the near field solution (3.129), it is possible to compute the crack closure integral (3.89), which leads to the relation

$$G = -\lim_{\Delta a \to 0} \frac{\Delta \Pi}{\Delta a} = \frac{K_1^2 + K_2^2}{E^* \cosh^2(\pi\epsilon)}.$$

$$(3.136)$$

Fortunately, the oscillations and the dependency on the reference length l disappear in the energetic view. What is left are the influence of the bi–material constants ϵ and the mode ratio K_2/K_1. Without knowing the phase angle ψ it is impossible to deduce both intensity factors from (3.136). Further information about energetic fracture criteria for interfaces and their experimental confirmations can be found in [33].

Theoretical investigations for interface cracks between *anisotropic* elastic materials were done by Qu and Bassani [34], Suo [35] and Beom and Atluri [36] on the basis of Stroh–Lekhnitski–formalism. The mathematical structure of the solutions is naturally more complicated as in the isotropic case and features mostly oscillating singularities. Nevertheless, anisotropy plays an important role for numerous applications on crystallographic interfaces, fiber–reinforced laminates and microelectronic coating systems.

3.2.9 Cracks in Plates and Shells

Thin-walled plate and shell structures occur particularly in light-weight and aerospace constructions where, due to the service load, fatigue cracks play a particular role. Besides the membrane stresses that generate a stress state at the crack tip such as in sheets, the bending and torsional moments cause an additional different near field at the crack tip.

Based on Kirchhoff's theory of thin shear–rigid plates (see explanation in the Appendix A.5.5), Williams [37] was able to calculate the eigenfunctions for a crack

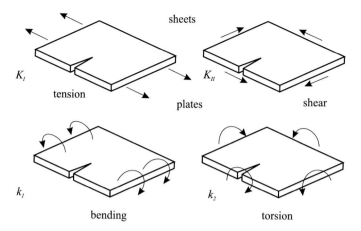

Fig. 3.21 Crack opening types in plane sheets, plates and shells due to membrane stresses, bending and torsional moments

in the infinite plate by using a series expansion for the deflection function $w(x_1, x_2)$. Later, Sih et al. [38] introduced the stress intensity factors k_1 and k_2 for plate bending/plate torsion. The form of representation is consistent with the crack loading by K_I and K_{II} in sheet problems in the way that it can be related to the normal and shear stresses σ_{22} or τ_{12}, respectively on the ligament ahead of the crack. Figure 3.21 illustrates all four crack opening types, which can occur in shells at the same time. If the coordinate in the direction of the plate thickness h is denoted by $z \triangleq x_3$, see Fig. 3.22, the asymptotic solution at the crack has the following form in cylindrical coordinates (r, θ, z)

$$
\left\{\begin{matrix} \sigma_{rr}^b \\ \sigma_{\theta\theta}^b \\ \tau_{r\theta}^b \end{matrix}\right\} = \frac{k_1}{(3+\nu)\sqrt{2r}} \frac{z}{2h} \left\{\begin{matrix} (3+5\nu)\cos\frac{\theta}{2} - (7+\nu)\cos\frac{3\theta}{2} \\ (5+3\nu)\cos\frac{\theta}{2} - (7+\nu)\cos\frac{3\theta}{2} \\ -(1-\nu)\sin\frac{\theta}{2} + (7+\nu)\sin\frac{3\theta}{2} \end{matrix}\right\}
$$

$$
+ \frac{k_2}{(3+\nu)\sqrt{2r}} \frac{z}{2h} \left\{\begin{matrix} -(3+5\nu)\sin\frac{\theta}{2} + (5+3\nu)\sin\frac{3\theta}{2} \\ -2(5+3\nu)\cos\frac{\theta}{2}\sin\theta \\ -(1-\nu)\cos\frac{\theta}{2} + (5+3\nu)\cos\frac{3\theta}{2} \end{matrix}\right\} \qquad (3.137)
$$

$$
\left\{\begin{matrix} \tau_{rz}^b \\ \tau_{\theta z}^b \end{matrix}\right\} = \frac{[1-(2z/h)^2]}{(3+\nu)(2r)^{\frac{3}{2}}} \frac{h}{2} \left\{\begin{matrix} -k_1\cos\frac{\theta}{2} + k_2\sin\frac{\theta}{2} \\ -k_1\sin\frac{\theta}{2} - k_2\cos\frac{\theta}{2} \end{matrix}\right\}. \qquad (3.138)
$$

The bending and shear stresses in the plane of the plate (x_1, x_2) or (r, θ) behave again singularly with $1/\sqrt{r}$. The $r^{-3/2}$-singularity of the shear stresses acting vertically to the plate plane is a consequence of the shear-rigid plate model, which fulfills the stress-free conditions at the crack faces only approximately by introducing a substitute shear-force. According to the plate theory, the bending stresses run across

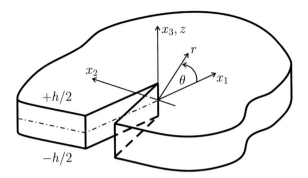

Fig. 3.22 Cylindrical coordinate system at the crack front in plates and shells

the thickness h linearly with z. Thus, the stresses vary from tension to compression along the crack front and assume $(z = \pm h/2)$ maxima with opposite signs at the upper and lower surfaces. In the neutral plane $(z = 0)$ the crack is not stressed at all. A possible contact of the crack faces however cannot be considered within the plate theory. The region, where the k-controlled near field dominates, is about $a/10$.

The deflection function $w(r, \theta)$ of the plate's mid-plane has the following asymptotics at the crack tip (Fig. 3.22):

$$w = \frac{(2r)^{\frac{3}{2}}(1-\nu^2)}{2Eh(3+\nu)}\left(k_1\left[\frac{1}{3}\frac{7+\nu}{1-\nu}\cos\frac{3\theta}{2} - \cos\frac{\theta}{2}\right] + k_2\left[\frac{1}{3}\frac{5+3\nu}{1-\nu}\sin\frac{3\theta}{2} - \sin\frac{\theta}{2}\right]\right).$$
$$(3.139)$$

Since the stresses and sectional variables m_{ij}, q_i are calculated from w by twofold derivation (see A.167), r needs to stand in the power of $3/2$. The differential equation of Kirchhoff's plate theory is very similar to the bi-potential equation of the sheet problem. For this reason the complex analysis methods from Sect. 3.2.2 can often be adopted to find solutions, see [38]. Thus, the stress intensity factors are obtained from the complex stress function ϕ of Eq. (A.171) by a limiting process for $z \to z_0$ towards the crack tip

$$k_1 - ik_2 = -\frac{\sqrt{2}Eh(3+\nu)}{1-\nu^2}\lim_{z \to z_0}\sqrt{z-z_0}\,\phi'(z).$$
$$(3.140)$$

As an example, for the infinite plate under a constant bending moment m_0 on all sides, the solution is

$$k_1 = \frac{6m_0}{h^2}\sqrt{a}, \qquad k_2 = 0.$$
$$(3.141)$$

Using the Reissner–theory for thick plates allowing for shear deformations [39], one gets, as expected, the same asymptotics as in a plane strain state, whereby the K_I-, K_{II}-factors run along the crack front linearly with z. Since this crack tip field is only valid in a region $r < h/10$, it lies embedded within the Kirchhoff–asymptotics and is uniquely defined by it [40]. That is why the Kirchhoff–theory is mostly sufficient for fracture–mechanical calculations of plates and shells. The theory is also preferred because of its lower effort necessary for discretization.

Hui and Zehnder [40] established the connection between the energy release rates and Kirchhoff's stress intensity factors

$$G_1 = \frac{k_1^2 \pi (1 + \nu)}{3E(3 + \nu)}, \quad G_2 = \frac{k_2^2 \pi (1 + \nu)}{3E(3 + \nu)}. \tag{3.142}$$

3.2.10 Fracture Mechanical Weight Functions

In this section a very useful, semi-analytical method to calculate stress intensity factors for linear–elastic, static crack problems will be introduced. Based on certain basic solutions for the relevant geometrical crack configuration, it is possible to find the K factors for further arbitrary load situations of the same crack configuration.

Behind this the fascinating fact is concealed that inside the solution of *one* special boundary value problem rests the variety of all kinds of possible solutions of the same crack configuration. The key for this approach lies in *crack weight functions*.

The Principle of Superposition

As for all boundary value problems of linear partial differential equation, the *principle of superposition* is valid in elastostatics, too. It implies that solutions for different boundary values can be combined additively to the total solution, which then repre-

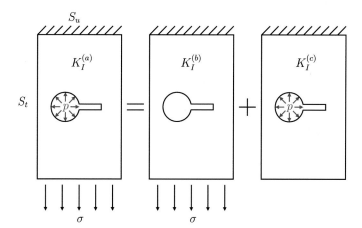

Fig. 3.23 Example to apply the principle of superposition

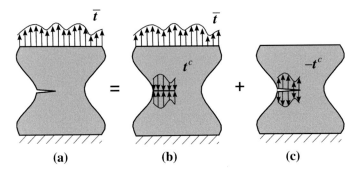

Fig. 3.24 Conversion of external stresses into equivalent crack face tractions

sents the exact solution of the boundary value problem for the sum of all boundary
values. The prerequisite for doing so is that the geometry of the body as well as
the subdivision of the boundary into parts, owing to Dirichlet S_u and Neumann S_t
conditions, are always the same, see Fig. A.16. Thereby it is possible to reduce complicated loading states of a crack configuration to the combination of a set of known
solutions with simpler loads. For example, let's consider the sheet containing a pre-
crack starting from a hole, as drafted in Fig. 3.23, which is loaded by tension σ at the
lower face and by a pressure p inside the hole. The resulting stress intensity factor
is obtained by superimposing both sub-problems $K_{\mathrm{I}}^{(\mathrm{a})} = K_{\mathrm{I}}^{(\mathrm{b})} + K_{\mathrm{I}}^{(\mathrm{c})}$. The principle
of superposition is also valid for thermal stresses and body forces.

The following technique is also very advantageous, combining the method of sections with the principle of superposition. Given a crack configuration under external
load, such as the edge-cracked sheet under tension shown in Fig. 3.24. The boundary
value problem can be split into a subproblem (b) without crack and a subproblem (c)
with pure crack face load. To do this, we calculate the sectional stresses at the position of the crack S_c in the uncracked configuration (b) from the stresses $t_i^c = \pm\sigma_{ij}n_j$.
Next we make notionally a body cut along the crack, but let the sectional stress act on
its faces so that the crack stays closed as before. Now in load case (c), the sectional
stresses t_i^c are applied exactly with the opposite signs. As it becomes evident from
Fig. 3.24, adding the boundary conditions (b) and (c) gives the original problem (a).
Since the subproblem (b) contains in fact no crack ($K^{(\mathrm{b})} = 0$), the stress intensity
factor of the considered problem (a) is identical to the one of the boundary value
problem (c)

$$K_{\mathrm{I}}^{(\mathrm{a})} = K_{\mathrm{I}}^{(\mathrm{c})} . \tag{3.143}$$

Using this method, any kind of loading (surface tractions, body forces, temperature fields) imposed onto a crack can be converted into equivalent crack
face tractions. This allows for systematic and unified calculations of the stress
intensity factors.

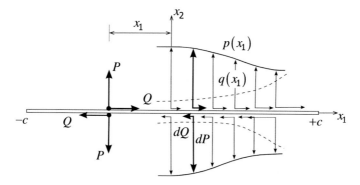

Fig. 3.25 Example of using superposition of crack face stresses

Weight Function for Crack Face Loads

The use of the superposition principle will be exemplified by means of the simple Griffith crack of length $2c$, see Fig. 3.25. Our starting point is the known solution (e.g. reference book [15]) for the K–factors due to a force couple P and Q (per thickness), imposed on the upper and lower crack face

$$
\begin{Bmatrix} K_{\mathrm{I}}^{\pm} \\ K_{\mathrm{II}}^{\pm} \end{Bmatrix} = \frac{1}{\sqrt{\pi c}} \sqrt{\frac{c \pm x_1}{c \mp x_1}} \begin{Bmatrix} P \\ Q \end{Bmatrix} \tag{3.144}
$$

From Fig. 3.25 it is evident that because of symmetry P creates a mode–I–loading and the antisymmetry of Q leads to mode II. The K factors and signs refer to the positive ($x_1 = +c$) and to the negative ($x_1 = -c$) crack tip.

This solution is also known as Green's function for crack faces, since it enables us to calculate the stress intensity factors for arbitrarily distributed line loads $p(x_1)$ or $q(x_1)$ respectively along the crack faces. To do this, the line loads are interpreted as continuous infinitesimal concentrated forces $dP = p(x_1)\,dx_1$ or $dQ = q(x_1)\,dx_1$ respectively, whose superposition leads to the following integral for the K–factors

$$
\begin{Bmatrix} K_{\mathrm{I}}^{\pm} \\ K_{\mathrm{II}}^{\pm} \end{Bmatrix} = \frac{1}{\sqrt{\pi c}} \int_{-c}^{+c} \sqrt{\frac{c \pm x_1}{c \mp x_1}} \begin{Bmatrix} p(x_1) \\ q(x_1) \end{Bmatrix} dx_1 . \tag{3.145}
$$

If for instance the K_{I}–factor for a constant compressive stress $p(x_1) = -\sigma_{\mathrm{F}}$ in the region of the crack tips $a \leq |x_1| \leq a + d = c$ is supposed to be determined (the result is needed in Sect. 3.3.3, see Fig. 3.34), the use of (3.145) results in

$$K_{\mathrm{I}}^{+} = K_{\mathrm{I}}^{-} = \frac{-\sigma_{\mathrm{F}}}{\sqrt{\pi c}} \left[\int_{-c}^{-a} \sqrt{\frac{c+x_1}{c-x_1}} \, dx_1 + \int_{a}^{c} \sqrt{\frac{c+x_1}{c-x_1}} \, dx_1 \right]$$

$$= \frac{-\sigma_{\mathrm{F}}}{\sqrt{\pi c}} \int_{a}^{c} \frac{2c}{\sqrt{c^2 - x_1^2}} \, dx_1 = -2\sigma_F \sqrt{\frac{c}{\pi}} \arccos\left(\frac{a}{c}\right). \tag{3.146}$$

General Weight Functions

The generalization of the explained calculation method to any kind of loading of a crack configuration comprising surface tractions $\bar{t}$ at the boundary S_t and volume loads $\bar{b}$ in the body V leads to the actual *fracture–mechanical weight functions*.

The fracture–mechanical weight function $H_i^{\mathrm{I}}(x, a)$ describes the effect of a concentrated force $F = F_i e_i$ of the value $|F| = 1$ at position x on the stress intensity factor $K_{\mathrm{I}}(a)$ for a crack of length a in the considered body. Thus, the stress intensity factor can be determined for this crack configuration under an arbitrary load $\bar{t}$ and $\bar{b}$ using a simple integration

$$K_{\mathrm{I}}(a) = \int_{S_t} \bar{t}_i(x) H_i^{\mathrm{I}}(x, a) \, dS + \int_{V} \bar{b}_i(x) H_i^{\mathrm{I}}(x, a) \, dV . \tag{3.147}$$

$H_i^{\mathrm{I}}(x, a)$ depends on the body and crack geometry, on the assignment of the boundary into S_t and S_u aa well as on the elastic material properties.

In the following, one method for calculating fracture–mechanical weight functions will be explained that traces back to Rice [41]. For that purpose, we consider a crack configuration under two different loading conditions (1) and (2), which is shown in Fig. 3.26 for an edge crack of size a. The load case (1) indicates that crack loading, for which we seek the K factors, while load case (2) represents an already known solution. The associated displacement fields $u_i^{(m)}$, boundary tractions $t_i^{(m)}$ (overbar is omitted from here on) and K factors $K_{\mathrm{I}}^{(m)}$, $K_{\mathrm{II}}^{(m)}$ of both load cases $m = 1$, 2 are marked by superscripted indexes. We now perform a virtual crack extension by the length Δa, where the boundary loads $t_i^{(m)}$ are kept constant. The displacement state in the body however changes as follows:

$$u_i^{(m)}(x, a + \Delta a) = u_i^{(m)}(x, a) + \Delta u_i^{(m)}, \quad \Delta u_i^{(m)} = \frac{\partial u_i^{(m)}}{\partial a} \Delta a . \tag{3.148}$$

According to Sect. 3.2.5 the potential energy released during this process matches half the work of the external loads $t_i^{(m)}$ to the displacement changes $\Delta u_i^{(m)}$ and is

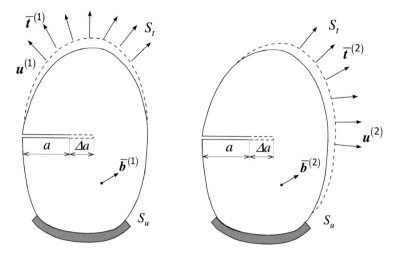

Fig. 3.26 Derivation of generalized fracture–mechanical weight functions

therefore equal to the energy release rate $G = \Delta W_{\text{ext}}/\Delta a$. On the other hand, the energy release rate G is linked with the intensity factors K_{I} and K_{II} by the relation (3.93). Thereby the result for both load cases is

$$\Delta W_{\text{ext}}^{(1)} = \frac{1}{2} \int_{S_t} t_i^{(1)} \Delta u_i^{(1)} dS = G^{(1)} \Delta a = \frac{1}{E'} \left[\left(K_{\text{I}}^{(1)} \right)^2 + \left(K_{\text{II}}^{(1)} \right)^2 \right] \Delta a$$

$$\Delta W_{\text{ext}}^{(2)} = \frac{1}{2} \int_{S_t} t_i^{(2)} \Delta u_i^{(2)} dS = G^{(2)} \Delta a = \frac{1}{E'} \left[\left(K_{\text{I}}^{(2)} \right)^2 + \left(K_{\text{II}}^{(2)} \right)^2 \right] \Delta a. \quad (3.149)$$

When superimposing both load cases (1) and (2), the K factors $K_L = K_L^{(1)} + K_L^{(2)}$ ($L = \text{I, II}$) are added. In the boundary integrals, the work of every other load case has to be considered, respectively. Therefore the energy release for the combined total state is

$$G^{(1+2)} \Delta a = \frac{1}{E'} \left[\left(K_{\text{I}}^{(1)} + K_{\text{I}}^{(2)} \right)^2 + \left(K_{\text{II}}^{(1)} + K_{\text{II}}^{(2)} \right)^2 \right] \Delta a$$

$$= \frac{1}{2} \int_{S_t} t_i^{(1)} \left(\Delta u_i^{(1)} + \Delta u_i^{(2)} \right) dS + \frac{1}{2} \int_{S_t} t_i^{(2)} \left(\Delta u_i^{(2)} + \Delta u_i^{(1)} \right) dS.$$

$$(3.150)$$

The difference of Eqs. (3.150) and (3.149) identifies the interaction energy between both load cases

$$\frac{2}{E'}\left[K_{\mathrm{I}}^{(1)}K_{\mathrm{I}}^{(2)} + K_{\mathrm{II}}^{(1)}K_{\mathrm{II}}^{(2)}\right]\Delta a = \frac{1}{2}\int_{S_t} t_i^{(1)}\Delta u_i^{(2)}\,dS + \frac{1}{2}\int_{S_t} t_i^{(2)}\Delta u_i^{(1)}\,dS. \quad (3.151)$$

According to the theorem of Betti (see for instance [42]) the work done by the boundary tractions of one load case to the displacements of the other load case are reciprocally identical

$$\int_{S_t} t_i^{(1)}u_i^{(2)}\,dS = \int_{S_t} t_i^{(2)}u_i^{(1)}\,dS, \quad (reciprocity\ theorem) \qquad (3.152)$$

which can also be applied this way to the situation after crack expansion $a + \Delta a$, too

$$\int_{S_t} t_i^{(1)}\left(u_i^{(2)} + \Delta u_i^{(2)}\right)dS = \int_{S_t} t_i^{(2)}\left(u_i^{(1)} + \Delta u_i^{(1)}\right)dS. \qquad (3.153)$$

The subtraction of (3.152) and (3.153) results in

$$\int_{S_t} t_i^{(1)}\Delta u_i^{(2)}\,dS = \int_{S_t} t_i^{(2)}\Delta u_i^{(1)}\,dS. \qquad (3.154)$$

Thereby, the 2nd integral can be replaced by the 1st integral in (3.151). Finally we put the differential quotient $\Delta u_i^{(2)}/\Delta a$ below the integral ($t_i^{(1)}$ does not depend on Δa) and form the limiting process $\Delta a \to 0$ using (3.148)

$$K_{\mathrm{I}}^{(1)}K_{\mathrm{I}}^{(2)} + K_{\mathrm{II}}^{(1)}K_{\mathrm{II}}^{(2)} = \frac{E'}{2}\int_{S_t} t_i^{(1)}\frac{\partial u_i^{(2)}}{\partial a}\,dS. \qquad (3.155)$$

This relation of general validly can be specialized in different aspects.

(a) Pure Mode I loading
In this special case of symmetrical geometry and load, $K_{\mathrm{II}}^{(1)} = K_{\mathrm{II}}^{(2)} = 0$ vanish. It is easy to solve (3.155) for the sought stress intensity factor of load case (1)

$$K_{\mathrm{I}}^{(1)}(a) = \frac{E'}{2K_{\mathrm{I}}^{(2)}(a)}\int_{S_t} t_i^{(1)}(x)\frac{\partial u_i^{(2)}(x, a)}{\partial a}\,dS. \qquad (3.156)$$

A comparison with (3.147) shows that the weight function H_i^{I} is exactly proportional to the change of the displacement field at position x on the boundary S_t during crack expansion

$$H_i^{\mathrm{I}}(x, a) = \frac{E'}{2K_{\mathrm{I}}^{(2)}(a)} \frac{\partial u_i^{(2)}(x, a)}{\partial a}. \tag{3.157}$$

Thus, the (same!) weight function can be calculated for this crack configuration from every known reference solution $u_i^{(2)}$ and $K_{\mathrm{I}}^{(2)}$.

(b) Pure crack face loading

Every external load $\bar{t}_i$ can be transformed into an equivalent crack face load t_i^c. Therefore it is allowed to put into Eq. (3.156) instead of S_t the entire crack surface S_c, using exactly the sectional stresses t_i^c acting there. This way the location $x \in S_c$, where the weight functions are to be determined and applied, are restricted to the crack itself → *crack face weight functions*.

(c) Mixed–mode loading

In this case two reference solutions (2) are necessary, which will be labeled (2a) and (2b). Applying (3.155) then provides a linear system of equations

$$\begin{aligned} K_{\mathrm{I}}^{(1)} K_{\mathrm{I}}^{(2a)} + K_{\mathrm{II}}^{(1)} K_{\mathrm{II}}^{(2a)} &= \tfrac{E'}{2} \int\limits_{S_t} t_i^{(1)} \frac{\partial u_i^{(2a)}}{\partial a} \, \mathrm{d}S \\ K_{\mathrm{I}}^{(1)} K_{\mathrm{I}}^{(2b)} + K_{\mathrm{II}}^{(1)} K_{\mathrm{II}}^{(2b)} &= \tfrac{E'}{2} \int\limits_{S_t} t_i^{(1)} \frac{\partial u_i^{(2b)}}{\partial a} \, \mathrm{d}S, \end{aligned} \tag{3.158}$$

whose solution gives the sought K–factors of the considered load case (1)

$$K_{\mathrm{I}}^{(1)} = \frac{E'}{2K^2} \left[K_{\mathrm{II}}^{(2a)} \int\limits_{S_t} t_i^{(1)} \frac{\partial u_i^{(2b)}}{\partial a} \, \mathrm{d}S - K_{\mathrm{II}}^{(2b)} \int\limits_{S_t} t_i^{(1)} \frac{\partial u_i^{(2a)}}{\partial a} \, \mathrm{d}S \right]$$

$$K_{\mathrm{II}}^{(1)} = \frac{E'}{2K^2} \left[K_{\mathrm{I}}^{(2b)} \int\limits_{S_t} t_i^{(1)} \frac{\partial u_i^{(2a)}}{\partial a} \, \mathrm{d}S - K_{\mathrm{I}}^{(2a)} \int\limits_{S_t} t_i^{(1)} \frac{\partial u_i^{(2b)}}{\partial a} \, \mathrm{d}S \right] \tag{3.159}$$

$$\text{with} \quad K^2 = K_{\mathrm{I}}^{(2b)} K_{\mathrm{II}}^{(2a)} - K_{\mathrm{I}}^{(2a)} K_{\mathrm{II}}^{(2b)}.$$

Both reference solutions must not be solely mode I or mode II, since the system of equations would then become indefinite. ($K = 0!$). Ideally, a pure mode I solution ($K_{\mathrm{II}}^{(2a)} = 0$) is used for (2a), and for (2b) a mode II case is chosen ($K_{\mathrm{I}}^{(2b)} = 0$), by which Eq. (3.158) is decoupling.

For mixed–mode loading of plane crack problems, the weight functions are gained from (3.159)

$$H_i^{\mathrm{I}}(x, a) = \frac{E'}{2K^2(a)} \left[K_{\mathrm{II}}^{(2\mathrm{a})}(a) \frac{\partial u_i^{(2\mathrm{b})}(x, a)}{\partial a} - K_{\mathrm{II}}^{(2\mathrm{b})}(a) \frac{\partial u_i^{(2\mathrm{a})}(x, a)}{\partial a} \right]$$

$$H_i^{\mathrm{II}}(x, a) = \frac{E'}{2K^2(a)} \left[K_{\mathrm{I}}^{(2\mathrm{b})}(a) \frac{\partial u_i^{(2\mathrm{a})}(x, a)}{\partial a} - K_{\mathrm{I}}^{(2\mathrm{a})}(a) \frac{\partial u_i^{(2\mathrm{b})}(x, a)}{\partial a} \right].$$
$$(3.160)$$

The application of these four functions H_i^L ($i = 1, 2;\ L = \mathrm{I,\ II}$) has the following form

$$K_{\mathrm{I}}^{(1)}(a) = \int_{S_t} t_i^{(1)}(x)\, H_i^{\mathrm{I}}(x, a)\, \mathrm{d}S\,, \quad K_{\mathrm{II}}^{(1)}(a) = \int_{S_t} t_i^{(1)}(x)\, H_i^{\mathrm{II}}(x, a)\, \mathrm{d}S\,. \quad (3.161)$$

Naturally, these weight functions can be restricted to the crack location $x \in S_c$. Instead of (or in addition to) the boundary loads $t^{(m)}$, the above derivation could have been performed for arbitrarily distributed volume loads $b^{(m)}$, whereof the weight functions $H_i^{\mathrm{I,II}}(x, a)$ for inner points $x \in V$ result, see Eq. (3.147).

Until now, identical displacement boundary conditions $\bar{u}$ in both load cases (1) and (2) have been implied at the surface part S_u. However, there are situations when the body is only stressed by boundary displacements $\bar{u}^{(m)}$ imposed on S_u, whose effect on K factors is of interest. For this situation, fracture mechanical weight functions can be derived in a complementary way, too [43]. Instead of (3.148), now the changes of reaction stresses $\Delta t_i^{(m)}$ occurring on the displacement boundary S_u have to be considered during a virtual crack expansion. Their work performed to the imposed displacements $u_i^{(m)}$ yields the energy release rate $G = -\Delta W_{\mathrm{int}}/\Delta a$, which corresponds to the loss of inner energy at fixed displacements. Therefore, the equivalent expression to (3.149) for $m = 1,\ 2$ is

$$- \Delta W_{\mathrm{int}}^{(m)} = -\frac{1}{2} \int_{S_u} u_i^{(m)} \Delta t_i^{(m)}\, \mathrm{d}S = G^{(m)} \Delta a\,. \quad (3.162)$$

Analogous considerations as mentioned above lead to weight functions G_i^L ($L = \mathrm{I,\ II}$) for displacement boundary conditions, whereof the sought stress intensity factors of the load case (1) can be calculated using a simple integration of the displacements $\bar{u}_i^{(1)} \triangleq \bar{u}_i$. Here, only the expressions equivalent to (3.147) and (3.157) are given for mode I

$$K_I(a) = \int_{S_u} \bar{u}_i(\boldsymbol{x})\, G_i^I(\boldsymbol{x}, a)\, \mathrm{d}S\,, \qquad G_i^I(\boldsymbol{x}, a) = \frac{E'}{2K_I^{(2)}(a)}\, \frac{\partial t_i^{(2)}(\boldsymbol{x}, a)}{\partial a}\,. \tag{3.163}$$

The generalization to mixed–mode loading similar to (3.160) and (3.161) is left to the reader as an exercise.

In the most general case, mixed boundary conditions $\boldsymbol{t}^{(m)}$ and $\boldsymbol{u}^{(m)}$ exist on the boundary parts S_t and S_u already indicated in Fig. 3.26. For this, only the result will be provided, which represents an extension of the relationship (3.155):

$$K_I^{(1)} K_I^{(2)} + K_{II}^{(1)} K_{II}^{(2)} = \frac{E'}{2} \left[\int_{S_t} t_i^{(1)} \frac{\partial u_i^{(2)}}{\partial a} \,\mathrm{d}S - \int_{S_u} u_i^{(1)} \frac{\partial t_i^{(2)}}{\partial a} \,\mathrm{d}S \right]\,. \tag{3.164}$$

Bueckner Singularity

A completely different approach to weight functions originates from Bueckner [44]. It makes use of »fundamental solution fields« of a crack problem, associated with a special load at the crack tip by a pair of forces. Starting points are again two load cases of exact the same (two-dimensional) crack configuration: load case (1) is again an arbitrary surface load $t_i^{(1)}$, for which the K factors are searched, see Fig. 3.27a. The corresponding displacement field $u_i^{(1)}(\boldsymbol{x})$ in the body is unknown. However it is known that the near field at the crack tip must exist and is defined by the yet unknown stress intensity factors $K_I^{(1)}$ and $K_{II}^{(1)}$. Using the Eqs. (3.16) for mode I and (3.25) for mode II, respectively, the crack face displacements can be found ($\theta = \pm\pi$)

$$u_1^{(1)}(r) = \pm\frac{\kappa+1}{2\mu}\sqrt{\frac{r}{2\pi}}\,K_{II}^{(1)}\,, \qquad u_2^{(1)}(r) = \pm\frac{\kappa+1}{2\mu}\sqrt{\frac{r}{2\pi}}\,K_I^{(1)}\,. \tag{3.165}$$

Load case (2) represents the solution for a force pair $\pm\boldsymbol{F} = \pm Q\boldsymbol{e}_1 \pm P\boldsymbol{e}_2$ that acts on the crack faces. Contrary to Fig. 3.25, here a crack in the finite body is examined according to Fig. 3.27b. With the help of Dirac's delta function, these point forces (per thickness) at distance $r = d$ to the crack tip, can be rewritten as traction vector of the load case (2) by $\boldsymbol{t}^{(2)} = t_i^{(2)}\boldsymbol{e}_i$

$$\boldsymbol{t}^{(2)}(\boldsymbol{x}) = \delta(r-d)\boldsymbol{F}\,, \quad t_1^{(2)} = \delta(r-d)Q\,, \quad t_2^{(2)} = \delta(r-d)P\,. \tag{3.166}$$

The corresponding displacement field is $u_i^{(2)}(\boldsymbol{x})$.

We will now again make use of Betti's theorem for these two load cases, i. e. the reciprocal interaction energies are equated

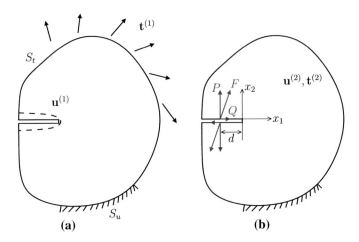

Fig. 3.27 Derivation of the weight functions according to Bueckner

$$\int\limits_S t_i^{(2)} u_i^{(1)} \mathrm{d}S = \int\limits_S t_i^{(1)} u_i^{(2)} \mathrm{d}S . \tag{3.167}$$

The integral on the left-hand side consists of known functions (3.165) and (3.166) that can be calculated along the boundary S including the two crack faces

$$\int\limits_S \delta(r-d) F_i u_i^{(1)} \mathrm{d}S = 2 F_i u_i^{(1)}(d) = 2 \frac{\kappa+1}{2\mu} \sqrt{\frac{d}{2\pi}} \left[P K_{\mathrm{I}}^{(1)} + Q K_{\mathrm{II}}^{(1)} \right] . \tag{3.168}$$

Equalizing it with the right-hand side of (3.167) gives

$$\frac{\sqrt{d}}{\pi} \left[P K_{\mathrm{I}}^{(1)} + Q K_{\mathrm{II}}^{(1)} \right] = \frac{2\mu}{\kappa+1} \frac{1}{\sqrt{2\pi}} \int\limits_S t_i^{(1)} u_i^{(2)} \mathrm{d}S . \tag{3.169}$$

For an easier understanding, we will confine ourselves furthermore to mode I, which means $Q = 0$ and $K_{\mathrm{II}}^{(1)} = 0$ disappears. Next, the force pair $\pm P$ is displaced directly into the crack tip by the limiting process $d \to 0$ that creates a special singularity. Its intensity should remain unchanged, which is why the quantity

$$B_{\mathrm{I}} = \lim_{d \to 0} \left(\frac{P\sqrt{d}}{\pi} \right) = \mathrm{const} \tag{3.170}$$

is introduced. Thereby, (3.169) can be rearranged for the sought stress intensity factor $K_{\mathrm{I}}^{(1)}$

$$K_I^{(1)} = \frac{2\mu}{\kappa + 1} \frac{1}{\sqrt{2\pi}} \frac{1}{B_I} \int_S t_i^{(1)} u_i^{(2)} \, dS. \tag{3.171}$$

A comparison with (3.147) reveals the structure of the weight function

$$H_i^I(\mathbf{x}, a) = \frac{2\mu}{\kappa + 1} \frac{1}{\sqrt{2\pi}} \frac{u_i^{(2)}(\mathbf{x})}{B_I}, \tag{3.172}$$

which is calculated from the displacement field $u_i^{(2)}(\mathbf{x})$. It is generated by the force pair at the crack tip. Displacement fields like this were denoted as *fundamental fields* by Bueckner (Not to be confused with the fundamental solution of a differential equation!).

What is the structure of this fundamental displacement field? From Eq. (3.144) and Fig. 3.25 it is obvious that the stress intensity factors become infinitely large, when the force pair P or Q is located directly at the crack tip so that $d = c - x_1 \to 0$. Therefore, an even stronger stress singularity develops as $1/\sqrt{r}$! Fundamental fields have been analytically calculated for simple two-dimensional [44, 45] and three-dimensional [46, 47] crack configurations. For the semi-infinite crack in the infinite sheet (see Fig. 3.7, right-hand side), eigenfunctions (3.38)–(3.41) have been derived in Sect. 3.2.2. The Bueckner singularity complies just with that eigenfunction belonging to the eigenvalue $\lambda = -1/2$ or $n = -1$. The displacements at the crack tip thereby become singular with $r^{-1/2}$ and the stresses with $r^{-3/2}$. The strain energy as well becomes singular in a finite area because of $u = \frac{1}{2}\sigma_{ij}\varepsilon_{ij} \sim r^{-3}$ (compare with Sect. 3.3.6). For these reasons Bueckner's singularity must not be understood and approved as a real physical solution. But it constitutes a mathematically correct solution of the boundary value problem and is fully legitimated as a weight function. The coefficient of the (-1)st eigenfunction is directly correlated to the intensity of the Bueckner singularity $\Re A_{(-1)} = -B_I$. Thus, the appropriate displacement and stress fields for mode I in a plane strain state are obtained from (3.41) with $n = -1$

$$\begin{Bmatrix} u_1 \\ u_2 \end{Bmatrix} = \frac{B_I}{\mu\sqrt{r}} \begin{Bmatrix} \cos\frac{\theta}{2}\left[(2\nu - 1) + \sin\frac{\theta}{2}\sin\frac{3\theta}{2}\right] \\ \sin\frac{\theta}{2}\left[(2 - 2\nu) - \cos\frac{\theta}{2}\cos\frac{3\theta}{2}\right] \end{Bmatrix} \tag{3.173}$$

$$\begin{Bmatrix} \sigma_{11} \\ \sigma_{22} \\ \tau_{12} \end{Bmatrix} = B_I r^{-\frac{3}{2}} \begin{Bmatrix} \cos\frac{3\theta}{2} - \frac{3}{2}\sin\theta\sin\frac{5\theta}{2} \\ \cos\frac{3\theta}{2} + \frac{3}{2}\sin\theta\sin\frac{5\theta}{2} \\ \frac{3}{2}\sin\theta\cos\frac{5\theta}{2} \end{Bmatrix}. \tag{3.174}$$

By inserting (3.173) as $u_i^{(2)}$ into (3.172), the weight functions at the crack faces $(\theta = \pm\pi)$ are directly identified

$$H_1^I(r, \pm\pi) = 0\,s, \quad H_2^I(r, \pm\pi) = \frac{\pm 1}{\sqrt{2\pi r}}. \tag{3.175}$$

Since the tangential force Q creates no intensity factor K_I, function H_1^1 has to be zero. H_2^1 embodies exactly the effect on K_I by a vertical concentrated load P in the distance of r. Weight functions of the Bueckner type can be extended to mixed–mode problems and mixed boundary value problems as well.

Comparing the terms for the weight functions according to Bueckner (3.172) with the ones developed by Rice (3.157), the difference becomes clear: In the former we use a fundamental singular displacement field, whereas a regular displacement solution is differentiated in the latter. For the above-mentioned semi-infinite crack, the near field (3.16) for Rice's method could be used. In fact, the differentiation of (3.16) with $\frac{\partial}{\partial a} = -\frac{\partial}{\partial x} = -\cos\theta\frac{\partial}{\partial r} + \frac{\sin\theta}{r}\frac{\partial}{\partial\varphi}$ provides just that fundamental field (3.173).

> In conclusion, it has to be highlighted that especially the weight functions for crack faces offer a very efficient calculation method for K factors, since the sectional stresses at the crack location can be obtained from any conventional stress analysis of the considered component. Therefore, an analysis with an explicit crack is not necessary. For this reason, the numerical computation of weight functions by means of FEM is dealt with in detail in Sect. 5.6.

The book by Fett and Munz [48] and the articles regarding mixed–mode loading [43, 45, 49] and three-dimensional crack configurations [46, 47, 50, 51] are recommended as additional literature.

3.2.11 Thermal and Electric Fields

Today, there is increasing interest in technical problems where cracks are not exclusively a mechanical phenomenon but where they are exposed to important impacts from other physical fields. The problem of thermally induced stresses in components with cracks, which develop due to inhomogeneous temperature fields, is well known. They play an important role especially in facilities and components of power plants as well as in cast parts. Recently, questions have arisen as to how a crack influences an electric field (for instance in capacitors, electric conductors or microelectronic components), or which magnetic field concentrations (in motors, transformers) are caused by material defects. Especially the usage of new multifunctional materials with piezoelectric, magnetostrictive (and other) properties in mechatronics, adaptronics and microsystem technology raises new questions of strength and reliability. Their solution requires assessment of cracks under coupled thermal, electric, magnetic and mechanic fields. In the spirit of this book, two simple field problems with cracks will be covered for a first understanding.

Crack in a Stationary Temperature Field

The *temperature field* $T(x_1, x_2)$ in a plane isotropic body with crack is obtained from the solution of a stationary thermal conduction problem. The heat flux $\boldsymbol{h}$ in the body is proportional to the negative temperature gradient (from hot to cold) $\boldsymbol{g} = -\nabla T$. This is described by Fourier's law employing the thermal conduction coefficient k

$$\boldsymbol{h} = -k\nabla T \quad \text{or} \quad h_i = -k\frac{\partial T}{\partial x_i} . \tag{3.176}$$

According to the 1st law of thermodynamics, the divergence of the heat flux vector $\boldsymbol{h}$ needs to be zero if no internal heat sources are present

$$\nabla \cdot \boldsymbol{h} = -k\,\Delta T(x_1, x_2) = 0 \quad \text{or} \quad -k\left(\frac{\partial^2 T}{\partial x_1^2} + \frac{\partial^2 T}{\partial x_2^2}\right) = 0 . \tag{3.177}$$

Therefore, the sought temperature field $T(x_1, x_2)$ in the crack plane has to obey the Laplace equation (potential function). If a certain heat flux $\bar{h}$ is prescribed on the boundary (normal vector n_i), the thermal balance postulates

$$h_i n_i = -\bar{h} . \tag{3.178}$$

In our specific case we assume that the body is exposed to a constant thermal flux in the direction of the x_2 axis due to a temperature gradient, which means

$$h_2 = \bar{h}_2 = h , \quad h_1 = 0 \quad \text{bei} \quad |z| \to \infty . \tag{3.179}$$

The surface of the crack is thermally isolated so that the flux lines have to circumvent the crack, as shown in Fig. 3.28. This leads to a concentration of field lines at the crack tip. The thermal boundary condition with $n_i = \mp e_2$ at the crack faces is

$$h_i n_i = \mp h_2 = 0 . \tag{3.180}$$

This boundary value problem, defined for the temperature field T, is mathematically absolutely identical with the one for the displacement field u_3 in anti-plane shear stress (compare Appendix A.5.4 and Sect. 3.2.1). In both cases the Laplace equation has to be fulfilled. The heat flux here corresponds to the shear stress component $h_i \,\hat{=}\, \tau_{i3}$. The boundary conditions (3.27) are analogous as well. Therefore, the solution can be expressed by the same complex functions as in (A.165):

$$T(x_1, x_2) = -\Re\Omega(z)/k , \quad h_1 - ih_2 = \Omega'(z). \tag{3.181}$$

In analogy to (3.28)–(3.32) we obtain the following thermal solution at the crack tip:

Fig. 3.28 Concentration of thermal and electric fields at a crack

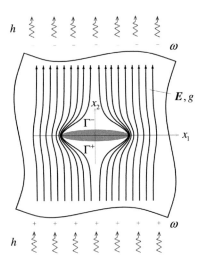

$$T(r, \theta) = h\sqrt{\pi a}\left(-\tfrac{2}{k}\right)\sqrt{\tfrac{r}{2\pi}}\,\sin\tfrac{\theta}{2}$$

$$\left\{\begin{array}{c} h_1 \\ h_2 \end{array}\right\} = h\sqrt{\pi a}\frac{(-1)}{\sqrt{2\pi r}}\left\{\begin{array}{c} -\sin\tfrac{\theta}{2} \\ +\cos\tfrac{\theta}{2} \end{array}\right\}. \qquad (3.182)$$

One can see that at the crack tip the temperature field (like the displacements in mechanics) complies with $\sqrt{r}$, and the heat flux (analog to the stresses) becomes singular with $1/\sqrt{r}$! The coefficient depends on thermal load and crack length. It plays the same role as K_{III} and should be denoted as »heat flux intensity factor« K_{h}.

$$h\sqrt{\pi a} = K_{\mathrm{h}} \quad [\mathrm{W\ m^{-3/2}}] \qquad (3.183)$$

Crack in an Electrostatic Field

The *electric field* $E(x_1, x_2)$ in a body with cracks made of dielectric isotropic material is studied. The approach is basically identical with the previously used thermal example. The primary field quantity in electrostatics is the electric potential $\varphi(x_1, x_2)$, wherefrom the vector of the electric field strength is calculated as a negative gradient $E = -\nabla\varphi$. The dielectric material law correlates E to the dielectric displacement vector D

$$D = \epsilon E \quad \text{bzw.} \quad D_i = \epsilon E_i \quad (\epsilon \text{ dielectricity constant}). \qquad (3.184)$$

Gauss's law requires the balance of the electric charge density in the volume

$$\nabla \cdot \boldsymbol{D} = -\epsilon \Delta \varphi = 0 \quad \text{bzw.} \quad -\epsilon \left(\frac{\partial^2 \varphi}{\partial x_1^2} + \frac{\partial^2 \varphi}{\partial x_2^2} \right) = 0 \qquad (3.185)$$

and at the surface, if a surface charge density $\bar{\omega}$ is specified

$$D_i n_i = -\bar{\omega} . \qquad (3.186)$$

It is evident that we can use the same approach for the electrostatic boundary value problem

$$\varphi(x_1, x_2) = -\Re \Omega(z)/\epsilon , \quad D_1 - iD_2 = \Omega'(z) . \qquad (3.187)$$

We now consider a crack in the infinite plane (Fig. 3.28), exposed to an external vertical electric field given by the appropriate charge density

$$D_2 = \omega . \qquad (3.188)$$

The crack is assumed to be impermeable for the electric field, which is expressed by a vanishing electric charge $D_i n_i = \mp \omega = 0$ at the crack faces. The mathematical solution occurs entirely analogous to the previous thermal example and we obtain the electric crack tip field

$$\varphi(r, \theta) = K_D \left(-\frac{2}{\epsilon} \right) \sqrt{\frac{r}{2\pi}} \sin \frac{\theta}{2}$$

$$\left\{ \begin{array}{c} D_1 \\ D_2 \end{array} \right\} = \epsilon \left\{ \begin{array}{c} E_1 \\ E_2 \end{array} \right\} = K_D \frac{(-1)}{\sqrt{2\pi r}} \left\{ \begin{array}{c} -\sin \dfrac{\theta}{2} \\ +\cos \dfrac{\theta}{2} \end{array} \right\} . \qquad (3.189)$$

Obviously, a singularity of the electric fields $\boldsymbol{D}$ and $\boldsymbol{E}$ originates at the crack tip, which is quantified by an »intensity factor of dielectric displacement« K_D.

$$K_D = \omega \sqrt{\pi a} \quad [\text{C m}^{-3/2}] \qquad (3.190)$$

The parallels between the mechanical (anti-plane shear), thermal and electric field problem is summarized once again in Table 3.1. This leads to the conclusion that all available mechanical solutions for cracks under mode III loading may be converted with the indicated correlations to boundary value problems of stationary thermal conduction or electrostatics.

In the mentioned examples the corresponding field problems themselves were isolated. The problem becomes more interesting when a direct coupling between different physical fields comes into existence due to the material laws. For instance, mechanical loading in piezoelectric material causes electric field singularities at the crack tip and vice versa. A treatise on fracture mechanics for piezoelectrics and their numerical analysis can be found for instance in Qin [52] and Kuna [53].

Table 3.1 Analogy of mechanical, thermal and electric field variable for fracture mechanical problems

	Mechanics	Thermal conduction	Electrostatics
Primary field variable	Displacement u_3	Temperature T	el. potential φ
Derived field variable	Strains	Temperature gradient	el. field strength
	$\gamma_{i3} = u_{3,i}$	$g_i = T_{,i}$	$E_i = -\varphi_{,i}$
Dual field variable	Stresses	Heat flux	el. displacement
Material law	$\tau_{i3} = \mu\gamma_{i3}$	$h_i = -kg_i$	$D_i = \epsilon E_i$
Balance solution in V	Equilibrium	Thermal energy	Charge density
	$\tau_{i3,i} = 0$	$h_{i,i} = 0$	$D_{i,i} = 0$
Balance equation on S	$\tau_{i3}n_i = \bar{t}_3$	$h_i n_i = -\bar{h}$	$D_i n_i = -\bar{\omega}$

3.3 Elastic-Plastic Fracture Mechanics

3.3.1 Introduction

Many construction materials (metals, plastics, and others) show elastic-plastic deformation behavior. Therefore the application of linear-elastic fracture mechanics has its limitations. Due to the stress concentration at the crack tip, here the yield stress of the material is already exceeded at low external loads and actually a small *plastic zone* is formed. When the load increases, the plastic zone in the body expands further. It causes a redistribution of the stress and strain fields which leads to a blunting of the crack tip. In the beginning, the plastified zone is surrounded by an elastic region. In the theory of plasticity, this state is called »constrained plastic flow«. In the further course, the plastic regions can reach the boundaries of the body and the result is the »fully plastic state«. In an ideally plastic material, the *plastic limit load* F_L of the structure would be reached, which means unlimited plastic deformations would happen at this load level. Real materials possess further strength reserves due to their hardening behavior.

These stages of plastification of a body with crack are schematically pictured in Fig. 3.29. With increasing plastification, the non-linearity of the global force–displacement–curve is enhanced. How big a plastic zone can grow before the crack is initiated and the fracture process begins, depends on the material properties and the load situation. The higher the ratio of fracture toughness and yield stress of the material, the higher is the extent of plastification before fracture. Besides that, it is important to note that the plastic deformation is associated with a significant amount of energy *dissipation* in the body, which can get rather large compared to the energy *consumption* during crack propagation, and is to be distinguished clearly from it.

It is obvious that the plastic deformations influence the situation at the crack and in the body considerably. Thus, special failure criteria need to be established for fracture phenomena that are preceded or accompanied by elastic-plastic deformations. This task is pursued in *Elastic-plastic fracture mechanics(EPFM)* also designated as *ductile fracture mechanics*.

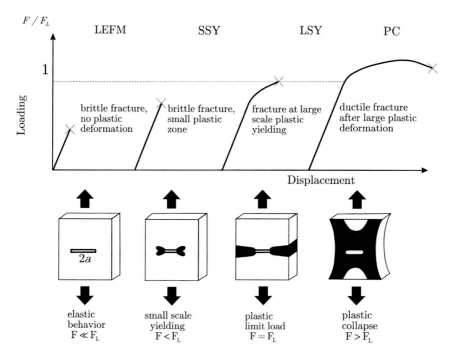

Fig. 3.29 Stages of plastic deformation in a body with crack

Because of the non-linear, load path-dependent material behavior, the solution of boundary value problems of the theory of plasticity proves rather difficult for cracked bodies. Therefore, the analytical solution methods of elastic-plastic fracture mechanics are very limited and restricted to simple material models, plane crack configurations and mostly monotonic loads. Only since powerful numerical solution methods have became available new possibilities for analyzing elastic-plastic fracture problems opened up. The most important fracture-mechanical parameters and concepts of ductile fracture mechanics, which have proven effective until now, will be presented in the following sections. However, their application to real crack configurations in structural components requires in most cases appropriate numerical calculations of the parameters..

3.3.2 Small Plastic Zones at the Crack

Estimating Size and Form of Plastic Zones

If the size of the plastic zone is small compared to the length of the crack and all other dimensions of the structure, it is considered as *small scale yielding*, commonly abbreviated as SSY. This model is based on the idea that the plastic zone is situated

inside the elastic crack tip solution known from Sect. 3.2.2. That means the plastification at the crack tip is controlled by stresses and deformations of the surrounding
elastic fields, which again are defined by the stress intensity factors. The essential
condition for this is that the radius of the plastic zone r_p stays considerably smaller
than the radius r_K of validity of the near field solution, pictured in Fig. 3.30. Since
the range of validity $r_K \approx 0,02 - 0,10\,a$ itself is only a fraction of the crack length,
the SSY assumption thus requires very small plastic zones. Still, the model of small
scale yielding leads to the first interesting findings. The explanations in this section
will focus on the mode I load but could logically be transferred to the other two crack
opening types (see Sähn & Göldner [54]) as well.

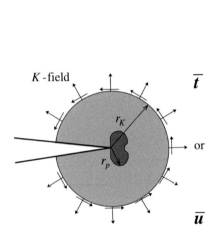

Fig. 3.30 Model of small scale yielding
SSY

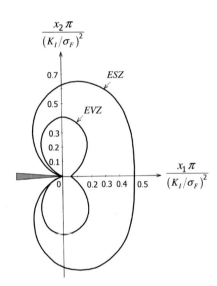

Fig. 3.31 Shape of the plastic zone in small
scale yielding

As the simplest case, ideal-plastic material behavior with an initial yield stress of
σ_F is assumed. In order to calculate the plastic zones to a first approximation, the
stresses of the elastic crack tip solution (3.12) are inserted into the yield criterion
according to v. Mises (A.102). For this purpose, we determine the maximum principal normal stresses in the (x_1, x_2) plane and the corresponding principal direction
angle θ_0 from (3.12) according to Appendix A.3.3:

$$\begin{Bmatrix} \sigma_\mathrm{I} \\ \sigma_\mathrm{II} \end{Bmatrix} = \frac{K_\mathrm{I}}{\sqrt{2\pi r}} \cos\frac{\theta}{2} \begin{Bmatrix} 1 + \sin\frac{\theta}{2} \\ 1 - \sin\frac{\theta}{2} \end{Bmatrix}, \quad \theta_0 = \pm\frac{\pi}{4} + \frac{3}{4}\theta. \tag{3.191}$$

The third principal stress is σ_{33}, see Sect. A.5.2:

$$\sigma_{\text{III}} = 0 \text{ (plane stress)}, \quad \sigma_{\text{III}} = \nu(\sigma_{\text{I}} + \sigma_{\text{II}}) = 2\nu \frac{K_{\text{I}}}{\sqrt{2\pi r}} \cos \frac{\theta}{2} \text{ (plane strain)}.$$
(3.192)

Inserting into the yield criterion (A.102) provides the radius $r_{\text{p}}(\theta)$ of the plastic zone as a function of the polar angle θ. It is distinguished between the models of plane strain and plane stress.

$$r_{\text{p}}(\theta) = \frac{1}{2\pi} \left(\frac{K_{\text{I}}}{\sigma_{\text{F}}} \right)^2 \cos^2 \frac{\theta}{2} \begin{cases} 3 \sin^2 \frac{\theta}{2} + 1 & \text{plane stress} \\ 3 \sin^2 \frac{\theta}{2} + (1 - 2\nu)^2 & \text{plane strain}. \end{cases}$$
(3.193)

Figure 3.31 illustrates the resulting shapes of the plastic zones for $\nu = 1/3$. The plastic zone for plane strain state is considerably smaller than those for plane stress. Its form stretches laterally in the crack direction, whereas it is oriented more straight ahead in plane stress.

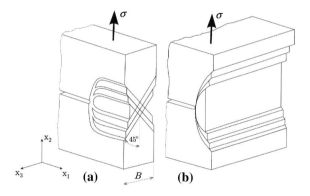

Fig. 3.32 Glide planes of maximum shear stress **a** plane stress **b** plane strain

Plastic deformations in metals take place in slip bands, which are formed on planes of maximum shear stress. In order to determine the orientation of the glide processes at the crack tip, we calculate the principal shear stresses $\tau_{\text{max}} = (\sigma_{\text{max}} - \sigma_{\text{min}})/2$ using relation (A.64) for the angular range $\theta \approx \pm 45°$ in front of the crack tip with the help of the near field solution (3.191), (3.192). There are significant differences between the models of plane stress and plane strain

$$\text{plane stress:} \quad \tau_{\text{max}} = \frac{\sigma_{\text{I}} - \sigma_{\text{III}}}{2} = \frac{\sigma_{\text{I}}}{2}$$

$$\text{plane strain:} \quad \tau_{\text{max}} = \frac{\sigma_{\text{I}} - \sigma_{\text{II}}}{2} = \frac{K_{\text{I}}}{\sqrt{2\pi r}} \cos \frac{\theta}{2} \sin \frac{\theta}{2}.$$
(3.194)

In plane stress, τ_{max} occurs in section planes, which are inclined by 45° with regard to the (x_1, x_2) plane. The assumption of plane stress is valid for thin-walled structures so that the slip bands run slanted across the thickness direction, see Fig. 3.32a. This causes a necking of the cross section along a small strip in front of the crack tip. Under plane strain conditions, the greatest shear stresses are caused by the principal stresses in the (x_1, x_2) plane. Therefore, the plastic slip takes place in planes that lie parallel to the x_3 axis, just as pictured by Fig. 3.32b. This hinge-like plastic deformation leads to a blunting of the originally sharp crack tip. These essential differences in the plastic deformation kinematics are confirmed by experimental findings in thin (plane stress) and thick (plane strain) structures, whereby the amount of constraint strain in x_3–direction does substantially depend on the size of the plastic zone r_p relative to thickness B:

$$\text{plane stress: } r_p \gg B \qquad \text{plane strain: } r_p \ll B. \qquad (3.195)$$

Irwin's crack length correction for small plastic zones

These considerations are based on stresses at the crack ligament $(r, \; \theta = 0)$, whose values can be determined from the crack tip solution (3.12) as follows

$$\sigma_{11} = \sigma_{22} = \frac{K_I}{\sqrt{2\pi r}}, \qquad \tau_{12} = 0, \qquad \sigma_{33} = \begin{cases} 0 & \text{plane stress} \\ 2\nu\sigma_{22} & \text{plane strain.} \end{cases} \qquad (3.196)$$

The function of the crack-opening stress $\sigma_{22}(r)$ is depicted in Fig. 3.33. The shape of the plastic zone in an ideally-plastic material has been calculated in (3.193) by means of the V. Mises's yield criterion. Its extension r_F along the x_1 axis is

$$r_F = r_p(\theta = 0) = \frac{1}{2\pi}\left(\frac{K_I}{\sigma_F}\right)^2 \begin{cases} 1 & \text{plane stress} \\ (1-2\nu)^2 & \text{plane strain.} \end{cases} \qquad (3.197)$$

In plane stress, r_F is determined just through the intersection of the σ_{22} curve and the yield stress σ_F since $\sigma_v = \sigma_{22}(r_F, 0) = \sigma_F$ is valid. For plane strain conditions, the yield criterion decreases to $\sigma_v = (1 - 2\nu)\sigma_{22}(r_F, 0) = \sigma_F$ because of the stress component σ_{33}, i. e. the acting normal stress σ_{22} must be greater by the factor $1/(1 - 2\nu)$ (about 3 times for $\nu = 1/3$). This increase of stress in consequence of the constraint deformation (in multiaxial stress state) is quantified by the *plastic constraint factor*

$$\alpha_{cf} = \frac{\sigma_{22}}{\sigma_F} = \begin{cases} 1 & \text{plane stress} \\ 1/(1-2\nu) \approx 3 \text{ during } \nu = 1/3 & \text{plane strain.} \end{cases} \qquad (3.198)$$

By cutting off the stress level at $\sigma = \sigma_F$ (hatched area in Fig. 3.33) however, the resulting force in the x_2–direction is being falsified. In order to restore the balance of forces, these stresses have to be reallocated to the ligament, leading to a greater size d_p of the plastic zone. From the condition that the area underneath the elastic curve (dashed line) is equivalent to the area under the elastic-plastic curve (full line), the length $d_p = 2r_F$ can be calculated. Using (3.197), the size of the hatched surface is

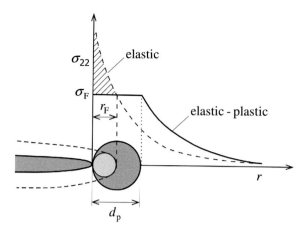

Fig. 3.33 Correction of the plastic zone at a crack tip by Irwin

$$\int_0^{r_F} \frac{K_I}{\sqrt{2\pi r}}\, dr - \sigma_F\, r_F = \sigma_F r_F \overset{!}{=} \sigma_F\,(d_p - r_F). \tag{3.199}$$

According to Irwin [55], this area is exactly compensated if the crack is effectively increasing by r_F (dashed crack shape), which unloads half of the area $\sigma_F d_p$. This gives the correction of the plastic zone size to the doubled value (large circle in Fig. 3.33)

$$d_p \overset{\frown}{=} 2r_p = 2r_F, \qquad d_p = \frac{1}{\pi}\left(\frac{K_I}{\sigma_F}\right)^2 \begin{cases} 1 & \text{plane stress} \\ (1-2\nu)^2 & \text{plane strain.} \end{cases} \tag{3.200}$$

The effective crack length and the corresponding stress intensity factor (3.64) are

$$a_{\text{eff}} = a + r_p, \qquad K_{\text{Ieff}} = \sigma\sqrt{\pi a_{\text{eff}}}\, g\left(\frac{a_{\text{eff}}}{w}\right). \tag{3.201}$$

By means of the plastic zone size, the range of validity for a linear-elastic assessment of cracks can be defined. In fracture-mechanical test procedures [20, 56] it is demanded that d_p must be considerably smaller than all relevant dimensions of specimens and components. According to that, specimen thickness B, crack length a and ligament size $(w - a)$ must fulfill the following requirements:

$$B,\, a,\, (w - a) \geq 2,5\left(\frac{K_{\text{Ic}}}{\sigma_F}\right)^2. \tag{3.202}$$

The expression on the right is proportional to the plastic zone size at fracture. If one of these criteria is violated, the application of LEFM becomes questionable. That means, fracture toughness K_{Ic} values, measured by tests outside this validity range, do not correspond to the lower conservative bound value K_{Ic} defined for plane strain.

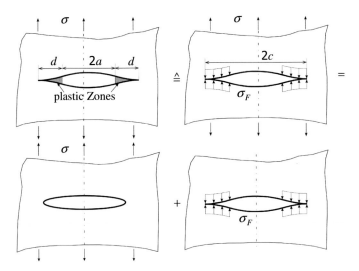

Fig. 3.34 Dugdale's model for strip yield zone

3.3.3 The Dugdale Model

Observing strip-like plastic zones in front of the crack tip in thin metal sheets during tensile tests, Dugdale [57] was inspired to create this model. As explained above, under plane stress conditions, necking occurs by plastic flow on 45° planes that restricts the height of the plastic zone to about the thickness B. The following assumptions underlie this model:

- The entire plastic deformation is concentrated on one strip (mathematical line) of the length d.
- The material inside the yield strip behaves ideal-plastically. Plane stress is valid so that yielding begins at $\sigma_{22} = \sigma_F$. Applying the model to plane strain, the effective yield stress to initiate plastic flow has to be modified by the constraint factor (3.198) to $\sigma_{22} = \alpha_{cf}\, \sigma_F$.
- The problem is simplified to a boundary value problem of a hypothetical crack of length of $2(a+d) = 2c$ in an elastic body. The crack model can then be imagined as superposition of the following two loading conditions (see Fig. 3.34):

(1) Crack in an infinite plane under constant tension σ.
(2) The supporting effect of the plastified material inside the yield zone $a \leq |x_1| \leq a + d$ is accounted for by tractions $t_2^c = \sigma_F$, which compress the crack.

For problem (1), the stress intensity factor is known from Sect. 3.2.2:

$$K_I^{(1)} = \sigma\sqrt{\pi(a+d)}. \tag{3.203}$$

For load case (2), the K–factor has been calculated in Sect. (3.2.10):

$$K_I^{(2)} = -2\sigma_F \sqrt{\frac{a+d}{\pi}} \arccos\left(\frac{a}{a+d}\right). \tag{3.204}$$

It is demanded that no stress singularities appear at the ends of the hypothetical crack $|x_1| = \pm c$. Therefore, the stress intensity factors of both subproblems (1) and (2) have to cancel each other out:

$$K_I = K_I^{(1)} + K_I^{(2)} = 0. \tag{3.205}$$

From this relation we can deduce the length d of the plastic zone

$$\frac{a}{a+d} = \cos\left(\frac{\pi\sigma}{2\sigma_F}\right), \quad d = a\left[\left(\cos\frac{\pi\sigma}{2\sigma_F}\right)^{-1} - 1\right]. \tag{3.206}$$

As expected, according to this relation no plastic zone appears without load ($\sigma = 0$), $d = 0$. However, it is interesting that d can grow to infinity if the stress approaches the yield level $\sigma \to \sigma_F$. In this case, the plastic limit load of the sheet is attained and the entire net cross-section flows plastically.

For small-scale yielding $\sigma \ll \sigma_F$, the cosine function can be approximated by

$$\left(\cos\frac{\pi\sigma}{2\sigma_F}\right)^{-1} \approx 1 + \frac{1}{2}\left(\frac{\pi\sigma}{2\sigma_F}\right)^2 \tag{3.207}$$

and using the relation $\sigma\sqrt{\pi a} = K_I$, one can write

$$d \approx \frac{a\pi^2\sigma^2}{8\sigma_F^2} = \frac{\pi}{8}\left(\frac{K_I}{\sigma_F}\right)^2. \tag{3.208}$$

Comparing the result of the Dugdale model (3.208) with the plastic zone correction by Irwin (3.200), it becomes obvious that both models provide similar relations, differing only in the prefactors $1/\pi = 0,318$ and $\pi/8 = 0,392$ respectively.

3.3.4 Crack Tip Opening Displacement (CTOD)

If fracture mechanics specimens made of ductile materials are loaded, it can be observed that the tip of the originally sharp crack undergoes wide stretching and blunting due to plastic deformation, even before the crack initiates, Fig. 3.35. This irreversible opening displacement of crack faces exceeds by far that crack opening due to purely elastic deformation. Therefore, it can be considered as a local measure of the plastic strains around the crack tip. This parameter δ_t is called *crack tip opening displacement* CTOD. Wells [58] and Burdekin & Stone [59] suggested a fracture concept that employs the crack tip opening displacement δ_t as characteristic parameter.

> The CTOD-criterion states that in ductile materials crack initiation starts, if the crack tip opening displacement δ_t exceeds a critical, material specific limit value δ_{tc}.
>
> $$\delta_t = \delta_{tc} \qquad (3.209)$$

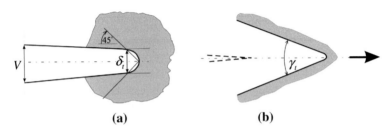

(a)   (b)

Fig. 3.35 Definition of the crack tip opening displacement δ_t (CTOD) at stationary cracks and the crack tip opening angle γ_t (CTOA) at moving cracks

In order to realize this concept, a quantitative relationship between δ_t and the external loads is required. The Dugdale-model offers one opportunity for computing this. Here, the crack opening displacement is identified at the tip of the physical crack at $|x_1| = a$, see Fig. 3.34, and is at this point

$$\delta_t = (u_2^+ - u_2^-) = 2u_2(a) = \frac{8\sigma_F a}{\pi E} \ln \left(\cos \frac{\pi \sigma}{2\sigma_F} \right)^{-1}. \qquad (3.210)$$

For very small loads we use again the series expansion (3.207), establishing this way a correlation to K_I under SSY-conditions.

$$\delta_t = \frac{K_I^2}{\sigma_F E}. \qquad (3.211)$$

Also the plastic zone correction by Irwin yields a value for the crack tip opening displacement. To do this, δ_t is calculated at that position, where the plastic zone touches the faces of the effective crack, i.e. at $x_1 = -r_p$ in Fig. 3.33. Inserting the displacement field (3.16) of linear-elastic solution at $(\theta = \pi)$ we obtain

$$u_2(r_p) = \frac{4}{E} K_I \sqrt{\frac{r_p}{2\pi}} \qquad \text{(plane stress)}, \qquad (3.212)$$

and using the known value at r_p from (3.200) results in

$$\delta_t = 2u_2 = 2\frac{4}{E}K_I\sqrt{\frac{K_I^2}{(2\pi)^2\sigma_F^2}} = \frac{4}{\pi}\frac{K_I^2}{E\,\sigma_F} = \frac{4}{\pi}\frac{G}{\sigma_F}. \tag{3.213}$$

At small plastic zones (SSY) we also find the correlation between the crack tip opening displacement δ_t and the energy release rate G. Depending on the amount of constraint in the real component and the material's hardening, an empirical approach can be made:

$$\delta_t = \frac{K_I^2}{m\,\sigma_F\,E} \quad \text{for plane stress and} \quad \delta_t = \frac{K_I^2(1-\nu^2)}{m\,\sigma_F\,E} \quad \text{for plane strain,} \quad (3.214)$$

at which the factor $m \approx 1$ is valid for plane stress state, and amounts to $m \approx 2$ for plane strain.

A generally accepted definition of the crack tip opening displacement δ_t that does not depend on a special model, is shown in Fig. 3.35. Hereby, two secants are placed in an angle of $\pm 45°$ at the blunted crack tip, and δ_t is identified from the intersection points with both crack faces. This definition is well suited for interpreting numerical analyses. The experimental determination of the crack tip opening displacement δ_t at a specimen or at a component turns out more complicated, since measurements directly in the crack tip region are difficult to handle. Because of this, usually the *crack opening displacement* COD $\hat{=} V$ is measured at the specimen's surface and extrapolated to the crack tip by means of geometrical assumptions (e.g. intercept theorem for rotation around a plastic hinge). A physically substantiated determination of δ_t is possible with the help of stereoscopic measurements of the »stretched zone height« SZH, although only after fracture. The interested reader is referred to the literature [21, 60] and official testing standards [61, 62].

3.3.5 Failure Assessment Diagram (FAD)

On the basis of the Dugdale crack model, a method for engineering-based evaluation of component safety was developed by Harrison [63] in 1976. Therefore, the English term *Failure Assessment Diagram*—or »Two–Criteria–Method« was coined. This concept combines both limiting cases of failure by brittle fracture on one hand and plastic collapse if the plastic limit load is attained on the other hand. For the transition region comprising elastic-plastic failure of cracked components in between, a geometry independent failure limit curve is formulated from both concepts. In the following, the basic idea shall be explained.

For large amounts of plastification up to the plastic limit load, the Dugdale-model (3.210) yields a suitable basis in combination with the CTOD-criterion (3.209)

$$a\frac{8\sigma_F}{\pi E}\ln\left(\cos\frac{\pi\sigma}{2\sigma_F}\right)^{-1} = \delta_t \stackrel{!}{=} \delta_{tc}. \tag{3.215}$$

If this equation is converted for σ, the critical stress for a given crack length a results in

$$\sigma_c = \frac{2}{\pi}\sigma_F \arccos\left[\exp\left(-\frac{\pi E \delta_{tc}}{8\sigma_F a}\right)\right]. \tag{3.216}$$

For vanishingly small cracks $a \rightarrow 0$ this gives $\sigma_c = \sigma_F$, i.e. the strength is controlled by the plastic collapse—which is the case of very large plastic zones, denoted by *large scale yielding* (LSY).

In the opposite extreme of LEFM or SSY, the Dugdale-model (3.211) provides a relation to K_I

$$\frac{K_I^2}{E\sigma_F} = a\frac{\pi\left(\sigma_c^{LEFM}\right)^2}{E\sigma_F} = \delta_t \overset{!}{=} \delta_{tc}, \tag{3.217}$$

from which the critical stress σ_c^{LEFM} is obtained using $K_I = \sigma\sqrt{\pi a}$

$$\sigma_c^{LEFM} = \sqrt{\frac{E\sigma_F\delta_{tc}}{\pi a}}. \tag{3.218}$$

This equation predicts an infinitely high strength $\sigma_c^{LEFM} \rightarrow \infty$ for small cracks $a \rightarrow 0$. By equating (3.217) and (3.215) the crack length is eliminated:

$$\left(\frac{\sigma_c^{LEFM}}{\sigma_F}\right)^2 = \frac{8}{\pi^2}\ln\left[\cos\frac{\pi\sigma}{2\sigma_F}\right]^{-1}. \tag{3.219}$$

Finally we introduce load parameters K_r and S_r normalized to the interval [0,1]:

$$\frac{\sigma}{\sigma_c^{LEFM}} \;\widehat{=}\; \frac{K_I}{K_{Ic}} = K_r \qquad \text{for brittle fracture LEFM}$$

$$\frac{\sigma}{\sigma_F} \;\widehat{=}\; \frac{F}{F_L} = S_r \qquad\qquad \text{for plastic failure} \tag{3.220}$$

$$K_r = S_r\left[\frac{8}{\pi^2}\ln\left(\cos\frac{\pi}{2}S_r\right)^{-1}\right]^{-1/2}. \tag{3.221}$$

In the FAD-diagram the K_r value is plotted on the ordinate and S_r in the abscissa, see Fig. 3.36. This way, Eq. (3.221) represents a failure limit curve that interpolates between brittle and ductile failure. For a given structural component with crack a under the load σ, the point $P(K_r, S_r)$ can be assigned in the FAD. If this point lies within the limit curve, safety is ensured. An increasing loading σ effects a proportional rise of K_r and S_r, so that the point is shifted along a radial line outward. Failure occurs if the limit curve is attained. In order to apply the FAD, besides the material parameters K_{Ic} and σ_F analytical or numerical solutions are required for the stress intensity factor K_I and the plastic limit load F_L of the component with crack. Although the FAD-concept has been derived by the Dugdale-model for a crack

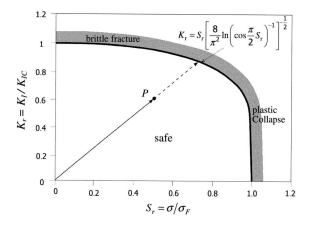

Fig. 3.36 Failure limit curve of the Failure Assessment Diagram FAD

in a sheet, the generalization of this concept to arbitrary crack configurations was proven. However, this limit curve is specific for a material ! Meanwhile, the concept is extended in revised versions [64] to materials with hardening, for secondary stresses, crack growth a. o. A contemporary overview of fracture mechanics assessment rules can be found in the European guidelines SINTAP [65] and FITNET [66].

3.3.6 Crack Tip Fields

In the framework of plastic flow theory no analytical solutions could be found for crack problems, not even for the simple case of a crack in the infinite plane. Closed-form solutions are only known for the asymptotic limit $r \to 0$ at the crack tip under simplified assumptions regarding the plastic material behavior. This means, we investigate the stress state in the interior of the plastic zone at a crack tip in an infinite domain. Information about the size and shape of the plastic zone is thus not possible. In the following, the most importation plastic crack tip fields for the crack opening mode I shall be described.

Ideal-Plastic Material

Under the assumption of a rigid-perfectly-plastic material the crack-tip field can be found using the so-called slip line theory. Thereby, a hyperbolic-ODE-system is set-up using the two-dimensional equilibrium conditions and the yield condition, whose characteristics are the slip lines. The orthogonal grid of slip lines in each point represents the directions of principal shear stresses and plastic shear strains. The plastic material is assumed as incompressible, i.e. Poisson's ratio $\nu = 1/2$ and $\sigma_{33} = (\sigma_{11} + \sigma_{22})/2$ for plane strain. For the plane strain state the slip line solution around the crack tip consists of three regions A, B and C shown in Fig. 3.37, where the following stress state exists (see e.g. [54, 67]):

$$A: \quad \sigma_{11} = \tau_F \, \pi, \quad \sigma_{22} = \tau_F(2 + \pi), \quad \tau_{12} = 0$$

$$B: \quad \sigma_{rr} = \sigma_{\theta\theta} = \tau_F \left(1 + \frac{3}{2}\pi - 2\theta\right), \quad \tau_{r\theta} = \tau_F$$

$$C: \quad \sigma_{11} = 2\tau_F, \quad \sigma_{22} = \tau_{12} = 0 \text{ (crack-face boundary consitions).} \quad (3.222)$$

Figure 3.38 depicts this stress distribution graphically. All stresses are proportional to the shear flow stress $\tau_F = \sigma_F/\sqrt{3}$. In regions A and C constant stress states prevail. In front of the crack (A), the normal stress amounts to $\sigma_{22} \approx 3\sigma_F$ and the triaxiality is $\hbar = \sigma^H/\sigma_v = (1+\pi)/\sqrt{3} \approx 2, 4!$ In the region B the slip lines run into the crack tip in a fan-shaped manner, inducing a singular behavior of the plastic shear strains $\varepsilon_{r\theta}$

$$\varepsilon_{r\theta} \sim \frac{1}{r}f(\theta). \quad (3.223)$$

Further slip line solutions for plane stress and mode II can be found in [68].

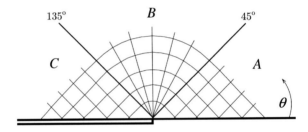

Fig. 3.37 Slip line field at the crack tip for a rigid-ideal-plastic material (plane strain)

Power Law Hardening

Often, the stress-strain curve of elastic-plastic materials can be represented in good approximation by the Ramberg-Osgood (A.118) law

$$\frac{\varepsilon}{\varepsilon_0} = \frac{\varepsilon^e}{\varepsilon_0} + \frac{\varepsilon^p}{\varepsilon_0} = \frac{\sigma}{\sigma_0} + \alpha \left(\frac{\sigma}{\sigma_0}\right)^n, \quad (3.224)$$

using the material parameter α, the hardening exponent n, the reference stress σ_0 ($\approx$ initial yield stress σ_{F0}) and the reference strain $\varepsilon_0 = \sigma_0/E$. By choosing n, the

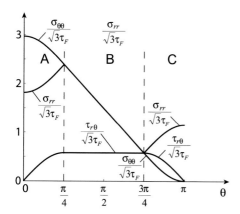

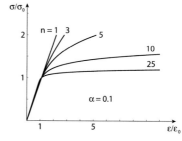

Fig. 3.38 Stress distribution at the crack tip for a rigid-ideal-plastic material (plane strain)

Fig. 3.39 Ramberg-Osgood Power-law hardening

material behavior can be varied in the whole range from linear-elastic ($n = 1$) to ideal-plastic ($n \to \infty$), see Fig. 3.39.

Assuming the plastic deformation theory and this power law, Hutchinson [69, 70] and Rice & Rosengreen [71] derived the relevant crack tip fields . This solution is named in fracture mechanics after the authors as an HRR-field, see [54, 67]. The plastic deformation theory (also referred to as total strain theory) and its limitations are discussed in detail in Appendix A.4.2.

The way to achieve this solution will only be outlined in the following. However, it reveals important relationships and insight. As usual in near tip field analyses, a polar coordinate system (r, θ) is placed at the tip of a crack in the infinite plane, see Fig. 3.7. According to (A.135), the plastic part of the constitutive law has the form

$$\frac{\varepsilon_{ij}^{\mathrm{p}}}{\varepsilon_0} = \frac{3}{2}\alpha \left(\frac{\sigma_v}{\sigma_0}\right)^{n-1} \frac{\sigma_{ij}^{\mathrm{D}}}{\sigma_0} \approx \frac{\varepsilon_{ij}}{\varepsilon_0}. \qquad (3.225)$$

With respect to the asymptotic behavior of the solution against the crack tip $r \to 0$, it is well justified to assume the non-linear plastic strains $\varepsilon_{ij}^{\mathrm{p}}$ much larger than the elastic parts $\varepsilon_{ij}^{\mathrm{e}}$, i.e. the latter can be neglected. The v. Mises equivalent stress is computed by the stress components in polar coordinates in plane strain state ($\sigma_{33} = (\sigma_{rr} + \sigma_{\theta\theta})/2$) as

$$\sigma_v = \sqrt{\frac{3}{4}(\sigma_{rr} - \sigma_{\theta\theta})^2 + 3\tau_{r\theta}^2}. \qquad (3.226)$$

Moreover, the stress state has to fulfill the equilibrium conditions, which is easiest realized by introducing the Airy's stress function (A.153) $F(r, \theta)$, from where the stress components in polar coordinates can be derived as follows:

$$\sigma_{rr} = \frac{1}{r}F_{,r} + \frac{1}{r^2}F_{,\theta\theta}, \quad \sigma_{\theta\theta} = F_{,rr}, \quad \tau_{r\theta} = -\left(\frac{1}{r}F_{,\theta}\right)_{,r}. \tag{3.227}$$

(it holds that $(\cdot)_{,r} = \frac{\partial(\cdot)}{\partial r}$, $(\cdot)_{,\theta} = \frac{\partial(\cdot)}{\partial\theta}$.) Furthermore, the pure plastic incompressible strain state at the crack tip has to obey the compatibility conditions (A.140), expressed in polar coordinates for plane strain

$$\frac{1}{r}(r\varepsilon_{\theta\theta})_{,rr} + \frac{1}{r^2}\varepsilon_{rr,\theta\theta} - \frac{1}{r}\varepsilon_{r,r} - \frac{2}{r^2}(r\varepsilon_{r\theta,\theta})_{,r} = 0. \tag{3.228}$$

Because of the asymptotic approach the near field can be separated multiplicatively into radial and angular functions to ensure the self-similarity for $r \to 0$ this way. Hence we make the separation ansatz for the Airy stress function

$$F(r, \theta) = Ar^s \tilde{F}(\theta) \tag{3.229}$$

with unknown factor A, exponent s and angular function $\tilde{F}(\theta)$. By using (3.227), we get expressions for stresses of the form

$$\sigma_{ij}(r, \theta) = A_\sigma r^{s-2}\tilde{\sigma}_{ij}(\theta), \tag{3.230}$$

with the stress coefficient A_σ and similar terms for the deviators σ_{ij}^D as well as the equivalent stress σ_v (3.226). By means of the material law (3.225) one finds the structure of the strain field including its angular distributions $\tilde{\varepsilon}_{ij}(\theta)$ and pre-factors

$$\varepsilon_{ij}(r, \theta) = \alpha\varepsilon_0 \left(\frac{A_\sigma}{\sigma_0}\right)^n r^{(s-2)n}\tilde{\varepsilon}_{ij}(\theta). \tag{3.231}$$

The relationship between the exponent s in (3.229) and the hardening exponent n can be established by the following considerations. The J-integral (3.100) is computed on a circular integration path with radius $r = \text{const}$, arc length $ds = r\,d\theta$ and $dx_2 = \cos\theta\,ds$

$$J = \int_{-\pi}^{+\pi} [U\cos\theta - \sigma_{ij}u_{i,1}]r\,d\theta = \int_{-\pi}^{+\pi} I(r, \theta)r\,d\theta. \tag{3.232}$$

The strain energy density U (A.136) and the term $\sigma_{ij}u_{i,1}$ have the dimension of $\sigma_{ij}\varepsilon_{ij}$, i.e. the kernel $I(r, \theta)$ of the integral has the radial dependency

$$I(r, \theta) \sim \sigma_{ij}\varepsilon_{ij} \sim \alpha\varepsilon_0 \frac{A_\sigma^{n+1}}{\sigma_0^n}r^{(s-2)(n+1)}\tilde{I}(\theta, n). \tag{3.233}$$

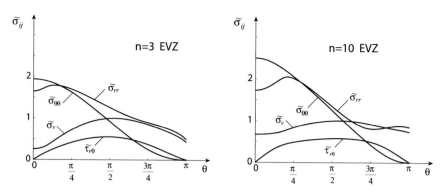

Fig. 3.40 Stress distribution at the crack tip according to HRR-solution for the hardening exponents $n = 3$ (*left*) and $n = 10$ (*right*) under plane strain (EVZ)

In the frame of plastic deformation theory, which is in essence equivalent to non-linear elasticity, the J-integral has to be independent of the integration path Γ. Its value must not depend on the chosen radius r! To ensure this property, the term $I(r, \theta)$ has to behave exactly as r^{-1}, from which follows $(s - 2)(n + 1) = -1$ and

$$s = \frac{2n + 1}{n + 1}. \tag{3.234}$$

From this result we can deduce the radial behavior of the respective field variables

$$\sigma_{ij} \sim r^{-\frac{1}{n+1}}, \quad \varepsilon_{ij} \sim r^{-\frac{n}{n+1}}, \quad U \sim r^{-1}. \tag{3.235}$$

To identify the unknown angular functions $\tilde{F}(\theta)$, $\tilde{\sigma}_{ij}(\theta)$ and $\tilde{\varepsilon}_{ij}(\theta)$ of the expressions (3.229), (3.230) and (3.231), we first plug the stress terms (3.227) into the material law (3.225) and then apply to it the compatibility conditions (3.228). This yields a non-linear ordinary differential equation for $\tilde{F}(\theta)$ (see [69, 71]) that has to be solved numerically. Thereby, the tractions-free boundary conditions on the crack faces $\tilde{\sigma}_{\theta\theta}(\pm\pi) = \tilde{\tau}_{r\theta}(\pm\pi) = 0$ and the symmetry of $\tilde{\sigma}_{\theta\theta}, \tilde{\sigma}_{rr}$ and the antimetry of $\tilde{\tau}_{r\theta}$ with respect to θ are taken into account. By inserting of $\tilde{F}(\theta)$ into (3.227) and (3.225) we finally gain the angular functions $\tilde{\sigma}_{ij}(\theta)$ and $\tilde{\varepsilon}_{ij}(\theta)$, respectively. The resulting stresses are illustrated in polar coordinates in Fig. 3.40. As an example the hardening exponents $n = 3$ and $n = 10$ are selected. By comparing with Figs. 3.5 and 3.38, one can quite clearly see the transition from linear-elastic to ideal-plastic near fields in the shape of angular functions.

In the end the question for the unknown coefficients A and A_σ remains. For this, we take advantage of the path-independence of the J-integral. To compute J, we choose at first an integration path Γ far remote from the crack tip and outside the plastic zone. A second circular-shaped path Γ_ε is placed into the validity region of the HRR-near field at $r \ll r_p$, see Fig. 3.18. Taking into account (3.232), (3.233) and (3.234) yields

$$J(\Gamma) = J(\Gamma_\varepsilon) = \alpha\varepsilon_0\sigma_0 \left(\frac{A_\sigma}{\sigma_0}\right)^{n+1} \underbrace{\int_{-\pi}^{+\pi} \tilde{I}(\theta, n)\mathrm{d}\theta.}_{I_n} \qquad (3.236)$$

The integral is only composed from the angular functions $\tilde{\sigma}_{ij}$, $\tilde{\varepsilon}_{ij}$ and the exponent n, delivering a constant $I_n(n)$

$$I_n(n) = 5.188 + 0.611n - 0.240n^2 + 0.027n^3 \qquad \text{(plane strain)}. \qquad (3.237)$$

The conversion of (3.236) yields the sought stress coefficient A_σ, so that by means of the J-value the stresses (3.230) and strains (3.231) can be written as follows:

$$\sigma_{ij} = \sigma_0 \left[\frac{J}{\alpha\varepsilon_0\sigma_0 I_n}\frac{1}{r}\right]^{\frac{1}{n+1}} \tilde{\sigma}_{ij}(\theta, n) \qquad (3.238a)$$

$$\varepsilon_{ij} = \alpha\varepsilon_0 \left[\frac{J}{\alpha\varepsilon_0\sigma_0 I_n}\frac{1}{r}\right]^{\frac{n}{n+1}} \tilde{\varepsilon}_{ij}(\theta, n) \qquad (3.238b)$$

and after integration of ε_{ij} the displacements

$$u_i = \alpha\varepsilon_0 \left[\frac{J}{\alpha\varepsilon_0\sigma_0 I_n}\right]^{\frac{n}{n+1}} r^{\frac{1}{n+1}} \tilde{u}_i(\theta, n) \qquad (3.238c)$$

These expressions describe the *HRR-near field* for an elastic-plastic material with the hardening exponent $1 \leq n \leq \infty$. The radial behavior of stresses and strains is singular at $r \to 0$. The »intensity« of the crack tip fields is quantified by the value of the J-integral.

The shape of the opened crack $u_2(r)$ equals a root-function of $(n+1)$th degree and gets more blunted with increasing n. An examination of the crack opening displacement δ_t from the HRR-field (3.238c) using the definition of Fig. 3.35 provides a linear relationship to the J-integral.

$$\delta_t = (\alpha\varepsilon_0)^{1/n}\frac{D_n}{\sigma_0} J. \qquad (3.239)$$

The known constant $D_n(n)$ has values between 1.72 and 0.79 for plane strain. The angular functions $\tilde{\sigma}_{ij}$, $\tilde{\varepsilon}_{ij}$ and $\tilde{u}_i$ in (3.238) were approximated by [72] as Fourier-series and are thereby available in a simple manageable form.

Extension by Higher Order Terms

The HRR-solution represents the strongest singularity of the elastic-plastic near field and embodies thus a theory of 1st order. Due to the non-linearity it is much more complicated than in LEFM to find the next terms of a series expansion (see Sect. 3.3.2). On the other hand it became evident that particularly in EPFM the fracture behavior of unlike crack configurations may differ substantially, although in all cases the same intensity of the HRR-field exists. The reason for that lies in different plastic zones, the size and form of which is rather sensitive to *triaxiality* of the stress state, see Fig. 3.41.

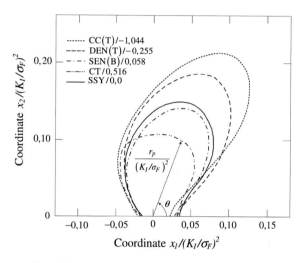

Fig. 3.41 Influence of triaxiality on the shape of plastic zones at the crack tip in various specimens [73]. Biaxial parameter $\beta_T = T_{11}\sqrt{\pi a}/K_I$

The triaxiality parameter $\hbar$ is commonly defined as the ratio of hydrostatic stress σ^H (A.66) and v. Mises equivalent stress σ_v (A.101):

$$\hbar = \frac{\sigma^H}{\sigma_v} \qquad \text{triaxiality parameter} \qquad (3.240)$$

From classical theory of strength it is already well known that with increasing triaxiality the tendency of a material for brittle fracture is enhanced (Hencky's diagram) [74]. Whereas in brittle materials the maximum of crack opening stress σ_{22} is deciding, the failure behavior of ductile materials is considerably effected by the hydrostatic stress. Therefore it is necessary to quantify the stress triaxiality at the crack by additional terms compared to the HRR-solution.

Fig. 3.42 Comparison of
numerical crack tip-solutions
for small and large deformations with the HRR-field,
plane strain, $n = 10$ [76]

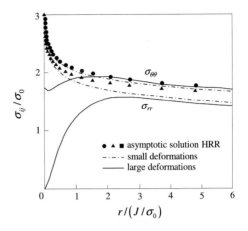

Furthermore it is important to note that the HRR-field was derived under the
assumption of infinitesimal strains. FEM-calculations by McMeeking & Parks [75]
assuming *finite* deformations have shown for the first time that the stresses at the
crack tip stay bounded, since the singularity is relieved due to the more realistically
modeled crack blunting, whereas the strains remain infinitely large. This effect has
been confirmed by numerous FEM-analyses of other authors who investigated the
near field further. Figure 3.42 presents the stresses σ_{rr} and $\sigma_{\theta\theta}$ on the x_1-axis ahead
of the crack as function of the dimensionless distance $r/(J/\sigma_0)$ by Yuan [76]. The
FEM-solution for small deformations agrees well with the HRR-field (symbols) at
$r \to 0$. For large deformations the stresses fall off sharply below $r < 2J/\sigma_0$, σ_{rr}
has even to go to zero at the notch surface. The size of this region, where large
deformations have essential influence, is understandably correlated with the CTOD-
value $\delta_t \approx 0, 5J/\sigma_0$ of (3.239). Above $r > 4\delta_t \approx 2J/\sigma_0$ the results for large and
small deformations hardly differ anymore.

Various approaches were pursued in order to extent the HRR-solution and to
account for the triaxiality. In case of small scale yielding SSY the plastic zone
is controlled by the K-factors and the T-*stresses* according to (3.60) and (3.61).
For plane crack problems only the T_{11}-component parallel to the crack exists. Its
magnitude depends on the crack configuration and the loading type. Meanwhile
T-data are documented in handbooks [77, 78]. For the Griffith's crack under tension
σ_{22}^∞, the lateral $T_{11} \equiv \sigma_{11} = -\sigma_{22}^\infty < 0$ corresponds to an equal-sized compression
(see (3.6)). On the contrary, tensile stresses $T_{11} > 0$ occur parallel to the crack in
bending-dominated specimens. The influence of the T_{11}-stress on the plastic zones in
different specimens was detected by FEM-computations of [73, 79]. Despite identical
K_I-values the shapes vary as shown in Fig. 3.41.

Compared with the singular K_I-field (3.12) (3.60), the hydrostatic stress at the
crack is changed by T_{11} as follows:

$$\sigma^{\mathrm{H}} = \frac{1}{3}\sigma_{kk} = \frac{K_{\mathrm{I}}}{\sqrt{2\pi r}}\frac{1}{3}f_{kk}^{\mathrm{I}}(\theta) + \frac{1}{3}T_{11}. \tag{3.241}$$

Considering the stress state (3.12) in front of the crack ($\theta = 0$), we have

$$\frac{1}{3}f_{kk}^{\mathrm{I}}(0) = \begin{cases} \frac{2}{3} & \text{plane stress} \\ \frac{2}{3}(1+\nu) & \text{plane strain} \end{cases} \tag{3.242}$$

and at the border r_{p} of the plastic zone (3.200) (with $\sigma_{\mathrm{F}} \mathrel{\widehat{=}} \sigma_0$) the triaxiality attains the following values:

$$\hbar(r_{\mathrm{p}}, 0) = \frac{\sigma^{\mathrm{H}}}{\sigma_0} = \frac{1}{3}\frac{T_{11}}{\sigma_0} + \frac{2}{3}\begin{cases} 1 & \text{plane stress} \\ (1+\nu)/(1-2\nu) & \text{plane strain} \end{cases} \tag{3.243}$$

Due to the *deformation constraint* under plane strain ($\varepsilon_{33} = 0, \sigma_{33} > 0$) the triaxiality increases substantially, which is also called *out-of-plane* constraint-effect, whereas the *in-plane* constraint is characterized by T_{11}. The extended fracture-mechanical concept suggested by Betegon & Hancock [80] relies on the T_{11}-stresses.

In case of large plastic zones (LSY) the T_{11}-stresses lose their validity, which is why extensions compared with the HRR-solution were sought on the basis of deformation plasticity and power-law hardening. Sharma and Aravas [81] found a 2nd term of the near-field in the form

$$\frac{\sigma_{ij}(r, \theta)}{\sigma_0} = \left(\frac{J}{\alpha\varepsilon_0\sigma_0 I_n r}\right)^{\frac{1}{n+1}}\tilde{\sigma}_{ij}(\theta) + Q\left(\frac{r\sigma_0}{J}\right)^q\hat{\sigma}_{ij}(\theta), \tag{3.244}$$

whereby the exponent $q\,(> 0)$ and the angular functions $\hat{\sigma}_{ij}(\theta)$ depend on the hardening exponent n. The factor Q is now a second geometry dependent loading parameter apart from J ! A series expansion up to the 3rd term [82, 83] provided no remarkable improvement and could be reduced to one additional parameter A_2 similar to Q. In Fig. 3.43 the angular stress distributions $\sigma_{ij}(\theta)$ are drawn for the various approaches and compared with FEM-computations [83]. In the angular range $|\theta| < \pi/2$ all solutions agree quite well, whereas the HRR-field deviates by a nearly constant value. These findings proven by detailed FEM-analyses, brought O'Dowd & Shih [84, 85] to suggest a simplification of the 2nd term. Thereby the weak r-dependency in (3.244) was set to a constant value $q \approx 0$ and the angular functions $\hat{\sigma}_{ij}(\theta)$ were approximated by the unity tensor δ_{ij}

$$\sigma_{ij}(r, \theta) = \left[\sigma_{ij}(r, \theta)\right]^{\mathrm{HRR}} + Q\sigma_0\delta_{ij} \quad \text{at } |\theta| < \pi/2. \tag{3.245}$$

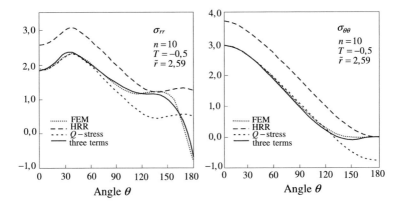

Fig. 3.43 Comparison of the higher order completion terms with the HRR-field, plane strain, $n = 10, \hbar = -0.5$ [83]

In order to determine the *Q- parameter* it was specified to calculate the difference between true (FEM)-stresses and HRR-solution ahead of the crack in a distance r_0

$$Q = \frac{\sigma_{\theta\theta}^{FEM} - \sigma_{\theta\theta}^{HRR}}{\sigma_0} \quad \text{at } \theta = 0, \quad r_0 = \frac{2J}{\sigma_0} \approx 4\delta_t. \quad (3.246)$$

This fixed distance $r_0 = 2J/\sigma_0$ was chosen purposely to circumvent the influence of the crack blunting. Thus also FEM-computations using infinitesimal deformations would be sufficient to determine Q, compare Fig. 3.42. According to this simplified definition the *Q-parameter* represents an averaged hydrostatic stress (related to the yield stress σ_0) in front of the crack, which quantifies the deviation of the stress state in a real crack configuration to those of the HRR-field in 1st approximation. Therefore, Q characterizes the geometry- and load-depending stress triaxiality in fracture test specimens or structural components under large plastic zones. Inserting of $r_0 = 2J/\sigma_0$ into the HRR-solution makes clear again that the dimensionless parameter Q directly indicates the change of $\hbar$ through (3.240)

$$\hbar(r_0, 0) = \frac{\sigma^H}{\sigma_0} = \frac{1}{3}\left(\frac{1}{2\alpha\varepsilon_0 I_n}\right)^{\frac{1}{n+1}} \tilde{\sigma}_{kk}(0) + Q. \quad (3.247)$$

Based on these investigations an assessment concept of ductile fracture was developed in the USA by Shih, O'Dowd, Dodds and Anderson [85–88] that relies on the two loading parameters J and Q. Forthcoming references on this topic can be found in [76, 89].

3.3.7 The J-Integral Concept

J as Fracture Parameter

The *J-integral* provides suitable prerequisites to formulate a fracture criterion in the framework of elastic-plastic fracture mechanics (EPFM). According to Sect. 3.3.6 the magnitude of J identifies the intensity of the crack tip fields in a hardening elastic-plastic material (with variable exponent) in the interior of the plastic zone. All stresses, strains and displacements are specified through (3.238) by the HRR-solution and are proportional to the J-value. Therefore, the parameter J controls the mechanical loading situation at the crack tip, where the micromechanical failure mechanisms proceed in the fracture process zone. In this sense, the quantity J plays a comparable role in EPFM as the stress intensity factor K_I does in LEFM. A necessary condition for this is again that the HRR-field dominates in a sufficiently large region around the crack tip with a radius r_J. Thereby all phenomena not captured by the plastic deformation theory as large strains, local unloading or physical cracking mechanisms, can be assigned to a small »process zone« of size $r_B \ll r_J$. Beyond the radius r_J higher order terms of plastic crack-fields gain in importance and at even greater distance the influence of geometric borders and boundary conditions effects the plastic solution. From this reason an upper bound is set $r_J \ll a, B, (w - a)$.

By far more serious restrictions on using J as a fracture criterion result from its theoretical fundamentals–the plastic deformation theory. The limitations of the deformation theory to capture real plastic flow processes are pointed out in the Appendix A.4.2. Hence the J-integral loses its validity in the strong sense, if the loading rises not monotonously on a proportional path. Any redistribution of the stress state or even more an unloading lead to a loss of path independence and uniqueness of J. At least local unloading processes occur at the crack tip during crack propagation or alternating loads, however. Therefore, the application of the J-concept has to be restricted to stationary cracks under monotonous loading. Under these conditions a deviation from the strong proportionality has a minor effect on the path independence of J, which has been verified by numerous FEM-analyses of crack configurations using plastic flow theory.

The *J-integral-concept* as fracture criterion in EPFM:
- For stationary cracks under monotonous loading the J is an established fracture mechanical parameter that describes quantitatively the loading situation at the crack tip in elastic-plastic materials.
- If the loading in the crack tip region attains a critical, material specific value J_{Ic}, then crack initiation occurs. The fracture criterion reads

$$J = J_{Ic}. \tag{3.248}$$

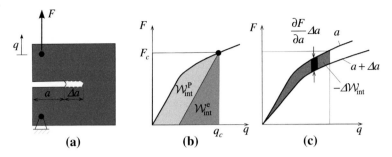

Fig. 3.44 Experimental determination of J_{Ic} by means of single and multiple specimen method

For (non-linear) elastic material behavior the J-integral is identical with the potential energy release rate, see Sect. 3.2.6. Theoretically, this holds as well for the plastic deformation theory but practically the part of plastic strain energy is dissipated and hence no longer available as a crack driving force. Therefore, this energetic interpretation and meaning of J are lost in EPFM.

How does one get the value of the loading parameter J for a given crack configuration during the course of loading for a specific elastic-plastic material law? Despite very simple geometries and estimations on the basis of ideal-plastic limit-load solutions, for this purpose the application of numerical methods, first of all FEM, is indispensable. This issue will be addressed in detail in Chap. 7. Partly, tabulated solutions are available in handbooks [90]. The essential advantage of computational methods is that the real definition of J as contour integral (3.100) can be employed, see Chap. 6.

Experimental Determination of Fracture Toughness J_{Ic}

In order to determine the critical material parameter J_{Ic} at ductile fracture, the same test specimens are used as for the measurement of K_{Ic} in LEFM, see Fig. 3.12. The testing procedure and result interpretation are subjects of standardized regulations [18–20]. The load-displacement curve F-q of the load point is recorded during monotonic loading of the specimen. Due to large plastification this curve is extremely non-linear, which is exemplified in Fig. 3.44 for a CT-specimen. At the point (F_c, q_c) of crack initiation, at the crack tip $J = J_{Ic}$ is reached. Since this point is not so easy to identify from the measurements, a certain amount of stable ductile crack growth is allowed for and recorded as *crack growth resistance curve* $J_R(\Delta a)$.

For interpretation of the experiment one needs a correlation between the load-displacement curve and the value of J. Therefore, we make use of the energetic meaning of J-integrals in the framework of plastic deformation theory. The area under the F-q-curve corresponds to the external work $\mathcal{W}_{ext}$ done on the specimen that is transformed into internal strain energy $\mathcal{W}_{int}$, which means potential energy Π_{int} in this model

$$\Pi_{int}(q, a) = \mathcal{W}_{int}(q, a) = \int_0^q F(\bar{q}, a)\mathrm{d}\bar{q}. \tag{3.249}$$

This energy is a function of crack length a. Under the assumption of fixed displacements (grips) ($q = \mathrm{const.}$) the imposed force F performs no work during a change of crack length by $\mathrm{d}a$ ($\mathrm{d}\Pi_{ext} = 0$). Thus the energy release rate is written as

$$J = -\frac{\mathrm{d}\Pi}{\mathrm{d}a} = -\frac{\partial\Pi_{int}}{\partial a}\bigg|_q = -\int_0^q \frac{\partial F(\bar{q}, a)}{\partial a}\bigg|_{\bar{q}} \mathrm{d}\bar{q}. \tag{3.250}$$

This integral describes the released internal energy for a crack extension and corresponds to the area between two load-displacement curves for the crack lengths a and $a + \Delta a$ as shown in Fig. 3.44c. It presents the non-linear pendant to the triangular areas depicted in Fig. 3.14. In a real elastic-plastic material however, at first unloading occurs during crack extension and second the work of deformation contains dissipative terms—which hence are *not* available as »potential crack driving energy « ($\mathcal{W}_{int} \neq \Pi_{int}$). Thus the relation (3.249) may only be interpreted as difference of work of deformation $\Delta \mathcal{W}_{int}$ between two identical specimens that have different initial crack lengths a and $a + \Delta a$ and are loaded monotonically.

This idea formed the basis of the original method suggested by [91] for determining J_{Ic}. Evaluating a test series of specimens with different crack lengths a_j, the differential areas of the load-displacement curves $F(q, a_j)$ are determined as functions of deflections q. From the function $J(q, a_j)$ one obtains the *fracture toughness* J_{Ic} by inserting the critical values at crack initiation. This so-called » multi-specimen method« did not gain acceptance because of its high experimental effort.

Instead of it, alternative methods have been elaborated [92], where the results measured at *only one single* specimen are sufficient to determine J_{Ic}. The underlying idea is to create a correlation between J and the work of deformation $\mathcal{W}_{int}$ itself (area under the F-q-curve in Fig. 3.44b).

With the help of numerous analytical and numerical calculations, formulas have been delivered to evaluate the elastic J_e and plastic J_p part of J-integral

$$J = J_e + J_p = \frac{K_I^2(F, a)}{E'} + \frac{\eta}{B(w - a)}\int_0^{q_p} F(\bar{q}_p, a)\,\mathrm{d}\bar{q}_p. \tag{3.251}$$

The two terms represent the elastic and plastic work of deformation $\mathcal{W}_{int}^e$ and $\mathcal{W}_{int}^p$, respectively, related to the cross sectional area $B(w - a)$ of the ligament, see Fig. 3.44b. They are determined from the reversible q_e and irreversible q_p displacements of the load application point. This way J_{Ic} can be gained in a simple manner from one single experiment by taking (3.251) at the initiation point (F_c, a). Correlations of the form (3.251) can be found for most types of specimens and components. Apart from the K-factor solution the correction functions $\eta(a/w, n)$ are needed which depend on geometry and hardening. Meanwhile, such functions η have been computed by extensive elastic-plastic FEM-calculations for the most important specimen types, supplying the basis for standardized J_{Ic}-procedures [18–20].

Similar to LEFM the J_{Ic}-determination requires compliance with certain size relations, to ensure that the J-controlled near-field (expressed in $\delta_{\text{tc}} \sim J_{\text{Ic}}/\sigma_{\text{F}}$) is small compared to all dimensions of the specimen so that the tests yield real, geometry-independent material parameters

$$B,\ (w - a) \geq 25 J_{\text{Ic}}/\sigma_{\text{F}}. \tag{3.252}$$

This means in practice a reduction of specimen size of 1–2 orders of magnitude compared to the conditions (3.202) at a K_{Ic}-test. Finally it has to be emphasized again that the justification to use J in EPFM is solely based on its relevance as an intensity parameter but not on its energetic meaning, although the experimental methods give such an impression!

Engineering Approach by EPRI

The *engineering approach* is a practice-orientated concept for the application of EPFM to structural components with cracks, which was developed at Electric Power Research Institute in the USA by [90]. This concept is based on the J-integral as fracture parameter. The critical material parameters are accepted from the J_{Ic}-test guidelines explained in the previous section. A substantial advantage consists in delivery of the loading parameter J in an easy form provided to all users together with an integrated assessment concept. The motivation was that numerical FEM-analyses at first can not be carried out by everybody and second require a considerable effort of manpower and costs.

Therefore, a catalog of solutions was compiled for a number of important crack configurations under fully-plastic conditions. These handbook-solutions were created by systematic FEM-calculations assuming for simplicity the deformation theory of plasticity and a power-law hardening according to (3.224). This has the advantage of parameterizing the (non-linear-elastic) solution with respect to the loading, i. e. only *one* fully-plastic solution is necessary instead of the whole load history. If P denotes a global load parameter (force, moment, surface loads, ...) at monotonic loading, so the respective stress and strain fields are given by the following scaling:

$$\sigma_{ij}(P) = \frac{P}{P_L}\sigma_{ij}^*(P_L), \quad \varepsilon_{ij}(P) = \left(\frac{P}{P_L}\right)^n \varepsilon_{ij}^*(P_L). \tag{3.253}$$

Here, σ_{ij}^* and ε_{ij}^* are the reference solutions if the plastic limit load P_L is attained. Having in mind (3.238a) the J-integral is scaled as

$$J(P) = \alpha\varepsilon_0\sigma_0 I_n r \left(\frac{\sigma_{ij}}{\sigma_0}\right)^{n+1} \tilde{\sigma}_{ij}^{n+1}(\theta; n) = \left(\frac{P}{P_L}\right)^{n+1} J^*(P_L). \tag{3.254}$$

In this way for most common crack configurations the plastic term of J-integral J_p and the global deformation variable q_p were calculated and cataloged [90], whereby the functions h_i comprise the results depending on geometry and hardening exponent n

$$J_p = \alpha \varepsilon_0 b \, h_1 \left(\frac{a}{w}, n\right) \left(\frac{P}{P_L}\right)^{n+1} \quad , \quad q_p = \alpha \varepsilon_0 a \, h_2 \left(\frac{a}{w}, n\right) \left(\frac{P}{P_L}\right)^{n} . \tag{3.255}$$

The elastic part J_e is calculated as in (3.251) via the K_I-factor of the considered crack configuration $K_I(P, a) \sim \sqrt{a}\, P\, g(a/w)$ using the tabulated geometry function g

$$J_e = \varepsilon_0 \sigma_0 a \left(\frac{P}{P_L}\right)^2 g^2 \left(\frac{a}{w}\right) . \tag{3.256}$$

Using Irwin's plastic crack length correction a_{eff} (3.200) the J-integral value is obtained by adding the elastic and plastic parts

$$J\left(\frac{P}{P_L}, \frac{a}{w}, n\right) = J_e(a_{\text{eff}}) + J_p\left(\frac{a}{w}, n\right) . \tag{3.257}$$

Figure 3.45 demonstrated once again graphically that the superposition of elastic and fully-plastic solutions according to the Engineering Approach yields J-values, which agree very well with sophisticated elastic-plastic FEM-calculations.

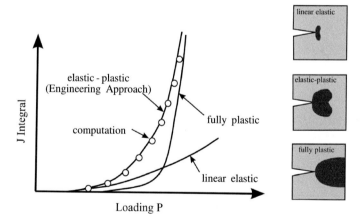

Fig. 3.45 J-estimation by the engineering approach

3.3.8 Ductile Crack Growth

Crack Growth Resistance Curves

Due to their high toughness, most ductile materials don't fail spontaneously at the critical J_{Ic}-value, but after crack initiation they build up a considerable resistance against further crack propagation. This means, a subsequent stable ductile crack growth can only be achieved by an enhancement of the loading. In an experiment typically a ductile *crack growth resistance curve* can be observed, see Fig. 3.46.

$$J = J_R(\Delta a), \tag{3.258}$$

In its initial phase the J_R-curve runs very steep and almost linear. Crack tip blunting induces a slight crack growth characterized by the so-called » blunting line«. If one supposes a circular-shaped rounding by plastic deformation (Fig. 3.35), the crack growth amounts to $\Delta a \sim \delta_t/2$ and from (3.239) a linear relation follows to the J-integral

$$J(\Delta a) \approx 2\sigma_F \Delta a \quad \textit{blunting line.} \tag{3.259}$$

With rising load the real physical crack initiation happens at J_i by fracturing in front of the blunted crack. This point is difficult to record during the measurements. Therefore, a minimum amount of crack propagation is needed to identify crack initiation (e.g. $\Delta a = 0.2$mm). Then, an engineering value of J_{Ic} is defined by the intersection point of the $J_R(\Delta a)$-curve with the parallel shifted blunting line, see Fig. 3.46. This pragmatic specification has a similarity with the R_{p02}-definition of the yield strength. Above $J > J_{Ic}$ a distinct crack growth resistance curve arises. According to the ASTM-standard [20] this curve will be simply fitted by a power law $J = C_1(\Delta a)^{C_2}$ and brought to intersection with the shifted blunting line. In order to determine the current crack length $a = a_0 + \Delta a$ in the specimen during the experiment, several measurement methods (partial unloading technique, electric potential-drop method) have been developed. Finally, J is evaluated using (3.251) from the work of deformation attained up to the current load level divided by the ligament area.

Since the fracture criterion $J = J_R(\Delta a)$ should be always fulfilled during stable crack propagation, the resistance curve is actually supposed to be a characteristic of the specific material. Unfortunately, this statement is only valid with restrictions:

Firstly we know that during crack growth parts of the plastic zone behind the crack tip are unloaded and that the region in front of the crack tip undergoes a non-proportional loading, which invalidates the meaning of J-integral as fracture parameter. To warrant nevertheless a crack growth controlled by J, the regions of elastic unloading and non-proportional plastic loading—that is the crack increment Δa—must be embedded in the zone of J-Dominanz $\Delta a < r < r_J$. Under SSY-conditions the plastic Zone is controlled by K_I and equivalently also by J, hence in this case the $J_R(\Delta a)$-curve represents a geometry independent materials

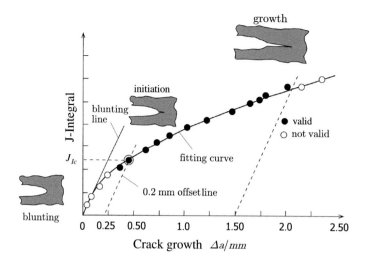

Fig. 3.46 Sketch of the ductile crack growth resistance curve

characteristic. Unfortunately, the SSY-assumption only seldom applies to ductile materials in common specimen sizes. In the case of LSY it was tried by [93] to estimate the conditions for J-controlled crack growth from the HRR-solution. To expose a point in the vicinity of the crack tip to a proportional stress increase, the increment dJ must have dominant effect over the unloading term due to crack extension da. This leads to the condition $dJ/J \gg da/r$, whereof a limiting relation is derived for fully plastified specimens with $r \approx w - a$

$$\frac{w - a}{J}\frac{dJ}{da} = \varpi \gg 1. \tag{3.260}$$

FEM-analyses by [94] have revealed that in specimens with dominating bending stresses a J-controlled crack growth Δa up to 6 % of ligament size $(w - a)$ is admissible ($\varpi = 10$). For this reason, the valid data-range for evaluating J_R-curves is narrowed down to a maximum crack growth of $\Delta a_{max} \leq 0.10(w - a)$ in the guidelines, as indicated in Fig. 3.46.

Secondly, it has been found that crack resistance curves are strongly influenced by the geometry (shape, thickness, crack size) of the specimens whereby they lose their meaning as materials characteristics. In the same way, this queries the transferability to cracks in components, too. Figure 3.47 represents $J_R(\Delta a)$-curves of the same material. The crack initiation values J_i are in all specimens equal, but afterwords the slopes of the curves dJ/da differ considerably. Therefore it has been suggested to take the physical initiation value J_i as obligatory material parameter. This would lead to a rather conservative safety assessment neglecting the reserves contained in the rising crack resistance. The reasons for the geometry dependence of crack resistance curves lie in the influence of *triaxiality* $\hbar$ of the stress state. The higher the triaxiality

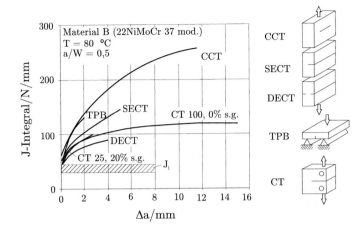

Fig. 3.47 Crack resistance curves and initiation values obtained by various specimen geometries for steel 22NiMoCr37 [95]

the flatter are the J_R-curves, i.e. ductile fracture is facilitated. At high triaxiality more potential elastic Energy $\mathcal{W}_{int}^e$ is available from the incoming external work into the structure (J-value), whereas at low triaxiality a substantial part is consumed by plastic work of deformation $\mathcal{W}_{int}^p$. This effect is enhanced by materials behavior since the mechanism of ductile dimple fracture is favored by high triaxiality, which reduces the fracture resistance. Indeed for every material a whole set of resistance curves $J_R(\Delta a, \hbar)$ would be needed as functions of $\hbar$. Therewith the ductile crack growth in another structure could be predicted if besides the loading J the existing value $\hbar$ of triaxiality would be known and the corresponding J_R-curve would be chosen. Owing to different strain constraints, the triaxiality varies along a crack front. Because of that this method of adjusted resistance curves has to be applied locally. In this way, very good predictions could be achieved in [96] by FEM-calculations simulating the propagation of a semi-elliptical surface crack in a tensile specimen of ductile steel.

In Sect. 3.3.6 the T- and Q-stresses have been introduced as additional terms to complete the elastic-plastic crack tip fields. They essentially influence the local stress triaxiality via (3.243) and (3.247). Hence they have similar consequences for the resistance curves as $\hbar$ has. Therefore, it is an urgent goal of numerical analyses to determine both parameters in an efficient way for a given crack configuration.

Crack Tip Fields for Stationary Crack Propagation

For a crack that propagates quasi-statically with a constant velocity $\dot{a}$ in an ideal-plastic material, the asymptotic stress and strain fields were found by Slepyan [97] (Tresca's yield condition, plane strain), Drugan, Rice and Sham [98] and Castaneda [99] (v.- Mises yield condition, plane stress). The solutions for mode I and II are obtained by means of the slip line theory and are subdivided into various sectors, similar to the stationary crack (see Figs. 3.37 and 3.38). The stresses are everywhere bounded, whereas the strains show a logarithmic singularity for $r \to 0$ in the central sector B (reference length $R \approx$ plastic zone size)

$$\varepsilon_{r\theta}(r, \theta) = \frac{2(1 - \nu^2)\sigma_F}{\sqrt{3}E} \ln \frac{R}{r} \tilde{\varepsilon}_{r\theta}(\theta). \qquad (3.261)$$

Under SSY-conditions the solution gives a relation between the rates of the crack opening displacement $\dot{\delta}_t$, the far field loading $\dot{J}$ and the crack velocity $\dot{a}$ with some constants c_1 and c_2

$$\dot{\delta}_t = c_1 \frac{\dot{J}}{\sigma_F} + c_2 \frac{\sigma_F}{E} \dot{a} \ln(\frac{R}{r}). \qquad (3.262)$$

For steady-state conditions at the moving crack the integration provides a crack opening displacement that is proportional to the distance r from the crack tip. Thus the crack faces run as straight lines with a *crack tip opening angle* $\gamma_t = $ CTOA as depicted in Fig. 3.35

$$\frac{\delta_t}{r} = \frac{c_1}{\sigma_F} \frac{dJ}{da} + \frac{c_2 \sigma_F}{E} \ln(e\frac{R}{r}) = \tan \gamma_t. \qquad (3.263)$$

Also in experiments with large amounts of ductile crack growth the formation of a constant angle CTOA$_c$ is observed, which is why the following crack propagation criterion has been suggested:

$$\arctan \frac{\delta_{tc}}{r_t} = \text{CTOA}_c = \text{const}. \qquad (3.264)$$

Energy Considerations

The energy balance during ductile crack propagation $a(t)$ will be studied for a two-dimensional crack problem. We assume that at the crack tip a fracture process zone A_B is formed which moves with the crack. The process zone may have a shape and structure typical for the material. $\dot{W}_{ext}$ denotes the mechanical power of the external loads $\bar{t}_i$ acting on the boundary S_t of the body (for simplicity let $\bar{b}_i = 0$). This power is transformed into internal energy W_{int} and supplies the energy $\mathcal{D}$ consumed in the process zone, see Fig. 3.48 left.

$$\dot{W}_{ext} = \int_{S_t} \bar{t}_i \dot{u}_i \, ds \qquad (3.265)$$

Contrary to LEFM (Sect. 3.2.5) the internal energy (3.71) is now composed from elastic and plastic work of deformation W_{int}^e and W_{int}^p

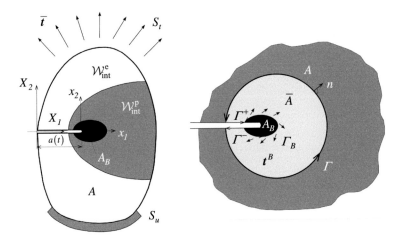

Fig. 3.48 Energy balance during ductile crack propagation

$$\mathcal{W}_{\text{int}} = \mathcal{W}_{\text{int}}^{\text{e}} + \mathcal{W}_{\text{int}}^{\text{p}} = \int_V (U^{\text{e}} + U^{\text{p}}) \, dV \,, \quad U^{\text{e}} + U^{\text{p}} = \int_0^{\varepsilon_{kl}^{\text{e}}} \sigma_{ij} \, d\varepsilon_{ij}^{\text{e}} + \int_0^{\varepsilon_{kl}^{\text{p}}} \sigma_{ij} \, d\varepsilon_{ij}^{\text{p}}.$$

$$(3.266)$$

Since the plastic work $\mathcal{W}_{\text{int}}^{\text{p}}$ is mainly dissipated as heat or microstructural defects, only the potential energy of the elastic part $\mathcal{W}_{\text{int}}^{\text{e}}$ is available for crack driving. Therefore, the global energy balance during crack propagation per time t or crack increment $da = \dot{a} \, dt$ is written as

$$\dot{\mathcal{W}}_{\text{ext}} = \dot{\mathcal{W}}_{\text{int}}^{\text{e}} + \dot{\mathcal{W}}_{\text{int}}^{\text{p}} + \dot{\mathcal{D}}. \qquad (3.267)$$

If all potential energy terms $\dot{\Pi}_{\text{ext}} = -\dot{\mathcal{W}}_{\text{ext}}$ and $\dot{\Pi}_{\text{int}} = \dot{\mathcal{W}}_{\text{int}}^{\text{e}}$ are brought on the left side, the extension of the Griffith's energy release rate (3.76) to ductile fracture is obtained

$$G = -\frac{d\Pi}{da} = \frac{d\mathcal{W}_{\text{ext}}}{da} - \frac{d\mathcal{W}_{\text{int}}^{\text{e}}}{da} = \frac{d\mathcal{W}_{\text{int}}^{\text{p}}}{da} + \frac{d\mathcal{D}}{da} \,. \qquad (3.268)$$

The released potential energy G is therefore not only spent for separation of material as a specific fracture energy $d\mathcal{D}/da = 2\gamma$ or crack resistance curve $d\mathcal{D}/da = J_{\text{R}}(\Delta a)$, but is predominantly consumed as plastic deformation. The greatest problem in EPFM is precisely that both contributions are difficult to distinguish in experiments. As a consequence the complete right-hand side of (3.268), the dissipation rate during crack propagation, is interpreted as fracture toughness. For the same reason, the crack resistance curves contain the geometry dependent plastic work in the specimen

$$J = -\frac{d\Pi}{da} = \frac{d\mathcal{W}_{\text{int}}^{\text{p}}}{da} + \frac{d\mathcal{D}}{da} = J_{\text{R}}(\Delta a) \,. \qquad (3.269)$$

In particular, Turner [100], Brocks [101] and Cotterell and Atkins [102] have pointed out this conflict.

For a better theoretical understanding it is useful to split the energetic considerations into a »continuum mechanical« part for the body A and a »material specific« part for the process zone A_B. Thereby the energy balance reads

$$\text{body } A: \quad \dot{\mathcal{W}}_{\text{ext}} + \overline{\dot{\mathcal{W}}}_B = \dot{\mathcal{W}}^e_{\text{int}} + \dot{\mathcal{W}}^p_{\text{int}} \qquad (3.270)$$

$$\text{process zone } A_B: \quad \dot{\mathcal{W}}_B = -\overline{\dot{\mathcal{W}}}_B = \dot{D}. \qquad (3.271)$$

$\dot{\mathcal{W}}_B$ describes the energy flux from the body across the boundary Γ_B into the process zone, which corresponds to the power performed by the tractions $t^B_i = \sigma_{ij} n_j$ with the displacement velocities

$$\dot{\mathcal{W}}_B = \int_{\Gamma_B} t^B_i \dot{u}_i ds. \qquad (3.272)$$

How this energy is converted for materials separation inside of A_B, remains a »black box«. The process zone extracts from the body the same amount of energy that can vice versa be considered as additional external work $\overline{\mathcal{W}}_B$ along the boundary Γ_B with opposite sign. Of course, the sum of both balances (3.270) and (3.271) results again in (3.267). The *energy flux* $d\mathcal{W}_B$ per crack increment da into the process zone will further on be denoted by $\mathcal{F}$. Using (3.270) gives

$$\mathcal{F} = \frac{d\mathcal{W}_B}{da} = \frac{1}{\dot{a}} \dot{\mathcal{W}}_B = \frac{1}{\dot{a}} \left[\int_{S_t} \bar{t}_i \dot{u}_i ds - \frac{d}{dt} \int_A U dA \right]. \qquad (3.273)$$

Now we introduce a coordinate system (x_1, x_2) that is moving together with the crack tip in the sense of an Eulerian description. On the other hand we denote by (X_1, X_2) a Lagrangeian coordinate system fixed with the material body, see Fig. 3.48.

$$x_1 = X_1 - a(t), \qquad x_2 = X_2 \qquad (3.274)$$

The material time derivative of an arbitrary field variable $f[x(X, t), t]$ associated with the particle X is calculated by the well–known rule

$$\dot{f} = \frac{df}{dt}\Big|_X = \frac{\partial f}{\partial t}\Big|_x + \frac{\partial x}{\partial t} \frac{\partial f}{\partial x} \quad \text{mit} \quad \frac{\partial x}{\partial t} = v = -\dot{a} e_1. \qquad (3.275)$$

In the special case of a steady state in the moving coordinate system the partial time derivative is zero, i.e. any temporal change is proportional to the gradient

$$\dot{f} = -\dot{a} \frac{\partial f}{\partial x_1}. \qquad (3.276)$$

The velocity $\dot{u}_i$ on Γ_B is to be transformed by (3.276) into the moving coordinate system $\boldsymbol{x}$. If, as we may assume, the displacement field $u_i(\boldsymbol{x}, t)$ is smooth and bounded at the crack tip and only its gradients (strains) become singular, so in (3.275) the convective 2. term dominates, which is called »local stationarity«.

$$\frac{\mathrm{d}u_i(\boldsymbol{x}, t)}{\mathrm{d}t} = \frac{\partial u_i}{\partial t} - \dot{a}\frac{\partial u_i}{\partial x_1} \approx -\dot{a}\, u_{i,1}. \tag{3.277}$$

When performing the material time differentiation of the 2. integral of (3.273), one must account for the temporal change of the domain $A(t)$ between the external boundary S and the moving process zone Γ_B. In an Eulerian coordinate system $\boldsymbol{x}$ this requires the use of Reynold's transport theorem (see [103, 104]) with $\boldsymbol{v} = -\dot{a}\boldsymbol{e}_1$

$$\frac{\mathrm{d}}{\mathrm{d}t}\int_A U\mathrm{d}A = \int_A \frac{\partial}{\partial t}U\mathrm{d}A - \dot{a}\int_{\Gamma_B} Un_1\mathrm{d}s. \tag{3.278}$$

In analogy to Sect. 3.2.6 the area integral on the right-hand side can be converted into a line integral along the closed contour $C = S + \Gamma^+ + \Gamma^- - \Gamma_B$. To do this, we rearrange $\partial U/\partial t = \sigma_{ij}\dot{\varepsilon}_{ij} = \sigma_{ij}\dot{u}_{i,j}$ into $\partial U/\partial t = (\sigma_{ij}\dot{u}_i)_{,j} - \sigma_{ij,j}\dot{u}_i$, whereby the 2. term vanishes due to the equilibrium equations $\sigma_{ij,j} = 0$. The 1st term is modified by the Gaussian integral theorem

$$\begin{aligned}
\frac{\mathrm{d}}{\mathrm{d}t}\int_A U\,\mathrm{d}A &= \int_A (\sigma_{ij}\dot{u}_i)_{,j}\,\mathrm{d}A - \dot{a}\int_{\Gamma_B} U\delta_{1j}n_j\,\mathrm{d}s \\
&= \int_S \sigma_{ij}\dot{u}_i n_j\,\mathrm{d}s - \int_{\Gamma_B}\sigma_{ij}\dot{u}_i n_j\,\mathrm{d}s - \dot{a}\int_{\Gamma_B} U\delta_{1j}n_j\,\mathrm{d}s.
\end{aligned} \tag{3.279}$$

The stress-free crack faces Γ^+ and Γ^- do not contribute because of $\sigma_{ij}n_j = t_i = 0$. Equation (3.277) is used for the velocity on Γ_B. After inserting these into relation (3.273) the integral over S cancels out, so that we get the final expression for the energy flux into the process zone

$$\mathcal{F} = \int_{\Gamma_B}[U\delta_{1j} - \sigma_{ij}u_{i,1}]n_j\,\mathrm{d}s = \int_{\Gamma_B}[Un_1 - t_i u_{i,1}]\,\mathrm{d}s. \tag{3.280}$$

This *energy flux integral* has a conspicuous similarity with the J-integral (3.100) but there are two essential differences: Firstly, now U includes the work of deformation for any elastic-plastic material law. Secondly, this integral is tied to the boundary of the process zone, upon which yet no specific assumptions were made.

In order to achieve (at least formally) the path-independency of the integral, an arbitrary closed contour Γ is introduced around the crack tip outside of the process zone A_B, see Fig. 3.48 right. The regular domain $\bar{A}$ between Γ and Γ_B is enclosed by a continuous line $\bar{C} = \Gamma + \Gamma^+ + \Gamma^- - \Gamma_B$. Applying the Gauss theorem in the opposite direction, whereby [ ] denotes the integrand of (3.280), we obtain, for vanishing crack face terms Γ^+, Γ^-,

$$\mathcal{F} = \int_{\Gamma_B} [\ ]n_j\, ds = \int_\Gamma [\ ]n_j\, ds - \int_{\Gamma - \Gamma_B} [\ ]n_j\, ds = \int_\Gamma [\ ]n_j\, ds - \int_{\bar{A}} \frac{\partial}{\partial x_j}[\ ]\, dA, \quad (3.281)$$

$$\mathcal{F} = \int_\Gamma [U\delta_{1j} - \sigma_{ij}u_{i,1}]n_j\, ds - \int_{\bar{A}} [U_{,1} - \sigma_{ij}u_{i,j1}]\, dA. \quad (3.282)$$

This result means: For an elastic-plastic material behavior the energy flux into the process zone can *not* be expressed by a path-independent contour integral alone (1. term), moreover an additional domain integral is required. Relationship (3.282) provides thus a modified opportunity to calculate the original energy flux integral along Γ_B, which now does not depend on the choice of the contour Γ and the enclosed domain $\bar{A}$. At least for the numerical computation this offers an advantage. Of course the expression merges for $\Gamma \to \Gamma_B$ into (3.280) since the domain integral disappears. By using (3.275) it is easy to show that for steady state conditions at the propagating crack tip the domain integral becomes zero.

In the end we focus on the modeling of the process zone itself:
As most simple variant the region A_B can be imagined as a point shrunk onto the crack tip and embedded by a continuum solution, which corresponds to the near field (3.261) treated in the previous section. Hereby the weaker logarithmic singularity of strains ($\varepsilon_{ij} \sim 1/\ln r$) prevails compared to those for a stationary crack ($\varepsilon_{ij} \sim 1/r$), which yields at bounded stresses $\sigma_{ij} \sim \sigma_F$ to a vanishing energy flux integral (3.280)! This leads to the contradictory result that in an ideal-plastic material no energy is transported into the crack tip [105]. However, the real reason is to be found in a too much simplified process zone. The correct conclusion is: *Without a better and more realistic modeling of the fracture process zone it is impossible to gain knowledge about ductile crack propagation in EPFM.* This experience was the starting point to develop cohesive zone models and damage mechanics approaches for the process zone. Further information can be found in [67, 102, 104, 106].

3.4 Fatigue Crack Propagation

Under alternating loads cracks may propagate stably although the stress intensity factor is far below the static fracture toughness. This phenomenon of subcritical crack propagation is called *fatigue crack growth*. Fatigue crack growth is the most common cause of failure in machines, vehicles, aircraft and other constructions that are exposed to time-varying operating loads. Depending on their temporal course one can distinguish between periodic (cyclic) or stochastic loads with constant or variable amplitudes.

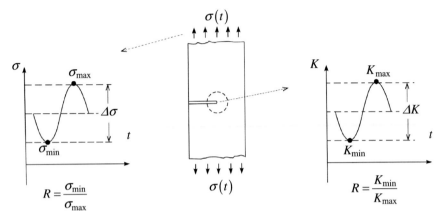

Fig. 3.49 Relation between alternating load and stress intensity factor

During fatigue of materials the following damage mechanism is observed. As a result of alternating micro-plastic deformations (dislocations, slip bands) first microcracks are formed on the surface or on microstructural inhomogeneities (inclusions, grain boundaries) in the interior. The propagation behavior of the result-ing *microstructurally short cracks* is very strongly influenced by the surrounding microstructure and is subject to special rules. Only above a crack length of about 10 grain diameters do we call it a *macroscopic initial crack*. Its behavior can now be described with methods of classical fracture mechanics. From a macroscopic view very small negligible plastic deformations occur and fatigue crack growth takes place in the K-controlled near-field, so that the LEFM is applicable.

3.4.1 Constant Amplitude Loading

At cyclic loading both the external loads, the stress distribution at the crack and the stress intensity factor are time-dependent. Because of their linear relationship they all run synchronously with the same time function

$$K_{\mathrm{I}}(t) = \sigma(t)\sqrt{\pi a}\, g(a, w), \quad \sigma_{ij}(t) = \frac{K_{\mathrm{I}}(t)}{\sqrt{2\pi r}} f_{ij}^{I}(\theta). \tag{3.283}$$

Cyclic loading $\sigma(t)$ of a structural component with constant frequency is characterized by its *range of stresses* $\Delta\sigma$ (double amplitude), the *mean stress* σ_m and the
stress ratio R:

$$\Delta\sigma = \sigma_{max} - \sigma_{min}, \quad \sigma_m = (\sigma_{max} + \sigma_{min})/2, \quad R = \sigma_{min}/\sigma_{max}. \quad (3.284)$$

Hence the range ΔK_I follows, also known as *cyclic stress intensity factor* (to
remain in general, the index for the crack opening mode is omitted)

$$\Delta K = K_{max} - K_{min} = \Delta\sigma\sqrt{\pi a}\, g\,(a, w). \quad (3.285)$$

This relationship is depicted in Fig. 3.49. ΔK is written with the help of stress
ration R as

$$\Delta K = (1 - R)K_{max}, \quad R = K_{min}/K_{max}. \quad (3.286)$$

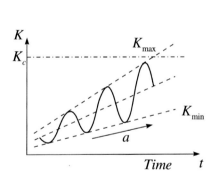

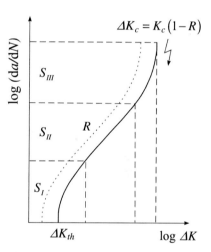

Fig. 3.50 Course of stress intensity factor
with increasing crack length a as function of
time t or load cycles N

Fig. 3.51 Crack growth velocity as function
of cyclic stress intensity factor

Even if the stress range $\Delta\sigma$ and the mean stress σ_m of external loading do not
change, nevertheless the stress intensity factor will commonly rise because of the
growing crack length $\sqrt{\pi a}\, g(a, w)$ in (3.285), which results in the course shown in
Fig. 3.50. If K_{max} reaches the critical value K_c, unstable forced fracture happens.

Fatigue crack growth is quantified by the crack extension da per load cycle. The
crack growth velocity or *crack growth rate* is defined as the ratio da/dN, where N is

the number of cycles. Plotting the experimentally determined crack velocity da/dN on a double logarithmic scale as a function of cyclic stress intensity factor ΔK, we obtain the *crack growth curve*. Its typical behavior is shown in Fig. 3.51. The lower limit of the curve is identified by the *threshold value* of the stress intensity factor ΔK_{th}. If the cyclic stress intensity factor ΔK is below this threshold value then the fatigue crack can not propagate. K_{th} quantifies so to say a » fracture mechanical endurance limit « and is about $K_{\text{c}}/10$ for metals. Whereas the initial stage S_{I} is sensitive to microstructural effects, in the subsequent range S_{II} the influence of loading dominates. In this range a linear relationship of the crack growth curve is observed that can be approximated by the law of Paris-Erdogan [107],

$$\frac{da}{dN} = C(\Delta K)^m \quad Paris\text{-law} \tag{3.287}$$

The exponent m and the coefficient C are characteristics of the material. C has a relatively strange dimension depending on the exponent m of the crack growth law. For metals the exponent is about $m \approx 2 - 7$, while for ceramic materials it attains much higher values ($m \approx 20 - 100$). The range S_{III} marks the transition to brittle forced fracture. The upper limit ΔK_{c} indicates that loading from here on, the crack propagation becomes unstable. The criterion $K_{\text{max}} = K_{\text{c}}$ and $K_{\text{max}} = \Delta K_{\text{c}}/(1 - R)$ respectively is valid. The behavior of the crack growth curve depends on the specific material and is influenced by manifold factors as microstructure, temperature, environmental medium or R-ratio. With increasing R-ratio, the crack growth velocity da/dN and the threshold value K_{th} commonly diminish.

Erdogan and Ratwani [108] have first proposed a formula for describing all three ranges of the curve

$$\frac{da}{dN} = \frac{C(\Delta K - \Delta K_{\text{th}})^m}{(1 - R)K - \Delta K}. \tag{3.288}$$

Meanwhile, there are many variants of *crack growth laws*. As an example, the very sophisticated version of the code ESACRACK [109] is given here that is based on an extended Paris equation. It covers the entire crack growth curve as function of stress ratio R at a constant load amplitude:

$$\frac{da}{dN} = C^* \left[\left(\frac{1 - \chi}{1 - R} \right) \Delta K \right]^{m^*} \frac{(1 - \Delta K_{\text{th}}/\Delta K)^p}{(1 - K_{\text{max}}/K_c)^q}$$

$$\chi = \frac{K_{\text{op}}}{K_{\text{max}}} = \begin{cases} \max(R, A_0 + A_1 R + A_2 R^2 + A_3 R^3) & \text{for } 0 \le R \\ A_0 + A_1 R & \text{for } -1 \le R < 0 \end{cases}$$

$$A_0 = (0.825 - 0.34\alpha_{\text{cf}} + 0.05\alpha_{\text{cf}}^2) \left[\cos \left(\frac{\pi \sigma_{\max}}{2\sigma_{\text{F}}} \right) \right]^{1/\alpha_{\text{cf}}}$$

$$A_1 = (0.415 - 0.071\alpha_{\text{cf}})\sigma_{\max}/\sigma_{\text{F}}$$

$$A_2 = 1 - A_0 - A_1 - A_3, \quad A_3 = 2A_0 + A_1 - 1. \tag{3.289}$$

Here, p and q are parameters to fit the transition from range II into ranges I and III, respectively. C^* and m^* are modified constants of the Paris Eq. (3.287) which follows from (3.289) by reducing $p = 0$ and $q = 0$ as well as $\chi = R$. χ means the R-dependent crack opening function according to Newman [110] that indicates the fraction between stress opening intensity K_{op} and the maximum stress intensity K_{max} during one cycle, see (3.294). The crack opening function χ accounts additionally for a dependence of the crack growth curve on the stress ratio R. The constants A_0 to A_3 are at first functions of the plastic zone size that is estimated by the Dugdale-model using the ratio of maximum nominal tensile stress σ_{max} and yield stress σ_F. Secondly, they are influenced by the strain constraint which is taken into account by the constraint factor α_{cf}. According to (3.198) the value of α_{cf} varies between 1 (plane stress) and 3 (plane strain) and is a function of the component's thickness. For steel the best agreement with experimental data is achieved using the values $\alpha_{cf} = 2.5$ and $\sigma_{max}/\sigma_F = 0.3$.

With the help of such crack growth laws it is possible to determine by integration how much a crack extends from its initial length a_0 at a given number of load cycles N. Applying to the simple Paris-Erdogan-law, one gets at constant amplitude

$$\Delta a = a(N) - a_0 = C \int_0^N [\Delta K(a)]^m \, dN = C \, \Delta\sigma^m \int_0^N \left[\sqrt{\pi a} \, g(a, w)\right]^m \, dN.$$
$$\tag{3.290}$$

By inverting the crack growth law to dN and integrating with respect to da, we can calculate vice versa the number of load cylces N required to attain the crack length a,

$$N(a) = \frac{1}{C\Delta\sigma^m} \int_{a_0}^a \frac{d\bar{a}}{\left[\sqrt{\pi\bar{a}} \, g(\bar{a}, w)\right]^m}$$
$$\approx \frac{2a_0}{C(m-2)\left[\sqrt{\pi a_0} \, g(a_0)\Delta\sigma\right]^m} \left[1 - \left(\frac{a_0}{a}\right)^{\frac{m}{2}-1}\right] \quad \text{for } m \neq 2. \tag{3.291}$$

If we neglect for simplicity that the geometry function depends on the crack length, the indicated approximative (non-conservative!) formula is received. Inserting for a the critical crack length a_c resulting from a fracture criterion, we obtain from (3.291) the number of load cycles N_B until fracture. In this way for a structural component with crack under in-service loading, the residual life time or the usable number of load cycles can be predicted until the crack is grown to its critical size a_c. Such considerations are essential ingredients of a damage tolerant dimensioning, where unavoidable crack-like defects in a component are admitted but are detected and monitored by means of non-destructive inspection methods in sufficient time periods. With the help of corresponding fracture mechanical assessments it is thus guaranteed that these cracks never become critical or indicated above which size they must be removed.

3.4.2 Stress State at the Crack Tip

In the following the characteristics of the stress situation at fatigue cracks will be discussed, which are basically different from the case of monotonic loading and are the reason for sequence effects if the amplitude of loading is changed.

Alternating Plastification

At cyclic loading, alternating plastic deformations occur in the small plastic zone at the tip of a fatigue crack. Each material point is undergoing a σ-ε-hysteresis as depicted schematically in Fig. A.10. The range of strain $\Delta\varepsilon$ corresponding to a given range of stress $\Delta\sigma$ depends essentially on the hardening behavior and the loading history. In fatigue crack growth a primary, secondary and cyclic plastic zone can be distinguished. The primary plastic zone is formed if the maximum load is attained the first time. The secondary or »reversed zone« is generated at the minimum load and the cyclic plastic zone is reached after a number of load cycles. Estimates of the plastic zone sizes have been presented in Sect. 3.3 for SSY.

Based on the Dugdale-model, Rice [111] developed an estimate of plastic zones under fatigue loading. Hereby, an elastic-ideal plastic material behavior is assumed, i. e. the yield stress has the same absolute magnitude of $\pm\sigma_F$ in the tensile and compression range. During the first time loading up to K_{max} the primary plastic zone is generated of size $r_{p\,max}$ according to formula (3.193) or (3.197) as shown in Fig. 3.52 (curve 1). The subsequent unloading by ΔK to K_{min} (Fig. 3.52 curve 2) is regarded as loading into reverse direction. Because of load reversal we have to take the doubled yield limit to attain compressive yielding (from $+\sigma_F$ to $-\sigma_F$) determining this way by (3.193) the size $r_{p\,min}$ of the secondary plastic zone.

$$r_{p\,max} = \frac{\pi}{8}\left(\frac{K_{max}}{\sigma_F}\right)^2, \quad r_{p\,min} = \frac{\pi}{8}\left(\frac{-\Delta K}{2\sigma_F}\right)^2 \implies \frac{r_{p\,min}}{r_{p\,max}} = \frac{1}{4}(1-R)^2.$$

(3.292)

By superimposing the static solutions for K_{max} and $-\Delta K$ the stress distribution at K_{min} is obtained as presented in Fig. 3.52 (curve 1+2). As can be seen, the secondary plastic zone is created by negative stresses resulting from the reverse deformation (compression) of the positive plastic elongations arisen during first tensile loading K_{max}. This does not necessarily require K_{min} to be in the negative compressive region ! By means of (3.292) we get the size relation of plastic zones at K_{min} and K_{max} taking into account the R-ratio. In this simple ideal-plastic model the stress state will not change at further load cycles so that the result of Fig. 3.52 corresponds to the cyclic plastic zone as well.

If isotropic hardening were assumed the elastic region would increase during the first load cycles until no further plastic alternating deformations appear (elastic shake down). According to the kinematic hardening model, a stabilized hysteresis will soon be adjusted, which is passed through in each cycle and leads to periodic plastic strains

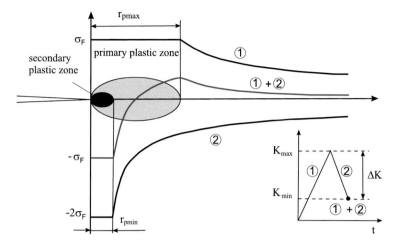

Fig. 3.52 Formation of plastic zones at the tip of a fatigue crack

$\pm \Delta \varepsilon_{ij}^{p}$ accumulating in the equivalent plastic strain ε_{v}^{p}. The true situation is by far more complex, since a multiaxial stress state and a combined hardening rule prevail in the plastic zone. Moreover, the plastic zone is permanently moving into new material regions.

Residual Stresses

The considerations of the previous section have shown that compressive stresses appear as a consequence of alternating plastic deformations directly ahead of the crack tip in fatigue cracking (Fig. 3.52). At a certain distance to the crack tip on the ligament the residual stresses alter from compression to tension.

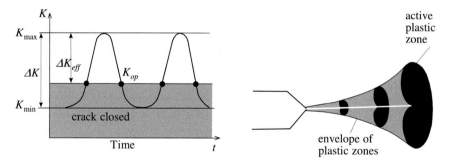

Fig. 3.53 Crack closure effect (*left*) caused by plastic deformations of crack faces (*right*)

By means of (3.292) it is obvious that $r_{p\,max}$ increases quadratically with the maximum K_{max}-factor. If for instance one single overload of size $K_{ol} = \beta K_{max}$ happens, then a plastic zone is formed with a β^2 times larger size $r_{p\,ol}$ and a corresponding region of compressive residual stresses. During the subsequent regular load amplitudes, this causes a retardation of fatigue crack growth until this compressive region has been crossed [112].

Compressive residual stresses are not only observed ahead of the crack tip, but as well as along the crack faces close to the tip, which was verified by experimental and numerical investigations.

Crack Closure Effect

The *crack closure effect* was discovered first by Elber [113]. Investigating fatigue cracks at cyclic tensile loading ($R = 0$) with constant amplitude, he observed that during unloading the crack was already closed before the minimum load was attained or that the crack keeps closed during re-loading up to a distinct level-the *crack opening intensity* K_{op}, see Fig. 3.53 (left).

The mechanism of crack closure has the consequence that not the entire loading ΔK contributes to crack propagation, but only an *effective cyclic stress intensity* ΔK_{eff} operating during the actual opening phase of the crack

$$\Delta K_{eff} = K_{max} - K_{op} \leq \Delta K \tag{3.293}$$

The quotient between effective and apparent range of external K-factor is described by the empirical crack opening function U or the ratio χ,

$$U = \frac{\Delta K_{eff}}{\Delta K}, \quad \chi = \frac{K_{op}}{K_{max}} = 1 - (1 - R)\,U. \tag{3.294}$$

The sources of crack closure may be quite different [114]. Plasticity-induced crack closure is regarded as the most important mechanism, which takes place in stage S_{II} of crack growth curve. The premature crack closure is caused by plastically elongated material along the crack faces, which is originated inside the plastic zones emerging permanently at the fatigue crack. This material is cut through during crack propagation, see Fig. 3.53 (right). The plastically stretched regions avoid both crack faces being closed in a compatible manner to each other so that a complete reverse displacement is impeded by mismatch at unloading.

Plasticity-induced crack closure is even intensified at variable-amplitude loading, leading to larger plastic zones and to a higher level of plastification, which affect once more crack closure and hence ΔK_{eff}.

Premature crack closure is also induced by small particles or fluids in the crack gap, by the roughness of the fracture surfaces or by phase transformations, respectively.

3.4.3 Variable-Amplitude Loading

The case of fatigue crack growth at constant load amplitude is only rarely found in real engineering practice. Machinery, components and vehicles in operational use are frequently exposed to time-varying loads. The load regimes with variable amplitude can be divided into three categories:

- single over- or under-loads
- stepwise change of amplitudes: block loads (e.g. engines, turbines)
- stochastic loads: load spectra of (e.g. vehicles, airplanes).

In addition to time-varying load histories the type of loading on a structure (tension, shear, bending, torsion) can also change during service. For a crack this results in an alteration of opening modes I, II and III, which is why the study of mixed-mode loads is of particular importance.

Contrary to the fatigue crack growth at constant amplitude, the crack propagation at variable amplitude does not only depend on the current load ΔK and R, but is determined by the temporal course of loading. This phenomenon is called *sequence effect*. Due to this fact variable-amplitude loads can lead both to accelerated and delayed crack growth (Fig. 3.54), i. e. they affect the lifetime either by reducing or

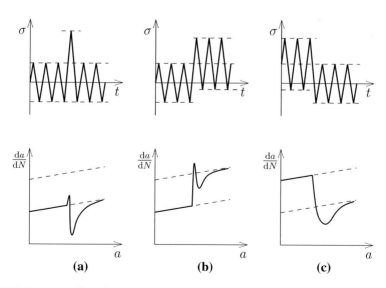

Fig. 3.54 Sequence effect after **a** a single overload, **b** a low-high block sequence and **c** a high-low block sequence

prolonging it. Whereas at constant loading parameters the lifetime can be obtained simply by integrating the crack growth curve, sequence effects have absolutely to be taken into account in case of variable loads, which comes along with severe difficulties.

One important prerequisite is a complete knowledge of the service loads at all. For its quantitative analysis, so-called in service load spectra are measured, whereby the load-time history of a component is recorded under real operating conditions in dependence on its profile of usage. Meanwhile, there exist approved standardized load spectra for many components (especially in automotive and aerospace industry). Instead of the field experiments, these data bases allow one to examine constructions or components either on test benches in the laboratory or by numerical stress analyses.

Such measurements of load spectra provide a necessary input-information for the fracture mechanical assessment of the crack behavior, too. In order to reduce the enormous amount of data, they are classified by common counting methods (e.g. rainflow-method) as used in service strength theory [115].

The simplest method to evaluate fatigue fracture under variable amplitudes consists in reducing all statistical data of a load spectrum to a single averaged value of the cyclic stress intensity factor. Barsom (see [116]) made the proposal to create an effective integral value ΔK_{rms} calculated as root mean square of the ranges ΔK_n of all N recorded load cycles

$$\Delta K_{\text{rms}} = \sqrt{\frac{1}{N} \sum_{n=1}^{N} (\Delta K_n)^2}. \tag{3.295}$$

The effective rms-value is put in a crack growth law for constant amplitudes, which results in an estimated averaged crack growth rate for this load spectrum. This so-called *global analysis* hence provides only integral values without temporal resolution and neglects any sequence effects on crack propagation.

An improved approach takes into account the load-time history of cycles. Provided the crack growth curves of a material are available for all amplitudes ΔK and stress ratios R, then the fatigue crack growth can be calculated and summed up *cycle by cycle*:

$$\frac{da}{dN} = f(\Delta K, R) \implies a = a_0 + \sum_{n=1}^{N} \Delta a_n, \quad \Delta a_n = f(\Delta K_n, R_n) \cdot 1. \tag{3.296}$$

For a block program, which consists of a sequence of load intervals $i = 1, 2, \ldots, I$ of ΔN_i cycles at the load levels ΔK_i and R_i, the crack growth can be accumulated from the contributions Δa_i of each interval as follows:

$$a = a_0 + \sum_{i=1}^{I} \Delta a_i, \quad \Delta a_i \approx f(\Delta K_i \, \bar{a}_i, R_i) \Delta N_i. \tag{3.297}$$

The approximation is that in each interval a mean constant crack length $\bar{a}_i$ is estimated. This approach is similar to the damage accumulation hypotheses used in service strength rules. After all it accounts for the temporal sequence of loads, however not for their influence among each other, which is insofar a simplification.

Different models have been proposed to capture more precisely the sequence effect during fatigue crack growth at variable amplitudes. Hereby, mainly the delaying effect of an overload or a step-load sequence is simulated, which are attributed either to residual stresses in the plastic zone (Wheeler [112], Willenborg, Gallagher [117]), plasticity induced crack closure (Onera [118], Corpus [119]), crack blunting or several factors.

The most advanced concept seems to be the *strip yield model* by Führing and Seeger [120], DeKoning [109] and Newman [110, 117]. It is an extension of the Dugdale model (Sect. 3.3.3). Assuming a set of bar elements bridging the crack faces, on the one hand the plastic deformations of the opened crack are taken into account and on the other hand the contact stresses during crack closure. This model allows us to calculate the crack opening stresses σ_{op}, at which the crack face contact just disappears and the crack is entirely opening. By means of (3.293) an effective stress intensity factor is obtained,

$$\Delta K_{eff} = (\sigma_{max} - \sigma_{op})\sqrt{\pi a}\, g(a, w),\qquad(3.298)$$

that enters the crack growth relation proposed by Newman [110] as follows:

$$\frac{da}{dN} = C_1(\Delta K_{eff})^{C_2}\left[1 - \left(\frac{\Delta K_0}{\Delta K_{eff}}\right)^2\right]\Bigg/\left[1 - \left(\frac{K_{max}}{C_5}\right)^2\right],\qquad(3.299)$$

whereby $\Delta K_0 = \Delta K_{th}\left(1 - \frac{\sigma_{op}}{\sigma_{max}}\right)/(1 - R)$ is set and the coefficients C_1, C_2, C_5 have to be fitted from experimental data. The crack opening stress σ_{op} is simulated by the strip yield model throughout the entire load history, in this way taking into account all sequence effects.

The above described models for predicting fatigue crack growth at constant and variable amplitudes have been implemented in various simulation tools as e.g. NASGRO [117] and ESACRACK [109] that enable assessment of the life time of components (especially in aerospace and astronautics).

Further literature on fatigue crack growth can be found in Schijve [121, 122] and the conference proceedings [123, 124].

3.4.4 Fracture Criteria at Mixed-Mode Loading

So far mainly crack problems for the opening mode I at symmetrical loading have been treated. The corresponding fracture criteria have been derived on the assumption that the crack continues to extend along its original line (2D) or plane (3D) in a straightforward manner on the ligament. If mode I is superimposed by mode II and/or III, the symmetry is violated and the situation is called *mixed-mode loading*. Mixed-mode loading occurs always when the sectional stresses t_i^c in the cutting plane of the body where the crack is located exhibit both normal and shear components as shown in Fig. 3.55 (left). In engineering practice, there are plenty of examples and reasons leading to mixed-mode loading of cracks:

- structural loads consisting of tension, shear, torsion or complex load cases
- oblique, curved, branched or kinked cracks
- time-varying, dynamic or thermal in service loads
- enforced crack propagation in a direction oblique to the principal stress because of preferred geometrical, material or technological orientations (interfaces, joints, anisotropy).

Furthermore, we will restrict ourselves to linear-elastic fracture mechanics at static and cyclic loading. In this case the stress state at the crack tip is clearly determined by the three intensity factors so that a generalized fracture criterion for mixed-mode loading would be set up in the form

$$B(K_I,\ K_{II},\ K_{III}) = B_c. \tag{3.300}$$

Loading of the crack is defined by the parameter B on the left side, while on the right the critical material parameter B_c characterizes crack initiation. In addition, now a decision has to be made about the direction of crack propagation which deviates from the original direction by an angle θ_c as shown in Fig. 3.55 (left). The propagation usually runs in that direction, where the parameter B reaches an extreme value

$$\max\{B(\theta)\} = B(\theta_c) \quad \text{or} \quad \min\{B(\theta)\} = B(\theta_c). \tag{3.301}$$

In the case of brittle instantaneous fracture it is sufficient for a strength evaluation to know the critical size and direction of crack loading. For subcritical and stable crack growth moreover a kinematic law is needed that quantifies the length of crack propagation under this stress situation

$$\frac{da}{dt} = f_1(B) \quad \text{or} \quad \frac{da}{dN} = f_2(\Delta B). \tag{3.302}$$

Especially in fatigue fracture, the crack is very sensitive to any variation in mixed-mode stress state and changes its direction accordingly, leading often to fancy curved crack paths and subtle fracture surfaces. To assess the residual life time it is especially important in this context that the geometric path of the crack can be predicted.

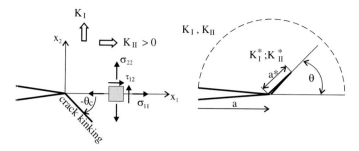

Fig. 3.55 Mixed-mode loading of a crack by mode I and II for plane problems

In the following the most approved and established fracture criteria for mixed-mode crack problems will be introduced from a variety of proposals (see e.g. [125]).

Criterion of Maximum Circumferential Stress

The *criterion of maximum circumferential stress* was suggested by Erdogan and Sih [126] and is based on the following assumptions:

- The crack extends radially from its tip with an angle θ_c in that direction, which is perpendicular to the maximum circumferential stress $\sigma_{\theta\theta\,\text{max}}$.
- Crack propagation initiates, if $\sigma_{\theta\theta\,\text{max}}$ (in a certain distance r_c) reaches a critical material constant σ_c that corresponds exactly to those under mode I at the ligament if $K_I = K_{Ic}$ is fulfilled.

From the near-tip solutions (3.17) and (3.24) we find

$$\sigma_{\theta\theta} = \frac{1}{4\sqrt{2\pi r_c}} \left[K_I \left(3\cos\frac{\theta}{2} + \cos\frac{3\theta}{2} \right) - K_{II} \left(3\sin\frac{\theta}{2} + 3\sin\frac{3\theta}{2} \right) \right] \quad (3.303)$$

and by applying the extremal condition

$$\left. \frac{\partial \sigma_{\theta\theta}}{\partial \theta} \right|_{\theta_c} = 0, \quad \frac{\partial^2 \sigma_{\theta\theta}}{\partial \theta^2} < 0 \quad \Rightarrow \quad K_I \sin\theta_c + K_{II}(3\cos\theta_c - 1) = 0 \quad (3.304)$$

the crack deflection angle θ_c is found to be

$$\theta_c = 2\arctan\left[\frac{1}{4}\frac{K_I}{K_{II}} - \frac{1}{4}\sqrt{\left(\frac{K_I}{K_{II}}\right)^2 + 8} \right]. \quad (3.305)$$

$\sigma_{\theta\theta}$ is a principal normal stress at $\theta = \theta_c$ and consequently the shear stress $\tau_{r\theta}$ vanishes. The second condition requires

$$\sigma_{\theta\theta}(\theta_c) = \sigma_c = \frac{K_{Ic}}{\sqrt{2\pi r_c}}, \qquad (3.306)$$

whereof by (3.303) the failure criterion is obtained. It can also be expressed by means of an *equivalent stress intensity factor* K_v on the radial ray θ_c,

$$K_v = \lim_{r \to 0} \left[\sigma_{\theta\theta}(\theta_c) \sqrt{2\pi r} \right] = \cos^2 \frac{\theta_c}{2} \left(K_I \cos \frac{\theta_c}{2} - 3 K_{II} \sin \frac{\theta_c}{2} \right) = K_{Ic}. \quad (3.307)$$

Of course, for opening mode-I collinear crack extension $\theta_c = 0°$ is recovered. For pure mode-II loading the deflection angle is $\theta_c = -70.5°$ and the critical loading amounts to $K_{II} = \sqrt{3}/2 K_{Ic}$.

Criterion of Maximum Energy Release Rate

For physical reasons, energetic considerations are favored to derive a suitable criterion for mixed-mode fracture. The Eq. (3.93) given in Sect. 3.2.5 holds however only for self-similar crack propagation along its original direction and is thus not applicable. Therefore, models and solutions have been developed by [127–130] whereby at the origin of the main crack a small kink crack of length $a^* \ll a$ and direction θ is assumed as illustrated in Fig. 3.55 (right). Hussain, Pu and Underwood [127] formulated the *criterion of maximum energy release rate* as follows:

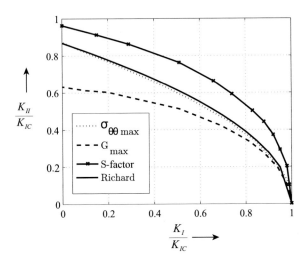

Fig. 3.56 Failure limit curves for mixed-mode loading by modes I and II

- The kink crack is formed under that angle θ_c, where the energy release rate $G^* = (K_I^{*2} + K_{II}^{*2})/E'$ attains a maximum.

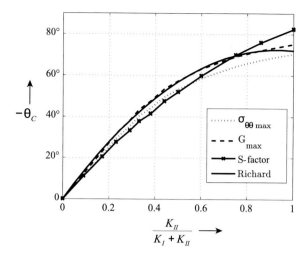

Fig. 3.57 Crack deflection angle for mixed-mode loading by modes I and II

$$\frac{\partial G^*(\theta)}{\partial \theta}\bigg|_{\theta_c} = 0, \quad \frac{\partial^2 G^*}{\partial \theta^2}\bigg|_{\theta_c} < 0 \qquad (3.308)$$

- Crack propagation initiates, if G^* reaches the critical material value G_c.

The loading situation at the small auxiliary crack a^* is controlled by the near-field of the main crack with its intensity factors K_I and K_{II}. Thereby the factors K_I^* and K_{II}^* can be calculated and the energy release rate is obtained in the limit $a^* \to 0$,

$$G^*(\theta) = \frac{4}{E'}\left(\frac{\pi - \theta}{\pi + \theta}\right)^{\theta/\pi}\frac{1}{(3 + \cos^2\theta)^2}\bigg[(1 + 3\cos^2\theta)K_I^2$$

$$+ (4\sin 2\theta)K_I K_{II} + (9 - 5\cos^2\theta)K_{II}^2\bigg] = G_c = K_{Ic}^2/E'. \qquad (3.309)$$

Here, we omit the derivation of $\partial G^*/\partial\theta$. Instead of it, the results for this failure limit curve and the corresponding deflection angle θ_c are presented in Figs. 3.56 and 3.57.

Criterion of Strain Energy Density

Sih [131] proposed the *criterion of strain energy density*, which is based on the angular dependence of the singular energy density U given by the two-dimensional near-tip field

$$U = \frac{1}{2}\sigma_{ij}\varepsilon_{ij} = \frac{S(\theta)}{r} = \frac{1}{r}\left(a_{11}K_I^2 + 2a_{12}K_I K_{II} + a_{22}K_{II}^2\right). \qquad (3.310)$$

The *strain energy density function* $S(\theta)$ is expressed by the K-factors

$$a_{11} = [(1 + \cos\theta)(\kappa - \cos\theta)]/16\pi\mu$$
$$a_{12} = \sin\theta[2\cos\theta - \kappa + 1]/16\pi\mu \qquad (3.311)$$
$$a_{22} = [(\kappa + 1)(1 - \cos\theta) + (1 + \cos\theta)(3\cos\theta - 1)]/16\pi\mu.$$

The following assumptions underlie the criterion:

- The crack extends radially in that direction θ_c, where the function $S(\theta)$ exhibits a minimum

$$\left.\frac{dS(\theta)}{d\theta}\right|_{\theta_c} = 0, \quad \left.\frac{d^2 S(\theta)}{d\theta^2}\right|_{\theta_c} > 0. \qquad (3.312)$$

- Crack propagation initiates, if $S(\theta_c)$ reaches a critical material value S_c calibrated at the mode-I case in plane strain

$$S(\theta_c) = S_c \,\hat{=}\, a_{11}(\theta_c = 0)K_{Ic}^2 = \frac{1 - 2\nu}{4\pi\mu}K_{Ic}^2. \qquad (3.313)$$

The deflection angle θ_c derived by (3.312) and the corresponding failure limit curve are included in Figs. 3.56 and 3.57 for plane strain with $\nu = 0.3$.

The S-criterion is physically motivated by the argument that in the direction of θ_c the part of volume-changing (dilatational) energy U_V predominates the shape-changing (deviatoric) part $U_G = U - U_V$, whereby brittle failure is favored. This idea is elaborated further in the criterion of Radaj and Heib [132] postulating a maximum of the function $S^*(\theta) = U_V(\theta)/U_G(\theta)$.

Criterion of J-Integral Vector

Another hypothesis can be formulated by means of the J-integral vector $\boldsymbol{J}$ that was explained in Sect. 6.1. Both planar components of $\boldsymbol{J}$ are shown in Fig. 3.58. Their values represent the energy release rate if the crack tip region enclosed by the integration path is virtually displaced by δl_1 or δl_2 in the corresponding coordinate direction. The computation is done as a path-independent integral according to the expressions (6.11) or (6.16). The x_1-component J_1 coincides geometrically with the crack extension in its original direction and is therefore equivalent to the energy release rate of (3.78), cf. Sect. 3.2.6. The situation is unlike with the J_2-component that corresponds geometrically with the shift of the entire considered crack tip region perpendicular to its original position. However, this kind of displacement δl_2 does not reflect the real physical crack propagation, where kinking of the crack takes place as shown in Fig. 3.55. In the frame of LEBM there exists, according to (6.12), the following relationship between both components of the J_k-integral vector (6.11) and the stress intensity factors

$$J_1 = G = \frac{1}{E'}\left(K_{\mathrm{I}}^2 + K_{\mathrm{II}}^2\right), \quad J_2 = -2K_{\mathrm{I}}K_{\mathrm{II}}/E'. \tag{3.314}$$

Based on the J-integral the criterion is now postulated:

- The crack extends radially in the direction of the vector $\boldsymbol{J}$, because here the configurational force is maximal (Fig. 3.58). The crack deflection angle is thus

$$\theta_c = \arctan\left(J_2/J_1\right). \tag{3.315}$$

- Crack initiation occurs if the absolute magnitude of $\boldsymbol{J}$ reaches a critical material value J_{Ic} that can be converted from the fracture toughness K_{Ic}.

$$|\boldsymbol{J}| = \sqrt{J_1^2 + J_2^2} \;=\; J_{\mathrm{Ic}} \mathrel{\hat=} K_{\mathrm{Ic}}^2 \frac{1-\nu^2}{E} \tag{3.316}$$

With the help of (3.314) the fracture criterion and the crack deflection angle can be likewise expressed by the stress intensity factors or their ratio $\varrho = K_{\mathrm{II}}/K_{\mathrm{I}}$:

$$K_{\mathrm{I}} \sqrt[4]{\left(1+\varrho^2\right)^2 + 4\varrho^2} - K_{\mathrm{Ic}} = 0, \quad \theta_c = \arctan\left(\frac{-2\varrho}{1+\varrho^2}\right). \tag{3.317}$$

The crack deflection angles deduced from the J-integral criterion are depicted in Fig. 3.59 as a function of mixed-mode loading ratio. For larger shear stresses $\varrho = K_{\mathrm{II}}/K_{\mathrm{I}} > 1$ the deflection angle decreases again and tends to $\theta_c = 0$. Due to (3.317), also the critical value at failure in pure shear loading is predicted to $K_{\mathrm{II}} = K_{\mathrm{Ic}}$ and not to $K_{\mathrm{II}} < K_{\mathrm{Ic}}$. Both statements are in contradiction to experimental observations and all other criteria (cf. Fig. 3.56). Therefore, a practical application of the J-integral vector seems to be useful only for $K_{\mathrm{II}} \le K_{\mathrm{I}}$.

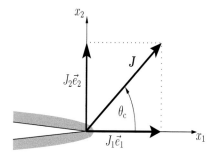

Fig. 3.58 J-integral vector at the crack tip

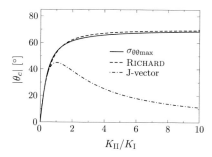

Fig. 3.59 Crack deflection angle for the J-integral

A comprehensive overview about other mixed-mode fracture criteria is given by Richard [125, 133] and Pook [134]. The investigation of three-dimensional crack

configurations at superposition of all three crack opening modes I, II and III is still
a subject of current research, see e.g. [135, 136].

3.4.5 Fatigue Crack Growth at Mixed-Mode Loading

As can be seen from Fig. 3.56, the various mixed-mode fracture criteria differ
considerably in their conclusions, especially if a high shear component K_{II} exists.
Experimental verifications require expensive and complicated tests. Due to manifold
influences of materials, side effects and scatterings, a universal criterion could not yet
be confirmed. Particularly difficult are measurements of the fracture toughness K_{IIc}
under pure shear loading. Note also that all of the above criteria are only applicable
if a minimum crack opening ($K_I > 0$) is present. Otherwise, both crack faces would
come into contact, which has to be taken into account by the analysis. A $K_I < 0$ is
practically impossible! A relative tangential movement of the crack faces at mode II
is impaired by friction.

Fortunately, for a low mode II component ($K_{II} \ll K_I$) all hypotheses provide in
a first approximation the same result for the crack deflection angle (cf. Fig. 3.57)

$$\theta_c \approx -2K_{II}/K_I. \tag{3.318}$$

A pragmatic approach was proposed amongst others by Richard [133] who introduced an *equivalent stress intensity factor* K_v in analogy to superimposed stress
states in classical theory of strength. It is calculated from K_I and K_{II} (and possibly K_{III}) and represents a measure of crack loading that corresponds to an equal
mode I case

$$K_v(K_I, K_{II}) = \frac{K_I}{2} + \frac{1}{2}\sqrt{K_I^2 + 4(\alpha_1 K_{II})^2} = K_{Ic}, \quad \alpha_1 = 1,155 \tag{3.319}$$

$$\theta_c = \mp \left[155, 5° \frac{|K_{II}|}{|K_I| + |K_{II}|} - 83, 4° \left(\frac{|K_{II}|}{|K_I| + |K_{II}|} \right)^2 \right], \tag{3.320}$$

whereby the signs hold for $K_{II} \gtrless 0$. These formulas are included in Figs. 3.56 and
3.57. The predictions are quite close to those of the maximum circumferential stress
criterion and were substantiated by experiments. Other empirical approaches describe
the failure limit curve by means of a generalized elliptic shape, the parameters of
which K_{Ic}, K_{IIc}, α and β, respectively have to be fitted for every material by mixed-
mode tests.

$$\left(\frac{K_I}{K_{Ic}} \right)^\alpha + \left(\frac{K_{II}}{K_{IIc}} \right)^\beta = 1 \tag{3.321}$$

Although these fracture criteria were derived for monotonic static loads until
catastrophic failure, they may with sufficient justification as well be applied to cyclic

loads within the scope of LEBM. In order to transfer this concept to fatigue cracks, instead of absolute values now the ranges ΔK_{I} and ΔK_{II} of the stress intensity factors are employed. Then the mixed-mode hypotheses are interpreted in the sense that the growth rate of subcritical crack propagation is quantified in dependence on the mode-ratio K_{I} over K_{II}. The best suited quantity for this purpose is the range of the equivalent stress intensity factor ΔK_{v} given e.g. in (3.307) or (3.319)

$$\Delta K_{\mathrm{v}}(\Delta K_{\mathrm{I}}, \Delta K_{\mathrm{II}}) = \frac{\Delta K_{\mathrm{I}}}{2} + \frac{1}{2}\sqrt{\Delta K_{\mathrm{I}}^2 + 4(\alpha_1 \Delta K_{\mathrm{II}}^2)}. \tag{3.322}$$

Hence, ΔK_{v} can be used in all crack growth laws, established in Sects. 3.4.1 and 3.4.3 for constant and variable amplitudes, instead of ΔK

$$\frac{\mathrm{d}a}{\mathrm{d}N} = f(\Delta K_{\mathrm{v}}, R) \quad \text{in the area } \Delta K_{\mathrm{th}} \leq \Delta K_{\mathrm{v}} \leq (1 - R)K_{\mathrm{c}}. \tag{3.323}$$

However, it must be provided that $K_{\mathrm{I}}(t)$ and $K_{\mathrm{II}}(t)$ oscillate synchronously and their ratio does not change.

3.4.6 Prediction of Crack Path and its Stability

To make a life time assessment for fatigue fracture in practice the essential question arises: Which geometrical path will the crack run under the given complex loading of the component? Let's assume the stress intensity factors K_{I} and K_{II} would be known for the current position of a fatigue crack in its local crack-tip coordinate system. Then, all mixed-mode criteria predict approximately the same alteration $\Delta\theta = -2K_{\mathrm{II}}/K_{\mathrm{I}}$ of the crack propagation direction. In addition two cases have to be distinguished:

- Alignment on K_{I} and $K_{\mathrm{II}} = 0$
  Fatigue cracks grow in an inhomogeneous stress field always in such a manner that they align themselves perpendicular to the local maximum principal stress, i.e. to get K_{I} dominant and let $K_{\mathrm{II}} \to 0$ fall to zero. As a result at the current crack tip a symmetric (mode I) stress state is adjusted. The direction of the crack is *continuously* turned and all criteria imply that $K_{\mathrm{II}} = 0$ must thereby vanish. Otherwise the change of direction $\Delta\theta$ would have been calculated falsely! This means in the frame of a 1st-order theory (only singular terms of the near field), the directional change is governed by the condition $K_{\mathrm{II}} = 0$. Prerequisite is however that the crack propagation is only controlled by the stress field and that the material is isotropic. This theory is confirmed by practical experience and countless failure cases.
- Real mixed-mode situation $K_{\mathrm{II}} \neq 0$
  Under those conditions mentioned in the introduction of Sect. 3.4.4 it happens that at the beginning of crack growth indeed real mode-II components exist or that they reappear spontaneously later again. Then in fact an *abrupt* kinking of the crack is observed according to (3.318).

In the numerical simulation of crack propagation by FEM one is forced to make finite crack increments so that a polygonal path with *discontinuous* changes in direction is obtained. Hereby, usually an explicit algorithm is applied for integrating the crack propagating path: The change of direction $\Delta\theta$ is derived from the K-factors at a_0 and not from the condition $K_{II}(a_0 + \Delta a) = 0$ at the end of the interval. This would require an implicit scheme. Thus a value $K_{II}(a_0 + \Delta a) \neq 0$ arises that is basically an approximation error which is believed to be corrected in the next increment. This presupposes a sufficiently fine increment Δa of a crack propagation. Extended criteria for the direction of curved crack propagation are worth being mentioned (see [137, 138]), whereby the T_{11}-stresses of 2nd order are accounted for in the near-field solution and the criterion of maximum circumferential stress is used.

Finally, the question of *directional stability* during crack propagation is to be discussed. This concerns the issue of how a crack reacts to small perturbations of its path due to material inhomogeneities, load fluctuation and more. Based on the solution for a kinked crack (Fig. 3.55) located inside the K-field of the main crack, Cotterell and Rice [139] have derived the following theory. The starting situation is a mixed-mode loading $K_I(a)$, $K_{II}(a)$ as depicted in Fig. 3.60, which causes a deflection angle θ_0 according to (3.318). The further propagation of the main crack $y(x)$ could approximately be calculated under the assumption that $K_{II}^*(a^*) = 0$ is kept, which corresponds to the maximum of $G^*(a^*)$. The T_{11}-stress at the main crack has significant influence on the course of the crack path. Hence, two branches of the solution were recognized, see Fig. 3.60:

$$y(x) \rightarrow \frac{\theta_0 K_I^2}{4T_{11}^2} \exp\left(\frac{8T_{11}^2}{K_I^2}x\right) \qquad \text{for } T_{11} > 0 \qquad (3.324)$$

$$y(x) \rightarrow \frac{\theta_0 K_I}{|T_{11}|} \sqrt{\frac{x}{2\pi}} \qquad \text{for } T_{11} < 0. \qquad (3.325)$$

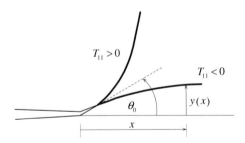

Fig. 3.60 Directional stability during crack propagation

The result implies: In case of positive stresses T_{11} parallel to the main crack the kink crack turns off upwards of the initial orientation θ_0, i. e. $y' \rightarrow \infty$–the direction becomes *unstable*. In case of compressive stresses $T_{11} < 0$ however, the crack is

forced to stay on its initial direction $\theta = 0$ because of $y' \to 0$—*stable* behavior.
(This can also be made clear by means of the principle stresses.) The predictions of
this model have been proved by experiments.

These investigations emphasize the importance of the T_{11}-stress or its dimensionless form, the *biaxial parameter*, in fracture mechanics

$$\beta_{\mathrm{T}} = \frac{T_{11}\sqrt{\pi a}}{K_{\mathrm{I}}} . \tag{3.326}$$

3.5 Dynamic Fracture Processes

3.5.1 Introduction

In fracture mechanics we speak about dynamic phenomena whenever the deformations in the pre-cracked body run with high velocity and if large inertial forces occur
as a result of accelerated mass particles. Both effects determine the stress situation
and the strength behavior significantly. High, undesirable dynamic loading of components with cracks are encountered in engineering practice at impact and shock
processes, crash events and explosions. Even earthquakes are wave propagations
caused by dynamic fracture processes in the earth crust. On the contrary, dynamic
fracture processes are consciously employed in mining (blasting), in geotechnical
engineering (hydrocracking) and in comminution technology. In principle, two questions have to be distinguished in dynamic fracture mechanics:

- The *stationary crack* under dynamic loading. The action of external loads is transferred by stress waves through the material to the crack.
- The *fast running crack*. Hereby elastodynamic waves are emitted by highly
  dynamic rupture processes from the crack itself.

Dynamic fracture processes are often more dangerous than stationary cracks for the
following reasons: Firstly, after crack initiation they almost always proceed in an
unstable manner, which may lead particularly in brittle materials to uncontrolled
failure of the whole component. After a short phase of acceleration, dynamic cracks
attain high speeds of propagation ranging in the order of sound wave speeds. Secondly, the high strain rates at the crack tip induce in many materials an embrittlement,
since viscoplastic effects reduce the energy absorption capacity of the fracture process
zone and the fracture toughness decreases.

Due to complicated elastodynamic wave phenomena such as reflection, superposition, dispersion and attenuation, in dynamic fracture processes no longer does
a proportional relationship exist between the temporal course of applied load $\sigma(t)$
and the stress state at the crack $K_{\mathrm{I}}(t)$. Principally, the following tendencies can be
identified at impact loading: In the beginning the wave phenomena dominate in a
short time range, as long as high kinetic energy is in the system. With increasing

time this energy is dissipated, and the waves are attenuated, scattered, and finally fade out.

The larger the crack is in proportion to the body, the more pronounced is the short-term effect, i. e. intense oscillations of $K_I(t)$ appear such as e. g. in the notch impact test. On the other hand a small crack in a huge component is hit only once by an incoming wave front that will subsequently be dispersed.

In the following some theoretical foundations of dynamic fracture mechanics are explained, restricted to brittle materials and two-dimensional problems. The time-dependent stress and deformation fields in the cracked body are thus calculated as solution of the related initial boundary value problems (IBVP) of the linear elasto-dynamic theory (Appendix A.5).

3.5.2 Fundamentals of Elastodynamics

The starting point is the PDE-system of Navier-Lamé, which expresses the linear elastodynamic problem in the form of displacement fields and has to be completed by corresponding initial and boundary conditions.

$$(\lambda + \mu)u_{j,ji} + \mu u_{i,jj} = \rho \ddot{u}_i \tag{3.327}$$

This relation is gained by inserting Hooke's law (A.88) into the equations of motion (A.70) (without body forces $b_i = 0$) and subsequent substitution of strains by the kinematic relations (A.29). In order to solve the PDE (3.327) the displacement field is represented by a scalar potential $\varphi(x, t)$ and a vector potential $\psi(x, t)$

$$\boldsymbol{u}(\boldsymbol{x}, t) = \nabla\varphi + \nabla \times \boldsymbol{\psi} \quad \text{or} \quad u_i = \varphi_{,i} + \epsilon_{ijk}\psi_{j,k} , \tag{3.328}$$

which results in Helmholtz wave equations for both potentials.

$$c_d^2 \Delta\varphi = \ddot{\varphi} , \qquad c_s^2 \Delta\psi = \ddot{\psi} \tag{3.329}$$

The scalar potential φ characterizes a volume change, whereas the vector potential ψ is associated with a pure shape change. Therefore, the constants c_d and c_s correspond to the speeds of dilatational (longitudinal) waves and shear (transversal) waves, respectively.

$$c_d = \sqrt{\frac{\lambda + 2\mu}{\rho}} , \qquad c_s = \sqrt{\frac{\mu}{\rho}} \tag{3.330}$$

Plane elastic waves that propagate in the direction of the normal vector $\boldsymbol{n}$ through an infinite 3D body can be represented as

$$\varphi = \varphi(\boldsymbol{x} \cdot \boldsymbol{n} - c_d t) \qquad \psi = \psi(\boldsymbol{x} \cdot \boldsymbol{n} - c_s t) . \tag{3.331}$$

The dilatational wave excites the particles in direction of the wave propagation, whereas the shear wave produces displacements along both transversal directions in the wave plane. In addition, surface waves (so–called Rayleigh waves) become important for dynamic crack problems, since they mainly propagate along free crack surfaces but fade rapidly away towards the interior. Their speed of propagation c_R is obtained from the root of the so-called Rayleigh's function $D(c)$.

$$D(c) = 4\alpha_d\alpha_s - (1 + \alpha_s^2)^2 \quad \rightarrow \quad D(c_R) = 0 \tag{3.332}$$

$$\alpha_d(c) = \sqrt{1 - \frac{c^2}{c_d^2}}, \qquad \alpha_s(c) = \sqrt{1 - \frac{c^2}{c_s^2}} \tag{3.333}$$

Wave velocities in engineering materials vary from minimum $900\,\mathrm{m/s}$ (plastics) up to maximum $11{,}000\,\mathrm{m/s}$ (ceramics). For steel and aluminum alloys they amount to about: $c_d \approx 5{,}900\,\mathrm{m/s}$, $c_s \approx 3{,}100\,\mathrm{m/s}$ and $c_R \approx 2{,}900\,\mathrm{m/s}$. The relation $c_R < c_s < c_d$ is valid.

3.5.3 Dynamic Loading of Stationary Cracks

At the tip of a dynamically loaded, stationary crack exactly the same near-tip fields arise as in the static case. There exist identical singularities and the same separation applies to the three opening modes I, II and III with the stress intensity factors as loading parameters. This way all asymptotic relations derived in Sect. 3.2 can be adopted for the stresses, strains and displacements at mode I ((3.12)–(3.16)), mode II ((3.23)–(3.26)) and mode III ((3.31)–(3.32)). The only but essential difference to the static case consists in that the K-factors in dynamics depend on time t.

Dynamic crack-tip field at a stationary crack:

$$\sigma_{ij}(r, \theta, t) = \frac{1}{\sqrt{2\pi r}} \left[K_I(t)f_{ij}^I(\theta) + K_{II}(t)f_{ij}^{II}(\theta) + K_{III}(t)f_{ij}^{III}(\theta) \right] \tag{3.334}$$

$$u_i(r, \theta, t) = \frac{1}{2\mu}\sqrt{\frac{r}{2\pi}} \left[K_I(t)g_i^I(\theta) + K_{II}(t)g_i^{II}(\theta) + K_{III}(t)g_i^{III}(\theta) \right] \tag{3.335}$$

This agreement can mathematically be explained by means of the PDE (3.327). If for the displacement field at the crack tip a non-singular ansatz is made $u_i = r^\lambda \tilde{g}(\theta, t)$ $(0 < \lambda < 1)$, then the stresses will behave like $r^{\lambda-1}$ and the second spatial derivative on the left-hand side of (3.327) like $r^{\lambda-2}$. Hence in the asymptotic limit $r \rightarrow 0$ the inertial forces $\rho\ddot{u}_i \sim r^\lambda$ vanish as terms of higher order and may be neglected.

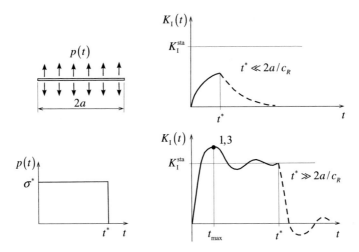

Fig. 3.61 Stress intensity factor as function of time if a crack is loaded by a rectangular impulse

The calculation of the K-factors as a function of transient load, component geometry, crack length and time requires solution of the elastodynamic IBVP. The few available analytical solutions for unbounded domains are mostly based on the Laplace integral transform, see Freund [140]. Some fundamental phenomena of dynamic crack analyses are exemplified and discussed in the 2D problem shown in Fig. 3.61. We consider a crack of length $l = 2a$ in the plane (plane strain state). Its faces are subjected to a sudden stress jump $p(t) = \sigma^*$ at $t = 0$ and are completely unloaded again after the time t^*. According to the principle of Huygen's, the impact loading generates elementary waves in every point of the crack faces and their envelope forms the current wave front. Far away from the crack tips, plane waves are emitted parallel to the crack faces. Around each crack tip however, two concentric circular wave fronts arise, associated with a dilatational and a shear wave, respectively, the radii of which increase with time $c_d t$ and $c_s t$. As long as these wave fronts have not reached the respective other crack tip, i.e. if the run time is $t < 2a/c_d$, each crack tip behaves autonomously like in the case of a semi-infinite crack. Because of self-similarity of the stress field $\sigma_{ij}(r, t)\sqrt{r} \sim \sigma^*\sqrt{c_d t}$, the stress intensity factor increases with time (Fig. 3.61, top right)

$$K_I(t) = 2\sigma^* \frac{\sqrt{1 - 2\nu}}{1 - \nu}\sqrt{\frac{c_d t}{\pi}} \qquad \text{foür } 0 < t < 2a/c_d. \qquad (3.336)$$

If the dilatational waves have attained the other crack tip, they generate there an additional tensile loading that fades away only when the Rayleigh wave impinges later at $\tau = 2a/c_R$. The occurring maximum of the dynamic stress intensity factor $K_{I\,max}$ exceeds the static solution $K_I^{sta} = \sigma^*\sqrt{\pi a}$ for this crack by about 30 % (for $\nu = 0.3$)! This effect is called *dynamic overshoot*. The enhancement factor $\varkappa > 1$ with respect to the static solution depends on the crack configuration and the edge steepness of the impulse $p(t)$.

$$\max_t K_{\mathrm{I}}(t) \approx K_{\mathrm{I}}(t_{\max} = 2a/c_{\mathrm{R}}) = \varkappa K_{\mathrm{I}}^{\mathrm{sta}} \qquad (3.337)$$

After the maximum, the course of $K_{\mathrm{I}}(t)$ is leveled off in several oscillations to its static value as depicted in Fig. 3.61 (right bottom). The amplitudes decay since the crack radiates permanently waves into the infinite domain. The behavior of $K_{\mathrm{I}}(t)$ after unloading at t^* is illustrated for both cases in Fig. 3.61.

It is apparent that the characteristic of the solution substantially depends on the ratio between pulse duration t^* and run time of a Rayleigh wave $\tau = 2a/c_{\mathrm{R}}$ along the crack length. By applying the K-concept $K_{\mathrm{I}}(t) = K_{\mathrm{Id}}$, the critical value σ_{c}^* of the rectangular impulse can be calculated for both cases from (3.336) and (3.337) (plane strain, $\nu = 0.3$)

$$\sigma_{\mathrm{c}}^* = 0.87 \frac{K_{\mathrm{Id}}}{\sqrt{\pi a}} \Big/ \sqrt{\frac{t^* c_{\mathrm{R}}}{2a}} \qquad \text{for } t^* \ll 2a/c_{\mathrm{R}}$$

$$\sigma_{\mathrm{c}}^* = \frac{1}{1.3} \frac{K_{\mathrm{Id}}}{\sqrt{\pi a}} = \text{const.} \qquad \text{for } t^* \gg 2a/c_{\mathrm{R}} \qquad (3.338)$$

For long pulses the critical loading behaves qualitatively similar as in the static case (3.79). However, at very short load impacts the tolerable stress (for constant crack length) is enhanced with $1/\sqrt{t^*}$, since the K-factor has to be built-up first. In fact these durations are extremely short. For a crack of size 10 mm in aluminum it is $\tau \approx 3.4\,\mu s$.

3.5.4 Dynamic Crack Propagation

What is the structure of the near-field at the tip of a crack that is propagating with a velocity $\dot{a}$? To find the solution, a coordinate system is attached at the moving crack tip $(x_1 = X_1 - \dot{a}t, x_2 = X_2)$, see Fig. 3.62. At first the mode III case is considered, followed by the opening modes I and II in the crack plane.

Antiplane Shear Mode III

In this case the Navier-Lamé equation (3.327) simplify to a wave equation for the longitudinal displacement field u_3 since $u_1 \equiv u_2 \equiv 0$, $\frac{\partial(\cdot)}{\partial x_3} \equiv 0$,

$$c_{\mathrm{s}}^2 u_{3,jj}(x_1, x_2) = \ddot{u}_3(x_1, x_2)\,, \qquad (3.339)$$

and due to $\frac{\partial(\cdot)}{\partial t} = -\dot{a}\frac{\partial(\cdot)}{\partial x_1}$ the following form is yielded

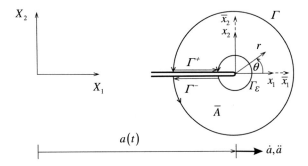

Fig. 3.62 Moving coordinate system with a fast running crack and integration paths

$$\frac{\partial^2 u_3}{\partial x_1^2} + \frac{1}{\alpha_s^2}\frac{\partial^2 u_3}{\partial x_2^2} = 0 \quad \text{with } \alpha_s = \sqrt{1 - \frac{\dot{a}^2}{c_s^2}}. \tag{3.340}$$

In order to transform this PDE into a potential equation, a modified (compressed) x_2-coordinate $\bar{x}_2 := \alpha_s x_2$ is introduced, which reads in complex notation

$$z_s = x_1 + i\alpha_s x_2 = r_s e^{i\theta_s} \quad \text{with} \quad r_s = \sqrt{x_1^2 + \alpha_s^2 x_2^2}, \quad \theta_s = \arctan\left(\frac{\alpha_s x_2}{x_1}\right). \tag{3.341}$$

Thus we get the equation

$$\Delta u_3(r_s, \theta_s) = 0, \tag{3.342}$$

which can be solved by using the complex functions method as in the static case (cf. Sect. 3.2.2)

$$\mu u_3 = \Re\Omega(z_s), \quad \tau_{13} - i\frac{\tau_{23}}{\alpha_s} = \Omega'(z_s). \tag{3.343}$$

The same series approach (3.51) for the stress function Ω satisfies the crack-face boundary conditions. Only the singular term $n = 1$ is evaluated $\Omega(z_s) = c_1 z_s^{1/2}$, whereby $c_1 = -K_{III}\sqrt{2/\pi}$ holds. Using (3.343) the stresses and displacements of the near-field are finally found

$$\begin{Bmatrix} \tau_{13} \\ \tau_{23} \end{Bmatrix} = \frac{K_{III}(t)}{\sqrt{2\pi r_s}} \begin{Bmatrix} -\dfrac{\sin(\theta_s/2)}{\alpha_s\sqrt{\gamma_s}} \\ \dfrac{\cos(\theta_s/2)}{\sqrt{\gamma_s}} \end{Bmatrix}, \quad u_3 = \frac{K_{III}(t)\sqrt{r}}{\sqrt{2\pi}}\frac{\sqrt{\gamma_s}}{\alpha_s}\sin\frac{\theta_s}{2}. \tag{3.344}$$

As in the static case, this solution is composed as the product of radial and angular functions. The stresses become singular as $r^{-1/2}$ and the displacement field is proportional to $r^{1/2}$. Of course now the angular functions of the fields depend via α_s and r_s on the crack velocity $\dot{a}$. This difference to the static problem is vanishing for $\dot{a} = 0$ ($\alpha_s = 1$, $r_s = r$, $\theta_s = \theta$), cf. (3.31) and (3.32). As abbreviation the ratio between the scaled and true radii is introduced

$$\gamma_s(\theta) = \frac{r_s}{r} = \sqrt{1 - \left(\frac{\dot{a} \sin \theta}{c_s}\right)^2}, \qquad \gamma_d(\theta) = \frac{r_d}{r} = \sqrt{1 - \left(\frac{\dot{a} \sin \theta}{c_d}\right)^2} \qquad (3.345)$$

In-Plane Tension (Mode I) and Shear (Mode II)

For plane problems (plane strain: $u_3 \equiv 0$, $: \frac{\partial (\cdot)}{\partial x_3} \equiv 0$) the representation of the solution (3.328) is reduced to two wave equations that identify the scalar potential and the $\psi = \psi_3$-component of the vector potential ($\psi_1 \equiv \psi_2 \equiv 0$)

$$c_d \, \varphi_{,jj} = \ddot{\varphi}, \qquad c_s \, \psi_{,jj} = \ddot{\psi}. \qquad (3.346)$$

Similar to the mode III problem, a transformation into a moving coordinate system (x_1, x_2) as depicted in Fig. 3.62 gives two time-independent PDE with the parameters

$$\frac{\partial^2 \varphi}{\partial x_1^2} + \frac{1}{\alpha_d} \frac{\partial^2 \varphi}{\partial x_2^2} = 0, \qquad \alpha_d = \sqrt{1 - \frac{\dot{a}^2}{c_d^2}}$$

$$\frac{\partial^2 \psi}{\partial x_1^2} + \frac{1}{\alpha_s} \frac{\partial^2 \psi}{\partial x_2^2} = 0, \qquad \alpha_s = \sqrt{1 - \frac{\dot{a}^2}{c_s^2}}. \qquad (3.347)$$

Inserting of the scaled coordinates (3.341) into the 2nd equation and declaring corresponding expressions

$$z_d = x_1 + i\alpha_d x_2 = r_d e^{i\theta_d} \quad \text{with} \quad r_d = \sqrt{x_1^2 + \alpha_d^2 x_2^2}, \quad \theta_d = \arctan\left(\frac{\alpha_d x_2}{x_1}\right) \qquad (3.348)$$

for the 1st equation results in the potential equations

$$\Delta\varphi(r_d, \theta_d) = 0, \qquad \Delta\psi(r_s, \theta_s) = 0. \qquad (3.349)$$

To get the solution, both functions are represented either as real or imaginary part of a complex analytical function, whereby again only the dominant terms of a series expansion will be used

$$\varphi = A \, \Re z_d^{3/2}, \quad \psi = B \, \Im z_s^{3/2} \quad \text{for symmetry (mode I)}$$

$$\varphi = A \, \Im z_d^{3/2}, \quad \psi = B \, \Re z_s^{3/2} \quad \text{for antimetry (mode II)}. \qquad (3.350)$$

The real constants A and $B = A\, 2\alpha_d/(1 + \alpha_s^2)$ are determined by employing the boundary conditions of stress-free crack faces ($\tau_{21} = \sigma_{22} = 0$ at $\theta = \pm\pi$). After extensive calculations, see [141], the stress and displacement fields are gained at the crack tip via the relationships

$$u_1 = \frac{\partial\varphi}{\partial x_1} + \frac{\partial\psi}{\partial x_2}, \qquad u_2 = \frac{\partial\varphi}{\partial x_2} - \frac{\partial\psi}{\partial x_1} \tag{3.351}$$

$$\sigma_{11} = \lambda\Delta\varphi + 2\mu\left[\frac{\partial^2\varphi}{\partial x_1^2} + \frac{\partial^2\psi}{\partial x_1\partial x_2}\right], \qquad \sigma_{22} = \lambda\Delta\varphi + 2\mu\left[\frac{\partial^2\varphi}{\partial x_2^2} - \frac{\partial^2\psi}{\partial x_1\partial x_2}\right]$$

$$\sigma_{12} = \mu\left[2\frac{\partial^2\varphi}{\partial x_1\partial x_2} + \frac{\partial^2\psi}{\partial x_2^2} - \frac{\partial^2\psi}{\partial x_1^2}\right]. \tag{3.352}$$

The stress intensity factors are defined as for static cracks

$$K_{\mathrm{I}}(t) = \lim_{r\to 0} \sqrt{2\pi r}\,\sigma_{22}(r, \theta = 0, t)$$

$$K_{\mathrm{II}}(t) = \lim_{r\to 0} \sqrt{2\pi r}\,\tau_{21}(r, \theta = 0, t) \tag{3.353}$$

$$K_{\mathrm{III}}(t) = \lim_{r\to 0} \sqrt{2\pi r}\,\tau_{23}(r, \theta = 0, t).$$

Dynamic crack tip field for mode I

$$\begin{Bmatrix} \sigma_{11} \\ \sigma_{22} \\ \tau_{12} \end{Bmatrix} = \frac{K_{\mathrm{I}}(t)}{\sqrt{2\pi r}D(\dot a)} \begin{Bmatrix} (1+\alpha_s^2)(1+2\alpha_d^2-\alpha_s^2)\dfrac{\cos(\theta_d/2)}{\sqrt{\gamma_d}} - 4\alpha_d\alpha_s\dfrac{\cos(\theta_s/2)}{\sqrt{\gamma_s}} \\[2mm] -(1+\alpha_s^2)^2\dfrac{\cos(\theta_d/2)}{\sqrt{\gamma_d}} + 4\alpha_d\alpha_s\dfrac{\cos(\theta_s/2)}{\sqrt{\gamma_s}} \\[2mm] 2\alpha_d(1+\alpha_s^2)\left(\dfrac{\sin(\theta_d/2)}{\sqrt{\gamma_d}} - \dfrac{\sin(\theta_s/2)}{\sqrt{\gamma_s}}\right) \end{Bmatrix}$$

$$\begin{Bmatrix} u_1 \\ u_2 \end{Bmatrix} = \frac{2K_{\mathrm{I}}(t)\sqrt{r}}{\mu\sqrt{2\pi}D(\dot a)} \begin{Bmatrix} (1+\alpha_s^2)\sqrt{\gamma_d}\cos\dfrac{\theta_d}{2} - 2\alpha_d\alpha_s\sqrt{\gamma_s}\cos\dfrac{\theta_s}{2} \\[2mm] -(1+\alpha_s^2)\alpha_d\sqrt{\gamma_d}\sin\dfrac{\theta_d}{2} + 2\alpha_d\sqrt{\gamma_s}\sin\dfrac{\theta_s}{2} \end{Bmatrix} \tag{3.354}$$

Dynamic crack tip field for mode II

$$
\begin{Bmatrix} \sigma_{11} \\ \sigma_{22} \\ \tau_{12} \end{Bmatrix} = \frac{K_{\mathrm{II}}(t)}{\sqrt{2\pi r}D(\dot{a})} \begin{Bmatrix} -2\alpha_{\mathrm{s}}(1+2\alpha_{\mathrm{d}}^2-\alpha_{\mathrm{s}}^2)\dfrac{\sin(\theta_{\mathrm{d}}/2)}{\sqrt{\gamma_{\mathrm{d}}}} + 2\alpha_{\mathrm{s}}(1+\alpha_{\mathrm{s}}^2)\dfrac{\sin(\theta_{\mathrm{s}}/2)}{\sqrt{\gamma_{\mathrm{s}}}} \\[2ex] 2\alpha_{\mathrm{s}}(1+\alpha_{\mathrm{s}}^2)\left(\dfrac{\sin(\theta_{\mathrm{d}}/2)}{\sqrt{\gamma_{\mathrm{d}}}} - \dfrac{\sin(\theta_{\mathrm{s}}/2)}{\sqrt{\gamma_{\mathrm{s}}}}\right) \\[2ex] 4\alpha_{\mathrm{d}}\alpha_{\mathrm{s}}\dfrac{\cos(\theta_{\mathrm{d}}/2)}{\sqrt{\gamma_{\mathrm{d}}}} - (1+\alpha_{\mathrm{s}}^2)^2\dfrac{\cos(\theta_{\mathrm{s}}/2)}{\sqrt{\gamma_{\mathrm{s}}}} \end{Bmatrix}
$$

$$
\begin{Bmatrix} u_1 \\ u_2 \end{Bmatrix} = \frac{2K_{\mathrm{II}}(t)\sqrt{r}}{\mu\sqrt{2\pi}D(\dot{a})} \begin{Bmatrix} 2\alpha_{\mathrm{s}}\sqrt{\gamma_{\mathrm{d}}}\sin\dfrac{\theta_{\mathrm{d}}}{2} - \alpha_{\mathrm{s}}(1+\alpha_{\mathrm{s}}^2)\sqrt{\gamma_{\mathrm{s}}}\sin\dfrac{\theta_{\mathrm{s}}}{2} \\[2ex] -2\alpha_{\mathrm{d}}\alpha_{\mathrm{s}}\sqrt{\gamma_{\mathrm{d}}}\cos\dfrac{\theta_{\mathrm{d}}}{2} + (1+\alpha_{\mathrm{s}}^2)\sqrt{\gamma_{\mathrm{s}}}\cos\dfrac{\theta_{\mathrm{s}}}{2} \end{Bmatrix}
\qquad (3.355)
$$

The dynamic crack tip fields have basically the same structure as in statics, cf. (3.12), (3.16) and (3.23), (3.25). But their magnitude and their angular distribution depend on the crack velocity $\dot{a}$, which affects the Rayleigh function (3.332) as well as the constants α_{d}, α_{s} (3.347) and γ_{d}, γ_{s} (3.345). For $\dot{a} \to 0$ the formulas (3.354) and (3.355) migrate into the well-known relations of statics. It can be shown that the acceleration $\ddot{a}$ of the crack doesn't influence the singular near-field solution but has an impact on higher order terms of the series expansion [141]. This means, the yet deduced relationships (3.344), (3.354) and (3.355) are valid for (arbitrarily) accelerated crack propagation, too. Thus, the K-factors and the velocity $\dot{a}$ control definitely the loading situation at a moving crack in an isotropic elastic material. The 2nd term of the series expansion describes the dynamic T-stresses:

$$
T_{11}^{\mathrm{dyn}} = T_{11}^{\mathrm{sta}}(\alpha_{\mathrm{d}}^2 - \alpha_{\mathrm{s}}^2), \qquad T_{22}^{\mathrm{dyn}} = T_{12}^{\mathrm{dyn}} = 0. \qquad (3.356)
$$

Finally the stress field (3.354) at a fast running crack under mode I should be discussed. From the fracture mechanics point of view the maximum circumferential stresses $\sigma_{\theta\theta}(r, \theta)$ are of most interest. Their angular distribution is plotted in Fig. 3.63 for different crack velocities. Up to a velocity of $\dot{a} < 0.6\,c_{\mathrm{s}}$ of the shear wave speed the maximum $\sigma_{\theta\theta}$ is located at $\theta = 0$, i.e. in the direction of crack extension. At higher crack velocities the position of the maximum is shifted towards an angle of $\theta = 60°$. Supposing the validity of the criterion of maximum circumferential stress (Sect. 3.4.4), the crack should deflect in this orientation or branch symmetrically at higher velocities $\dot{a} > 0, 6c_{\mathrm{s}}$. Even though crack branching is a frequently observed phenomenon at such velocities, no sufficient explanation is delivered from the stress state alone. Another important reason for the loss of directional stability at fast crack propagation lies in the excessive supply of kinetic energy that the crack can only relieve by branching.

If the ratio of stresses σ_{22}/σ_{11} is investigated at the ligament $\theta = 0$ as a function of crack velocity $\dot{a}$ ($\gamma_{\mathrm{d}} = \gamma_{\mathrm{s}} = 1$)

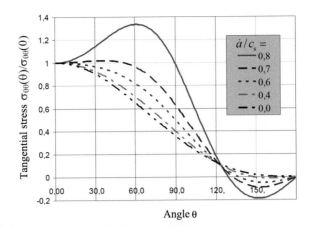

Fig. 3.63 Angular distribution of dimensionless circumferential stress as a function of relative crack velocity $\dot{a}/c_s$ for steel

$$\frac{\sigma_{22}}{\sigma_{11}} = \frac{-(1+\alpha_s^2)^2 + 4\alpha_d\,\alpha_s}{(1+\alpha_s^2)(1+2\alpha_d^2 - \alpha_s^2) - 4\alpha_d\,\alpha_s}, \tag{3.357}$$

a falling trend is obtained from the value 1 ($\dot{a} = 0$) down to zero ($\dot{a} = c_R$), since in the denominator stands the Rayleigh function $D(\dot{a} = c_R) = 0$. Therefore, at high crack velocities $\sigma_{11} > \sigma_{22}$, favoring a material rupture perpendicular to the crack direction. Moreover, the triaxiality $\hbar$ is considerably decreased, which explains the larger plastic deformation and enhancement of the fracture resistance with $\dot{a}$.

3.5.5 Energy Balance and J-Integrals

In the case of dynamic fracture processes the kinetic energy $\mathcal{K}$ (3.72) has to be included into the energy balance. We consider an elastic, thermally isolated body ($\dot{Q} = 0$) and employ the results from Sect. 3.2.5, Eq. (3.69). The external work $\mathcal{W}_{ext}$ (3.70) is converted into strain energy $\mathcal{W}_{int}$ (3.71) and kinetic energy $\mathcal{K}$ (3.72), the remaining amount of energy is available for crack propagation, i.e.

$$\frac{d\mathcal{W}_{ext} - d(\mathcal{W}_{int} + \mathcal{K})}{dA} = \frac{-d(\Pi + \mathcal{K})}{dA} = G = \frac{d\mathcal{D}}{dA} = 2\gamma. \tag{3.358}$$

The global energy release rate G quantifies the total mechanical energy provided by the system during an infinitesimal crack increment. For a stationary, dynamically loaded crack the same correlation exists to the stress intensity factors as in statics (3.93), since identical crack tip fields are present, however now time-dependent

$$G(t) = \frac{1}{E'}\left(K_I^2(t) + K_{II}^2(t)\right) + \frac{1+\nu}{E}\,K_{III}^2(t). \tag{3.359}$$

3.5 Dynamic Fracture Processes

To analyze the energy balance for a fast moving crack, we refer to the considerations made in Sect. 3.3.8 for ductile crack growth, see Fig. 3.48. We restrict the derivation to elastodynamic material behavior and concentrate the process zone A_B to a point at the crack tip enclosing the dominant singularity, i. e. $\Gamma_B = \Gamma_\varepsilon \to 0$. The energy flux (3.273) is now extended by the kinetic energy of the body, $\dot{\mathcal{W}}_B = \dot{\mathcal{W}}_{\text{ext}} - \dot{\mathcal{W}}_{\text{int}} - \dot{\mathcal{K}}$, and has the meaning of a real energy release rate since $\mathcal{W}_{\text{int}}$ does not contain dissipative plastic terms.

$$\mathcal{F} = G^{\text{dyn}} = \frac{d\mathcal{W}_B}{da} = \frac{\dot{\mathcal{W}}_B}{\dot{a}} = \frac{1}{\dot{a}} \left[\int_{S_t} \bar{t}_i \dot{u}_i \, ds - \frac{d}{dt} \int_A \left(U + \frac{\rho}{2} \dot{u}_i \dot{u}_i \right) dA \right] \quad (3.360)$$

Again, we switch over to the spatial coordinates (x_1, x_2) moving with the crack (Figs. 3.62 and 3.48), which requires us to treat the material time derivative in (3.360) with the help of the Reynold's transport theorem

$$\frac{d}{dt} \int_A \left[U + \frac{\rho}{2} \dot{u}_i \dot{u}_i \right] dA = \int_A \left[\frac{\partial U}{\partial t} + \rho \dot{u}_i \ddot{u}_i \right] dA - \dot{a} \int_{\Gamma_\varepsilon} \left[U + \frac{\rho}{2} \dot{u}_i \dot{u}_i \right] n_1 \, ds \quad (3.361)$$

Taking into account the equations of motion $\sigma_{ij,j} = \rho \ddot{u}_i$ and the relation $\partial U / \partial t = \sigma_{ij} \dot{u}_{i,j}$, the kernel of the 1st integral can be converted into $(\sigma_{ij} \dot{u}_i)_{,j}$. Applying Gauss's theorem delivers a contour integral over $C = S + \Gamma^+ + \Gamma^- - \Gamma_\varepsilon$ as visible in Fig. 3.48, which simplifies for load-free crack faces and with $\dot{u}_i = 0$ on S_u to the contributions along S_t and Γ_ε only. Inserting these expressions in (3.360) yields in the end

$$\mathcal{F} = \frac{1}{\dot{a}} \left[\int_{\Gamma_\varepsilon} t_i \dot{u}_i \, ds + \dot{a} \int_{\Gamma_\varepsilon} \left(U + \frac{\rho}{2} \dot{u}_i \dot{u}_i \right) n_1 \, ds \right]. \quad (3.362)$$

The displacement velocity on $\Gamma_\varepsilon \to 0$ is obtained by the local stationarity condition (3.277) $\dot{u}_i = -\dot{a} u_{i,1}$.

Thus the energy flux into the crack tip or the energy release rate of the system at elastodynamic self-similar crack propagation amounts to

$$\mathcal{F} = G^{\text{dyn}} = \lim_{\Gamma_\varepsilon \to 0} \int_{\Gamma_\varepsilon} \left[\left(U + \dot{a}^2 \frac{\rho}{2} u_{i,1} u_{i,1} \right) n_1 - \sigma_{ij} n_j u_{i,1} \right] ds \quad (3.363)$$

$$G^{\text{dyn}} = \lim_{\Gamma_\varepsilon \to 0} \int_{\Gamma_\varepsilon} \left[\left(U + \frac{\rho}{2} \dot{u}_i \dot{u}_i \right) \delta_{1j} - \sigma_{ij} u_{i,1} \right] n_j \, ds. \quad (3.364)$$

For the numerical computation of G^{dyn} it is more comfortable to replace the near-tip integral (3.364) along $\Gamma_\varepsilon \to 0$ by a line integral along an arbitrary remote path Γ, which requires an additional integral over the enclosed area $\bar{A}$, see Fig. 3.62. By analogy to (3.279), this conversion of (3.364) results in the path-independent contour-area integral

$$\mathcal{F} = G^{\text{dyn}} = \int_\Gamma \left[\left(U + \frac{\rho}{2} \dot{u}_i \dot{u}_i \right) \delta_{1j} - \sigma_{ij} u_{i,1} \right] n_j \, \mathrm{d}s + \int_{\bar{A}} \left(\rho \ddot{u}_i u_{i,1} - \rho \dot{u}_i \dot{u}_{i,1} \right) \mathrm{d}A \ .$$

$$(3.365)$$

In the special case of a constant crack propagation speed $\dot{a}$ and stationary conditions in $\bar{A}$, the relations $\dot{u}_i = -\dot{a} u_{i,1}$, $\dot{u}_{i,1} = -\dot{a} u_{i,11}$ and $\ddot{u}_i = -\dot{a}^2 u_{i,1}$ apply making the area integral to zero. Under such conditions the energy release rate is exclusively represented by the path-independent contour integral along Γ. However, at crack propagations in finite structures these circumstances are rarely met.

For stationary cracks ($\dot{a} = 0$) the above expression of energy release rate simplifies to:

$$G^{\text{dyn}} = \int_\Gamma \left[U \delta_{1j} - \sigma_{ij} u_{i,1} \right] n_j \, \mathrm{d}s + \int_{\bar{A}} \rho \ddot{u}_i u_{i,1} \, \mathrm{d}A \qquad (3.366)$$

In the pure static case even the inertia forces $\rho \ddot{u}_i$ in $\bar{A}$ are omitted, which is consequently leading back to the classical J-integral (3.100).

The relationship between energy release rate G^{dyn} and stress intensity factors can be realized as in statics either by the energy flux integral (3.363) or by the dynamic analogue of the crack closure integral (3.89), inserting there the dynamic crack tip fields of Sect. 3.5.4 [140].

Elastodynamic energy release rate as a function of stress intensity factors and crack velocity under mixed-mode loading (plane strain):

$$G^{\text{dyn}}(t) = G_{\text{I}}(t) + G_{\text{II}}(t) + G_{\text{III}}(t)$$

$$= \frac{1}{2\mu} \left[A_{\text{I}}(\dot{a}) K_{\text{I}}^2(t) + A_{\text{II}}(\dot{a}) K_{\text{II}}^2(t) + A_{\text{III}}(\dot{a}) K_{\text{III}}^2(t) \right] \qquad (3.367)$$

$$A_{\text{I}}(\dot{a}) = \frac{\dot{a}^2 \alpha_{\text{d}}}{c_{\text{s}}^2 D(\dot{a})}, \quad A_{\text{II}}(\dot{a}) = \frac{\dot{a}^2 \alpha_{\text{s}}}{c_{\text{s}}^2 D(\dot{a})}, \quad A_{\text{III}}(\dot{a}) = \frac{1}{\alpha_{\text{s}}}$$

This relation is only valid for straight crack propagation in its original direction (along the x_1-axis), but holds for any transient course $\dot{a}(t)$. For $\dot{a} = 0$ Eq. (3.367) merges into the static relation (3.93). The functions A_{I} and A_{II} become singular at $\dot{a} \to c_{\text{R}}$ ($D(c_{\text{R}}) = 0$) and the function A_{III} at $\dot{a} \to c_{\text{s}}$. To keep $G^{\text{dyn}}(= 2\gamma_{\text{D}})$ bounded, the K-factors must tend to zero. This means, the Rayleigh speed is the upper limit of propagation velocity $\dot{a}_{\max} = c_{\text{R}} \approx 0.57\sqrt{E/\rho}$ for cracks under mode I or II,

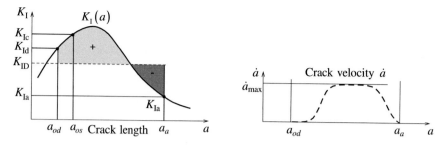

Fig. 3.64 Initiation, acceleration and arrest at dynamic crack propagation

which has been evidenced by experimental observations [141, 142]. For mode III the shear wave speed is the upper limit.

3.5.6 Fracture Criteria

The loading situation at the crack tip is controlled by the elastodynamic stress intensity factors. Therefore, the K-concept applies with the same arguments as in the static case. We restrict ourselves to the most important opening mode I. For the phase of *crack initiation* the following criterion can be postulated:

$$K_{\mathrm{I}}(t) = K_{\mathrm{Id}}(\dot{K}_{\mathrm{I}}, t^*) \qquad (3.368)$$

The right–hand side K_{Id} represents the *dynamic initiation toughness* that depends on the velocity of crack tip loading $\dot{K}_{\mathrm{I}}$. For metallic materials it lies below the static fracture toughness $K_{\mathrm{Id}} < K_{\mathrm{Ic}}$. In practice, typical impact loads occur with values of about $\dot{K}_{\mathrm{I}} \approx 10^3\text{--}10^7 \, \text{MPa} \sqrt{\text{m}}/\text{s}$. According to Sect. 3.5.2, at short-time load impulses the interval length t^* during which the crack tip field is built-up plays an important role.

After a dynamic fracture process is started, the further course of crack propagation depends on the supply and consumption of energy. The stress field at a moving crack and the energy release rate are determined by the dynamic stress intensity factor K_{I} and the crack velocity $\dot{a}$. One the other hand the fracture energy 2γ dissipated by the material is influenced by the velocity, too. Therefore, the fracture criterion for a moving crack is formulated as:

$$K_{\mathrm{I}}(t, \dot{a}) = K_{\mathrm{ID}}(\dot{a}, \dot{K}_{\mathrm{I}}, T) \quad \text{or} \quad G_{\mathrm{I}}^{\mathrm{dyn}}(t, \dot{a}) = 2\gamma_{\mathrm{D}}(\dot{a}, \dot{K}_{\mathrm{I}}, T) . \qquad (3.369)$$

The index D denotes the *dynamic crack growth toughness* that is principally different from the initiation quantity K_{Id}. Expensive experiments are necessary to determine K_{ID} as a function of crack velocity $\dot{a}$. Nearly all materials show an enormous increase of fracture toughness K_{ID} with $\dot{a}$ due to various mechanisms (crack branching, strain rate dependency, adiabatic heating, et al).

If the energy supply falls off, the crack velocity is slowed down and the crack finally stops, which is called *crack arrest*. Then the condition applies:

$$K_I(t, \dot{a}) \leq K_{Ia} . \tag{3.370}$$

The *crack arrest toughness* K_{Ia} characterizes the ability of a material to absorb a running crack. Because of inertial effects, the crack arrest is not yet simply the reverse of crack initiation! Therefore the relation holds $K_{Ia} < K_{ID}(\dot{a} \to 0) < K_{Id}$. In this context the *crack arrest concept* should be mentioned. This concept is applied in fracture mechanical safety design (e.g. of reactor pressure vessels) to ensure that a crack will come to a standstill if it runs in a region of the component with tougher material $K_{Ia} > K_I$.

In the end, the typical scenario of a dynamic fracture process should be outlined. Figure 3.64 depicts the temporal course of the stress intensity K_I for a crack in a structural component as a function of crack length a. Under static loading the crack length a_{0s} would be necessary to initiate fracture at $K_I = K_{Ic}$. If the same K_I-factor is now imposed by dynamic loading, then already a crack size $a_{0d} < a_{0s}$ becomes critical. After the crack is started it gains more energy (dashed area), since $K_I(a)$ is further increasing and its velocity speeds up. Let's assume for simplicity a constant dynamic fracture toughness K_{ID}. If the driving load $K_I(t)$ falls below the K_{ID} value, the remaining available energy causes the fracture process to continue, leading to a much later crack arrest at $K_I(a_a) \leq K_{Ia}$.

References

1. Muskhelishvili NI (1971) Einige Grundaufgaben zur mathematischen Elastizitätstheorie. Fachbuchverlag Leipzig
2. Williams ML (1957) On the stress distribution at the base of a stationary crack. J Appl Mech 24:109–114
3. Gross B, Srawley JE (1965) Stress-intensity factors for single-edge-notch specimens in bending or combined bending and tension by boundary collocation of a stress function. Technical Report NASA Technical Note D-2603, NASA, Lewis Research Center
4. Xiao QZ, Karihaloo BL (2002) Coefficients of the crack tip asymptotic field for a standard compact tension specimen. Int J Fract 118:1–15
5. Sneddon LN, Lowengrub M (1969) Crack problems in the classical theory of elasticity. Wiley, New York
6. Fabrikant VI (2001) Application of potential theory in mechanics: a selection of new results. Kluwer Academic Publishers, Dordrecht
7. Hartranft RJ, Sih GC (1969) The use of eigenfunction expansions in the general solution of three-dimensional crack problems. J Math Mech 19(2):123–138
8. Hartranft RJ, Sih GC (1977) Stress singularity for a crack with an arbitrary curved crack front. Eng Fract Mech 9:705–718
9. Sih GC (1973) Methods of analysis and solutions of crack problems. Mechanics of fracture, vol. 1. Noordhoff International Publisher, Leyden

10. Tamuz V, Romalis N, Petrova V (2000) Fracture of solids with microdefects. Nova Science Publishers, New York

11. Erdogan F, Gupta GD, Cook TS (1973) Numerical solution of singular integral equations. In: Sih GC (ed) Methods of analysis and solutions of crack problems. Mechanics of fracture, vol. 1. Noordhoff, Leyden, pp. 368–425

12. Kassir MK, Sih GC (1975) Three dimensional crack problems. Mechanics of fracture, vol. 2. Noordhoff Internatioanl Publisher, Leyden

13. Murakami Y (1987) Stress intensity factors handbook, vol. 1–5. Pergamon Press, Oxford

14. Rooke DP, Cartwright DJ (1976) Compendium of stress intensity factors. Her Majesty's Stationary Office, London

15. Tada H, Paris P, Irwin G (1985) The stress analysis of cracks handbook, 2nd edn. Paris Production Inc., St. Louis

16. Theilig H, Nickel J (1987) Spannungsintensitätsfaktoren. Fachbuchverlag, Leipzig

17. Irwin GR (1957) Analysis of stresses and strains near the end of a crack traversing a plate. J Appl Mech 24:361–364

18. ISO12135 (2002) Metallic materials—Unified method of test for the quasitatic fracture toughness. Technical Report, International Organization for Standardization, Genf

19. ESIS P2 (1992) Procedure for determining the fracture behaviour of materials. Technical Report, European Structural Integrity Society

20. ASTM-E 1820 (2007) Standard test method for measurement of fracture toughness. Technical Report, American Society for Testing and Materials, West Conshohocken

21. Blumenauer H, Pusch G (1993) Technische Bruchmechanik, 3rd edn. Deutscher Verlag für Grundstoffindustrie, Leipzig

22. Griffith AA (1921) The phenomena of rupture and flow in solids. Philos Trans Ser A 221: 163–198

23. Irwin GR (1958) Fracture. In: Flügge S (ed) Handbuch der Physik. Band 6, Engineering fracture mechanics, Springer, Berlin, pp 551–590

24. Cherepanov G (1967) Rasprostranenie trechin v sploshnoi srede (about crack advance in the continuum). Prikladnaja Matematika i Mekhanica 31:478–488

25. Rice J (1968) A path independent integral and the approximate analysis of strain concentration by notches and cracks. J Appl Mech 35:379–386

26. Altenbach H, Altenbach J, Rikards R (1996) Einführung in die Mechanik der Laminatwerkstoffe. Deutscher Verlag für Grundstoffindustrie, Stuttgart

27. Lekhnitskii SG (1981) Theory of elasticity of an anisotropic body. Mir Publisher, Moscow

28. Stroh AN (1962) Steady state problems on anisotropic elasticity. J Math Phys 41:77–103

29. Sih GC, Paris PC, Irwin GR (1965) On cracks in rectilinear anisotropic bodies. Int J Fract Mech 1:189–203

30. Comninou M (1990) An overview of interface cracks. Eng Fract Mech 37:197–208

31. Rice JR, Suo Z, Wang JS (1990) Mechanics and thermodynamics of brittle interface failure in bimaterial systems. In: Rühle M, Evans A, Ashby M, Hirth J (eds) Metal-Ceramic Interfaces. Pergamon Press, Oxford, pp 269–294

32. Rice JR (1988) Elastic fracture mechanics concepts for interfacial cracks. Trans ASME 55:98–103

33. Banks-Sills L, Ashkenazi D (2000) A note on fracture criteria for interface fracture. Int J Fract 103:177–188

34. Qu J, Bassani JL (1993) Interfacial fracture mechanics for anisotropic bimaterials. J Appl Mech 60:422–431

35. Suo Z (1990) Singularities, interfaces and cracks in dissimilar anisotropic media. Proc Roy Soc London A 427:331–358

36. Beom HG, Atluri SN (1995) Dependence of stress on elastic constants in an anisotropic bimaterial under plane deformation; and the interfacial crack. Comput Mech 16:106–113

37. Williams ML (1961) The bending stress distribution at the base of a stationary crack. J Appl Mech 28:78–82

38. Sih GC, Paris PC, Erdogan F (1962) Crack-tip stress-intensity factors for plane extension and plate bending problems. J Appl Mech 29:306–312

39. Hartranft RJ, Sih GC (1968) Effect of plate thickness on the bending stress distribution around through cracks. J Math Phys 47:276–291

40. Hui CY, Zehnder AT (1993) A theory for the fracture of thin plates subjected to bending and twisting moments. Int J Fract 61:211–229

41. Rice JR (1972) Some remarks on elastic crack-tip stress fields. Int J Solid Struct 8:751–758

42. Stein E, Barthold FJ (1996) Werkstoffe - Elastizitätstheorie. In: Mehlhorn G (ed) Der Ingenieurbau, vol 4. Verlag Ernst & Sohn, Berlin, pp 165–425

43. Chen YZ (1989) Weight function technique in a more general case. Eng Fract Mech 33:983–986

44. Bueckner HF (1970) A novel principle for the computation of stress intensity factors. Zeitschrift für Angewandte Mathematik und Mechanik 50:529–546

45. Paris PC, McMeeking RM, Tada H (1976) The weight function method for determining stress intensity factors. Cracks and Fracture, STP 601, American Society for Testing of Materials pp 471–489

46. Bueckner HF (1987) Weight functions and fundamental fields for the penny-shaped and the half-plane crack in three-space. Int J Solid Struct 23(1):57–93

47. Bueckner HF (1989) Observations of weight functions. Eng Anal Bound Elem 6:3–18

48. Fett T, Munz D (1997) Stress intensity factors and weight functions. Computational Mechanics Publications, Southampton, Boston

49. Bortmann Y, Banks-Sills L (1983) An extended weight function method for mixed-mode elastic crack analysis. J Appl Mech 50:907–909

50. Gao H, Rice JR (1987) Somewhat circular tensile cracks. Int J Fract 33:155–174

51. Rice JR (1985) First order variation in elastic fields due to variation in location of a planar crack front. J Appl Mech 52:571–579

52. Qin QH (2001) Fracture mechanics of piezoelectric materials. WIT Press, Southampton, Boston

53. Kuna M (2006) Finite element analyses of cracks in piezoelectric structures: a survey. Arch Appl Mech 76:725–745

54. Sähn S, Göldner H (1993) Bruch- und Beurteilungskriterien in der Festigkeitslehre. Fachbuchverlag, Leipzig-Köln

55. Irwin GR (1956) Onset of fast crack propagation in high strength steel and aluminum alloys. In: Sagamore research conference proceedings, vol. 2, pp 289–305

56. BS 5447 (1974) Methods of testing for plane-strain fracture toughness (K_{Ic}) of metallic materials. Technical Report, British Standards

57. Dugdale D (1960) Yielding of steel sheets containing slits. J Mech Phys Solid 8:100–104

58. Wells AA (1961) Unstable crack propagation in metals: cleavage and fast fracture. In: Proceedings of the crack propagation symposium, vol. 1, Paper 84, Cranfield, UK

59. Burdekin FM, Stone DEW (1966) The crack opening displacement approach to fracture mechanics in yielding materials. J Strain Anal 1:145–153

60. Schwalbe KH (1980) Bruchmechanik metallischer Werkstoffe. Carl Hanser Verlag, München

61. ASTM-E 1290–93 (1993) Fracture toughness measurement crack-tip opening displacement (CTOD). Technical Report, American Society for Testing and Materials, Philadelphia

62. BS 5762 (1979) Methods for crack opening displacements (COD) testing. Technical Report, British Standards

63. Harrison RPEA (1976) Assessments of the integrity of structures containing defects. CEGB-Report R/H.R6

64. Milne I, Ainsworth RA, Dowling AR, Stewart AT (1991) Assessments of the integrity of structures containing defects. British Energy-Report, R6-Revision 3

65. Zerbst U, Wiesner C, Kocak M, Hodulak L (1999) Sintap: Entwurf einer vereinheitlichten europäischen fehlerbewertungsprozedur - eine einführung. Technical Report, GKSS-Forschungszentrum, Geesthacht

66. Kocak M, Ainsworth RA, Dowling AR, Stewart AT (2008) FITNET Fitness-for-Service, Fracture-Fatigue-Creep-Corrosion, vol. I+II. GKSS Research Centre
67. Gross D, Seelig T (2001) Bruchmechanik: Mit einer Einführung in die Mikromechanik. Springer Verlag, Berlin
68. Shih CF (1973) Elastic-plastic analysis of combined mode crack problems. Ph.D. thesis, Harvard University Cambridge, Massachusetts
69. Hutchinson JW (1968) Plastic-stress and strain fields at a crack tip. J Mech Phys Solid 16:337–347
70. Hutchinson JW (1968) Singular behavior at the end of a tensile crack tip in a hardening material. J Mech Phys Solid 16:13–31
71. Rice JR, Rosengren GF (1968) Plain strain deformation near a crack tip in a power-law hardening material. J Mech Phys Solid 16:1–12
72. Uhlmann W, Knésl Z, Kuna M, Bilek Z (1976) Approximate representation of elastic-plastic small scale yielding solution for crack problems. Int J Fract 12:507–509
73. Larsson SG, Carlsson AJ (1973) Influence of non-singular stress terms and specimen geometry on small-scale yielding at track tips in elastic-plastic materials. J Mech Phys Solid 21:263–277
74. Issler L, Ruoß H, Häfele P (2003) Festigkeitslehre - Grundlagen. Springer, Berlin
75. McMeeking RM, Parks DM (1979) On criteria for J-dominance of crack-tip fields in large-scale yielding. In: Landes JD, Begley JA, Clarke GA (eds) Elastic-plastic fracture, vol. ASTM STP 668, ASTM, pp. 175–194
76. Yuan H (2002) Numerical assessments of cracks in elastic-plastic materials. Springer, Berlin
77. Fett T (1998) A compendium of T-stress solutions. Technical Report FZKA 6057, Forschungszentrum Karlsruhe, Technik und Umwelt
78. Sherry AH, France CC, Goldthorpe MR (1995) Compendium of T-stress solutions for two and three dimensional cracked geometries. Fatigue Fract Eng Mater Struct 18:141–155
79. Rice JR (1974) Limitations to the small-scale yielding approximation for crack tip plasticity. J Mech Phys Solid 22:17–26
80. Betegón C, Hancock JW (1991) Two-parameter characterization of elastic-plastic crack-tip fields. J Appl Mech 58:104–110
81. Sharma SM, Aravas N (1991) Determination of higher-order terms in asymptotic crack tip solutions. J Mech Phys Solid 39(8):1043–1072
82. Yang S, Chao YJ, Sutton MA (1993) Higher order asymptotic fields in a power-law hardening material. Eng Fract Mech 45:1–20
83. Nikishkov GP (1993) Three-term elastic-plastic asymptotic expansion for the description of the near-tip stress field. Technical Report, University of Karlsruhe, Institute for Reliability and Failure Analysis
84. O'Dowd NP, Shih CF (1991) Family of crack-tip fields characterized by a triaxiality parameter: I structure of fields. J Mech Phys Solid 39:989–1015
85. O'Dowd NP, Shih CF (1992) Family of crack-tip fields characterized by a triaxiality parameter: II fracture applications. J Mech Phys Solid 40:939–963
86. Dodds RHJ, Tang M, Anderson TL (1994) Effects of prior ductile tearing on cleavage fracture toughness in the transition region. In: Kirk M, Bakker A (eds) Constraint effects in fracture—theory and applications, vol. ASTM STP 1244. ASTM
87. Dodds RH Jr, Shih CF, Anderson TL (1993) Continuum and micromechanics treatment of constraint in fracture. Int J Fract 64(2):101–133
88. Kirk MT, Koppenhoefer KC, Shih CF (1993) Effect of constraint on specimen dimensions needed to obtain structurally relevant toughness measures. Constraint effects in fracture, ASTM STP 1171, American Society for Testing and Materials, pp 79–103
89. Brocks W, Schmitt W (1994) The second parameter in J-R curves: constraint or triaxiality? Second symposium on constraint effects, ASTM STP 1244. In: Kirk MT, Bakker A (eds), American society for testing and materials
90. Kumar V, German MD, Shih CF (1981) An engineering approach for elastic-plastic fracture analysis. EPRI-Report NP-1931

91. Begley JA, Landes JD (1972) The J-integral as a fracture criterion. ASTM STP 514, American Society of Testing and Materials, pp 1–20
92. Rice JR, Paris PC, Merkle JG (1973) Some further results of J-integral analysis and estimates. ASTM STP 536, American society of testing and materials, pp 231–245
93. Hutchinson JW, Paris PC (1979) Stability analysis of J-controlled crack growth. In: Landes J, Begley J, Clarke G (eds) Elastic-plastic fracture, vol. ASTM STP 668, ASTM, pp 37–64
94. Shih CF, deLorenzi HG, Andrews WR (1979) Studies on crack initiation and stable crack growth. Elastic-Plastic Fracture, pp 65–120
95. Kussmaul K, Roos E, Föhl J (1997) Forschungsvorhaben Komponentensicherheit (FKS). Ein wesentlicher Beitrag zur Komponentensicherheit. In: Kussmaul K (ed) 23. MPA Seminar Sicherheit und Verfügbarkeit in der Anlagentechnik, vol. 1, MPA, Stuttgart, pp 1–20
96. Kordisch H, Sommer E, Schmitt W (1989) The influence of triaxiality on stable crack growth. Nucl Eng Des 112:27–35
97. Slepyan LI (1974) Growing crack during plane deformation of an elastic-plastic body. Mekh. Tverdogo Tela 9:57–67
98. Drugan WJ, Rice JR, Sham TL (1982) Asymptotic analysis of growing plane strain tensile cracks in elastic-ideally plastic solids. J Mech Phys Solid 30:447–473
99. Castañeda PP (1987) Asymptotic fields in steady crack growth with linear strain-hardening. J Mech Phys Solid 35:227–268
100. Turner CE (1990) A re-assessment of ductile tearing resistance. In: Firrao D (ed) Fracture behaviour and design of materials and structures, vol. 2, EMAS, Warley, pp. 933–949, 951–968
101. Memhard D, Brocks W, Fricke S (1993) Characterization of ductile tearing resistance by energy dissipation rate. Fatigue Fract Eng Mater Struct 16:1109–1124
102. Cotterell B, Atkins AG (1996) A review of the J and I integrals and their implications for crack growth resistance and toughness in ductile fracture. Int J Fract 81:357–372
103. Belytschko TKLW, Moran B (2001) Nonlinear finite elements for continua and structures. Wiley, New York
104. Moran B, Shih CF (1987) Crack tip and associated domain integrals from momentum and energy balance. Eng Fract Mech 27:615–642
105. Nguyen QS (1991) An energetic analysis of elastic-plastic fracture. In: Blauel JG, Schwalbe KH (eds) Defect assessment in components—fundamentals and applications. Mechanical Engineering Publications, London, pp 75–85
106. Yuan H, Brocks W (1991) On the J-integral concept for elastic-plastic crack extension. Nucl Eng Des 131:157–173
107. Paris P, Erdogan F (1963) A critical analysis of crack propagation laws. J Basic Eng 85:528–534
108. Erdogan F, Ratwani M (1970) Fatigue and fracture of cylindrical shells containing a circumferential crack. Int J Fract Mech 6:379–392
109. ESA (2000) ESACRACK User's manual. European Space Research and Technology Centre (ESTEC)
110. Newman JC (1981) A crack-closure model for predicting fatigue crack growth under aircraft spectrum loading. In: Chang JB, Hudson CM (eds) Methods and models for predicting fatigue crack growth under random loading. ASTM STP 748, pp 53–84
111. Rice JR (1967) Mechanics of crack tip deformation and extension by fatigue. In: Fatigue crack propagation, vol. ASTM STP 415, American Society for Testing and Materials, London, pp 247–309
112. Wheeler OE (1972) Spectrum loading and crack growth. J Basic Eng 94:181–186
113. Elber W (1970) Fatigue crack closure under cyclic tension. Eng Fract Mech 2:37–45
114. Suresh S, Ritchie RO (1984) Propagation of short fatigue cracks. Int Metall Rev 29:445–476
115. Westermann-Friedrich A, Zenner (Verfasser) H (1999) Zählverfahren zur Bildung von Kollektiven aus Zeitfunktionen - Vergleich der verschiedenen Verfahren und Beispiele. Technical Report, Forschungsvereinigung Antriebstechnik. FVA-Merkblatt Nr. 0/14

116. Dominguez J (1994) Fatigue crack growth under variable amplitude loading. In: Carpinteri A (ed) Handbook of fatigue crack propagation in metallic structures, Elsevier Science, pp 955–997
117. NASA (2000) Fatigue crack growth computer program "NASGRO" version 3.0. JSC-22267B, Johnson Space Center, Texas
118. de Koning AU (1981) A simple crack closure model for prediction of fatigue crack growth rates under variable-amplitude loading. In: Fracture mechanics, vol. ASTM STP 743, American Society for Testing and Materials, pp 63–85
119. Padmadinata UH (1990) Investigation of crack-closure prediction models for fatigue in aluminum sheet under flight-simulation loading. Ph.D. Thesis, Delft University of Technology
120. Führing H, Seeger T (1979) Dugdale crack closure analysis of fatigue cracks under constant amplitude loading. Eng Fract Mech 11:99–122
121. Schijve J (2001) Fatigue of structures and materials. Kluwer Academic Publisher, Dordrecht
122. Schijve J (1979) Four lectures on fatigue crack growth. Eng Fract Mech 11:176–221
123. Miller KJ, de los Rios ER (1992) Short fatigue cracks, vol. ESIS 13. Mechanical Engineering Publications, London
124. Ravichandran KS, Ritchie RO, Murakami Y (1999) Small fatigue cracks. mechanics, mechanisms and applications. Elsevier, Amsterdam
125. Richard HA (1985) Bruchvorhersagen bei überlagerter Normal- und Schubbeanspruchung von Rissen. VDI-Forschungsheft 631
126. Erdogan F, Sih GE (1963) On the crack extension in plates under plane loading and transverse shear. J Basic Eng 85:519–527
127. Hussain MA, Pu SL, Underwood J (1974) Strain energy release rate for a crack under combined Mode I and Mode II. ASTM STP 560:2–28
128. Ichikawa M, Tanaka S (1982) A critical analysis of the relationship between the energy release rate and the stress intensity factors for non-coplanar crack extension under combined mode loading. Int J Fract 18:19–28
129. Lo KK (1978) Analysis of branched cracks. J Appl Mech 45:797–802
130. Nuismer RJ (1975) An energy release rate criterion for mixed mode fracture. Int J Fract 11:245–250
131. Sih GC (1974) Strain energy density factor applied to mixed mode crack problems. Int J Fract 10:305–321
132. Radaj D, Heib M (1978) Energy density fracture criteria for cracks under mixed mode loading. Materialprüfung 20:256–62
133. Richard HA, Fulland M, Sander M (2005) Theoretical crack path prediction. Fatigue Fract Eng Mater Struct 28:3–12
134. Pook LP (2002) Crack paths. WIT Press, Boston
135. Schöllmann M, Richard HA, Kullmer G, Fulland M (2002) A new criterion for the prediction of crack development in multiaxially loaded structures. Int J Fract 117:129–141
136. Richard HA, Kuna M (1990) Theoretical and experimental study of superimposed fracture modes I. II and III. Eng Fract Mech 35(6):949–960
137. Chen CS, Wawrzynek PA, Ingraffea AR (1997) Methodology for fatigue crack growth and residual strength prediction with applications to aircraft fuselages. Comput Mech 19:527–532
138. Sumi Y, Nemat-Nasser S, Keer LM (1985) On crack path instability in a finite body. Eng Fract Mech 22:759–771
139. Cotterell B, Rice JR (1980) Slightly curved or kinked cracks. Int J Fract 16:155–169
140. Freund LB (1998) Dynamic fracture mechanics. Cambridge University Press, Cambridge
141. Ravi-Chandar K (2003) Dynamic fracture. In: Milne I, Ritchie RO, Karihaloo B (eds) Comprehensive structural integrity—fundamental theories and mechanisms of failure, vol 2. Elsevier, Oxford, pp 285–361
142. Miannay DP (2001) Time-dependent fracture mechanics. Springer, New York

Chapter 4
Finite Element Method

The *finite element method* (FEM) is currently one of the most efficient and universal methods of numerical calculation for solving partial differential equations from engineering and scientific fields. The basic mathematical concepts are based on the work of Ritz, Galerkin, Trefftz and others at the beginning of the twentieth century. With the advance of modern computer science in the 1960s, these approaches of numerical solution could be successfully implemented with FEM. This development was motivated to an enormous extent by tasks of structural analyses in aviation, construction and mechanical engineering. The formulation of the finite element method in its current standard was developed thanks to the pioneer work of (among others) Argyris, Zienkiewicz, Turner, and Wilson. Therein, the system of differential equations is converted into an equivalent variational problem (weak formulation), mostly utilizing mechanical principles or weighted residual methods. To solve a boundary value problem (BVP), trial functions are created for limited subdomains—the so-called » finite elements « –, the free variables of which are finally determined numerically by solving an algebraic system of equations.

In this chapter at first a few principles of continuum mechanics will be introduced in order to clarify the basic theoretical foundations of FEM. This will be followed by a concise presentation of discretization techniques and the numerical realization of FEM. These explanations are required to understand the following chapters, in which the application of FEM in fracture mechanics will then be treated in detail. There is also a large selection of related textbooks on FEM, which are recommended to supplement the reader's knowledge of the subject.

4.1 Spatial and Temporal Discretization of Boundary Value Problems

Section A.5 provides a compilation of the basic relations of solid mechanics necessary for formulating an (initial) boundary value problem (IBVP) (cf. Fig. A.16 and equation set (A.139)). The finite element method is a technique used to approximate

M. Kuna, *Finite Elements in Fracture Mechanics*, Solid Mechanics and Its
Applications 201, DOI: 10.1007/978-94-007-6680-8_4,
© Springer Science+Business Media Dordrecht 2013

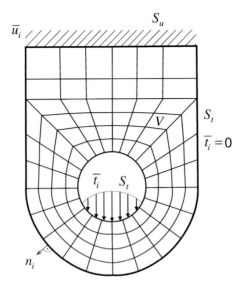

$S = S_u \cup S_t$ $\bar{t}_i$ – given boundary tractions on S_t
$\bar{u}_i$ – given boundary displacement on S_u

Fig. 4.1 Finite element discretization: a tension strap

the solution of an IBVP. It is based on the numerical implementation of energy principles from mechanics, which are for this purpose discretized in space and time.

To this end, the domain V of the body under consideration is divided into a number n_E of finite subdomains V_e, *finite elements*, and simplified formulations are made for these. The finite elements are numbered with the running index $e = 1, \ldots, n_E$. Figure 4.1 shows the example of a spatial discretization of the boundary value

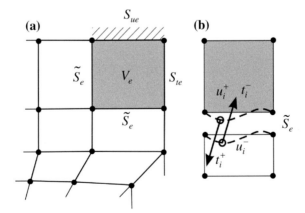

Fig. 4.2 a Detailed view of a finite element, **b** transition conditions to the neighboring element

problem of a tension strap. In the case of initial boundary value problems or non-linear problems, the transient temporal progression $[t_0 \leq t \leq t_{end}]$ is generally discretized with a sequence of time steps or load increments Δt_i.

Figure 4.2 a shows a detail of Fig. 4.1, in which the element e has been highlighted. Belonging to element e are its volume V_e and its boundary $S_e = S_{ue} \cup S_{te} \cup \tilde{S}_e$. S_{ue} and S_{te} represent the intersections of S_e with the displacement or traction boundaries S_u or S_t of the IBVP. Moreover, every element adjoins to neighboring elements on the part $\tilde{S}_e$ of its boundary. The sum of these interelement boundaries is designated as $\tilde{S}$. All external, given field quantities are represented by an overbar. A tilde indicates quantities that are only defined on the boundary.

To solve a BVP, the field quantities in the finite elements have to fulfill the following basic relations:

1. The displacement field u_i is a continuous function of location. By gradient formation, the strains ε_{ij} inside each element are derived by:

$$\varepsilon_{ij} = \frac{1}{2}(u_{i,j} + u_{j,i}) \quad \text{in } V_e \quad \text{(compatibility conditions).} \quad (4.1)$$

2. The displacement field $\bar{u}_i$ must take on prescribed values $\bar{u}_i$ on section S_{ue} of the boundary:

$$u_i = \bar{u}_i \qquad \text{on } S_{ue} \quad \text{(essential boundary conditions).} \quad (4.2)$$

3. Balance of the stress state with the body forces $\bar{b}_i$ inside the element:

$$\sigma_{ij,j} + \bar{b}_i = 0 \qquad \text{in } V_e \quad \text{(equilibrium conditions).} \quad (4.3)$$

4. Tractions on boundary section S_{te} correspond to the external given values $\bar{t}_i$:

$$\sigma_{ij} n_j = t_i = \bar{t}_i \qquad \text{on } S_{te} \quad \text{(natural boundary conditions).} \quad (4.4)$$

5. Continuity of displacements at the interelement boundaries $\tilde{S}_e$, i.e. the value must be the same if approached from both elements (symbolized by $+$ and $-$):

$$u_i^+ = u_i^- \qquad \text{on } \tilde{S}_e . \quad (4.5)$$

6. Reciprocity of the traction vectors (actio = reactio) on the interelement boundaries:

$$t_i^+ = -t_i^- \qquad \text{on } \tilde{S}_e . \quad (4.6)$$

7. The stress-strain law. In the following we will first assume linear-elastic material properties:

$$\sigma_{ij} = C_{ijkl}\varepsilon_{kl} , \quad \varepsilon_{ij} = S_{ijkl}\sigma_{kl} , \quad U = \widehat{U} = \frac{1}{2}\sigma_{ij}\varepsilon_{ij} . \quad (4.7)$$

Every displacement field that fulfills Eqs. (4.1) and (4.2) is called a *kinematically admissible displacement field* u_i^{kin}. Every stress field that obeys (4.3) and (4.4) is designated as a *statically admissible stress field* σ_{ij}^{sta}. The last two conditions for interelement boundaries are clarified in Fig. 4.2 b.

The *true* solution of the BVP must satisfy all static and kinematic conditions (4.1)–(4.6) and the material law (4.7) in the entire domain V *exactly* (see Sect. A.5.2). With FEM however, an approximation is made for the primary field quantities. This approximation needs to fulfill a priori only a part of the above listed basic equations for the element interior and continuity requirements on the boundary. The remaining, basic equations that are not satisfied from the outset, usually for the derived dual field quantities, will be obtained as Euler equations of a variational principle and are realized only approximately in terms of weighted residuals.

4.2 Energy Principles of Continuum Mechanics

The starting point of our examination will be the *generalized principle of work*. It states that the internal work $\mathcal{W}_{\text{int}}$ of a statically admissible stress field σ_{ij}^{sta} performed with a kinematically admissible displacement field $\varepsilon_{ij}^{\text{kin}}$ is equal to the work $\mathcal{W}_{\text{ext}}$ done by external stresses $\bar{t}_i^{\text{sta}}$ and $\bar{b}_i^{\text{sta}}$ with associated displacement field u_i^{kin}:

$$\mathcal{W}_{\text{int}} \mathrel{\hat{=}} \int_V \sigma_{ij}^{\text{sta}} \varepsilon_{ij}^{\text{kin}} dV = \int_{S_t} \bar{t}_i^{\text{sta}} u_i^{\text{kin}} dS + \int_V \bar{b}_i^{\text{sta}} u_i^{\text{kin}} dV + \int_{S_u} t_i^{\text{sta}} \bar{u}_i^{\text{kin}} dS \mathrel{\hat{=}} \mathcal{W}_{\text{ext}} .$$
(4.8)

We can also apply (4.8) to the true stresses $\sigma_{ij}^{\text{sta}} := \sigma_{ij}$ of the BVP and the true increments of the kinematic quantities $u_i^{\text{kin}} := du_i$ and $\varepsilon_{ij}^{\text{kin}} := d\varepsilon_{ij}$ of a non-linear analysis. Integration over all load steps from the undeformed initial state to the final state then results–for any material law–in the identity of internal work (work of deformation) and the work of external forces:

$$\mathcal{W}_{\text{int}} \mathrel{\hat{=}} \int_V \int_0^{\varepsilon_{ij}} \sigma_{ij} d\varepsilon_{ij} dV = \int_S \int_0^{u_i} \bar{t}_i du_i dS + \int_V \int_0^{u_i} \bar{b}_i du_i dV \mathrel{\hat{=}} \mathcal{W}_{\text{ext}} .$$ (4.9)

4.2.1 Variation of Displacement Field

We assume a state of stress in the body that is in equilibrium with the external system of forces, i.e. (4.3) and (4.4) are satisfied. In addition to the actual displacements u_i now a virtual displacement δu_i is carried out. This is defined as a continuous function of x that has the following properties:

(a) infinitesimal (compared to u_i),
(b) imagined, i.e. not really existing,

(c) kinematically admissible, i.e. $\delta u_i = 0$ on S_u, so that the boundary condition $u_i = \bar{u}_i$ is not violated.

These virtual displacements lead to virtual deformations:

$$\delta\varepsilon_{ij} = \frac{1}{2}\delta(u_{j,i}) = \frac{1}{2}\left[(\delta u_i)_{,j} + (\delta u_j)_{,i}\right] . \tag{4.10}$$

Now the principle of work (4.8) is applied, whereby $\sigma_{ij}^{\mathrm{sta}} := \sigma_{ij}$ corresponds to the state of equilibrium and the virtual displacements are used for $u_1^{\mathrm{kin}} := \delta u_i$ (the integral over S_u must disappear according to presumption (c)):

$$\int_V \sigma_{ij}\delta\varepsilon_{ij}\mathrm{d}V = \int_{S_t} \bar{t}_i\delta u_i\mathrm{d}S + \int_V \bar{b}_i\delta u_i\mathrm{d}V + \int_{S_u} t_i\delta u_i\mathrm{d}S$$

$$\delta\mathcal{W}_{\mathrm{int}} = \delta\mathcal{W}_{\mathrm{ext}} . \tag{4.11}$$

Principle of virtual displacements:

A deformable body is exactly in the state of equilibrium if the work of applied external forces is equal to the deformation work of the internal forces with an arbitrary kinematically admissible virtual displacement field δu_i, or the total virtual work is equal to zero. This is true for every material law.

$$\delta\mathcal{W} = \delta\mathcal{W}_{\mathrm{int}} - \delta\mathcal{W}_{\mathrm{ext}} = 0 \tag{4.12}$$

Equation (4.11) represents the integral form of equilibrium conditions (4.3) and (4.4). It is also called the *weak formulation* because the order of the differentiation of σ_{ij} is one less than in the differential Eq. (4.3). The function δu_i in (4.11) can be mathematically regarded as a *test function* or *weighting function*, which corresponds to the method of weighted residuals (Galerkin).

For the principle of virtual displacements, only the current state of stress is important, even if it is a functional of the entire deformation history as with non-linear material behavior. For this reason, the principle can also be applied to every load step Δt of a non-linear analysis in order to fulfill equilibrium conditions. For this purpose, we insert in (4.11) instead of the total kinematic quantities their increments $\Delta u_i = \dot{u}_i\Delta t$ or $\Delta\varepsilon_{ij} = \dot{\varepsilon}_{ij}\Delta t$:

$$\int_V \sigma_{ij}\delta\Delta\varepsilon_{ij}\,\mathrm{d}V = \int_{S_t} \bar{t}_i\delta\Delta u_i\,\mathrm{d}S + \int_V \bar{b}_i\delta\Delta u_i\,\mathrm{d}V . \tag{4.13}$$

Dividing by Δt, we thus obtain the relation in rate form, which is also called the *principle of virtual velocities*:

$$\int_V \sigma_{ij} \delta \dot{\varepsilon}_{ij} \, dV = \int_{S_t} \bar{t}_i \delta \dot{u}_i \, dS + \int_V \bar{b}_i \delta \dot{u}_i \, dV \,. \qquad (4.14)$$

This principle can also be applied to large deformations (see Sect. A.2). In the current configuration, the internal work is calculated with the Cauchy stress tensor σ_{ij} according to (A.42) and the Euler-Almansi strain tensor η_{ij} (A.18), which has a non-linear relation with the displacements (A.20). The integration extends over the current volume v and the surface a

$$\delta W_{\text{int}} \mathrel{\widehat{=}} \int_v \sigma_{ij} \delta \eta_{ij} \, dv = \int_v \rho \bar{f}_i \delta u_i \, dv + \int_a \bar{t}_i \delta u_i \, da \mathrel{\widehat{=}} \delta W_{\text{ext}} \,. \qquad (4.15)$$

The representation in the reference configuration utilizes the 2nd Piola-Kirchhoff stress tensor T_{IJ} (A.51) and the Green-Lagrange strain tensor E_{IJ} (A.19) with the kinematic relation (A.20). Correspondingly, the boundary traction vector $\hat{\bar{T}}$ according to (A.50) and the volumes V and areas A of integration must be related to the reference state

$$\delta W_{\text{int}} \mathrel{\widehat{=}} \int_V T_{IJ} \delta E_{IJ} \, dV = \int_V \rho_0 \bar{f}_i \delta u_i \, dV + \int_A \hat{\bar{T}}_i \delta u_i \, dA \mathrel{\widehat{=}} \delta W_{\text{ext}} \,. \qquad (4.16)$$

The volume forces $\bar{b}_i$ are obtained from the force vector per mass $\bar{f}_i$ such that $\rho_0 \bar{f}_i \, dV = \rho \bar{f}_i \, dv$, i. e. by accounting for the different densities ρ_0 and ρ in both configurations.

The principle of virtual displacements can be converted into a variational formulation under two conditions. Firstly, hyperelastic material behavior is required so that the work of deformation constitutes an internal potential strain energy (see Sect. A.4.1).

$$\Pi_{\text{int}} = \int_V U \, dV = W_{\text{int}} \qquad \sigma_{ij} = \frac{\partial U}{\partial \varepsilon_{ij}} \qquad \left(U(\varepsilon_{ij}) = \frac{1}{2} \sigma_{ij} \varepsilon_{ij} \quad \text{linear-elastic} \right) \qquad (4.17)$$

Secondly, the external forces must be conservative (independent of path), i. e. be derivable from a potential Π_{ext} (gravity, elastic spring force, …).

$$\Pi_{\text{ext}} = -W_{\text{ext}} = - \int_V \bar{b}_i u_i \, dV - \int_{S_t} \bar{t}_i u_i \, dS \qquad (4.18)$$

From (4.12) thus follows $\delta W_{\text{int}} - \delta W_{\text{ext}} = \delta \Pi_{\text{int}} + \delta \Pi_{\text{ext}} = \delta \Pi_P = 0$ the stationarity of the total potential

$$\Pi_P(u_i) = \Pi_{\text{int}} + \Pi_{\text{ext}} = \int_V U(\varepsilon_{ij}) \, dV - \int_{S_t} \bar{t}_i u_i \, dS - \int_V \bar{b}_i u_i \, dV \,. \qquad (4.19)$$

The variation of $\delta \Pi_P$ with respect to δu_i that must be kinematically admissible, results in:

$$\delta \Pi_P = \int_V \frac{\partial U}{\partial \varepsilon_{ij}} \delta \varepsilon_{ij} \, \mathrm{d}V - \int_{S_t} \bar{t}_i \delta u_i \, \mathrm{d}S - \int_V \bar{b}_i \delta u_i \, \mathrm{d}V = 0 . \tag{4.20}$$

With (4.17) and $\sigma_{ij} \delta \varepsilon_{ij} = \sigma_{ij} \delta u_{i,j} = (\sigma_{ij} \delta u_i)_{,j} - \sigma_{ij,j} \delta u_i$ we obtain

$$\delta \Pi_P = \int_V (\sigma_{ij} \delta u_i)_{,j} \, \mathrm{d}V - \int_V \left[\sigma_{ij,j} + \bar{b}_i \right] \delta u_i \, \mathrm{d}V - \int_{S_t} \bar{t}_i \delta u_i \, \mathrm{d}S , \tag{4.21}$$

and the application of Gauss's theorem and the Cauchy's formula lead to

$$\delta \Pi_P = \int_{S_t} \left[\sigma_{ij} n_j - \bar{t}_i \right] \delta u_i \, \mathrm{d}S - \int_V \left[\sigma_{ij,j} + \bar{b}_i \right] \delta u_i \, \mathrm{d}V = 0 . \tag{4.22}$$

Due to the fundamental law of variational calculus (δu_i is an arbitrary test function) the expressions in brackets must disappear. We obtain exactly (4.3) and (4.4) as Euler equations, i.e. the differential equation and the boundary conditions for the stress tensor σ_{ij}.

Principle of minimum potential energy:
  Of all kinematically admissible displacement fields, the true displacements, which also correspond to the state of equilibrium, render the potential energy a minimum value. Therefore, the true solution of the BVP can be found by choosing an admissible displacement approach with free parameters using the variational principle of minimum potential energy (Ritz method, Galerkin's method of weighted residuals → FEM)

4.2.2 Variation of Forces

The *principle of virtual forces* can also be derived from the general principle of work (4.8). However, we now assume a kinematically admissible state of displacement $u_i(\boldsymbol{x})$ and strain $\varepsilon_{ij}(\boldsymbol{x})$ in the body, which satisfies (4.1) and (4.2) and should correspond to the quantities u_i^{kin} and $\varepsilon_{ij}^{\text{kin}}$ in (4.8). In addition to the actual stresses, now a virtual change of internal and external forces is applied, which are identified with the stresses $\sigma_{ij}^{\text{sta}} := \delta \sigma_{ij}(\boldsymbol{x})$, boundary stresses $\bar{t}_i^{\text{sta}} := \delta t_i$ and body forces $\bar{b}_i^{\text{sta}} := \delta b_i$ of the principle of work. These *virtual forces* form a system of equilibrium with the properties:

(a) infinitely small,
(b) imagined, i.e. not really existing,

(c) statically admissible, i.e. $\delta\sigma_{ij,j} + \delta\bar{b}_i = 0$ and $\delta\bar{t}_i = 0$ on S_t.

Inserting these field quantities into the principle of work (4.8) results in

$$\int_V \delta\sigma_{ij}\varepsilon_{ij}\,\mathrm{d}V = \int_{S_t} \delta\bar{t}_i u_i\,\mathrm{d}S + \int_{S_u} \delta t_i u_i\,\mathrm{d}S + \int_V \delta b_i u_i\,\mathrm{d}V$$
$$\delta\widehat{\mathcal{W}}_{\text{int}} = \delta\widehat{\mathcal{W}}_{\text{ext}}. \tag{4.23}$$

The left side is called *complementary internal work* $\widehat{\mathcal{W}}_{\text{int}}$ and $\widehat{\mathcal{W}}_{\text{ext}}$ designates *complementary external work*. Expressed verbally, the principle of virtual forces means:

> The displacements and strains of a deformable body are kinematically compatible with each other and with the boundary conditions only if, for every arbitrary statically admissible system of virtual forces and stresses, the virtual complementary internal work and external work are equal, or the total virtual complementary work is equal to zero. This principle is valid for all material laws.
>
> $$\delta\widehat{\mathcal{W}} = \delta\widehat{\mathcal{W}}_{\text{int}} - \delta\widehat{\mathcal{W}}_{\text{ext}} = 0. \tag{4.24}$$

The *principle of minimum complementary energy* can analogously be derived from the principle of virtual forces like the derivation of the minimum potential energy in Sect. 4.2.1. Here too, hyperelastic material behavior and a conservative external load system have to be presupposed. Then the complementary work of deformation

$$\widehat{\mathcal{W}}_{\text{int}} = \int_V \widehat{U}(\sigma_{ij})\,\mathrm{d}V, \quad \widehat{U}(\sigma_{ij}) = \int_0^{\sigma_{ij}} \varepsilon_{kl}\,\mathrm{d}\sigma_{kl} \tag{4.25}$$

represents an internal complementary potential $\widehat{\Pi}_{\text{int}}$ and the external complementary work $\widehat{\mathcal{W}}_{\text{ext}}$ is performed by a corresponding potential-$\widehat{\Pi}_{\text{ext}}$. Both quantities together make up the total complementary potential

$$\Pi_{\text{C}}(\sigma_{ij}) = \widehat{\Pi}_{\text{int}} + \widehat{\Pi}_{\text{ext}} = \int_V \widehat{U}(\sigma_{ij})\,\mathrm{d}V - \int_{S_u} u_i t_i\,\mathrm{d}S - \int_V u_i b_i\,\mathrm{d}V, \tag{4.26}$$

which should be conceived as a function of the statically admissible stress field $\sigma_{ij}(\boldsymbol{x})$. The variational principle $\delta\Pi_{\text{C}} = \delta\widehat{\Pi}_{\text{int}} + \delta\widehat{\Pi}_{\text{ext}} = 0$ then leads precisely to the relation (4.23) and the kinematic relations (4.1) are thus obtained as the Euler equations.

> Of all forces and stresses that satisfy equilibrium conditions, those that also correspond to the true kinematically admissible deformation state reduce the complementary potential energy to a minimum.

4.2.3 Mixed and Hybrid Variational Principles

Fundamentals

The principle of minimum potential energy, in which we assume kinematically admissible displacement functions, is the foundation of the most prevalent FEM variant – the *displacement method*. If the principle of complementary energy is used on the other hand, one works with the variation of equilibrated stress functions, which leads to the *force method* of FEM. If we try to satisfy all basic equations inside the body (compatibility conditions, stress equilibrium) and all boundary conditions exclusively via the variational principle, we then obtain *generalized energy principles*. The additional requirements are then introduced into the functional as auxiliary conditions with the help of *Lagrange multipliers*. If for example, in the case of the variational principle of complementary energy (4.26), we would like to allow for an arbitrary function for the stress state, the equilibrium conditions (4.3) in V and (4.4) on S_t must be satisfied by the requirements

$$\int_V (\sigma_{ij,j} - \bar{b}_i)\lambda_i \, dV = 0 \qquad \int_{S_t} (\sigma_{ij} n_j - \bar{t}_i)\lambda_i \, dS = 0. \qquad (4.27)$$

The Lagrange multiplier thus turns out to be a variation of the displacement field $u_i(x)$, which should be regarded as independent. We arrive in this way at the Hellinger-Reissner principle $\Pi_R(u_i, \sigma_{ij})$–a two-field functional which is the foundation of the *mixed finite element formulations* .

Hybrid element formulations differ from ordinary displacement, stress or mixed element functions in that we dispense a priori with satisfying the continuity requirement at the element boundaries for the displacements (4.5) or tractions (4.6), i.e. a reduced compatibility of elements is permitted. This has the advantage of higher flexibility in element formation, since we can now select different displacement and stress functions that do *not conform* with the neighboring element. In the hybrid model, conditions of interelement continuity are enforced by introducing separate displacements $\tilde{u}_i$ and tractions $\tilde{t}_i$ on the interelement boundaries $\tilde{S}_e$, the selection of which is *independent* of the internal shape functions. The boundary displacements $\tilde{u}_i$ are equally valid for two adjacent elements and are most advantageously defined by nodal variables. In contrast, the boundary stresses $\tilde{t}_i$ are applied separately in each element. The continuity requirement in the transition from one element to the other is incorporated into the variational principle as an additional auxilary condition, whereby the dual field quantity on the boundary usually serves as the Lagrange parameter. Only through extending the variational principle in this way to a hybrid functional, the continuity of the field quantities can be achieved approximately in integral form.

The book by Washizu [1] provides a comprehensive view of all conventional variational principles of solid mechanics and their hybrid modifications, also referring to the various finite element formulations. The most essential principles and their correlations are compiled in a schematic overview in Fig. 4.3. The two classical minimum

Conventional variational principles	Modified hybrid variational principles
Principle of minimum potential energy	Modified principle of potential energy
$\Pi_P(u_i)$ $\Rightarrow$	$\Pi_{PH}(u_i, \tilde{u}_i, \tilde{t}_i)$
COMPATIBLE DISPLACEMENT ELEMENTS	HYBRID DISPLACEMENT ELEMENTS
Hellinger-Reissner principle	Modified principle of Hellinger-Reissner
$\Pi_R(u_i, \sigma_{ij})$ $\Rightarrow$	$\Pi_{GH}(u_i, \sigma_{ij}, \tilde{u}_i)$
MIXED ELEMENTS	MIXED HYBRID ELEMENTS
Principle of minimum complementary energy	Modified principle of complementary energy
$\Pi_C(\sigma_{ij})$ $\Rightarrow$	$\Pi_{CH}(\sigma_{ij}, \tilde{u}_i)$
COMPATIBLE STRESS ELEMENTS	HYBRID STRESS ELEMENTS

Fig. 4.3 Diagram of conventional and hybrid variational principles of elasticity theory

principles of potential and complementary energy can be seen as counterpoles. Since hybrid element formulations are particularly effective in the development of special crack tip elements, the three most important hybrid variational principles will be described in detail below.

The Hybrid Stress Model

The *hybrid stress model* is based on the principle of minimum complementary energy Π_C (4.26), which has been modified by an additional displacement function $\tilde{u}_i$ on the entire element boundary S_e (Pian and Tong [2]):

$$\Pi_{CH}(\sigma_{ij}, \tilde{u}_i) = \sum_{e=1}^{n_E} \left\{ \int_{V_e} \left[\frac{1}{2} \sigma_{ij} S_{ijkl} \sigma_{kl} \right] dV - \int_{S_e} t_i \tilde{u}_i \, dS + \int_{S_{te}} \bar{t}_i \tilde{u}_i \, dS \right\}.$$

(4.28)

The only a priori prerequisite for the stress ansatz is to

- satisfy the equilibrium conditions (4.3) in V_e.

But, as opposed to conventional stress elements, no stress equilibrium (4.4), neither on the traction boundary S_{te} nor on the interelement boundaries $\tilde{S}_e$ (4.6) is needed. The boundary displacements are required (4.2) to

- assume given boundary values $\tilde{u}_i = \bar{u}_i$ on S_{ue}.

The functional Π_{CH} is varied with respect to both field quantities $\delta\sigma_{ij}$ (and associated $\delta t_i = \delta\sigma_{ij}n_j$) and $\delta\tilde{u}_i$. It has to take on a stationary value $\delta\Pi_{CH} = 0$.

$$\delta\Pi_{CH} = \sum_{e=1}^{n_E} \left\{ \int_{V_e} \delta\sigma_{ij}S_{ijkl}\sigma_{kl}\,\mathrm{d}V - \int_{S_e} \delta t_i \tilde{u}_i\,\mathrm{d}S \right.$$
$$\left. - \int_{S_e} t_i\delta\tilde{u}_i\,\mathrm{d}S + \int_{S_{te}} \bar{t}_i\delta\tilde{u}_i\,\mathrm{d}S \pm \int_{S_e} \delta t_i u_i\,\mathrm{d}S \right\} = 0. \qquad (4.29)$$

The zero complement in the final term is on the one hand subtracted from the first term and converted by the divergence theorem into an integral over V_e, and on the other hand it is added to the second term:

$$\delta\Pi_{CH} = \sum_{e=1}^{n_E} \left\{ \int_{V_e} \delta\sigma_{ij}\left[S_{ijkl}\sigma_{kl} - u_{i,j}\right]\,\mathrm{d}V \right.$$
$$\left. - \int_{S_e} \delta t_i\left[\tilde{u}_i - u_i\right]\,\mathrm{d}S - \int_{\tilde{S}_e} t_i\delta\tilde{u}_i\,\mathrm{d}S + \int_{S_{te}} \left[\bar{t}_i - t_i\right]\delta\tilde{u}_i\,\mathrm{d}S \right\} = 0. $$
$$(4.30)$$

Setting the expressions in square brackets equal to zero provides the missing basic relations:

- compatibility (4.1) of the strains in V_e resulting from σ_{ij}:
  $S_{ijkl}\,\sigma_{kl} = \varepsilon_{ij} = \frac{1}{2}\left(u_{i,j} + u_{j,i}\right)$
- satisfaction of the stress boundary conditions (4.4) on S_{te}
- continuity of displacements $u_i^+ = \tilde{u}_i = u_i^-$ on the interelement boundaries (4.5)
- stress reciprocity (4.6) on the interelement boundaries $\tilde{S}_e$. When summing the third term, every interelement boundary appears twice (from both adjacent elements), which ensures the condition $t_i^+ = -t_i^-$

The Hybrid Displacement Model

In contrast to conventional displacement elements, the *hybrid displacement model* is based on displacement functions that only need to be continuous *in* the element, but not on the intersection to neighboring elements. The hybrid variational principle for displacement elements is created by extending the principle of minimum potential energy $\Pi_P(u_i)$ with independent functions for the displacements $\tilde{u}_i$ and tractions $\tilde{t}_i$ on the boundary:

$$\Pi_{\mathrm{PH}}(u_i,\, \tilde{u}_i,\, \tilde{t}_i) = \sum_{e=1}^{n_{\mathrm{E}}} \left\{ \int_{V_e} \left[\frac{1}{2}\, \varepsilon_{ij}\, C_{ijkl}\, \varepsilon_{kl} - \bar{b}_i\, u_i \right] \mathrm{d}V - \int_{S_{te}} \bar{t}_i\, u_i\, \mathrm{d}S \right.$$

$$\left. - \int_{S_{ue}} t_i\, (u_i - \bar{u}_i)\, \mathrm{d}S - \int_{\tilde{S}_e} \tilde{t}_i\, (u_i - \tilde{u}_i)\, \mathrm{d}S \right\}. \qquad (4.31)$$

We need only to assume a priori the

- compatibility conditions (4.1) in V_e.

When the functional is varied with respect to three variables δu_i, $\delta \tilde{u}_i$ and $\delta \tilde{t}_i$, the condition of stationarity $\delta \Pi_{\mathrm{PH}} = 0$ yields the following Euler equations:

- equilibrium conditions (4.3) in V_e
- satisfaction of displacement boundary conditions (4.2) on S_{ue}
- satisfaction of traction boundary conditions (4.4) on S_{te}
- displacement compatibility (4.5) between the elements on $\tilde{S}_e$
- stress reciprocity (4.6) on the interelement boundaries $\tilde{S}_e$

Different variants of hybrid displacement models can be found in the literature, often without separate boundary displacements $\tilde{u}_i$. Their explicit inclusion has the essential advantage that, by a suitable choice of $\tilde{u}_i$, special elements can be designed whose boundary displacements are fully compatible with those of conventional isoparametric element types (see [3, 4]).

The Simplified Mixed Hybrid Model

In the development of hybrid special elements in fracture mechanics, we can often utilize closed-form solutions existing from elasticity theory for the crack or crack tip near field, which are then set up inside the element. Such functions fulfill a priori *both* basic Eqs. (4.1) and (4.3) *in* the domain. With the help of Hooke's law, the compatibility conditions (4.1) and the Gauss theorem, the volume integral in the hybrid stress model (4.28) can be transformed into a boundary integral over S_e:

$$\int_{V_e} \frac{1}{2}\sigma_{ij} S_{ijkl} \sigma_{kl}\, \mathrm{d}V = \int_{V_e} \frac{1}{2}\sigma_{ij}\varepsilon_{ij}\, \mathrm{d}V = \int_{V_e} \frac{1}{2}\sigma_{ij}u_{i,j}\, \mathrm{d}V = \int_{S_e} \frac{1}{2}t_i u_i\, \mathrm{d}S. \qquad (4.32)$$

We thereby obtain the functional $\Pi_{\mathrm{MH}*}$ for the *simplified mixed hybrid model* first introduced by Tong et al. [5]:

$$\Pi_{\mathrm{MH}*}(u_i, \tilde{u}_i) = \sum_{e=1}^{n_E} \left\{ \int_{S_e} t_i \, \tilde{u}_i \, \mathrm{d}S - \frac{1}{2} \int_{S_e} t_i \, u_i \, \mathrm{d}S - \int_{S_{te}} \bar{t}_i \, \tilde{u}_i \, \mathrm{d}S \right\} . \quad (4.33)$$

We must now merely presuppose the:

- displacement boundary conditions by selecting $\tilde{u}_i = \bar{u}_i$ (4.2) on S_{ue}

The following relations are satisfied by the variational principle:

- compatibility of internal displacements u_i with the boundary functions $\tilde{u}_i$, i.e. (4.5) on S_e
- traction boundary conditions (4.4) on S_{te}
- reciprocity of tractions (4.6) on $\tilde{S}_e$.

Thus, this simplified variant also ensures the compatibility of complete analytical solutions with boundary displacements $\tilde{u}_i$ typical of FEM, i.e. it is ideally suited for embedding known solutions into special finite elements. It must be stressed here that, (4.33) only contains boundary integrals, which is in contrast to all previously introduced variational principles.

4.2.4 Hamilton's Principle

When dealing with dynamic problems, we typically utilize *Hamilton's variational principle* to derive the weak formulation of the field problem. The starting point is the Lagrange function $\mathcal{L}$, which depends on the displacement field u_i, the velocities $\dot{u}_i$ and time t. It is defined by means of kinetic energy $\mathcal{K}$ and total potential Π_{P}:

$$\mathcal{L}(u_i, \dot{u}_i, t) = \mathcal{K}(\dot{u}_i) - \Pi_{\mathrm{P}}(u_i, t) . \quad (4.34)$$

We consider the so-called principle function of $\mathcal{L}$ over a time span of t_0 to t. According to Hamilton's principle it takes on an extreme value, which is why the first variation $\delta\,()$ of the integral has to disappear:

$$\delta \int_{t_0}^{t} \mathcal{L}(u_i, \dot{u}_i, t) \, \mathrm{d}t = \int_{t_0}^{t} (\delta\mathcal{K} - \delta\Pi_{\mathrm{P}}) \, \mathrm{d}t = 0 . \quad (4.35)$$

According to the product rule, the variation of kinetic energy with respect to $\delta\dot{u}_i$ yields:

$$\delta \mathcal{K} = \delta \left(\frac{1}{2} \int_V \rho \dot{u}_i \dot{u}_i \, dV \right) = \int_V \rho \dot{u}_i \delta \dot{u}_i \, dV . \qquad (4.36)$$

If we calculate the time integral and invert the sequence of spatial and temporal integration, the following conversion can be made by partial integration:

$$\int_V \int_{t_0}^t \rho \dot{u}_i \delta \dot{u}_i \, dt \, dV = \int_V \left[(\rho \dot{u}_i \delta u_i) \big|_{t_0}^t - \int_{t_0}^t \rho \ddot{u}_i \delta u_i \, dt \right] dV = - \int_{t_0}^t \int_V \rho \ddot{u}_i \delta u_i \, dV \, dt .$$

$$(4.37)$$

The first integrand becomes zero since, according to the assumptions, the variations $\delta u_i = 0$ must disappear at the bounds t_0 and t. The variation of the total potential $\delta \Pi_P$ was already examined in (4.20). In sum, Hamilton's principle then reads:

$$\int_{t_0}^t \int_V \left[-\rho \ddot{u}_i \delta u_i - \sigma_{ij} \delta \varepsilon_{ij} \right] dV \, dt + \int_{t_0}^t \int_V \bar{b}_i \delta u_i \, dV \, dt + \int_{t_0}^t \int_{S_t} \bar{t}_i \delta u_i \, dS \, dt = 0. \qquad (4.38)$$

Conversion of (4.38) analogous to that made in Sect. 4.2.1 leads to

$$\int_{t_0}^t \int_V \left[-\rho \ddot{u}_i + \sigma_{ij,j} + \bar{b}_i \right] \delta u_i \, dV \, dt + \int_{t_0}^t \int_{S_t} \left[\bar{t}_i - \sigma_{ij} n_j \right] \delta u_i \, dS \, dt = 0. \quad (4.39)$$

In order to satisfy this relation for all times and all δu_i, the bracketed expressions must be equal to zero, i.e. we obtain as Euler equations the equations of motion (A.70) and the natural boundary conditions (4.4) on S_t. In this way, this variational formulation describes completely the initial boundary value problem of elastodynamics for a conservative system of forces.

4.3 Basic Equations of FEM

In the following, the basic equations of FEM will be briefly presented for the *displacement method*, which is considered to be the most successful and most common FEM variant. In general, it is based on the principle of virtual work or, in the particular case of hyperelastic conservative systems on the principle of minimum total potential energy. As an example, the relations will first be derived for the two-dimensional BVP of elastostatics (A.5) in order to move onto the non-linear and time-dependent BVP later. In the following, *matrices* will be written as *bold, upright letters* to differentiate them from vectors and tensors.

4.3.1 Constructing Stiffness Matrix for One Element

In FEM, the body is first subdivided into a number of simple geometrical subdomains, which is shown in Figs. 4.1 and 4.2 with quadrilateral elements. In each of these finite elements a separate mathematical function is chosen for the displacement field which must be continuous inside the element and on its boundary to neighboring elements. This is best realized by introducing suitable grid points – the n_K nodes – which usually lie on the element boundary.

The displacements $\mathbf{u}(\mathbf{x})$ in the element are expressed by interpolation functions N_a with the help of associated values $\mathbf{u}^{(a)}$ at the nodes $a = 1, 2, \dots, n_K$. $N_a(\boldsymbol{\xi})$ are the so-called *ansatz functions* or *shape functions*. The variables $\boldsymbol{\xi}$ form the *natural coordinates* of the element in a uniform parameter space, which will be treated in detail in Sect. 4.4. This is illustrated for a quadrilateral element in Fig. 4.5. Accordingly, we obtain for the element-related displacement values the approximation:

$$\begin{bmatrix} u_1(\mathbf{x}) \\ u_2(\mathbf{x}) \end{bmatrix} = \mathbf{u}(\mathbf{x}) = \sum_{a=1}^{n_K} N_a(\boldsymbol{\xi})\mathbf{u}^{(a)} = \mathbf{N}\mathbf{v}. \tag{4.40}$$

In the column vector $\mathbf{v}$, the components of the displacement vectors $\mathbf{u}^{(a)}$ of all nodes of the element are summarized:

$$\mathbf{v} = [\mathbf{u}^{(1)} \dots \mathbf{u}^{(a)} \dots \mathbf{u}^{(n_K)}]^{\mathrm{T}}, \qquad \mathbf{u}^{(a)} = [u_1^{(a)} u_2^{(a)}]^{\mathrm{T}} \tag{4.41}$$

The matrix $\mathbf{N}$ thus has the structure:

$$\mathbf{N} = \begin{bmatrix} N_1 & 0 & N_a & 0 & \dots & N_{n_K} & 0 \\ 0 & N_1 & 0 & N_a & \dots & 0 & N_{n_K} \end{bmatrix}. \tag{4.42}$$

According to (4.1) we obtain the strains from the spatial derivative of the displacement function (4.40), for which the differentiation matrix $\mathbf{D}$ is used:

$$\begin{bmatrix} \varepsilon_{11} \\ \varepsilon_{22} \\ \gamma_{12} \end{bmatrix} = \boldsymbol{\varepsilon} = \mathbf{D}\mathbf{u}(\mathbf{x}) \quad \text{with } \mathbf{D} = \begin{bmatrix} \dfrac{\partial}{\partial x_1} & 0 \\ 0 & \dfrac{\partial}{\partial x_2} \\ \dfrac{\partial}{\partial x_2} & \dfrac{\partial}{\partial x_1} \end{bmatrix}. \tag{4.43}$$

The *strain-displacement matrix* $\mathbf{B}$ finally relates displacements at the nodes with deformations in the element:

$$\varepsilon = \mathbf{DNv} = \mathbf{Bv} \quad \text{with } \mathbf{B} = \mathbf{DN},$$

$$\mathbf{B} = [\mathbf{B}_1 \dots \mathbf{B}_a \dots \mathbf{B}_{n_K}], \quad \mathbf{B}_a = \begin{bmatrix} N_{a,1} & 0 \\ 0 & N_{a,2} \\ N_{a,2} & N_{a,1} \end{bmatrix}. \tag{4.44}$$

The stress distribution in the element is obtained with the material law, which here for clarity's sake is set as isotropically elastic (plane stress) with thermal or other initial strains ε^*:

$$\begin{bmatrix} \sigma_{11} \\ \sigma_{22} \\ \tau_{12} \end{bmatrix} = \sigma(\mathbf{x}) = \mathbf{C}\left(\varepsilon(\mathbf{x}) - \varepsilon^*(\mathbf{x})\right) = \mathbf{C}\left(\mathbf{Bv} - \varepsilon^*\right) \tag{4.45}$$

$$\mathbf{C} = \frac{E}{1-\nu^2} \begin{bmatrix} 1 & \nu & 0 \\ \nu & 1 & 0 \\ 0 & 0 & \frac{1-\nu}{2} \end{bmatrix}. \tag{4.46}$$

Now the same function according to (4.40) is used for the virtual displacements and associated strains as for the displacement fields themselves (Galerkin's method of weighted residuals):

$$\delta\mathbf{u} = \mathbf{N}\delta\mathbf{v}, \quad \delta\varepsilon = \mathbf{B}\delta\mathbf{v}. \tag{4.47}$$

The relations (4.43)–(4.47) are inserted into the principle of virtual work (4.11). Together with boundary conditions $\bar{\mathbf{t}} = \begin{bmatrix} \bar{t}_1 & \bar{t}_2 \end{bmatrix}^{\mathrm{T}}$ and volume loads $\bar{\mathbf{b}} = \begin{bmatrix} \bar{b}_1 & \bar{b}_2 \end{bmatrix}^{\mathrm{T}}$ it reads in matrix form:

$$\begin{aligned} \delta W_{\text{int}} - \delta W_{\text{ext}} &= \int_V \delta\varepsilon^{\mathrm{T}}\sigma\,\mathrm{d}V - \int_{S_t} \delta\mathbf{u}^{\mathrm{T}}\bar{\mathbf{t}}\,\mathrm{d}S - \int_V \delta\mathbf{u}^{\mathrm{T}}\bar{\mathbf{b}}\,\mathrm{d}V \\ &= \delta\mathbf{v}^{\mathrm{T}}\left\{ \int_V \mathbf{B}^{\mathrm{T}}\mathbf{C}(\mathbf{Bv} - \varepsilon^*)\,\mathrm{d}V - \int_{S_t} \mathbf{N}^{\mathrm{T}}\bar{\mathbf{t}}\,\mathrm{d}S - \int_V \mathbf{N}^{\mathrm{T}}\bar{\mathbf{b}}\,\mathrm{d}V \right\} = 0 \end{aligned} \tag{4.48}$$

The expression in curly brackets must disappear for every variation $\delta\mathbf{v}^{\mathrm{T}}$, from which we obtain the FEM system of equations on the element level:

$$\mathbf{kv} = \mathbf{f}, \quad \mathbf{f} = \mathbf{f}_t + \mathbf{f}_b + \mathbf{f}_\varepsilon. \tag{4.49}$$

All node-related quantities $\mathbf{v}$ can be pulled in front of the integrals so that the following matrices and vectors can be set up:

$$\text{stiffness matrix: } \mathbf{k} = \int_{V_e} \mathbf{B}^{\mathrm{T}} \mathbf{C} \mathbf{B} \, dV \tag{4.50}$$

Force vectors due to volume loads, surface loads and initial strains:

$$\mathbf{f}_b = \int_{V_e} \mathbf{N}^{\mathrm{T}} \bar{\mathbf{b}} \, dV, \quad \mathbf{f}_t = \int_{S_{te}} \mathbf{N}^{\mathrm{T}} \bar{\mathbf{t}} \, dS, \quad \mathbf{f}_\varepsilon = \int_{V_e} \mathbf{B}^{\mathrm{T}} \mathbf{C} \, \varepsilon^* \, dV. \tag{4.51}$$

4.3.2 Assembly and Solution of the Total System

So far only a single finite element (index e) was considered in the derivation of the stiffness relations. Application of the principle of virtual work requires the summation of the contributions of all finite elements $e = 1, 2, \ldots, n_E$ of the total structure.

$$\delta W_{\mathrm{ext}} - \delta W_{\mathrm{int}} = \sum_{e=1}^{n_E} \left(\delta W_{\mathrm{ext}}^{(e)} - \delta W_{\mathrm{int}}^{(e)} \right) \tag{4.52}$$

The system matrices are assembled by addition of the single element matrices into the corresponding positions of the system matrices, whereby the assignment takes place by means of the respective *node variables*. The topological correlation of the finite elements in the total structure is described with the help of a so-called assignment or *incidence matrix* $\mathbf{A}_e$. This Boolean matrix (only elements $\{0,1\}$) determines at which location the nodal variables of element e are assigned in the vector $\mathbf{V}$ of all *nodal degrees of freedom* of the system.

$$\mathbf{v}_e = \mathbf{A}_e \mathbf{V} \tag{4.53}$$

$\mathbf{A}_e$ has the dimension $[n_K \cdot n_D, N_K \cdot n_D]$, if N_K is the number of all nodes in the system and n_D denotes the degree of freedom per node (in case of displacement functions the geometric dimension $n_D = 1, 2, 3$). In practice, this is often managed by means of a global numbering of all N_K nodes and a pointer vector on the local element nodes.

We thereby obtain the *stiffness relation of the total system* from those of all elements

$$\left(\sum_{e=1}^{n_\mathrm{E}} \mathbf{A}_e^\mathrm{T} \mathbf{k}_e \mathbf{A}_e\right) \mathbf{V} = \sum_{e=1}^{n_\mathrm{E}} \mathbf{A}_e^\mathrm{T} \mathbf{f}_e \quad \Rightarrow \quad \mathbf{KV} = \mathbf{F}. \tag{4.54}$$

This is a system of $[N_\mathrm{K} \cdot n_\mathrm{D}]$ equations for determining the vector $\mathbf{V}$, which contains the displacement degrees of freedom of all nodes of the structure. On the right hand side the contributions of all elements (4.51) are compiled to the corresponding nodal forces of the total structure in the *external load vector* $\mathbf{F}$. The *stiffness matrix* $\mathbf{K}$ of the total system is put together in the same way. This process is called the *assembly* of the total system and will be abbreviated in the following with the set symbol:

$$\mathbf{K} = \bigcup_{e=1}^{n_\mathrm{E}} \mathbf{k}_e, \quad \mathbf{V} = \bigcup_{e=1}^{n_\mathrm{E}} \mathbf{v}_e, \quad \mathbf{F} = \bigcup_{e=1}^{n_\mathrm{E}} \mathbf{f}_e \quad \text{etc.} \tag{4.55}$$

Assembly is obviously only allowed if all local and global nodal degrees of freedom are defined in the same (global) coordinate system. Otherwise a prior transformation of $\mathbf{k}_e$, $\mathbf{v}_e$ and $\mathbf{f}_e$ is required. As we can see in (4.50), the stiffness matrices of finite elements are symmetrical in the displacement formulation. Due to their energetic meaning, they are also positively definite, i.e. multiplication with an arbitrary state of motions $\mathbf{v}_e$ results in a positive deformation energy $\mathcal{W}_\mathrm{int} = \frac{1}{2} \mathbf{v}_e^\mathrm{T} \mathbf{k}_e \mathbf{v}_e > 0$. Naturally, it is presumed that possible rigid body motions are prevented by a statically determined support.

The same properties apply to the system stiffness matrix $\mathbf{K}$. It has only a small amount of elements and has a band structure at optimal nodal numbering. With linear material behavior, the FEM system therefore represents a linear algebraic equation system for the nodal variables having a unique solution. The *system stiffness matrix* thus embodies all the mechanical properties of a structure, which incorporate the selected structural model (e. g. a plate), the geometry and the material properties.

A huge number of mathematical methods exists with indirect or iterative methods for the numerical solution of the *FEM system of equations*. Due to the symmetry of the stiffness matrix, Cholesky's method is suitable in addition to the classical Gaussian elimination technique. However, iterative conjugated gradient methods with preconditioning are also very effective. Because of the very large systems of equations involved, fast computational algorithms for storage, processing and structuring of matrices are of considerable importance for improving efficiency. Techniques such as bandwidth minimization, frontal solution method and parallel processing on several processors, all fall into this category. For details, the reader should refer to the copious technical literature [6].

4.4 Numerical Realization of FEM

4.4.1 Selection of Displacement Functions

For the *shape functions* N_a (4.40), simple mathematical functions are chosen as a rule. These must however obey certain requirements so that the approximate numerical solution converges towards the exact solution with increasing mesh refinement:

1. Firstly, the displacement functions may not lead to deformations $\varepsilon \neq 0$ in the case of rigid body motion.
2. Secondly, they must be capable of realizing a constant state of deformation $\varepsilon = $ const in the entire element.
3. Thirdly, the shape functions should in general continuously proceed across the element boundaries so that under loading no discontinuities (gaps, overlapping) occur (C^0-continuity of displacement functions in case of the continuum elements considered here).
4. Moreover, the shape functions should correctly reproduce a constant function value (e. g. the element area)

$$\sum_{a=1}^{n_K} N_a(\xi) = 1 \tag{4.56}$$

and map the function value at the respective node ξ_a exactly and solely

$$N_a(\xi_b) = \delta_{ab} = \begin{cases} 1 & a = b \\ 0 & a \neq b \end{cases}. \tag{4.57}$$

4.4.2 Isoparametric Element Family

Depending on the element geometry, the structural model and the accuracy requirements, there are a number of possible ways to formulate the shape functions. For a detailed presentation of finite element types, see the relevant FEM literature (e. g. [7–9]). For most problems, the use of *isoparametric elements* has proved advantageous. This concept involves the interpolation of both the geometry and all primary field quantities in the element by the same functions $N_a(\xi)$.

The relationship between natural and local coordinate system is made by the shape functions. In this way, for each element a transformation is created, which maps the local geometrical coordinates $\mathbf{x} = [x_1\ x_2\ x_3]^T$ onto the natural parametric coordinates $\boldsymbol{\xi} = [\xi_1\ \xi_2\ \xi_3]^T$:

$$x_i = \sum_{a=1}^{n_K} N_a(\boldsymbol{\xi})x_i^{(a)} \quad \text{or} \quad \mathbf{x}(\boldsymbol{\xi}) = \mathbf{N}(\boldsymbol{\xi})\hat{\mathbf{x}}, \quad \hat{\mathbf{x}} = \left[x_1^{(1)}\ x_2^{(1)}\ \cdots\ x_1^{(n_K)}\ x_2^{(n_K)}\right]^T.$$

$$(4.58)$$

By means of this universal mapping, it is possible to convert the arbitrarily shaped, different geometries of all elements into a consistent system of coordinates ξ_i. Figs. 4.4–4.6 illustrate this for various element types. A further advantage of this representation method is that we can always use the same algorithms for all mathematical operations on the element level.

To assemble the $\mathbf{B}$ matrices from (4.43) and (4.44), we need the partial derivatives of the functions N_a with respect to the coordinates x_j, which are formed by the chain rule

$$\frac{\partial N_a}{\partial x_j}(\boldsymbol{\xi}) = \frac{\partial N_a}{\partial \xi_1}\frac{\partial \xi_1}{\partial x_j} + \frac{\partial N_a}{\partial \xi_2}\frac{\partial \xi_2}{\partial x_j} + \frac{\partial N_a}{\partial \xi_3}\frac{\partial \xi_3}{\partial x_j} \Rightarrow \frac{\partial N_a}{\partial \mathbf{x}} = \mathbf{J}^{-1}\frac{\partial N_a}{\partial \boldsymbol{\xi}}. \quad (4.59)$$

$\mathbf{J}$ represents the Jacobian functional matrix, which associates the line elements of the natural coordinates $\mathrm{d}\xi_i$ with those of the local coordinates $\mathrm{d}x_j$:

$$\mathbf{J} = \begin{bmatrix} \dfrac{\partial x_1}{\partial \xi_1} & \dfrac{\partial x_2}{\partial \xi_1} & \dfrac{\partial x_3}{\partial \xi_1} \\[2mm] \dfrac{\partial x_1}{\partial \xi_2} & \dfrac{\partial x_2}{\partial \xi_2} & \dfrac{\partial x_3}{\partial \xi_2} \\[2mm] \dfrac{\partial x_1}{\partial \xi_3} & \dfrac{\partial x_2}{\partial \xi_3} & \dfrac{\partial x_3}{\partial \xi_3} \end{bmatrix}. \quad (4.60)$$

To ensure that the inverse of the Jacobian matrix $\mathbf{J}^{-1}$ exists, the transformation of $\boldsymbol{\xi}$ to $\mathbf{x}$ must be uniquely invertible. The elements of the Jacobian matrix are calculated from the local coordinates $x_j^{(a)}$ of the particular element nodes by differentiating the shape functions

$$J_{ij} = \frac{\partial x_j}{\partial \xi_i} = \sum_{a=1}^{n_K} \frac{\partial N_a}{\partial \xi_i}x_j^{(a)}. \quad (4.61)$$

Isoparametric Triangular Elements

The geometrical description of a triangle is done best by its area coordinates $\boldsymbol{\xi} \triangleq (L_1, L_2, L_3)$. The cartesian coordinates $\boldsymbol{x}$ of a point P in the triangle are obtained from L_i via linear interpolation with the corner points $x_i^{(a)}$ $(a = 1, 2, 3)$

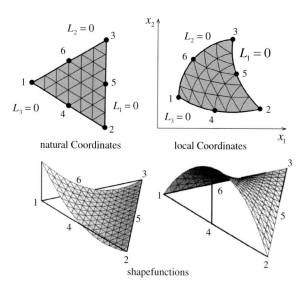

Fig. 4.4 Isoparametric triangular element mit with quadratic shape functions, $n_K = 6$ nodes

$$x_1 = L_1 x_1^{(1)} + L_2 x_1^{(2)} + L_3 x_1^{(3)}$$
$$x_2 = L_1 x_2^{(1)} + L_2 x_2^{(2)} + L_3 x_2^{(3)} \qquad (4.62)$$
$$1 = L_1 + L_2 + L_3.$$

At a corner point a exactly holds $L_a = 1$ and $L_b = 0$ ($b \neq a$). A point in the middle of a side (e. g. node 4) has the coordinates $(\frac{1}{2}, \frac{1}{2}, 0)$ etc. Lines $L_i = $ const. run parallel to the opposite side of the corner i, whereby the values decline linearly from $L_i = 1$ (corner) to $L_i = 0$ (opposite side) see Fig. 4.4. The sum of the triangle's coordinates corresponds exactly to the normalized surface area (last Eq. (4.62)).

(a) linear function for $n_K = 3$ corner nodes:

$$N_1 = L_1, \qquad N_2 = L_2, \qquad N_3 = L_3 \qquad (4.63)$$

(b) quadratic function for 3 corner and 3 mid-side nodes ($n_K = 6$, see Fig. 4.4):

$$N_1 = L_1(2L_1 - 1), \qquad N_2 = L_2(2L_2 - 1), \qquad N_3 = L_3(2L_3 - 1)$$
$$N_4 = 4L_1 L_2, \qquad N_5 = 4L_2 L_3, \qquad N_6 = 4L_3 L_1 \qquad (4.64)$$

Isoparametric Quadrilateral Elements

In the natural coordinate system ξ_i, these finite elements form a square in the interval $[-1, +1]$.

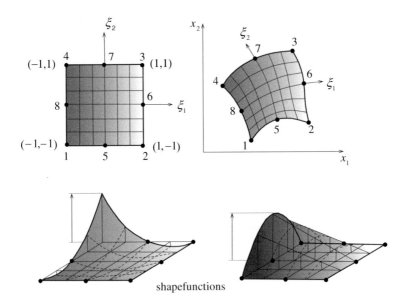

shapefunctions

Fig. 4.5 Isoparametric quadrilateral element with quadratic shape functions, $n_K = 8$ nodes

(a) linear function for $n_K = 4$ corner nodes:

$$N_a(\xi_1, \xi_2) = \frac{1}{4}(1 + \xi_1\xi_1^a)(1 + \xi_2\xi_2^a) \qquad (4.65)$$

(b) quadratic function for 4 corner nodes and 4 mid-side nodes ($n_K = 8$, Fig. 4.5):

- corner nodes: $N_a(\xi_1, \xi_2) = \frac{1}{4}(1 + \xi_1\xi_1^a)(1 + \xi_2\xi_2^a)(\xi_1\xi_1^a + \xi_2\xi_2^a - 1)$

- mid-side nodes: $\xi_1^a = 0 :\ N_a(\xi_1, \xi_2) = \frac{1}{2}(1 - \xi_1^2)(1 + \xi_2\xi_2^a)$

$$\xi_2^a = 0 :\ N_a(\xi_1, \xi_2) = \frac{1}{2}(1 + \xi_1\xi_1^a)(1 - \xi_2^2).$$

$$(4.66)$$

Isoparametric Hexahedron Elements

Every 3D element is mapped onto a regular unit cube $[-1 \le \xi_i \le +1]$. Again, linear or quadratic shape functions are customary.

(a) linear function for $n_K = 8$ corner nodes:

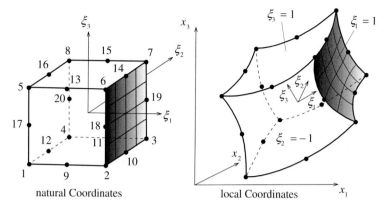

natural Coordinates   local Coordinates

Fig. 4.6 Isoparametric hexahedral elements with quadratic shape functions, $n_K = 20$ nodes

$$N_a(\xi_1, \xi_2, \xi_3) = \frac{1}{8}(1 + \xi_1\xi_1^a)(1 + \xi_2\xi_2^a)(1 + \xi_3\xi_3^a) \qquad (4.67)$$

(b) quadratic function for 8 corner nodes and 12 mid-side nodes: ($n_K = 20$, Fig. 4.6):

- corner nodes:
  $N_a(\xi_1, \xi_2, \xi_3) = \frac{1}{8}(1 + \xi_1\xi_1^a)(1 + \xi_2\xi_2^a)(1 + \xi_3\xi_3^a)(\xi_1\xi_1^a + \xi_2\xi_2^a + \xi_3\xi_3^a - 2)$
- mid-side nodes:
  $\xi_1^a = 0$, $\xi_2^a = \pm 1$, $\xi_3^a = \pm 1$
  $N_a(\xi_1, \xi_2, \xi_3) = \frac{1}{4}(1 - \xi_1^2)(1 + \xi_2\xi_2^a)(1 + \xi_3\xi_3^a)$
  and analogously by permuting ξ_1, ξ_2 and ξ_3 for nodes on the
  surfaces $\xi_2^a = 0$ and $\xi_3^a = 0$.

$$(4.68)$$

4.4.3 Numerical Integration of Element Matrices

To assemble element matrices and load vectors, the volume or surface integrals
in Eqs. (4.50) and (4.51) have to be evaluated. Since in most cases an analytical
integration is not feasible, numerical integration techniques are employed that are
based on special quadrature formulae for different element shapes and functions.
In this way, the definite integral of an arbitrary function $f(\xi)$ can be calculated
approximately by the quadrature formula $\mathcal{Q}(f)$

$$\int_{-1}^{+1} f(\xi)\, d\xi \approx \mathcal{Q}(f) = \sum_{g=1}^{n_G} w_g\, f(\xi^g)\,, \qquad (4.69)$$

Table 4.1 Grid points and weights of Gaussian integration ($g = 1, \ldots, n_G$) in a one-dimensional element $[-1, +1]$

n_G	ξ^g			w_g		
1	0			2		
2	$-\dfrac{1}{\sqrt{3}}$	$+\dfrac{1}{\sqrt{3}}$		1	1	
3	$-\sqrt{\dfrac{3}{5}}$	0	$+\sqrt{\dfrac{3}{5}}$	$\dfrac{5}{9}$	$\dfrac{8}{9}$	$\dfrac{5}{9}$

whereby the natural coordinate $\xi \in [-1, +1]$ serves as an independent variable. ξ^g denotes the grid points in the interval $[-1, +1]$ and the coefficients w_g are the associated weights of the quadrature formula. For the Gaussian quadrature, that is most often used in FEM, the grid points and weights are available for any number of integration points n_G in tabular form [10]. Table 4.1 contains the data of the most important integration orders $n_G = 1, 2, 3$. It can be shown that the summation (4.69) is exact for polynomials up to the $(2n_G - 1)$th degree.

When calculating a domain integral over a finite element, a multiple integral has to be evaluated in accordance with the dimension n_D of the element type. For this, the quadrature formula (4.69) is applied to all n_D spatial directions in the natural coordinate system. In the function to be integrated $F(x)$, we can first substitute the independent variable x via the shape function (4.59) with the natural coordinates:

$$F[x(\xi)] = F[x_1(\xi_1, \xi_2, \xi_3), x_2(\xi_1, \xi_2, \xi_3), x_3(\xi_1, \xi_2, \xi_3)] = f(\xi_1, \xi_2, \xi_3). \quad (4.70)$$

Secondly, the volume element dV must be expressed in natural coordinates, which is achieved with the help of the determinant $J_V = |\mathbf{J}|$ of the Jacobian matrix:

$$I = \int_{V_e} F(x)\,dV = \int_{V_e} f(\xi_i)\,dV = \int_{-1}^{+1}\int_{-1}^{+1}\int_{-1}^{+1} f(\xi_i)\,|\mathbf{J}(\xi_i)|\,d\xi_1 d\xi_2 d\xi_3. \quad (4.71)$$

By applying the integration formula (4.69) to each of the integrals of (4.71), we obtain a ($n_D = 3$)-fold summation, in which the natural coordinates in every dimension $i = 1, 2, 3$ run through all grid points ξ_i^l ($l = 1, 2, \ldots, n_G$) and the weights are multiplied.

$$I = \sum_{k=1}^{n_G}\sum_{n=1}^{n_G}\sum_{m=1}^{n_G} f(\xi_1^m, \xi_2^n, \xi_3^k)w_m w_n w_k \left|\mathbf{J}(\xi_1^m, \xi_2^n, \xi_3^k)\right| = \sum_{g=1}^{m_G} f(\xi_1^g, \xi_2^g, \xi_3^g)\bar{w}_g \left|\mathbf{J}(\xi^g)\right| \quad (4.72)$$

In practical terms, this is realized as a sum over all $m_G = n_G^{n_D}$ grid points $\boldsymbol{\xi}^g = [\xi_1^g \; \xi_2^g \; \xi_3^g]$ with the weights $\bar{w}_g = w_m w_n w_k$. Figure 4.7 illustrates the integration orders $n_G = 2$ and $n_G = 3$ for a $n_D = 2$-dimensional element.

Furthermore, surface integrals must be evaluated over a partial area S of 3D elements in order for example to take surface loads according to (4.51) into account. An element surface (e. g. $\xi_1 = 1$ in Fig. 4.6) is represented by the natural coordinates

(ξ_2, ξ_3). Applying the rules for parameter integrals, we obtain the surface normal vector $\boldsymbol{n}$ from the directional derivative with respect to ξ_2 and ξ_3, and the surface area dS is to be transformed with the Jacobian determinant J_S.

$$\boldsymbol{n} = \left(\frac{\partial \boldsymbol{x}}{\partial \xi_2} \times \frac{\partial \boldsymbol{x}}{\partial \xi_3} \right) / J_S$$

$$dS = \left| \frac{\partial \boldsymbol{x}}{\partial \xi_2} \times \frac{\partial \boldsymbol{x}}{\partial \xi_3} \right| d\xi_2 d\xi_3 = J_S(\xi_2, \xi_3)\, d\xi_2 d\xi_3 \tag{4.73}$$

The surface integral is then written as follows and can be calculated by the 2D Gaussian integration rule.

$$I = \int_S f(\boldsymbol{x}) dS = \int_{-1}^{+1} \int_{-1}^{+1} f[\boldsymbol{x}(\xi_2, \xi_3)] J_S d\xi_2 d\xi_3 \tag{4.74}$$

Analogously, for the line integral along an element edge L (e.g. $\xi_1 = 1$ in Fig. 4.5) we get the calculation rules using the parameter ξ_2:

$$\boldsymbol{n} = \left(\frac{\partial x_2}{\partial \xi_2} \boldsymbol{e}_1 - \frac{\partial x_1}{\partial \xi_2} \boldsymbol{e}_2 \right) / J_L$$

$$ds = J_L(\xi_2) d\xi_2 = \sqrt{\left(\frac{\partial x_1}{\partial \xi_2} \right)^2 + \left(\frac{\partial x_2}{\partial \xi_2} \right)^2} \, d\xi_2 \tag{4.75}$$

$$I = \int_L f(\boldsymbol{x}) ds = \int_{-1}^{+1} f[\boldsymbol{x}(\xi_2)] J_L d\xi_2 . \tag{4.76}$$

4.4.4 Numerical Interpolation of Results

The primary result of a FEM calculation is the solution vector $\mathbf{V}$ (4.54), which contains the displacement vectors $\mathbf{u}^{(a)}$ of all nodes $\mathbf{x}^{(a)}$ of the mesh. From the nodal displacements $\mathbf{v}$ of each element, we can calculate the total strain ε by differentiating via the kinematic relation (4.44). For linear material laws, the stresses at each point $\boldsymbol{\xi}$ of the element are easily obtained from (4.45). This becomes more difficult in the case of inelastic material laws, where the stresses $\boldsymbol{\sigma}$, plastic strains ε^{p}, and hardening variables $\mathbf{h}$ can only be obtained as independent outcome variables via the complex solution of the entire IBVP. Usually, these secondary field quantities are merely calculated, stored and output at the integration points (IP). Solely the displacement solution is continuous across the element boundaries, whereas all other »derived« field quantities are discontinuous there (jump). For a fracture-mechanical evaluation, we need these field quantities and their spatial derivatives at an arbitrary point $\mathbf{x}$ of the component. Some useful techniques for this will be outlined using the example of quadrilateral elements.

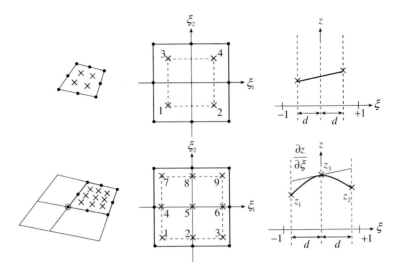

Fig. 4.7 Interpolation and extrapolation of FEM-results in an element

For the sake of simplicity, this procedure is exemplified with a scalar field quantity $z(\mathbf{x})$, which may represent any component of $\mathbf{u}$, σ, ε et al. For the considered point, at first we must find the associated finite element and then its natural coordinates $\boldsymbol{\xi} = [\xi_1 \, \xi_2]^{\mathrm{T}}$ inside this element. Since the isoparametric representation $\mathbf{x} = \sum \mathbf{N}_a(\boldsymbol{\xi}) \, \hat{\mathbf{x}}^{(a)}$ cannot be inverted analytically for $\boldsymbol{\xi}$, a suitable search algorithm (interval nesting or the Newtonian method) is required. If we have found the natural coordinates $\boldsymbol{\xi}$ and know the function values $z^{(a)}$ at the nodes $a = 1, 2, \ldots, n_{\mathrm{K}}$ of the element, we easily obtain $z(\mathbf{x}(\boldsymbol{\xi}))$ from the isoparametric shape functions:

$$z = \sum_{a=1}^{n_{\mathrm{K}}} N_a(\boldsymbol{\xi}) z^{(a)} \qquad \text{Interpolation rule.} \qquad (4.77)$$

In order to determine the gradients with respect to global coordinates (x_1, x_2), the inverse $\mathbf{J}^{-1}(\boldsymbol{\xi})$ of the transformation matrix (4.60)–(4.61) is needed as correlation to the natural coordinates (ξ_1, ξ_2):

$$
\begin{bmatrix} \dfrac{\partial z}{\partial x_1} \\[2ex] \dfrac{\partial z}{\partial x_2} \end{bmatrix}
=
\begin{bmatrix} \dfrac{\partial \xi_1}{\partial x_1} & \dfrac{\partial \xi_2}{\partial x_1} \\[2ex] \dfrac{\partial \xi_1}{\partial x_2} & \dfrac{\partial \xi_2}{\partial x_2} \end{bmatrix}
\begin{bmatrix} \dfrac{\partial z}{\partial \xi_1} \\[2ex] \dfrac{\partial z}{\partial \xi_2} \end{bmatrix}
= [\mathbf{J}^{-1}]
\begin{bmatrix} \dfrac{\partial z}{\partial \xi_1} \\[2ex] \dfrac{\partial z}{\partial \xi_2} \end{bmatrix}. \qquad (4.78)
$$

The natural derivatives are obtained directly with (4.77):

$$\frac{\partial z}{\partial \xi_i} = \sum_{a=1}^{n_K} \frac{\partial N_a}{\partial \xi_i} z^{(a)} .\tag{4.79}$$

If the nodal values are not available, like for the secondary field quantities, we must interpolate or extrapolate the IP values in a suitable way. In case of the isoparametric quadrilateral elements shown in Fig. 4.7, either the complete (3×3 IP) or reduced (2×2 IP) integration rule is used. The function $z(\xi_1, \xi_2)$ can be approximated very well by a product of two polynomials of first (2×2 IP) or second degree (3×3 IP) with respect to ξ_1 and ξ_2.

$$
\begin{aligned}
z(\xi_1, \xi_2) &= c_1 + c_2\xi_1 + c_3\xi_2 + c_4\xi_1\xi_2 && (2 \times 2) \\
z(\xi_1, \xi_2) &= c_1 + c_2\xi_1 + c_3\xi_2 + c_4\xi_1\xi_2 + c_5\xi_1^2 && (4.80) \\
&\quad + c_6\xi_2^2 + c_7\xi_1^2\xi_2 + c_8\xi_1\xi_2^2 + c_9\xi_1^2\xi_2^2 && (3 \times 3)
\end{aligned}
$$

To determine the unknown coefficients c_i, exactly 4 (2×2) or 9 (3×3) function values $z(\xi_1^g, \xi_2^g) = z^{(g)}$ are available at the IP. By inserting the grid points (ξ_1^g, ξ_2^g) (see Table 4.1) and values $z^{(g)}$ into (4.80), we can build a linear system of equations, the solution of which gives the sought c_i. In this way, the function $z(\xi_1, \xi_2)$ can be interpolated (4.80) and, by inserting the respective isoparametric coordinates ξ_i^a with (4.66), extrapolated to the nodes of the element. Since extrapolation towards a shared boundary node leads to different results in each element (Fig. 4.7, bottom left), usually a (weighted) averaging of different element values is carried out, which yields the value $\bar{z}^{(a)}$ at node a. For interpolation and differentiation, we can now also proceed in accordance with (4.77) and (4.79) for these variables.

Often, also the derivatives of the secondary variables are required at the IPs. For this purpose, the method shown in Fig. 4.7 (right) is recommended. The 2 or 3 function values at the IPs along one coordinate $\xi \in \{\xi_1, \xi_2\}$ are approximated as a straight line or a parabola:

$$
\begin{aligned}
z(\xi) &= \frac{z_2 - z_1}{2d}\xi + \frac{1}{2}(z_1 + z_2) && (2 \text{ IP}) \\
z(\xi) &= \frac{z_1 - 2z_3 + z_2}{2d^2}\xi^2 + \frac{z_2 - z_1}{2d}\xi + z_3 && (3 \text{ IP}), && (4.81)
\end{aligned}
$$

whose derivatives are

$$
\begin{aligned}
\frac{\partial z}{\partial \xi} &= (z_2 - z_1)/(2d) && (2 \text{ IP}) \\
\frac{\partial z}{\partial \xi} &= \frac{z_1 - 2z_3 + z_2}{d^2}\xi + \frac{z_2 - z_1}{2d} && (3 \text{ IP}). && (4.82)
\end{aligned}
$$

With the help of these derivatives w.r.t. ξ_1 and ξ_2 at the grid point (ξ_1, ξ_2), we go directly into (4.78) in order to obtain the global derivatives. This method is especially suitable for the differentiation of σ, ε^p or U^p at the IPs if the results must later be integrated over the element.

4.5 FEM for Non-Linear Boundary Value Problems

4.5.1 Basic Equations

There are essentially two factors leading to non-linear behaviour of BVP in solid mechanics:

(a) *Material non-linearity*
In the case of hyper-elastic, elastic-plastic or other non-linear materials, the state of stress occurring in an element depends on the strains and thus on displacements in a non-linear fashion: $\sigma(\mathbf{v}) = f\left(\varepsilon(\mathbf{v}), \mathbf{h}(\mathbf{v})\right)$. If the deformation process is irreversible, then σ is a function of the load history, which was explained in Sect. A.4.2 using the example of hardening history. Thus the current state of the material is described by several internal variables that are compiled in the matrix $\mathbf{h}$.

(b) *Geometrical non-linearity*
In the case of large, *finite deformations*, two non-linear phenomena occur. If there are *large displacements* with respect to the initial geometry, first the redistribution of the equilibrium of forces and the load boundary conditions upon the deformed structure must be taken into consideration. Secondly, in the case of *large strain*, there is a non-linear relation (A.20) between the strain tensors and the displacement gradients.

If we apply the principle of virtual displacements (4.12)–(4.13) to a non-linear BVP, then we obtain the following extensions in the FEM formulation compared to the linear relations from Sect. 4.3:

$$\delta \mathcal{W}_{\text{ext}} - \delta \mathcal{W}_{\text{int}} = \left[\mathbf{F}^{\text{ext}}(t) - \mathbf{F}^{\text{int}}\left(\mathbf{V}(t)\right) \right] \delta \mathbf{V} = 0 \qquad (4.83)$$

$$\mathbf{F}^{\text{ext}} = \bigcup_{e=1}^{n_E} \left[\int_{S_{te}} \mathbf{N}^{\text{T}} \bar{\mathbf{t}} \, dS_e + \int_{V_e} \mathbf{N}^{\text{T}} \bar{\mathbf{f}} \, dV_e \right] \qquad (4.84)$$

$$\mathbf{F}^{\text{int}}\left(\mathbf{V}(t)\right) = \bigcup_{e=1}^{n_E} \int_{V_e} \mathbf{B}^{\text{T}}(\mathbf{v}) \sigma(\mathbf{v}) \, dV_e . \qquad (4.85)$$

The external load vector $\mathbf{F}^{\text{ext}}(t)$ describes the system load vector $\mathbf{F}$ according to (4.55) as a result of temporally imposed surface and volume loads. The corresponding nodal force vector $\mathbf{F}^{\text{int}}$ of the internal forces is determined by the non-linear behavior of the material $\sigma(\mathbf{v})$ and the kinematics $\mathbf{B}(\mathbf{v})$.

In the case of non-linear problems, it is necessary to subdivide the load history into a number n_t of finite time steps Δt_n. (For scleronomous material behavior, we may also use a monotonously increasing load parameter instead of time → load steps.)

$$t_{n+1} = t_n + \Delta t_{n+1}, \qquad n = 0, 1, 2, \ldots, n_t - 1 \qquad (4.86)$$

The external load is thus applied as a temporal sequence of load steps. For each point in time we must find via iteration the state of equilibrium (4.83) with the internal forces at the end of the increment. This is also called an *incremental-iterative algorithm*.

$$\mathbf{F}^{\text{ext}}(t_{n_t}) = \sum_{n=1}^{n_t} \Delta \mathbf{F}_n^{\text{ext}}, \qquad \Delta \mathbf{F}_{n+1}^{\text{ext}} = \mathbf{F}^{\text{ext}}(t_{n+1}) - \mathbf{F}^{\text{ext}}(t_n) \qquad (4.87)$$

The increments of all other quantities arise in a corresponding manner, such as nodal displacements

$$\mathbf{V}(t_{n_t}) = \sum_{n=1}^{n_t} \Delta \mathbf{V}_n, \qquad \Delta \mathbf{V}_{n+1} = \mathbf{V}(t_{n+1}) - \mathbf{V}(t_n), \qquad (4.88)$$

stresses $\Delta \boldsymbol{\sigma}_n$, strain $\Delta \boldsymbol{\varepsilon}_n$, etc., whereby the subscript index n always marks the time step.

The incremental procedure is absolutely necessary in order to integrate the plastic material laws given in rate form over the load path. The incremental algorithm is also extremely useful for the numerical solution of a non-linear problem. Figure 4.8 shows this solution strategy (1 degree of freedom) in a simplified form. The true solution of the non-linear problem $\mathbf{F}^{\text{ext}}(\mathbf{V})$ is drawn as a continuous line. On the ordinate, a load increment $\Delta \mathbf{F}_{n+1}^{\text{ext}}$ is plotted as an example for which the associated displacement increment $\Delta \mathbf{V}_{n+1}$ is sought. To this end, the non-linear system of Eq. (4.83) must be solved in the $(n + 1)$-th load step.

$$\mathbf{F}_{n+1}^{\text{ext}} - \mathbf{F}_{n+1}^{\text{int}} = \mathbf{R}(\mathbf{V}) \rightarrow 0 \qquad (4.89)$$

The vector $\mathbf{R}(\mathbf{V})$ comprises the non-balanced »residual« nodal forces and must be reduced to zero to attain the true state of equilibrium. To calculate these zeros, often the Newton-Raphson method is used. In order to be able to use this iterative method, we first linearize the non-linear functions $\mathbf{F}^{\text{int}}$ with respect to the nodal displacements $\mathbf{V}$ (as a multi-dimensional Taylor expansion) at point $\mathbf{V}_n$:

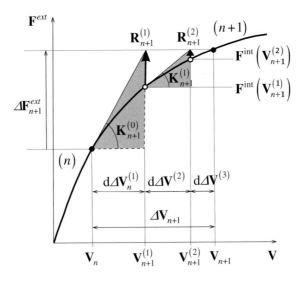

Fig. 4.8 Solution of the non-linear FEM system with the Newton-Raphson iteration method

$$\mathbf{F}^{int}\left(\mathbf{V}_n + \Delta\mathbf{V}_{n+1}\right) = \mathbf{F}^{int}\left(\mathbf{V}_n\right) + \left.\frac{\partial\mathbf{F}^{int}}{\partial\mathbf{V}}\right|_{\mathbf{V}_n}\Delta\mathbf{V} + \ldots$$

$$\left.\frac{\partial\mathbf{F}^{int}}{\partial\mathbf{V}}\right|_{\mathbf{V}_n} = \mathbf{K}(\mathbf{V}_n) . \tag{4.90}$$

The derivative in the last equation corresponds to the tangent $\mathbf{K}_n$ in Fig. 4.8, which is why this expression is also called a *tangential stiffness matrix*.

In the following, the Newton-Raphson method is explained for the $(n+1)$-th load step. The superscript index i denotes the current number of iteration.

(1) load step $n + 1$:
$$\mathbf{F}^{ext}_{n+1} = \mathbf{F}^{ext}_n + \Delta\mathbf{F}^{ext}_{n+1}$$

initial values ($i = 0$):
$$\mathbf{V}^{(0)}_{n+1} = \mathbf{V}_n, \quad \Delta\mathbf{V}^{(0)}_{n+1} = 0, \quad \mathbf{K}^{(0)}_{n+1} = \mathbf{K}_n, \quad \mathbf{R}^{(0)}_{n+1} = \Delta\mathbf{F}^{ext}_{n+1}$$

(2) iteration loop (i):

- calculation of tangential stiffness: $\mathbf{K}^{(i-1)}_{n+1}\left(\mathbf{V}^{(i-1)}_{n+1}\right)$
- solution of (4.89): $d\Delta\mathbf{V}^{(i)}_{n+1} = \left[\mathbf{K}^{i-1}_{n+1}\right]^{-1}\mathbf{R}^{(i-1)}_{n+1}$
- displacement increment: $\Delta\mathbf{V}^{(i)}_{n+1} = \Delta\mathbf{V}^{(i-1)}_{n+1} + d\Delta\mathbf{V}^{(i)}_{n+1}$
- total displacement: $\mathbf{V}^{(i)}_{n+1} = \mathbf{V}_n + \Delta\mathbf{V}^{(i)}_{n+1}$
- internal forces: $\mathbf{F}^{int} = \mathbf{F}^{int}\left(\mathbf{V}^{(i)}_{n+1}\right)$

- calculation of residual vector: $\mathbf{R}_{n+1}^{(i)} = \mathbf{F}_{n+1}^{\text{ext}} - \mathbf{F}^{\text{int}}\left(\mathbf{V}_{n+1}^{(i)}\right)$
- convergence test: If $\|\mathbf{R}_{n+1}^{(i)}\| >$ tolerance, then $i := i + 1$ and further with (2). Otherwise displacement solution $\mathbf{V}_{n+1} := \mathbf{V}_{n+1}^{(i)}$, next load step $n := n + 1$ and further with (1).

This algorithm is illustrated in Fig. 4.8 for $i = 3$ iterations. Because the numerical complexity for calculating and inverting the tangential stiffness matrix $\mathbf{K}_{n+1}^{(i)}$ in each iteration loop is very high, often the Newton-Raphson method is modified such that $\mathbf{K}_{n+1}$ is only updated after a certain number of iterations, or we use the first approximation $\mathbf{K}_{n+1}^{(1)}$ without any changes. This only slows convergence speed down, but not the accuracy of the solution.

4.5.2 Material Non-linearity

The explanations are limited to non-linear elastic-plastic material behavior with small strains as explained in Sect. A.4.2. The essential relations will be summarized again in matrix form. The stress-strain relation with the elasticity matrix $\mathbf{C}$ reads:

$$\boldsymbol{\sigma} = \mathbf{C}\left(\boldsymbol{\varepsilon} - \boldsymbol{\varepsilon}^{\mathrm{p}}\right). \tag{4.91}$$

The yield criterion is defined with the help of the current state of stress $\boldsymbol{\sigma}$ and the hardening variables $\mathbf{h} = [h_1 h_2 \ldots h_{n_{\mathrm{H}}}]^{\mathrm{T}}$, the number n_{H} and type of which is determined by the chosen hardening model. The following is valid for isotropic and kinematic hardening:

$$\Phi\left(\boldsymbol{\sigma}, \mathbf{h}\right) \begin{cases} < 0 & \text{elastic} \\ = 0 & \text{plastic} \end{cases} \tag{4.92}$$

The plastic strain velocities are calculated from the direction normal to the yield surface, whereby $\dot{\Lambda}$ signifies the plastic multiplier.

$$\dot{\boldsymbol{\varepsilon}}^{\mathrm{p}} = \dot{\Lambda} \frac{\partial \Phi}{\partial \boldsymbol{\sigma}} = \dot{\Lambda} \, \widehat{\mathbf{N}}(\boldsymbol{\sigma}, \mathbf{h}) \tag{4.93}$$

As a rule, the evolution equations for the hardening variables are given in the following form:

$$\dot{\mathbf{h}} = \dot{\Lambda} \, \mathbf{H}(\boldsymbol{\sigma}, \mathbf{h}). \tag{4.94}$$

The Eqs. (4.93) and (4.94) describe the development of plastic strains and hardening variables as a function of the current state. They are ordinary differential equations that are to be integrated over the load history. Both depend on the plastic multiplier Λ, the value of which is determined from the requirement that the yield criterion $\Phi(\boldsymbol{\sigma}, \mathbf{h}) = 0$ must be held with further hardening.

For non-linear FEM analysis, this means that at each integration point IP not only the stresses σ and total strains ε but also the plastic strains ε^p and hardening parameters $\mathbf{h}$ must be calculated, stored and updated. Since the load steps represent finite increments, suitable integration algorithms are required to satisfy the material equations. In principle, all explicit and implicit solution methods for ordinary differential equations can be utilized for this. In the following, the most important and most established method will be explained, which is called *radial return*.

Let's consider the $(n+1)$-th load step in an IP. At the starting point (n) the material equations are all satisfied and all quantities σ_n, ε_n, ε_n^p and $\mathbf{h}_n$ are known. As a result of global FEM analysis we now obtain the total strain $\varepsilon_{n+1} = \varepsilon_n + \Delta\varepsilon_{n+1}$ at the IP. Its increment consists of the elastic and plastic part $\Delta\varepsilon_{n+1} = \Delta\varepsilon_{n+1}^e + \Delta\varepsilon_{n+1}^p$.

The state of stress at the end of the increment is calculated from (4.91):

$$\sigma_{n+1} = \mathbf{C}(\varepsilon_{n+1} - \varepsilon_{n+1}^p) = \sigma_n + \mathbf{C}(\Delta\varepsilon_{n+1} - \Delta\varepsilon_{n+1}^p) = \sigma_{n+1}^{tr} - \mathbf{C}\Delta\varepsilon_{n+1}^p \quad (4.95)$$

whereby so far neither $\Delta\varepsilon_{n+1}^p$ nor $\Delta\mathbf{h}_{n+1}$ are known. For this reason, the method is split into an elastic predictor step (a) and a plastic corrector step (b) (operator split).

(a) *predictor step*
The plastic variables are first frozen to state (n) and the strain increment is understood purely elastically:

$$\Delta\varepsilon_{n+1}^p = \Delta\mathbf{h}_{n+1} = 0, \quad \mathbf{h}_{n+1}^{tr} = \mathbf{h}_n . \quad (4.96)$$

Equation (4.95) thus supplies an »elastic trial value« for the state of stress (superscript index tr):

$$\sigma_{n+1} = \sigma_{n+1}^{tr} \quad \text{mit } \sigma_{n+1}^{tr} = \sigma_n + \mathbf{C}\Delta\varepsilon_{n+1} . \quad (4.97)$$

Inserting into the yield criterion, we find out whether the load step is plastic or elastic:

$$\Phi\left(\sigma_{n+1}^{tr}, \mathbf{h}_{n+1}^{tr}\right) \begin{cases} < 0 & \text{elastic} \\ \geq 0 & \text{plastic} . \end{cases} \quad (4.98)$$

If the state of stress remains within the previous yield surface Φ_n (marked gray), shown in Fig. 4.9 as a dashed line, we have a purely elastic change in stress. The trial value σ_{n+1}^{tr} is already the correct solution, i.e. the algorithm is ended for this load step.

(b) *corrector step*
In the case of a plastic change of state (solid line in Fig. 4.9), the state of stress σ_{n+1}^{tr} according to (4.97) is located outside the yield surface and must be corrected to this. As a result of hardening $\Delta\mathbf{h}_{n+1} \neq 0$, the yield surface changes to Φ_{n+1} at the same time. In addition, now the evolution laws of the plastic variables must be integrated over the load step. With (4.93) we obtain for the plastic strain increment:

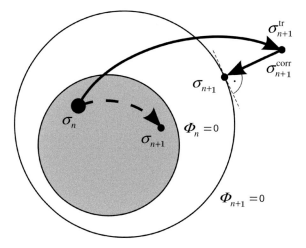

Fig. 4.9 Integration of elastic-plastic material laws with radial return

$$\Delta\varepsilon^p_{n+1} = \int_n^{n+1} \widehat{\mathbf{N}}(\boldsymbol{\sigma}, \mathbf{h})\, d\Lambda \approx \Delta\Lambda_{n+1}\widehat{\mathbf{N}}(\boldsymbol{\sigma}_{n+1}, \mathbf{h}_{n+1}),\qquad(4.99)$$

whereby the variables at the grid point $(n+1)$ at the end of the interval are used as an approximation. Since their values are still unknown, we are concerned with an implicit (backward Euler) method, which is very exact and absolutely stable. We proceed in the same manner with the evolution law (4.94) for the hardening variables:

$$\Delta\mathbf{h}_{n+1} = \mathbf{h}_{n+1} - \mathbf{h}_n = \int_n^{n+1} \mathbf{H}(\boldsymbol{\sigma}, \mathbf{h})\, d\Lambda \approx \Delta\Lambda_{n+1}\mathbf{H}(\boldsymbol{\sigma}_{n+1}, \mathbf{h}_{n+1}).\qquad(4.100)$$

Finally there is still the requirement that the true state of stress must lie on the yield surface, i.e.

$$\Phi_{n+1}(\boldsymbol{\sigma}_{n+1}, \mathbf{h}_{n+1}) = 0.\qquad(4.101)$$

Therefore, according to (4.95) the »trial stress« $\boldsymbol{\sigma}^{tr}_{n+1}$ must be reduced by the term $\boldsymbol{\sigma}^{corr}_{n+1} = \mathbf{C}\,\Delta\varepsilon^p_{n+1}$, i.e. with (4.99) holds

$$\boldsymbol{\sigma}_{n+1} = \boldsymbol{\sigma}^{tr}_{n+1} - \Delta\Lambda_{n+1}\mathbf{C}\widehat{\mathbf{N}}(\boldsymbol{\sigma}_{n+1}, \mathbf{h}_{n+1}),\qquad(4.102)$$

which is illustrated by the arrow $\boldsymbol{\sigma}^{corr}_{n+1}$ in Fig. 4.9. Since this correction stress has the direction of the normal $\widehat{\mathbf{N}}$ on the yield surface, we speak of »radial return«. The relations (4.100), (4.101) and (4.102) form a non-linear system of $(6 + n_H + 1)$ equations for the determination of the unknown quantities $\boldsymbol{\sigma}_{n+1}$, $\Delta\mathbf{h}_{n+1}$ and $\Delta\Lambda_{n+1}$.

This can again be solved iteratively with the Newton method by formulating the equations as residuals, the zeros of which must be found:

$$\mathbf{r}_\sigma = \sigma_{n+1} + \Delta\Lambda_{n+1}\widehat{\mathbf{C}\mathbf{N}}(\sigma_{n+1}, \mathbf{h}_{n+1}) - \sigma^{\mathrm{tr}}_{n+1} \qquad \to 0$$
$$\mathbf{r}_h = \Delta\mathbf{h}_{n+1} - \Delta\Lambda_{n+1}\mathbf{H}(\sigma_{n+1}, \mathbf{h}_{n+1}) \qquad \to 0$$
$$\mathbf{r}_\Phi = \Phi(\sigma_{n+1}, \mathbf{h}_{n+1}) \qquad \to 0 . \qquad (4.103)$$

For the treatment of large deformations and other non-linear material laws, the reader is referred to the relevant literature [11, 12].

4.5.3 Geometrical Non-linearity

In order to clarify the effects of large displacements and large strains on non-linear FEM relations, we will return to the basic equations of continuum-mechanics in the initial configuration, see Sects. A.2 and A.3. The starting point is the principle of virtual displacements in the Lagrangian form (4.16), which reads in symbolic notation ($: \widehat{=}$ doubled scalar product):

$$\delta\mathcal{W}_{\mathrm{int}} \widehat{=} \int_V \boldsymbol{T} : \delta\boldsymbol{E}\,\mathrm{d}V = \int_V \varrho_0\bar{\boldsymbol{f}} \cdot \delta\boldsymbol{u}\,\mathrm{d}V + \int_A \hat{\bar{\boldsymbol{T}}} \cdot \delta\boldsymbol{u}\,\mathrm{d}A \widehat{=} \delta\mathcal{W}_{\mathrm{ext}} . \qquad (4.104)$$

The variation $\delta\boldsymbol{E}$ of the Green-Lagrange strain tensor can be expressed according to (A.20) and (A.17) by the variations of the displacement vectors and deformation gradients:

$$\delta\boldsymbol{E} = \frac{1}{2}\left\{\nabla\delta\boldsymbol{u} + (\nabla\delta\boldsymbol{u})^{\mathrm{T}} + (\nabla\delta\boldsymbol{u}) \cdot (\nabla\boldsymbol{u}) + (\nabla\boldsymbol{u}) \cdot (\nabla\delta\boldsymbol{u})\right\}$$

$$\Delta\boldsymbol{E} = \frac{1}{2}\left\{\nabla\Delta\boldsymbol{u} + (\nabla\Delta\boldsymbol{u})^{\mathrm{T}} + (\nabla\Delta\boldsymbol{u}) \cdot (\nabla\boldsymbol{u}) + (\nabla\boldsymbol{u}) \cdot (\nabla\Delta\boldsymbol{u})\right\} \qquad (4.105)$$

$$\delta\boldsymbol{E} = \delta\left\{\frac{1}{2}\left(\boldsymbol{F}^{\mathrm{T}} \cdot \boldsymbol{F} + \boldsymbol{I}\right)\right\} = \frac{1}{2}\left\{\boldsymbol{F}^{\mathrm{T}} \cdot \delta\boldsymbol{F} + \delta\boldsymbol{F}^{\mathrm{T}} \cdot \boldsymbol{F}\right\} \qquad (4.106)$$

$$\delta\boldsymbol{F} = \delta(\nabla\boldsymbol{u} + \boldsymbol{I}) = \nabla(\delta\boldsymbol{u}) \qquad \text{or} \qquad \delta F_{iN} = \frac{\partial\delta u_i}{\partial X_N} \qquad (4.107)$$

Since the internal $\mathcal{W}_{\mathrm{int}}(\boldsymbol{u})$ and external work $\mathcal{W}_{\mathrm{ext}}(\boldsymbol{u})$ are now non-linear functions of the displacements $\boldsymbol{u}$ in a finite element, we carry out a linearization of the system at the previously reached already equilibrated state (index 0), i.e. we investigate the changes of all quantities in the functional (4.104) in the case of a displacement increment $\Delta\boldsymbol{u}$. For this, we use the linearization of a tensor-valued function $\boldsymbol{A}(\boldsymbol{a})$ at location $\boldsymbol{a}_0$ with respect to the vectorial variables $\Delta\boldsymbol{a}$ (directional derivation).

$$A(a_0 + \Delta a) = A(a_0) + \Delta A \qquad \text{mit } \Delta A = \frac{\mathrm{d}A}{\mathrm{d}a}\bigg|_{a_0} \cdot \Delta a \qquad (4.108)$$

For simplicity's sake, it is assumed that the external forces do not depend on the deformation (conservative, no contact, friction or co-moving forces), i. e. the external work δW_{ext} remains linear in (4.104) with respect to δu.

$$\Delta(\delta W_{\mathrm{int}}) = \int_V (\Delta T : \delta E + T : \Delta \delta E)\,\mathrm{d}V. \qquad (4.109)$$

In the 1st integral term, the 2nd Piola-Kirchhoff stress tensor (A.51) should be linearized with respect to Δu, which is done using the chain rule via the strain tensor ΔE and (4.105). The fourth-order tensor C^{ep} denotes the material tangent at this point.

$$\Delta T = \frac{\partial T}{\partial E} : \Delta E = C^{\mathrm{ep}} : \Delta E, \qquad C^{\mathrm{ep}} = \frac{\partial T}{\partial E}\bigg|_{T_0} \qquad (4.110)$$

Linearization of (4.106) yields

$$\Delta \delta E = \frac{1}{2}\left\{\Delta F^{\mathrm{T}} \cdot \delta F + \delta F^{\mathrm{T}} \cdot \Delta F\right\} \text{ or } \Delta \delta E_{MN} = \frac{1}{2}\left\{\Delta F_{iM}\delta F_{iN} + \delta F_{iM}\Delta F_{iN}\right\}. \qquad (4.111)$$

Due to the symmetry of the stress tensor T, the 2nd term of the integrand in (4.109) becomes

$$T_{MN}\Delta \delta E_{MN} = \delta F_{iM}T_{MN}\Delta F_{iN}, \qquad (4.112)$$

whereby we obtain the linearization of internal virtual work:

$$\Delta(\delta W_{\mathrm{int}}) = \int_V \left(\delta E : C^{\mathrm{ep}} : \Delta E + \delta F^{\mathrm{T}} \cdot T \cdot \Delta F\right)\mathrm{d}V. \qquad (4.113)$$

The FEM realization of this algorithm is briefly outlined using again the example of a two-dimensional element under plane stress assumption. To this end, we will switch to matrix notation. In contrast to Sect. 4.3.1, now the location vectors will be interpolated in spatial $\mathbf{x}$ and material $\mathbf{X}$ representation with the help of isoparametric variables $\boldsymbol{\xi}$ in the element, whereby $\hat{\mathbf{x}}$, $\hat{\mathbf{X}}$ and $\hat{\mathbf{u}} \equiv \mathbf{v}$ designate the respective nodal values:

$$\mathbf{x}(\boldsymbol{\xi}) = \mathbf{N}(\boldsymbol{\xi})\,\hat{\mathbf{x}}, \qquad \mathbf{X}(\boldsymbol{\xi}) = \mathbf{N}(\boldsymbol{\xi})\,\hat{\mathbf{X}}. \qquad (4.114)$$

From the kinematic relationships, we obtain the functions for the displacements $\mathbf{u}$ in the element as well as their variation $\delta \mathbf{u}$ and increment $\Delta \mathbf{u}$:

$$\hat{\mathbf{x}} = \hat{\mathbf{X}} + \hat{\mathbf{u}} = \hat{\mathbf{X}} + \mathbf{v} \tag{4.115}$$

$$\mathbf{u}(\boldsymbol{\xi}) = \mathbf{N}(\boldsymbol{\xi})\,\mathbf{v}, \quad \delta\mathbf{u}(\boldsymbol{\xi}) = \mathbf{N}(\boldsymbol{\xi})\,\delta\mathbf{v}, \quad \Delta\mathbf{u}(\boldsymbol{\xi}) = \mathbf{N}(\boldsymbol{\xi})\,\Delta\mathbf{v} \tag{4.116}$$

The three components of the strain tensor are varied in accordance with (4.105), whereby the linear term represents the known $\mathbf{B}$ matrix (4.44) and the quadratic terms lead to a non-linear relation $\mathbf{B}^{\mathrm{nlin}}(\mathbf{v})$:

$$\delta\mathbf{E} = \begin{bmatrix} \delta E_{11} \\ \delta E_{22} \\ \delta E_{12} \end{bmatrix} = \bar{\mathbf{D}}(\mathbf{u})\delta\mathbf{u} = \bar{\mathbf{B}}(\mathbf{v})\delta\mathbf{v} \quad \text{and} \quad \Delta\mathbf{E} = \bar{\mathbf{B}}(\mathbf{v})\Delta\mathbf{v}, \tag{4.117}$$

$$\bar{\mathbf{B}} = \mathbf{B} + \mathbf{B}^{\mathrm{nlin}}(\mathbf{v}), \tag{4.118}$$

$$\mathbf{B}^{\mathrm{nlin}} = [\mathbf{B}_1 \dots \mathbf{B}_a \dots \mathbf{B}_{n_{\mathrm{K}}}]$$

$$\mathbf{B}_a = \begin{bmatrix} u_{1,1}N_{a,1} & u_{2,1}N_{a,1} \\ u_{1,2}N_{a,2} & u_{2,2}N_{a,2} \\ u_{1,1}N_{a,2} + u_{1,2}N_{a,1} & u_{2,1}N_{a,2} + u_{2,2}N_{a,1} \end{bmatrix}. \tag{4.119}$$

In a similar way as shown in (4.59), we find the derivation of the isoparametric displacement functions with respect to the material coordinates in order to determine the variation $\delta\mathbf{F}$ and linearization $\Delta\mathbf{F}$ of the deformation gradient from (4.107) :

$$\delta\mathbf{F} = \sum_{a=1}^{n_{\mathrm{K}}} \frac{\partial N_a(\boldsymbol{\xi})}{\partial\mathbf{X}}\delta\mathbf{u}^{(a)} = \bar{\mathbf{H}}\delta\mathbf{v} \quad \text{and} \quad \Delta\mathbf{F} = \bar{\mathbf{H}}\Delta\mathbf{v}. \tag{4.120}$$

Inserting the relations (4.116)–(4.120) into (4.113) results in the virtual internal work for one increment $\Delta\mathbf{v}$ of the nodal displacements of an element V_e:

$$\Delta\left(\delta W_{\mathrm{int}}(\mathbf{v}, \delta\mathbf{v})\right) = \delta\mathbf{v}^{\mathrm{T}}\left\{ \underbrace{\int_{V_e} \bar{\mathbf{B}}^{\mathrm{T}} \mathbf{C}^{\mathrm{ep}} \bar{\mathbf{B}}\, dV_e}_{\mathbf{k}^{\mathrm{nlin}}} + \underbrace{\int_{V_e} \bar{\mathbf{H}}^{\mathrm{T}} \mathbf{T} \bar{\mathbf{H}}\, dV_e}_{\mathbf{k}^{\mathrm{sp}}} \right\}\Delta\mathbf{v}. \tag{4.121}$$

The expression in curly brackets represents the *tangential stiffness matrix* $\mathbf{k}(\mathbf{v})$ at point 0, which consists of two components. The first term $\mathbf{k}^{\mathrm{nlin}}$ is due to the non-linear displacement-strain relation $\bar{\mathbf{B}}$ and the material tensor $\mathbf{C}^{\mathrm{ep}}$, which in the case of non-linear material laws also depends on $\mathbf{v}$ (see previous Sect. 4.5.2). In addition, we get the so-called *geometric stiffness matrix* or *initial stress matrix* $\mathbf{k}^{\mathrm{sp}}$, which reflects the effect of the current state of stress $\mathbf{T}$ on the altered geometry $\bar{\mathbf{H}}$.

$$\mathbf{k}(\mathbf{v}) = \mathbf{k}^{\mathrm{nlin}}(\mathbf{v}) + \mathbf{k}^{\mathrm{sp}}(\mathbf{T}) \tag{4.122}$$

Upon assembling all element components (4.122) we obtain the tangential system stiffness matrix $\mathbf{K}$ for the linearized equation system of FEM.

$$\mathbf{K}(\mathbf{V}_0)\,\Delta\mathbf{V} = -\mathbf{R} := \mathbf{F}^{\text{ext}} - \mathbf{F}^{\text{int}} \qquad (4.123)$$

Its iterative solution is best carried out with the Newton- Raphson method (see Sect. 4.5.1).

The algorithm described here is referred to as the *total Lagrangian method* in FEM literature. Further explanations of FEM with physical and geometrical non-linearities can be found in [11–13].

4.6 Explicit FEA for Dynamic Problems

In order to analyse impact loading of cracks and highly dynamic crack propagation processes, which are mostly associated with contact problems, the technique of explicit time integration has become the most established one in FEM. As opposed to implicit time integration schemes (e. g. Newmark), the explicit method is numerically highly effective and robust. A disadvantage is its limited stability, which necessitates keeping to very small time steps $\Delta t_n \le \Delta t_c$.

With the first and second time derivation of the displacement functions (4.40), we obtain the velocities and accelerations from the nodal values $\mathbf{v}$:

$$\dot{\mathbf{u}} = \mathbf{N}\dot{\mathbf{v}}, \qquad \ddot{\mathbf{u}} = \mathbf{N}\ddot{\mathbf{v}}. \qquad (4.124)$$

For damping-free systems, the equations of motion follow from the Hamiltonian principle. After FEM-discretization according to Sect. 4.3, the stiffness relation (4.49) can be extended by the inertial term $\mathbf{m}\ddot{\mathbf{v}}$:

$$\mathbf{m}\ddot{\mathbf{v}}(t) + \mathbf{k}\mathbf{v}(t) = \mathbf{f}_b + \mathbf{f}_t + \mathbf{f}_\varepsilon = \mathbf{f}(t)$$

$$\mathbf{M}\ddot{\mathbf{V}}(t) + \mathbf{K}\mathbf{V}(t) = \mathbf{F}_b + \mathbf{F}_t + \mathbf{F}_\varepsilon = \mathbf{F}(t) \qquad (4.125)$$

$$\text{mass matrix: } \mathbf{m} = \int_{V_e} \mathbf{N}^{\text{T}}\rho\,\mathbf{N}\,\mathrm{d}V\,, \qquad \mathbf{M} = \bigcup_{e=1}^{n_{\text{E}}} \mathbf{m}_e\,. \qquad (4.126)$$

In principle, we can also integrate the element mass matrix (4.126) and assemble the system mass matrix $\mathbf{M}$ applying the algorithms shown in Sect. 4.3. Using the shape functions, we then obtain the so-called *consistent mass matrix*, which has symmetry and a band structure. This results in a relatively high computational effort in solving (4.125), since $\mathbf{M}$ must be inverted. This can be reduced considerably if we use a *lumped mass matrix* in diagonal form. In this case, inversion of $\mathbf{M}$ is trivial so that (4.125) can be solved directly for the sought nodal accelerations $\ddot{\mathbf{v}}$. Various interpolation and integration techniques exist for determining the diagonalized mass matrix [14, 15].

As a rule, the explicit central difference scheme is used to solve the 2nd order differential Eq. (4.125) with respect to time. Thereby, the dynamic quantities at the end of the $(n + 1)$th time step $t_{n+1} = t_n + \Delta t_{n+1}$ are calculated from the principle of linear momentum (4.125) at the beginning of the interval t_n. The velocities in the middle of the interval $t_{n+\frac{1}{2}} = t_n + \frac{1}{2}\Delta t_{n+1}$ are obtained from the differential quotient

$$\dot{\mathbf{v}}_{n+\frac{1}{2}} = (\mathbf{v}_{n+1} - \mathbf{v}_n)/\Delta t_{n+1}, \qquad \dot{\mathbf{v}}_{n-\frac{1}{2}} = (\mathbf{v}_n - \mathbf{v}_{n-1})/\Delta t_n, \qquad (4.127)$$

from which the acceleration at t_n follows:

$$\mathbf{a}_n = \ddot{\mathbf{v}}_n = (\dot{\mathbf{v}}_{n+\frac{1}{2}} - \dot{\mathbf{v}}_{n-\frac{1}{2}})/\left[\frac{1}{2}(\Delta t_{n+1} + \Delta t_n)\right]. \qquad (4.128)$$

Hence, the following algorithm is derived:

(0) determination of initial conditions $\mathbf{v}_0$ and $\dot{\mathbf{v}}_0$
(1) incrementing of time and external load $\mathbf{f}(t)$
(2) calculation of acceleration in time step Δt_{n+1} from (4.125):

$$\mathbf{a}_n = \mathbf{m}^{-1}(\mathbf{f} - \mathbf{k}\,\mathbf{v}_n)$$

(3) determination of velocity with (4.128):

$$\dot{\mathbf{v}}_{n+\frac{1}{2}} = \dot{\mathbf{v}}_{n-\frac{1}{2}} + \frac{\Delta t_{n+1} + \Delta t_n}{2}\mathbf{a}_n$$

(4) updating the displacements with (4.127):

$$\mathbf{v}_{n+1} = \mathbf{v}_n + \Delta t_{n+1}\dot{\mathbf{v}}_{n+\frac{1}{2}}$$

(5) next time step $n := n + 1$, go to (4.6)

The explicit algorithm requires neither iterations nor inversion of the stiffness matrix, even the assemblage of element matrices is superfluous. Material and geometrical non-linearities are taken into consideration simply with $\mathbf{k}(\mathbf{v}_n)$.

The maximum admissible time step length Δt_c is inversely proportional to the highest angular eigenfrequency $\omega_{\max}$ of the system. In practice, Δt_c is estimated by that time, which requires a dilation wave with a material-dependent propagation velocity c_d to pass through the smallest finite element of length $L_{\min}$, whereby b is still an empirical factor:

$$\Delta t \le \Delta t_c = b\frac{L_{\min}}{c_d}. \qquad (4.129)$$

4.7 Procedure of a Finite Element Analysis

In the following, the essential procedural steps are compiled necessary for the FEM calculation of a component. The special tasks needed for dealing with cracks are emphasized in *italic*.

4.7.1 PRE-processing

(a) generation of the FEM mesh (node coordinates, elements, topology)
(b) *special mesh generators for cracks (special elements, geometrical particularities)*
(c) specification of loading and bearing conditions
(d) input of material properties
(e) specification of the temporal load program
(f) supervision and visualization of the input model

4.7.2 FEM-processing

(a) assemblage and storage of the stiffness $\mathbf{K}(t)$ and mass matrix $\mathbf{M}$
(b) calculation of the right hand side $\mathbf{F}(t)$ from all loads
(c) *crack tip elements and special algorithms for cracks*
(d) incorporation of kinematic boundary conditions
(e) solution of the (non-)linear FEM system of equations
(f) storage of results (solution vector $\mathbf{V}(t)$ et al.)
(g) incremental increase of loads for the next time step for non-linear or dynamic analyses; if required, repetition of this loop

4.7.3 POST-processing

(a) calculation of pertinent field quantities (u_i, ε_{ij}, σ_{ij}, T) for desired locations and moments from the solution vector $\mathbf{V}(t)$
(b) nodal data: $\mathbf{v}^{(a)}$, $\dot{\mathbf{v}}^{(a)}$, $\ddot{\mathbf{v}}^{(a)}$, $\mathbf{f}$, T, h_R
(c) element data (integration points): ε_{ij}, σ_{ij}, ε_{max}, σ_{max}, ε_v, σ_v, U
(d) graphical representation: isoline images, temporal course, deformed structure, animation of motions etc.
(e) *specific analyses (EDI-integral, MCCI-closure integral, DIM etc.) to determine fracture mechanical loading parameters (K-factors, G, J)*

References

1. Washizu K (1975) Variational methods in elasticity and plasticity. Pergamon Press, Oxford
2. Pian THH, Tong P (1969) Basis of finite element methods for solid continua. Int J Numer Methods Eng 1:3–28
3. Atluri SN, Kobayashi AS, Nakagaki M (1975) An assumed displacement hybrid finite element model for linear fracture mechanics. Int J Fract 11:257–271
4. Atluri SN, Kartiresan K (1979) 3D analysis of surface flaws in thick-walled reactor pressure vessels using displacement-hybrid finite element method. Nucl Eng Des 51:163–176
5. Tong P, Pian THH, Lasry H (1973) A hybrid element approach to crack problems in plane elasticity. Int J Numer Methods Eng 7:297–308
6. Ueberhuber C (1995) Computer-Numerik. Springer, Berlin
7. Bathe KJ (1986) Finite-Elemente-Methoden. Matrizen und lineare Algebra. Springer, Berlin
8. Schwarz HR (1991) Methoden der finiten Elemente. Teubner Studienbücherei
9. Zienkiewicz OC, Taylor RL, Zhu JZ (2005) The finite element method: its basis and fundamentals, vol 1. Elsevier, Amsterdam
10. Stroud AH (1971) Approximate calculation of multiple integrals. Prentice Hall, Englewood Cliffs
11. Wriggers P (2001) Nichtlineare finite-elemente-methoden. Springer, Berlin
12. Simo JC, Hughes TJR (1998) Computational inelasticity. Springer, New York
13. Belytschko T, Liu WK, Moran B (2001) Nonlinear finite elements for continua and structures. Wiley, Chichester
14. Hughes TJR (1987) The finite element method. Prentice-Hall, Englewood Cliffs
15. Belytschko T, Hughes TJR (1983) Computational methods for transient analysis. North-Holland, Amsterdam

Chapter 5
FE-Techniques for Crack Analysis in Linear-Elastic Structures

The goal of a FEM analysis is the calculation of fracture-mechanical loading parameters for a crack in a structure (test piece, component, material's microstructure) in the case of linear-elastic (isotropic or anisotropic) material behavior. In Sect. 3.2 the relevant loading parameters of LEFM were introduced: the stress intensity factors K_I, K_{II}, K_{III} and the energy release rate $G \equiv J$. Their values depend on the geometry of the structure, its load, the length and shape of the crack and on the material's elastic properties.

Although FEM can be directly applied to solve a BVP, its use in crack problems involves a fundamental difficulty. This difficulty lies in the exact determination of the singularity at the crack tip with the help of a numerical approximation method such as FEM. Conventional finite element types only have regular polynomial functions for u_i, ε_{ij} and σ_{ij}. Therefore, they reproduce the crack singularity poorly. For this reason, special element functions, numerical algorithms or evaluation techniques are needed to obtain loading parameters from a FEM solution efficiently and accurately. In the following chapter, we will introduce the methods that have been developed for this, concentrating mainly on stationary cracks. The particularities of FEM techniques and meshes in analyzing unsteady cracks will be dealt with in Chap. 8.

5.1 Interpreting the Numerical Solution at the Crack Tip

If only regular standard elements (RSE) are available to us, a very fine mesh is required at the crack tip. One should again bear in mind that, with the stress intensity factor K_I, we are looking for the *coefficient of a singularity*! This means firstly that the finer we make the mesh, the greater ($\to \infty$) the stresses become. Secondly, the discretization has to be so fine that the field quantities within the near field solution are resolved with sufficient accuracy.

Figure 5.1 demonstrates a typical mesh for crack problems using the example of a Griffith crack in a finite plate under tensile load. Because of the symmetry of the

M. Kuna, *Finite Elements in Fracture Mechanics*, Solid Mechanics and Its Applications 201, DOI: 10.1007/978-94-007-6680-8_5,
© Springer Science+Business Media Dordrecht 2013

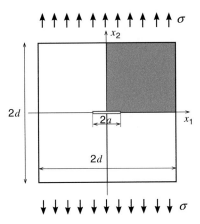

Fig. 5.1 Griffith crack in a finite plate under tensile load

problem with respect to the x_1 and x_2 axes, it suffices to model only one quadrant (shaded in gray). The normal displacements ($u_2 = 0$ and $u_1 = 0$) must be prevented on the axes of symmetry ($x_1 = 0$ and $x_2 = 0$). On the upper edge, the tensile stresses are applied with equivalent nodal forces, and the crack faces remain free of load. Figure 5.2 shows the FEM model using quadrilateral elements. For stationary cracks, a polar FEM mesh with a strong concentric refinement around the crack tip is effective (see detail right). The smallest elements at the crack tip must have a size L that is still considerably below the validity range r_K of the K_I-dominated near field. In order to reproduce properly the angular distribution, a sufficient number of elements must also be distributed over the circumference. With the estimation $r_K \approx a/50 \dots a/10$, the following relation should thus be observed:

- element size at the crack tip: $L < a/100 \dots a/20$
- number of elements/semicircle: $n > 6$ or $\Delta\theta < 30°$

The simplest method to determine the stress intensity factor K_I is directly to compare the FEM solution (3.12) and (3.16) with the near field solution for mode I:

$$u_i(r, \theta) = \frac{1}{2\mu}\sqrt{\frac{r}{2\pi}} K_I g_i^I(\theta), \quad \sigma_{ij}(r, \theta) = \frac{K_I}{\sqrt{2\pi r}} f_{ij}^I(\theta). \tag{5.1}$$

For a selected point (r^*, θ^*), we can thus calculate a value K_I^* either from the displacements u_i^{FEM} or the stresses σ_{ij}^{FEM} by converting (5.1):

$$K_I^*(r^*, \theta^*) = 2\mu\sqrt{\frac{2\pi}{r^*}} \frac{u_i^{FEM}(r^*, \theta^*)}{g_i^I(\theta^*)}, \quad K_I^*(r^*, \theta^*) = \frac{\sqrt{2\pi r^*}\,\sigma_{ij}^{FEM}(r^*, \theta^*)}{f_{ij}^I(\theta^*)}. \tag{5.2}$$

The displacements are best taken from values at the nodes, while the stresses are generally given at the integration points, having there the highest accuracy (see Fig. 5.3).

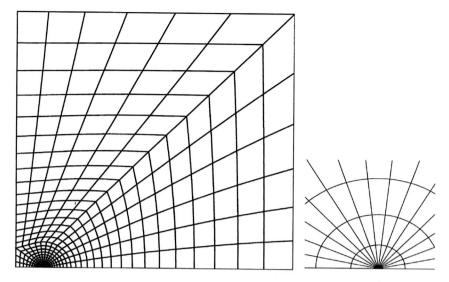

Fig. 5.2 FEM mesh with enlarged detail

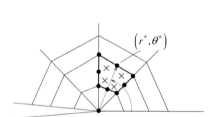

Fig. 5.3 Evaluation of the FEM solution at a node (r^*, θ^*)

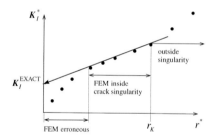

Fig. 5.4 Validity range of the FEM interpretation

Figure 5.4 shows a typical result for $K_I^*(r^*)$ along a radial ray $\theta^* = $ const. We can distinguish three domains here: The elements very close to the crack tip can only represent the singularity inaccurately, which is why K_I^* falls too short in comparison to the exact value of K_I. At the mid-range, the quality of the FEM solution is sufficient. Outside the dominance region $r^* > r_K$, Eq. (5.2) loses its justification, since further solution terms arise in addition to the singularity. Therefore, a linear extrapolation of the function $K_I^*(r^*)$ from the mid-range towards the crack tip $r^* \to 0$ is suggested as a pragmatic technique. This interpretation yields the best results at the following positions:

(a) values of crack-opening displacements at the crack face (relative to a possible displacement $u_2^{\mathrm{FEM}}(0)$ of the crack tip node)

$$K_{\mathrm{I}} = \lim_{r^* \to 0} u_2^{\mathrm{FEM}}(r^*, \theta = \pi) \frac{E'}{4} \sqrt{\frac{2\pi}{r^*}}. \tag{5.3}$$

$E' = E$ (plane stress)   and $E' = E/(1 - \nu^2)$ (plane strain)

(b) values of normal stresses on the ligament in front of the crack

$$K_{\mathrm{I}} = \lim_{r^* \to 0} \sigma_{22}^{\mathrm{FEM}}(r^*, \theta = 0) \sqrt{2\pi r^*}. \tag{5.4}$$

Occasionally, a doubled logarithmic plot of (5.4) is recommended

$$\ln[2\pi\sigma_{22}(r^*)] - \ln K_{\mathrm{I}} = -\frac{1}{2} \ln r^*, \tag{5.5}$$

whereby the reproduction quality of the crack singularity can be controlled by observing the slope (factor $-1/2$), and K_{I} results from the intersection with the ordinate.

The interpretation till now was focused on pure mode I loading. It can be applied analogously to two-dimensional crack geometries loaded exclusively in mode II or mode III (see Sect. 3.2.1 and Fig. 3.6). In these cases, the crack lies on a symmetry plane of the body (x_1-axis), so a half model is sufficient for the FEM. Due to its symmetry and anti-symmetry properties, the following boundary conditions are to be observed on the ligament ($|x_1| > 0$, $x_2 = 0$): mode II: $u_1 = \sigma_{22} = 0$, mode III: $u_3 = 0$. Using the asymptotic solutions (3.25) and (3.23) for mode II or (3.31) and (3.32) for mode III, we obtain formulae for the determination of the K-factors from the crack face displacements or ligament stresses:

$$K_{\mathrm{II}} = \lim_{r^* \to 0} u_1^{\mathrm{FEM}}(r^*, \pi) \frac{E'}{4} \sqrt{\frac{2\pi}{r^*}}, \qquad K_{\mathrm{II}} = \lim_{r^* \to 0} \tau_{21}^{\mathrm{FEM}}(r^*, 0) \sqrt{2\pi r^*} \tag{5.6}$$

$$K_{\mathrm{III}} = \lim_{r^* \to 0} u_3^{\mathrm{FEM}}(r^*, \pi) \frac{E}{4(1 + \nu)} \sqrt{\frac{2\pi}{r^*}}, \qquad K_{\mathrm{III}} = \lim_{r^* \to 0} \tau_{23}^{\mathrm{FEM}}(r^*, 0) \sqrt{2\pi r^*} \tag{5.7}$$

In the general loading case of a crack, all three modes I, II, and III superimpose, and all symmetry properties are lost. However, the modes of crack opening separate on the crack faces, so the K-factors can be determined from the relative displacements $\Delta u_i(r^*) = u_i(r^*, \theta = +\pi) - u_i(r^*, \theta = -\pi)$ of two opposing nodes on the crack faces

$$\begin{Bmatrix} K_{\mathrm{I}} \\ K_{\mathrm{II}} \\ K_{\mathrm{III}} \end{Bmatrix} = \lim_{r^* \to 0} \sqrt{\frac{2\pi}{r^*}} \begin{Bmatrix} \frac{E'}{8} \Delta u_2(r^*) \\ \frac{E'}{8} \Delta u_1(r^*) \\ \frac{E}{8(1+\nu)} \Delta u_3(r^*) \end{Bmatrix}. \tag{5.8}$$

This evaluation formula is also valid for three-dimensional crack configurations, whereby $u_i(r^*, \theta^*)$ then refers to a local coordinate system perpendicular to the considered point on the crack front (see Fig. 3.10), which requires a corresponding transformation from the global system.

All previously given relations are valid for an isotropic material, for which each stress intensity factor is associated exactly with the displacement component of the respective crack opening mode. In the case of anisotropic material however, the displacement components on the crack faces are linked with all K-factors, resulting in a linear system of equations. For the special case of orthotropy from Sect. 3.2.7, we find

$$\left\{ \begin{array}{c} K_I \\ K_{II} \\ K_{III} \end{array} \right\} = \frac{1}{4} \left\{ \begin{array}{ccc} H_{11} & H_{12} & 0 \\ H_{21} & H_{22} & 0 \\ 0 & 0 & H_{33} \end{array} \right\} \lim_{r^* \to 0} \sqrt{\frac{2\pi}{r^*}} \left\{ \begin{array}{c} \Delta u_1(r^*) \\ \Delta u_2(r^*) \\ \Delta u_3(r^*) \end{array} \right\}, \qquad (5.9)$$

where the components of the matrix H_{ij} depend on the elastic constants:

$$\begin{bmatrix} H_{11} & H_{12} \\ H_{21} & H_{22} \end{bmatrix} = \frac{1}{H_{11}H_{22} - H_{12}H_{21}} \begin{bmatrix} \Im\left(\dfrac{q_1 - q_2}{s_1 - s_2}\right) & \Im\left(\dfrac{p_2 - p_1}{s_1 - s_2}\right) \\ \Im\left(\dfrac{s_1 q_2 - s_2 q_1}{s_1 - s_2}\right) & \Im\left(\dfrac{s_2 p_1 - s_1 p_2}{s_1 - s_2}\right) \end{bmatrix}$$

$$H_{33} = \Im(c_{45} + s_3 c_{44}) \qquad (5.10)$$

This direct *displacement interpretation method* (DIM) and *stress interpretation method* (SIM) are the simplest techniques to determine the K-factors. Because of the relatively arbitrary extrapolation (see Fig. 5.4), they also have the lowest level of accuracy. Nonetheless, they are suited for a rough interpretation of directly available FEM results by simple manual calculation.

5.2 Special Finite Elements at the Crack Tip

5.2.1 Development of Crack Tip Elements

The unsatisfactory solution quality of regular elements was recognized already in the 1970s. This led to the development of special element formulations, in which the shape functions contain singular crack-specific functions the free parameters of which are related to the K factors. Special elements of this type are called *crack tip elements* (CTE). They are utilized to discretize the direct surroundings of the crack tip, while regular elements are then used to model the rest of the structure. These CTEs can embed a crack tip entirely if their shape functions describe complete crack tip fields in (r, θ) coordinates (see [1, 2].) However, we usually restrict ourselves to the reproduction of the radial $r^{-1/2}$ singularity, which is why the angular dependence θ must be modeled with fan-shaped element arrangements around the crack tip. The most important developments of two-dimensional [3–6] and three-dimensional crack tip elements [7, 8] should be cited.

The biggest problem with such element formulations is that the singular crack functions are not compatible with the regular shape functions on the boundaries to neighboring elements. Also, their shape functions often do not permit rigid body motions or constant states of strain, which is the prerequisite for the convergence of the solution. A further disadvantage of many crack tip elements is that they have not been incorporated into commercial FEM programs due to their algorithmic peculiarities and thus are only usable by a few specialists.

5.2.2 Modified Isoparametric Displacement Elements

Decisive progress was made in this respect by the discovery of so-called *quarter-point elements* (QPE) made independently by Henshell and Shaw [9] and Barsoum [10]. The basic idea consists in modifying isoparametric elements with a quadratic shape function such that the position of the mid-side nodes is altered. That is, we shift the coordinates of these nodes from the middle to the quarter-point position in the direction of the crack tip for all element edges that point to the crack tip. This effects a change of the displacement, strain and stress fields in the element into a form, which corresponds exactly to the radial function of the crack tip field. The emergence of a singular $1/\sqrt{r}$ behavior is caused by the non-linear mapping between the natural (isoparametric) and local (geometrical) coordinates $\xi \rightarrow x$. Therefore, we refer to them as »nodal-distorted« shape functions.

Before we look more closely at different types of quarter-point elements, the principle should be illustrated using a one-dimensional element 1D or one element edge.

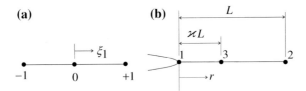

Fig. 5.5 One-dimensional quarter-point element: **a** natural coordinates, **b** local Cartesian coordinates

(a) One-Dimensional Quarter-Point Element

Figure 5.5b shows a 1D quarter-point element at the crack tip in geometrical space, the distance of the three nodes 1, 3 and 2 is given by the coordinate r. The position of the middle node 3 is controlled by the parameter $\varkappa$. Figure 5.5a refers to the parameter space $\xi\,(\widehat{=}\,\xi_1)$. The quadratic 1D displacement function of this element is:

$$u(\xi) = \sum_{a=1}^{3} N_a(\xi)\,u^{(a)} = \frac{1}{2}\xi(\xi-1)u^{(1)} + (1-\xi^2)u^{(3)} + \frac{1}{2}\xi(\xi+1)u^{(2)}$$

$$= u^{(3)} + \frac{1}{2}\left(u^{(2)} - u^{(1)}\right)\xi + \left[\frac{1}{2}\left(u^{(1)} + u^{(2)}\right) - u^{(3)}\right]\xi^2 \qquad (5.11)$$

The last equation was ordered by powers of ξ. Due to the isoparametric element formulation, the same interpolation function also holds for the coordinates, i.e. as well for the radius r with the nodal values $r^{(1)} = 0,\ r^{(3)} = \varkappa L,\ r^{(2)} = L$:

$$r(\xi) = \sum_{a=1}^{3} N_a(\xi)\,r^{(a)} = \varkappa L + \frac{1}{2}L\xi + \left(\frac{1}{2} - \varkappa\right)L\xi^2 \qquad (5.12)$$

In the case of a regular 1D element, $\varkappa = 1/2$ would be valid, and from (5.12) would then follow $\xi = 2r/L - 1$. Insertion of this linear relation into (5.11) results as well in a polynomial of the 2nd degree for the displacement $u(r)$. But if we shift node (3) to the quarter-point position $\varkappa = 1/4$, then (5.12) provides instead the relation

$$r = \frac{L}{4}(1+\xi)^2 \quad \Rightarrow \quad \xi = 2\sqrt{\frac{r}{L}} - 1, \qquad (5.13)$$

which yields with (5.11) the following radial dependence of the displacement and strain

$$u(r) = u^{(1)} + \left(-3u^{(1)} - u^{(2)} + 4u^{(3)}\right)\sqrt{\frac{r}{L}} + 2\left(u^{(1)} + u^{(2)} - 2u^{(3)}\right)\frac{r}{L} \qquad (5.14)$$

$$\varepsilon(r) = \frac{\partial u}{\partial r} = \left(-\frac{3}{2}u^{(1)} - \frac{1}{2}u^{(2)} + 2u^{(3)}\right)\frac{1}{\sqrt{Lr}} + 2\left(u^{(1)} + u^{(2)} - 2u^{(3)}\right)\frac{1}{L}. \qquad (5.15)$$

As we can see, the displacement function now contains besides a constant (rigid body displacement) and linear function also a $\sqrt{r}$ term, which reproduces exactly the displacement field at the crack tip. The strain in the quarter-point element exhibits the desired $1/\sqrt{r}$ singularity and also possesses the necessary constant term (patch test).

(b) Quadrilateral and Triangular Quarter-Point Elements 2D

To calculate two-dimensional crack problems, quarter-point elements are generated from isoparametric rectangular [9] or quadrilateral elements [10] by shifting the mid-side node along two edges as shown in Fig. 5.6a, b. The 1D consideration of the previous section can be applied directly to the edges 1–5–2 and 1–8–4 so that we obtain the desired crack-specific functions (5.14) and (5.15). These characteristics only exist in a narrow (gray) area on radial axes [11]. Moreover, it must be provided that all element edges are straight lines. Since the angular dependence of the near field solution can only be reproduced poorly with these crack elements (90° angle), they are used rather rarely.

   In this respect, quarter-point elements that are created from natural 6-noded triangular elements with quadratic shape functions have improved properties (see Fig. 5.6b). In the first place, we can lay many such elements sector-wise around the crack tip. Secondly, it has been shown [12] that the $r^{-1/2}$ singularity is reproduced in all directions within this element. Thus the nodal-distorted triangular elements can resolve the angular distribution around the crack tip much better, for which reason they are generally preferred. The edge 2–5–3 lying opposite to the crack tip may also be curved.

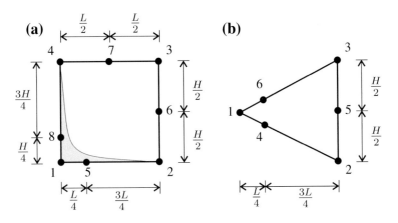

Fig. 5.6 Nodal-distorted isoparametric 8-noded quadrilateral element (**a**) and 6-noded triangular element (**b**)

(c) Collapsed Quadrilateral Elements

The probably most often used crack tip element is the isoparametric 8-noded element in which one side (e.g. $\xi_1 = -1$) is collapsed to a point so that the nodes have identical coordinates as shown in Fig. 5.7. According to Barsoum [13], this quadrilateral element, degenerated into a triangle, possesses several special properties that can be

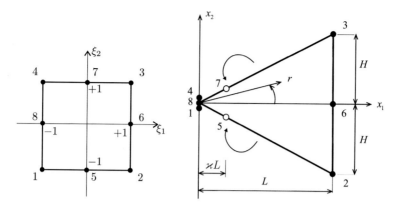

Fig. 5.7 Collapsed and distorted isoparametric 8-noded quadrilateral element

used advantageously in LEFM and EPFM. A group of these elements is arranged in a fan-shape around the crack tip, whereby the collapsed nodes all lie on the crack tip $(x_1 = 0, x_2 = 0)$. In addition, we allow for variable positioning of the middle nodes 5 and 7 with the parameter $\varkappa$. In accordance with Fig. 5.7, the nodal coordinates then read:

$$x_1^{(1)} = x_1^{(4)} = x_1^{(8)} = 0, \quad x_1^{(2)} = x_1^{(6)} = x_1^{(3)} = L, \quad x_1^{(5)} = x_1^{(7)} = \varkappa L$$

$$x_2^{(1)} = x_2^{(4)} = x_2^{(8)} = 0, \quad x_2^{(2)} = -H, \quad x_2^{(6)} = 0, \quad x_2^{(3)} = H, \quad x_2^{(5)} = -\varkappa H, \quad x_2^{(7)} = \varkappa H.$$
$$(5.16)$$

With the help of the shape functions (4.66) of this quadrilateral element we obtain the coordinates

$$x_1 = \frac{L}{2}\left[(1 + \xi_1)^2(1 - 2\varkappa) - (1 + \xi_1)(1 - 4\varkappa)\right]$$

$$x_2 = \frac{H}{2}\xi_2\left[(1 + \xi_1)^2(1 - 2\varkappa) - (1 + \xi_1)(1 - 4\varkappa)\right] \qquad (5.17)$$

and hence the distance r to the crack tip

$$r = \sqrt{x_1^2 + x_2^2} = \frac{1}{2}\sqrt{L^2 + H^2\xi_2^2}\left[(1 + \xi_1)^2(1 - 2\varkappa) - (1 + \xi_1)(1 - 4\varkappa)\right]. \qquad (5.18)$$

The special case of the quarter-point position $\varkappa = 1/4$ yields

$$x_1 = \frac{L}{4}(1 + \xi_1)^2, \quad x_2 = \frac{H}{4}\xi_2(1 + \xi_1)^2$$

$$r = \frac{1}{4}(1 + \xi_1)^2\sqrt{L^2 + H^2\xi_2^2} \quad \Rightarrow \quad (1 + \xi_1) = \frac{\sqrt{r}}{\frac{1}{2}\sqrt[4]{L^2 + H^2\xi_2^2}}. \qquad (5.19)$$

The elements of the Jacobian matrix are calculated from (5.17):

$$J_{11} = \frac{\partial x_1}{\partial \xi_1} = L\left[(1+\xi_1)(1-2\varkappa) - \frac{1}{2}(1-4\varkappa)\right]$$

$$J_{21} = \frac{\partial x_1}{\partial \xi_2} = 0$$

$$J_{12} = \frac{\partial x_2}{\partial \xi_1} = H\xi_2\left[(1+\xi_1)(1-2\varkappa) - \frac{1}{2}(1-4\varkappa)\right] = \frac{H}{L}\xi_2 J_{11}$$

$$J_{22} = \frac{\partial x_2}{\partial \xi_2} = \frac{H}{2}(1+\xi_1)\left[(1+\xi_1)(1-2\varkappa) - (1-4\varkappa)\right] = \frac{rH}{\sqrt{L^2 + H^2\xi_2^2}}.$$

$$(5.20)$$

From (5.20) we obtain with $\varkappa = 1/4$ the following dependence on the radius:

$$J_{11} = \frac{L}{2}(1+\xi_1) \sim \sqrt{r}, \qquad J_{21} = 0$$

$$J_{12} = \frac{H}{2}\xi_2(1+\xi_1) \sim \sqrt{r}, \qquad J_{22} = \frac{H}{4}(1+\xi_1)^2 \sim r \qquad (5.21)$$

$$J = \det |\mathbf{J}| = J_{11} J_{22} = \frac{LH}{8}(1+\xi_1)^3 \sim r^{3/2}.$$

The inverse of the Jacobian matrix is calculated via

$$\mathbf{J}^{-1} = \frac{1}{J}\begin{bmatrix} J_{22} & -J_{12} \\ 0 & J_{11} \end{bmatrix} \sim \begin{bmatrix} \frac{1}{\sqrt{r}} & \frac{1}{r} \\ 0 & \frac{1}{r} \end{bmatrix}. \qquad (5.22)$$

From here it becomes evident that singularities of type $r^{-1/2}$ and r^{-1} with respect to the radius to the crack tip arise, only by assigning node positions in the transformation ($\xi \leftrightarrow x$). The displacement function of the element obeys the isoparametric shape functions (4.66) independently of the nodal distortion. To calculate the strains, we have to form the displacement gradients:

$$\begin{bmatrix} \frac{\partial u}{\partial x_1} \\ \frac{\partial u}{\partial x_2} \end{bmatrix} = \mathbf{J}^{-1}\begin{bmatrix} \frac{\partial u}{\partial \xi_1} \\ \frac{\partial u}{\partial \xi_2} \end{bmatrix} \quad \text{with } J_{ij}^{-1} = \frac{\partial \xi_j}{\partial x_i}. \qquad (5.23)$$

Here, u stands for every component of the displacement vector u_i, e.g. u_1. With some computational effort, we find the derivatives with respect to the natural coordinates ξ_j sorted by powers of ξ_j:

$$\frac{\partial u}{\partial \xi_1} = \sum_{a=1}^{8} \frac{\partial N_a(\xi_1, \xi_2)}{\partial \xi_1} u^{(a)}$$

$$= \frac{1}{2}(1 + \xi_1)\left[\left(u^{(3)} + u^{(4)} - 2u^{(7)} + u^{(1)} + u^{(2)} - 2u^{(5)}\right)\right.$$

$$\left. + \xi_2\left(u^{(3)} + u^{(4)} - 2u^{(7)} - u^{(1)} - u^{(2)} + 2u^{(5)}\right)\right]$$

$$+ \frac{1}{4}(1 + \xi_2)\left[(\xi_2 - 2)u^{(3)} - (\xi_2 + 2)u^{(4)} + 4u^{(7)}\right] \tag{5.24}$$

$$+ \frac{1}{4}(1 - \xi_2)\left[(\xi_2 - 2)u^{(1)} - (\xi_2 + 2)u^{(2)} + 4u^{(5)}\right]$$

$$+ \frac{1}{2}(1 - \xi_2^2)\left[u^{(6)} - u^{(8)}\right]$$

$$= a_0 + a_1(1 + \xi_1) \quad \text{for } \xi_2 = \text{const.}$$

$$\frac{\partial u}{\partial \xi_2} = \frac{1}{4}(1 + \xi_1)^2\left[u^{(3)} + u^{(4)} - 2u^{(7)} - u^{(1)} - u^{(2)} + 2u^{(5)}\right]$$

$$+ \frac{1}{4}(1 + \xi_1)\left[(2\xi_2 - 1)u^{(3)} - (3 + 2\xi_2)u^{(4)} + (3 - 2\xi_2)u^{(1)}\right. \tag{5.25}$$

$$\left. + (2\xi_2 + 1)u^{(2)} + 4u^{(7)} + 4\xi_2 u^{(8)} - 4u^{(5)} - 4\xi_2 u^{(6)}\right]$$

$$+ \frac{1}{2}\left[(2\xi_2 + 1)u^{(4)} + (2\xi_2 - 1)u^{(1)} - 4\xi_2 u^{(8)}\right]^*$$

$$= b_0 + b_1(1 + \xi_1) + b_2(1 + \xi_1)^2 \quad \text{for } \xi_2 = \text{const.}$$

For a clearer view, the constants a_i and b_i were introduced in order to recognize the function of $(1 + \xi_1) \sim \sqrt{r}$. Analogously, we obtain the derivatives of $u_2 \hat{=} u$ from (5.24) and (5.25) with other constants:

$$\frac{\partial u_2}{\partial \xi_1} = c_0 + c_1(1 + \xi_1) \tag{5.26}$$

$$\frac{\partial u_2}{\partial \xi_2} = d_0 + d_1(1 + \xi_1) + d_2(1 + \xi_1)^2 \tag{5.27}$$

Via relation (5.23) we find with (5.22):

$$\varepsilon_{11} = \frac{\partial u_1}{\partial x_1} = J_{11}^{-1}\frac{\partial u_1}{\partial \xi_1} + J_{12}^{-1}\frac{\partial u_1}{\partial \xi_2}$$

$$= \frac{a_0 + a_1(1 + \xi_1)}{\sqrt{r}} + \frac{b_0 + b_1(1 + \xi_1) + b_2(1 + \xi_1)^2}{r} = \frac{b_0}{r} + \frac{e_1}{\sqrt{r}} + e_2 \tag{5.28}$$

$$\varepsilon_{22} = \frac{\partial u_2}{\partial x_2} = J_{21}^{-1}\frac{\partial u_2}{\partial \xi_1} + J_{22}^{-1}\frac{\partial u_2}{\partial \xi_2} = \frac{d_0}{r} + \frac{d_1}{\sqrt{r}} + d_2 \tag{5.29}$$

$$\varepsilon_{12} = \frac{1}{2}\left(\frac{\partial u_1}{\partial x_2} + \frac{\partial u_2}{\partial x_1}\right) = \frac{b_0 + d_0}{r} + \frac{f_1}{\sqrt{r}} + f_2 \qquad (5.30)$$

Using Eqs. (5.28)–(5.30) we determine the radial dependence of the strains in the element. The constants $a_i - f_i$ ($i = \{0, 1, 2\}$) depend on the actual nodal displacements and the second parameter ξ_2, which is constant on a radial ray. With the help of (5.25), we recognize that the expression in the marked brackets $[\ ]^* \stackrel{\wedge}{=} b_0 = 0$ vanishes if nodes 1, 4 and 8 possess identical displacements and are thus kinematically coupled. From this follows:

$$b_0 = d_0 = 0 \quad \text{for } u_i^{(1)} = u_i^{(4)} = u_i^{(8)}. \qquad (5.31)$$

Thus, in (5.28)–(5.30), the strong singular terms with $1/r$ are dropped, and the crack tip element possesses the desired $1/\sqrt{r}$-singularity of the elastic near field solution plus a constant strain term, which is essential for convergence behavior and the consideration of thermal expansions. Furthermore, the required continuity is satisfied with the neighboring quarter-point elements along edges 1–5–2 and 4–7–3 as well as with the regular elements of the next ring along 2–6–3. Also, the three degrees of freedom of rigid body motions are not restrained by this modification of the element. The edge 2–6–3 has to be straight.

> By collapsing one element edge to a node with common coupled displacements and by an additional quarter-point displacement, we obtain from an isoparametric 8-noded element a 2D crack tip element that possesses the required $1/\sqrt{r}$-singularity in ε_{ij} and σ_{ij} on all radial rays in the element.

(d) Three-Dimensional Quarter-Point Elements

The concept of quarter-point elements can be extended smoothly to three-dimensional crack problems by prismatically expanding the 2D elements introduced above along the crack front into the third dimension. In this way, nodal-distorted hexahedral elements and pentahedral elements arise that are grouped around every segment of the crack front as shown in Fig. 5.8. The singularity properties are then applicable to every element plane perpendicular to the crack front ($\xi_3 = $ const.). Figure 5.9a shows a pentahedral element with 15 nodes which lies with its edge 1–13–4 on the crack front. Nodes 7, 9, 10, and 12 are shifted into the quarter position, i.e. their coordinates $\mathbf{x}^{(i)} = [x_1^{(i)}, x_2^{(i)}, x_3^{(i)}]$ are calculated as:

$$\mathbf{x}^{(7)} = (3\mathbf{x}^{(1)} + \mathbf{x}^{(2)})/4, \qquad \mathbf{x}^{(9)} = (3\mathbf{x}^{(1)} + \mathbf{x}^{(3)})/4$$
$$\mathbf{x}^{(10)} = (3\mathbf{x}^{(4)} + \mathbf{x}^{(5)})/4, \qquad \mathbf{x}^{(12)} = (3\mathbf{x}^{(4)} + \mathbf{x}^{(6)})/4. \qquad (5.32)$$

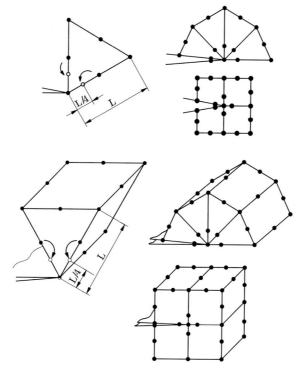

Fig. 5.8 Arrangement of different 2D and 3D quarter-point elements at the crack

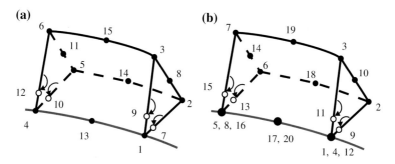

Fig. 5.9 3D-Quarter-point elements at the crack front: **a** pentahedral element, **b** collapsed hexahedral element

In order that the desired singular properties are satisfied along every element edge perpendicular to the crack front ($\xi_3 = $ const.), the following geometrical conditions must be fulfilled: The edges opposite to the crack must be straight lines:

$$\mathbf{x}^{(8)} = (\mathbf{x}^{(2)} + \mathbf{x}^{(3)})/2, \qquad \mathbf{x}^{(11)} = (\mathbf{x}^{(5)} + \mathbf{x}^{(6)})/2 . \qquad (5.33)$$

The mid-side nodes on the outer surface must be placed exactly such that their distance to the crack front corresponds to the arithmetic average of the distances on both front faces, i.e. with a given position of node 13 follows:

$$\mathbf{x}^{(14)} = \left[(\mathbf{x}^{(2)} - \mathbf{x}^{(1)}) + (\mathbf{x}^{(5)} - \mathbf{x}^{(4)}) + 2\mathbf{x}^{(13)}\right]/2$$
$$\mathbf{x}^{(15)} = \left[(\mathbf{x}^{(3)} - \mathbf{x}^{(1)}) + (\mathbf{x}^{(6)} - \mathbf{x}^{(4)}) + 2\mathbf{x}^{(13)}\right]/2 . \tag{5.34}$$

In the special case of a straight crack front $\mathbf{x}^{(13)} = (\mathbf{x}^{(1)} + \mathbf{x}^{(4)})/2$, we obtain the prismatic pentahedron with planar faces of Fig. 5.8.

In analogy to the collapsed quadrilateral elements, isoparametric hexahedral elements can be degenerated to pentahedral elements, whereby the collapsed surface coincides with the crack front (see Fig. 5.9b):

$$\mathbf{x}^{(1)} = \mathbf{x}^{(4)} = \mathbf{x}^{(12)}, \quad \mathbf{x}^{(17)} = \mathbf{x}^{(20)}, \quad \mathbf{x}^{(5)} = \mathbf{x}^{(8)} = \mathbf{x}^{(16)} . \tag{5.35}$$

The quarter-point nodes are located at

$$\mathbf{x}^{(9)} = (3\mathbf{x}^{(1)} + \mathbf{x}^{(2)})/4, \qquad \mathbf{x}^{(11)} = (3\mathbf{x}^{(1)} + \mathbf{x}^{(3)})/4$$
$$\mathbf{x}^{(13)} = (3\mathbf{x}^{(5)} + \mathbf{x}^{(6)})/4, \qquad \mathbf{x}^{(15)} = (3\mathbf{x}^{(5)} + \mathbf{x}^{(7)})/4 \tag{5.36}$$

and the nodes on the surface far from the crack have the positions

$$\mathbf{x}^{(10)} = (\mathbf{x}^{(2)} + \mathbf{x}^{(3)})/2, \qquad \mathbf{x}^{(14)} = (\mathbf{x}^{(6)} + \mathbf{x}^{(7)})/2$$
$$\mathbf{x}^{(18)} = \left[(\mathbf{x}^{(2)} - \mathbf{x}^{(1)}) + (\mathbf{x}^{(6)} - \mathbf{x}^{(5)}) + 2\mathbf{x}^{(17)}\right]/2$$
$$\mathbf{x}^{(19)} = \left[(\mathbf{x}^{(3)} - \mathbf{x}^{(1)}) + (\mathbf{x}^{(7)} - \mathbf{x}^{(5)}) + 2\mathbf{x}^{(17)}\right]/2 \tag{5.37}$$

In addition, all nodes on the same location of this element surface must be kinematically coupled with each other.

$$\mathbf{u}^{(1)} = \mathbf{u}^{(4)} = \mathbf{u}^{(12)}, \qquad \mathbf{u}^{(17)} = \mathbf{u}^{(20)}, \qquad \mathbf{u}^{(5)} = \mathbf{u}^{(8)} = \mathbf{u}^{(16)} \tag{5.38}$$

Detailed explanations of three-dimensional quarter-point elements can be found in Hussain et al. [14], Manu [15] and Banks-Sills and Sherman [16, 17].

The modeling of curved crack fronts deserves special attention. As long as they are approximated piecewise as polygonal lines and the crack tip elements retain two-dimensional surfaces, the properties discussed above are applicable without exception. If however we wish to take advantage of quadratic shape functions in order to reproduce curvilinear crack fronts with geometrically adjusted bent 3D elements, the above-mentioned geometrical restrictions should be observed in order to give the elements the best possible level of quality.

(e) Quarter-Point Elements for Plates and Shells

The quarter-point method unfortunately cannot be applied to finite element models for thin-walled Kirchhoff plates, as then a $r^{1/2}$-dependence would arise in accordance with (5.30) for the deflection function $w(r, \theta)$, but from (3.139) an asymptotic behavior of $r^{3/2}$ is required, see Sect. 3.2.9. Therefore, we must work with standard plate elements and use the displacements (DIM) or the crack closure integral (MCCI) for subsequent evaluation.

Alternatively, we can make use of the superior but more complex sixth order Reissner [18] theory for shear-deformable thick plates and shells. There are standard thick-walled, curved shell elements (see e.g. [19]) available that can be derived as a special case of three-dimensional 20-noded elements by degenerating them on the shell's mid-surface, setting a linear displacement function over thickness and neglecting all strain components perpendicular to the surface. The 8-noded shell element formed in this way has three displacement and two rotational degrees of freedom per node. On the other hand, Barsoum [20, 21] was able to prove that a quarter-point modification of these elements provides the required asymptotic crack behavior in the context of the Reissner theory, making it possible to achieve high levels of accuracy in the analysis of cracks in thick-walled plates and shells.

We should finally point out the possibility of meshing the crack tip region in plates and shells completely with 3D quarter-point elements in several layers across the thickness along the crack front. This mesh can be linked to the surrounding shell elements either with the substructure technique or as a submodel [22, 23]. This method is however the most costly in terms of discretization and computation time.

5.2.3 Computing Intensity Factors from Quarter-Point Elements

(a) Formulae for Plane Quarter-Point Elements

For two-dimensional crack problems, there is a simple formula for calculating the stress intensity factor from the results of the quarter-point elements. For this purpose, we interpret the displacements on the crack faces (see Fig. 5.10). No matter what type of quarter-point elements was used, on these edges the displacement course from (5.14) is valid. We introduce the general notations for the nodes at the crack tip $A(r = 0)$, quarter-point nodes $B(r = L/4, \theta = \pi)$ and $B'(r = L/4, \theta = -\pi)$ as well as the corner nodes $C(r = L, \theta = \pi)$ and $C'(r = L/4, \theta = -\pi)$. Comparing the $\sqrt{r}$ term from (5.14) and the near field solution for mode I (5.3) on the upper crack face $C-B-A$ yields:

$$u_2(r) = \frac{4}{E'} K_1 \sqrt{\frac{r}{2\pi}} \overset{!}{=} \left[-3u_2(r = 0) - u_2(r = L) + 4u_2(r = L/4) \right] \sqrt{\frac{r}{L}}$$

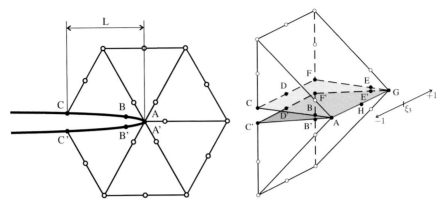

Fig. 5.10 Displacement interpretation for 2D-quarter-point elements

Fig. 5.11 Displacement interpretation for three-dimensional quarter-point elements on the crack surface

$$\Rightarrow \quad K_{\mathrm{I}} = \frac{E'}{4}\sqrt{\frac{2\pi}{L}}\left[4u_2(r = L/4) - u_2(r = L) - 3u_2(r = 0)\right]$$

$$= \frac{E'}{4}\sqrt{\frac{2\pi}{L}}\left[4u_2^B - u_2^C - 3u_2^A\right]. \tag{5.39}$$

For pure mode II or III loading, the displacements on the crack face also behave antisymmetrically, and we obtain the corresponding K-factors with similar considerations:

$$K_{\mathrm{II}} = \frac{E'}{4}\sqrt{\frac{2\pi}{L}}\left[4u_1^B - u_1^C - 3u_1^A\right] \tag{5.40}$$

$$K_{\mathrm{III}} = \frac{E}{4(1+\nu)}\sqrt{\frac{2\pi}{L}}\left[4u_3^B - u_3^C - 3u_3^A\right]. \tag{5.41}$$

In the general mixed-mode loading case of a crack, the relative displacements of the crack faces have to be evaluated in relation to each other with (5.8). The intensity factors K_{I}, K_{II} and K_{III} are only associated with the respective displacement directions u_2, u_1 and u_3 so that the following decoupled equations are obtained:

$$K_{\mathrm{I}} = \frac{E'}{8}\sqrt{\frac{2\pi}{L}}\left\{\left[4u_2(L/4, \pi) - u_2(L, \pi)\right] - \left[4u_2(L/4, -\pi) - u_2(L, -\pi)\right]\right\}$$

$$= \frac{E'}{8}\sqrt{\frac{2\pi}{L}}\left\{\left(4u_2^B - 4u_2^{B'}\right) - \left(u_2^C - u_2^{C'}\right)\right\} = \frac{E'}{8}\sqrt{\frac{2\pi}{L}}\left\{4\Delta u_2^B - \Delta u_2^C\right\}$$

$$\tag{5.42}$$

$$K_{II} = \frac{E'}{8} \sqrt{\frac{2\pi}{L}} \left\{ 4\Delta u_1^B - \Delta u_1^C \right\}$$

$$K_{III} = \frac{E'}{8(1+\nu)} \sqrt{\frac{2\pi}{L}} \left\{ 4\Delta u_3^B - \Delta u_3^C \right\}. \tag{5.43}$$

For the sake of brevity, the displacement jump over the crack face at the location of the node pair (e.g. BB') was denoted as follows:

$$\Delta u_i^B = u_i^B - u_i^{B'}. \tag{5.44}$$

Accordingly, we obtain the equation for determining the stress intensity factors for orthotropic material from (5.9)

$$\left\{ \begin{matrix} K_I \\ K_{II} \\ K_{III} \end{matrix} \right\} = \sqrt{\frac{\pi}{8L}} \begin{bmatrix} H_{11} & H_{12} & 0 \\ H_{21} & H_{22} & 0 \\ 0 & 0 & H_{33} \end{bmatrix} \left\{ \begin{matrix} 4\Delta u_1^B - \Delta u_1^C \\ 4\Delta u_2^B - \Delta u_2^C \\ 4\Delta u_3^B - \Delta u_3^C \end{matrix} \right\}. \tag{5.45}$$

(b) Formulae for Three-Dimensional Quarter-Point Elements

Also in the case of three-dimensional quarter-point elements nodal displacements are preferably interpreted on the crack surfaces because here (for isotropy) the crack modes are uncoupled. Regardless whether one uses hexahedral or pentahedral elements, always nodal-distorted 8-noded element faces lie on the crack (Fig. 5.11). If we again denote these nodes with letters A–H and their partners on the opposite crack faces with A'–H', then the following interpretation formulae [15, 24] for the K-factors are obtained.

Symmetry/Antisymmetry

If the crack lies on a symmetry plane of the structure under consideration, the FEM model can be reduced by half. Under symmetrical loading, only mode I occurs on the crack, and the normal displacements $u_2 \equiv 0$ disappear on the ligament, also at the crack front nodes A, H, G.

$$K_I(\xi_3) = \frac{E'}{4} \sqrt{\frac{2\pi}{L'}} \left\{ 2u_2^B - u_2^C + 2u_2^E - u_2^F + u_2^D + \frac{1}{2}\xi_3 \left(-4u_2^B + u_2^C + 4u_2^E - u_2^F \right) \right.$$
$$\left. + \frac{1}{2}\xi_3^2 \left(u_2^F + u_2^C - 2u_2^D \right) \right\} \tag{5.46}$$

Antisymmetrical loading leads to crack opening of mode II and/or III, which are generally coupled. Then the tangential displacements $u_1 \equiv u_3 \equiv 0$ must be prevented on the ligament, and the corresponding displacement components of the crack front nodes A, H, G are zero. From (5.46) we derive corresponding equations to determine

$K_{II}(\xi_3)$ and $K_{III}(\xi_3)$ if the respective displacement components u_1 or u_3 are evaluated instead of u_2. The stress intensity factors exhibit, like the displacement functions, a quadratic course along the crack front (ξ_3-coordinate) and are continuous at the transition from one crack element to the next.

General Case

With an arbitrary geometry and loading of the 3D-crack, we obtain all three intensity factors from the displacement differences of opposing nodes on the crack faces in accordance with (5.44), whereby the crack elements of course must be arranged mirror-symmetrically (see Fig. 5.11).

$$K_I(\xi_3) = \frac{E'}{8}\sqrt{\frac{2\pi}{L'}}\left\{2\Delta u_2^B - \Delta u_2^C + 2\Delta u_2^E - \Delta u_2^F + \Delta u_2^D \right.$$
$$+ \frac{1}{2}\xi_3\left(-4\Delta u_2^B + \Delta u_2^C + 4\Delta u_2^E - \Delta u_2^F\right) \qquad (5.47)$$
$$\left.+ \frac{1}{2}\xi_3^2\left(\Delta u_2^F + \Delta u_2^C - 2\Delta u_2^D\right)\right\}$$

$$K_{II}(\xi_3) = \frac{E'}{8}\sqrt{\frac{2\pi}{L'}}\left\{2\Delta u_1^B - \Delta u_1^C + 2\Delta u_1^E - \Delta u_1^F + \Delta u_1^D \right.$$
$$+ \frac{1}{2}\xi_3\left(-4\Delta u_1^B + \Delta u_1^C + 4\Delta u_1^E - \Delta u_1^F\right) \qquad (5.48)$$
$$\left.+ \frac{1}{2}\xi_3^2\left(\Delta u_1^F + \Delta u_1^C - 2\Delta u_1^D\right)\right\}$$

$$K_{III}(\xi_3) = \frac{E}{8(1+\nu)}\sqrt{\frac{2\pi}{L'}}\left\{2\Delta u_3^B - \Delta u_3^C + 2\Delta u_3^E - \Delta u_3^F + \Delta u_3^D \right.$$
$$+ \frac{1}{2}\xi_3\left(-4\Delta u_3^B + \Delta u_3^C + 4\Delta u_3^E - \Delta u_3^F\right)$$
$$\left.+ \frac{1}{2}\xi_3^2\left(\Delta u_3^F + \Delta u_3^C - 2\Delta u_3^D\right)\right\} \qquad (5.49)$$

The element quantity L' to be inserted into the above equations must be explained more precisely (see Fig. 5.12). For rectangular element surfaces ACFG, L' corresponds exactly to the edge lengths $L = \overline{AC} = \overline{GF}$. In the case of curved quarter-point elements with different edges $L_1 = \overline{AC} \neq L_2 = \overline{GF}$ which also form oblique angles to the crack front that deviate from the normal by γ_1 and γ_2, one must apply the interpolated perpendicular length [15]

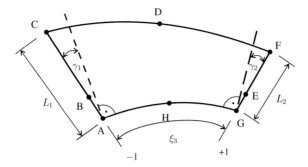

Fig. 5.12 Determination of the local element width L' in the case of curvilinear crack elements

$$L'(\xi_3) = -\frac{\xi_3 - 1}{2} L_1 \cos \gamma_1 + \frac{\xi_3 + 1}{2} L_2 \cos \gamma_2. \qquad (5.50)$$

In conclusion, the properties of quarter-point elements can be evaluated as follows. The important *advantage* of these crack tip elements is that they are easy to use and directly available. Almost all commercial finite element codes have isoparametric displacement elements with quadratic shape functions that can be converted to quite serviceable singular special elements for crack analysis simply by modifying the input values for the nodal coordinates and node numbers. Thus the user needs neither special fracture-mechanical expertise nor alteration of the FEM code. It only requires to comply with the above-mentioned conditions when setting up a mesh of the crack tip in the pre-processor as well as an uncomplicated interpretation of the result with clear formulae for the determination of stress intensity factors in the post-processor. There are quarter-point elements for two-dimensional, axially symmetric and three-dimensional crack problems as well as for thick-walled plates and shells.

Together with displacement interpretation DIM and virtual crack extension, quarter-point elements provide much more accurate results compared with standard elements and thus should generally be favored. One *disadvantage* is that only the singular solution of the crack tip fields is modeled, and of this only the radial function $r^{-1/2}$. In order to resolve the angular dependence with sufficient accuracy, an adequate number of quarter-point elements (at least 6 elements, preferably 12–16) must be arranged fan-wise over the semi-circle.

Since the validity range r_K of the crack singularity is narrowly limited, the radius of the quarter-point element must also be kept relatively small compared to the crack length. The guideline is: element size $L \approx a/20 \ldots a/10$.

5.3 Hybrid Crack Tip Elements

5.3.1 Development of Hybrid Crack Tip Elements

One particularly useful application area of the hybrid element formulations introduced in Sect. 4.2.3 is in the design of special crack tip elements. This procedure is illustrated schematically in Fig. 5.13. The hybrid technique enables us to apply known analytical crack solutions inside the element with free parameters and simultaneously to choose such boundary displacement functions that are compatible with those of the regular (e.g. isoparametric) neighboring elements. In this way, we can completely »embed« the singular asymptotic near field solutions for the crack tip region into a special element which is displacement-compatible with standard elements. As with ordinary displacement elements, the stiffness matrix is expressed by nodal variables at the boundary of the element. Thus, its assembly into the total FEM system can be carried out in the normal way. That is, its incorporation into a FEM program system requires only the particular integration routine for the stiffness matrix. Since the fracture parameters such as stress intensity factors are used as internal variables, they are obtained directly by the solution of the FEM system so that no interpretation or extrapolation techniques are needed for their determination. These advantages are characteristic of all hybrid crack tip elements, which is why they are much more effective compared to other crack tip elements. The price for this is a highly sophisticated theory as well as an increased implementation effort.

Fig. 5.13 Formulating hybrid finite elements

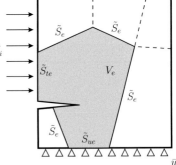

The first hybrid crack tip elements were developed for two-dimensional elastic crack problems on the basis of hybrid stress models Π_{CH} by Pian [25] and Schnack and Wolf [26]. Atluri et al. [27] were the first to apply the hybrid displacement model Π_{PH} with the three-variable function described in Sect. 4.2.3 to 2D crack problems. In these elements, the singular elastic crack tip solution of type $r^{-1/2}$ is included into the shape functions for the interior and partially for the boundary using the stress intensity factors K_I and K_{II} as free parameters. One group of hybrid elements encloses the crack tip. A significant step forward was the »superelement« devised by Tong et al. [28] which is based on the simplified mixed hybrid model $\Pi_{\text{MH}*}$ (Sect. 4.2.3). This element surrounds the crack tip completely (as in Fig. 5.13) and makes use of the eigenfunctions for the crack in the plane. Thus it contains besides the $r^{-1/2}$-singularity still higher order terms of the series expansion. This design principle is especially advantageous in combination with quadratic isoparametric elements [29] and will be described in detail in the next chapter. The simplified mixed hybrid model has in the meantime been applied to crack calculation in two-dimensional anisotropic materials [30], to 2D interface cracks [31] and to two-dimensional notch problems [32].

The development of three-dimensional (3D) hybrid crack tip elements mainly proceeded from pure hybrid displacement or stress models, since the lack of eigenfunctions for three-dimensional crack problems precludes the application of the simplified mixed principle, see the papers of Moriya and Pian [33], Kuna [34] (Sect. 5.3.3) and Atluri and Kartiresan [35]. In all elements mentioned the known asymptotic near field solution for a point on the 3D crack front was used either in the stress or displacement function for the element volume. The hybrid stress model has the advantage that we can do with two field quantities $(\sigma_{ij}, \tilde{u}_i)$ instead of three variables $(u_i, \tilde{u}_i, \tilde{T}_i)$ as in the hybrid displacement model. In exchange, the displacement model permits variability of the stress intensity factors along the crack front within the element, whereas the stress model only allows for a constant function. Since the approximations are only valid for subdomains around the crack tip, the crack front must again be surrounded by a group of hybrid crack tip elements, which necessitates volume integrals with removable singularities. In order to utilize the advantages of the simplified mixed method $\Pi_{\text{MH}*}$ for tackling three-dimensional crack problems as well, Kuna and Zwicke [36] have derived eigenfunctions for the special case of the straight 3D-crack and thus designed a hybrid 3D crack tip element that embeds the crack front segmentally.

Hybrid element formulations have also been used successfully for dealing with part-through cracks in plates, as can be seen in the papers of Rhee and Atluri [37] (hybrid stress model Π_{CH}) and Moriya [38] (simplified mixed model $\Pi_{\text{MH}*}$). The application of hybrid element concepts to elastic-plastic crack problems has been hindered especially by the lack of knowledge of corresponding analytical crack solutions. To date, no hybrid crack tip elements are known for the problem areas »dynamics of moving crack« or »dynamically loaded crack«. Overviews of the theory and application of hybrid crack tip elements are provided in [39–41].

5.3.2 2D Crack Tip Elements Based on Mixed Hybrid Model

Geometry and Concept of Crack Tip Elements

In principle, the crack tip element could be of any polygonal shape which surrounds
the crack tip as shown in Fig. 5.13. For the sake of simplicity however, a quadratic
or semi-quadratic shape was chosen as illustrated in Fig. 5.14. Variants (b) and (c)
are valid for the special case of purely symmetrical mode I crack loading with a left-
or right-hand crack. Along the boundary segments $\widetilde{S}_e$, the crack tip elements adjoin
to the neighboring standard elements, where they are coupled via $n_K = 17$ (variant
(a)) or $n_K = 9$ (variants (b) and (c)) nodes. On the crack itself there are no nodes.
The position of the crack tip in the element can be varied by setting the parameter
$\varkappa (-1 < \varkappa < +1)$. In Sect. 3.2.2, the eigenfunctions for a semi-infinite crack in
the infinite plane were derived for linear-elastic material behavior. They fulfill the
compatibility and equilibrium conditions in the domain as well as the traction-free
conditions on both crack faces. Thus they satisfy the prerequisites for the design of
a simplified mixed hybrid model (Sect. 4.2.3). The *eigenfunctions* form an infinite
series $n \in [1, \infty]$ with the complex coefficients $A_n = a_n + ib_n$ in the form $r^{\frac{n}{2}-1}$. The
associated angular functions of the stresses $\tilde{M}_{ij}^{(n)}$, $\tilde{N}_{ij}^{(n)}$ (3.41) and displacements $\tilde{F}_i^{(n)}$,
$\tilde{G}_i^{(n)}$ (3.43) were already indicated there. The first component $n = 1$ corresponds to
the known singular crack tip solution of type $r^{-1/2}$ (see (3.45)).

$$K_{\mathrm{I}} - iK_{\mathrm{II}} = \sqrt{2\pi}\,(a_1 + ib_1) \tag{5.51}$$

The real terms a_n should be assigned to the symmetrical solution component mode I,
while the imaginary parts b_n represent the crack opening mode II. The second eigenfunction merely represents a constant σ_{11}-stress:

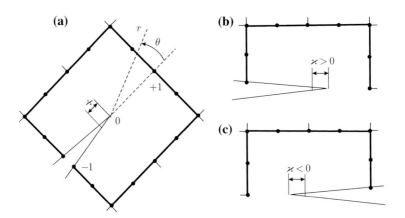

Fig. 5.14 Hybrid crack tip elements for two-dimensional tasks

$$T_{11} = \sigma_{11} = 4\,a_2, \quad \beta_T = \frac{T_{11}\sqrt{\pi a}}{K_I} \tag{5.52}$$

For the stresses and displacements in the hybrid element, N terms of the above-mentioned eigenfunctions are applied with the associated $2N$ free coefficients, which are compiled in the column vector

$$\boldsymbol{\beta} = [a_1\ a_2\ \ldots\ a_N\ b_1\ b_2\ \ldots\ b_N]^{\mathrm{T}}. \tag{5.53}$$

Written as a matrix, the displacements (3.43) then read

$$[u_1\ u_2]^{\mathrm{T}} = \mathbf{u}(r,\ \theta) = \mathbf{U}(r,\ \theta,\ n)\boldsymbol{\beta}. \tag{5.54}$$

The tractions $t_i = \sigma_{ij}\, n_j$ on S_e are obtained from (3.41) and the normal vector n_j on the boundary

$$[t_1\ t_2]^{\mathrm{T}} = \mathbf{t}(r,\ \theta) = \mathbf{R}(r,\ \theta,\ n)\boldsymbol{\beta}. \tag{5.55}$$

The displacements $\tilde{u}_i$ on the element boundary are expressed by the values $\mathbf{v}$ of the boundary nodes n_K:

$$[\tilde{u}_1\ \tilde{u}_2]^{\mathrm{T}} = \tilde{\mathbf{u}} = \mathbf{L}\mathbf{v}, \quad \mathbf{v} = \left[u_1^{(1)}\ u_2^{(2)}\ \ldots\ u_1^{(n_K)}\ u_2^{(n_K)}\right]^{\mathrm{T}}, \tag{5.56}$$

where $\mathbf{L}$ represents the matrix of the interpolation functions. $\mathbf{L}$ is now chosen such that the boundary displacements on every boundary segment are identical with those of the adjacent isoparametric elements. The combination of the crack element with quadratic isoparametric elements has led to convincing advantages compared with linear approaches. According to it, every boundary segment possesses three nodes (Fig. 5.14). By inserting the chosen displacement and stress functions (5.54)–(5.56) into the simplified mixed hybrid variational principle (4.33), we obtain for the crack tip element:

$$\Pi_{\mathrm{MH}*}(\boldsymbol{\beta}, \mathbf{v}) = \boldsymbol{\beta}^{\mathrm{T}}\mathbf{G}\mathbf{v} - \frac{1}{2}\boldsymbol{\beta}^{\mathrm{T}}\mathbf{H}\boldsymbol{\beta} - \mathbf{v}^{\mathrm{T}}\mathbf{f}$$

$$\mathbf{H} = \frac{1}{2}\int_{S_e}(\mathbf{R}^{\mathrm{T}}\mathbf{U} + \mathbf{U}^{\mathrm{T}}\mathbf{R})\mathrm{d}s, \quad \mathbf{G} = \int_{S_e}\mathbf{R}^{\mathrm{T}}\mathbf{L}\,\mathrm{d}s, \quad \mathbf{f} = \int_{S_{te}}\mathbf{L}^{\mathrm{T}}\bar{\mathbf{t}}\,\mathrm{d}s. \tag{5.57}$$

The second term of (5.57) is the strain energy of the element. The first term represents the interaction of the internal stress function with the independent boundary displacements. Variation of $\delta\boldsymbol{\beta}$ provides a relation between the internal coefficients $\boldsymbol{\beta}$ and the boundary displacements $\mathbf{v}$:

$$\boldsymbol{\beta} = \mathbf{H}^{-1}\mathbf{G}\mathbf{v} = \tilde{\mathbf{B}}\mathbf{v}. \tag{5.58}$$

In this way, $\boldsymbol{\beta}$ can be eliminated from (5.57). The variation $\delta \Pi_{MH*} = 0$ with respect to $\delta \mathbf{v}$ provides the stiffness relation

$$\mathbf{kv} = \mathbf{f}, \tag{5.59}$$

where $\mathbf{k}$ is the sought stiffness matrix of the hybrid element and $\mathbf{f}$ denotes the load vector due to the boundary loads $\bar{t}_i$.

$$\mathbf{k} = \mathbf{G}^T \mathbf{H}^{-1} \mathbf{G} \tag{5.60}$$

Numerical Implementation

The great advantage of the simplified mixed hybrid model is that we need only to integrate over the element boundary S_e to calculate the element matrices (5.57)–(5.59). Moreover, interpretation of singular integral terms typical of cracks is avoided. Thus, we can work with common 6th order Gaussian integration formulae and a subinterval technique in order to integrate exactly the more strongly oscillating higher eigenfunctions. The symmetrical, positive definite matrix $\mathbf{H}$ can be inverted easily. The likewise symmetrical stiffness matrix $\mathbf{k}$ is only positive semi-definite for hybrid elements, which is a result of the weakened continuity requirements of hybrid variational principles. The matrix $\mathbf{k}$ is incorporated in the same way as the stiffness matrices of the surrounding displacement elements into the total FEM system. After solving the FEM system of equations, we obtain from the nodal displacements of the crack element the values of the internal coefficients $\boldsymbol{\beta}$ (5.53) via (5.58). With (5.51) and (5.52) we thus directly obtain the sought stress intensity factors K_I and K_{II} as well as the T_{11}-stress.

The number $2N$ of coefficients to be used should correspond approximately to the number of nodal degrees of freedom $2n_K$ minus rigid body motions of the crack element. For the hybrid elements of Fig. 5.14, this means eigenfunctions up to an order of $N = 9$ (elements (b) and (c)) or $N = 17$ in the full element (a).

Examples

Let's take the CT-specimen (Fig. 3.12) as an example. Figure 5.15 shows the coarsely meshed discretization of the upper half (sufficient due to symmetry). Hybrid element type (b) was set on the crack tip, and the rest of the geometry was filled out with six isoparametric standard elements. Within the very large crack tip element (40 % of the crack length), the crack length was varied from $a = 0.35\,w$ to $a = 0.65\,w$ by changing the $\varkappa$-parameter. Instead of elaborate mesh alterations, here we need only to modify one input value! With these hybrid crack tip elements, the otherwise usual high mesh refinement at the crack tip becomes superfluous because, by using numerous higher eigenfunctions, the solution can be correctly represented even at a greater distance to the singularity. The accuracy of the calculated K_I-factors is also good.

Figure 5.16 shows the geometry function $g(a/w) = K_{\mathrm{I}}(a)Bw/\left(F\sqrt{\pi a}\right)$ in comparison to the reference solution [42]. In the crack length range $0.35 < a/w < 0.55$, the agreement is better than 0.5 %. Only for very deep cracks do the deviations go up to 3 %. We obtain just as easily with (5.52) the normalized T_{11}-stress $\beta_{\mathrm{T}} = 0.564$, which agrees to 3.5 % with the reference solution [43].

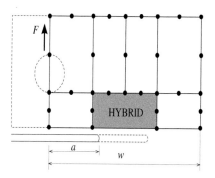

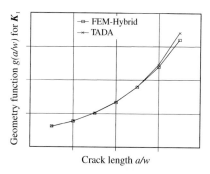

Fig. 5.15 FEM mesh of the CT-specimen using hybrid crack tip element type b

Fig. 5.16 Geometry function $g(a/w)$ for the CT-specimen as a function of the crack length

One way to investigate crack behavior under mixed mode loading experimentally is to use a disc-shaped tensile specimen. Hereby, the tensile force can be applied at an angle α inclined to the crack, which can be adjusted by the choice of bore holes. The test geometry and FEM modeling are shown in Fig. 5.17. To discretize the crack, the complete hybrid crack element type (a) was employed. By varying the direction of the applied pair of forces, crack loading can be set from pure mode I ($\alpha = 90°$) via various $K_{\mathrm{I}}/K_{\mathrm{II}}$-combinations up to the case of mode II ($\alpha = 0°$). This was reproduced in the FEM analysis. Evaluation of the hybrid element with (5.51) immediately supplied both K-factors, which are compiled in normalized form in Table 5.1.

Table 5.1 Dimensionless stress intensity factors g_{I} and g_{II} for the disc-shaped mixed mode specimen with $a = R/2$ and $R = 8\,\mathrm{cm}$

Angle α	90°	67.5°	45°	22.5°	0°
g_{I}	1.403	1.322	1.063	0.714	0.041
g_{II}	0.006	0.321	0.552	0.632	1.152

$K_L = \frac{F}{BR}\sqrt{\pi a}\, g_L(a/R), \quad L = \mathrm{I}, \mathrm{II}$

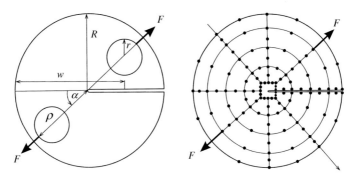

Fig. 5.17 Disc-shaped fracture test for mixed crack loading (*left*) and FEM discretization (*right*) with 32 quadrilateral elements, 1 hybrid element and 121 nodes

5.3.3 3D Crack Tip Elements Based on Hybrid Stress Model

Geometry and Concept

The three-dimensional crack tip elements have the shape of a hexahedron with planar element surfaces and $n_K = 20$ nodes. The path of the crack front through the body is approximated by a polygon, whereby a group of four crack tip elements is arranged around every straight line in accordance with Fig. 5.18. The rest of the body is meshed with isoparametric standard elements. Each of the qualitatively different hybrid elements (type A, A′, B, B′) has one edge in common with the crack front. To formulate the shape functions and for the sake of numerical integration, the element is transformed into the unit cube ($-1 < \xi_1, \xi_2, \xi_3 < 1$) with the help of the quadratic isoparametric shape functions (4.68) (see Fig. 5.19). At the crack front, we again introduce a local system of cylindrical coordinates ($r, \theta, z = x_3$) and Cartesian coordinates (x_1, x_2, x_3). The asymptotic solutions for the stresses (3.60) and displacements (3.62) are known from analytical investigations (see Sect. 3.2.3).

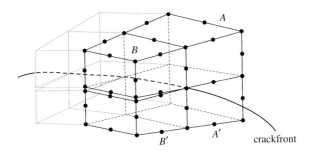

Fig. 5.18 Arrangement of hybrid crack tip elements around the crack front

Fig. 5.19 Systems of coordinates for the crack tip elements

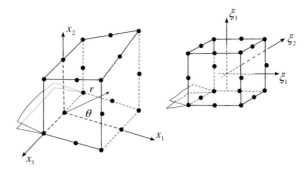

To design the crack elements, the hybrid stress model (4.28) from Sect. 4.2.3 was used. Application of this principle Π_{CH} requires a stress function in the element volume that satisfies a priori the equilibrium conditions and traction boundary conditions:

$$\boldsymbol{\sigma} = \begin{bmatrix} \sigma_{11} & \sigma_{22} & \sigma_{33} & \tau_{12} & \tau_{23} & \tau_{31} \end{bmatrix}^T = \mathbf{P}\boldsymbol{\beta} + \mathbf{P_B}\boldsymbol{\beta_B}, \tag{5.61}$$

whereby we can calculate the tractions with the normal vector

$$\mathbf{t} = \begin{bmatrix} t_1 & t_2 & t_3 \end{bmatrix}^T = \mathbf{R}\boldsymbol{\beta} + \mathbf{R_B}\boldsymbol{\beta_B}. \tag{5.62}$$

The stress functions contain the n_A unknown coefficients $\boldsymbol{\beta}$. The second term $\mathbf{P_B}\boldsymbol{\beta_B}$ takes into consideration particular solutions due to volume forces or boundary tractions. At present, the terms $\mathbf{P_B}$ are omitted in the crack tip elements. The chosen stress function $\mathbf{P}$ consist of n_R standard polynomial terms $\mathbf{P_R}(x_1, x_2, x_3)$ with the unknown coefficients b_n and the crack-specific part $\mathbf{P_S}$, which contains the crack tip solution (3.60) with the stress intensity factors as coefficients:

$$\mathbf{P}\boldsymbol{\beta} = \begin{bmatrix} \mathbf{P_R} & \mathbf{P_S} \end{bmatrix} \begin{bmatrix} b_1 & b_2 & \ldots & b_{n_R} & K_I & K_{II} & K_{III} \end{bmatrix}^T. \tag{5.63}$$

To constitute $\mathbf{P_R}$, for each stress component an incomplete 3rd order polynomial in (x_1, x_2, x_3) was employed, which is reduced to $n_R = 54$ (for element types A and A') coefficients by the equilibrium condition [34]. For element types B and B' only those states of stress were chosen that do not cause crack face tractions ($n_R = 60$). In this context, we must skip the extensive description of matrix $\mathbf{P_R}$. The stress ansatz (5.63) satisfies not only the equilibrium but also the stress-free conditions on the crack surfaces. It also satisfies the requirement that the number of coefficients $n_A = n_R + 3$ is larger than the number of nodal degrees of freedom $3n_K = 60$ minus the rigid body degrees of freedom $n_F = 6$.

Displacements $\tilde{\mathbf{u}}$ on the element surface are expressed by means of the interpolation function $\mathbf{L}$ by the nodal displacements $\mathbf{v}$.

$$\tilde{\mathbf{u}} = \begin{bmatrix} \tilde{u}_1 & \tilde{u}_2 & \tilde{u}_3 \end{bmatrix}^T = \mathbf{L}\mathbf{v} \quad \text{on } S_e \tag{5.64}$$

When specifying the shape functions, we distinguish between three surface types, which will be denoted in the following by the direction of their surface normals in Fig. 5.19:

(a) **surfaces** $\xi_1 = 1$ **and** $\xi_3 = 1$:
On these surfaces, the crack tip elements are adjacent to standard elements, so the usual quadratic isoparametric shape functions in (ξ_1, ξ_2, ξ_3) for an 8-noded surface are used for **L**. Only the crack tip elements themselves are connected via the other four surfaces. Their interpolation functions are adjusted to the displacement field (3.62) of the crack tip solution.

(b) **surfaces** $\xi_2 = 1$ **and** $\xi_2 = -1$:
These surfaces intersect the crack front at one point so that here the complete (r, θ)-dependence from (3.62) is used in addition to six polynomial terms

$$\tilde{u}_1 = a_1 + a_2\xi_1 + a_3\xi_3 + a_4\xi_1\xi_3 + a_5\xi_1^2 + a_6\xi_3^2$$
$$+ a_7\sqrt{r}\, g_1^{\mathrm{I}}(\theta) + a_8\sqrt{r}\, g_1^{\mathrm{II}}(\theta) \tag{5.65}$$
$$\tilde{u}_2 = a_9 + a_{10}\xi_1 + a_{11}\xi_3 + a_{12}\xi_1\xi_3 + a_{13}\xi_1^2 + a_{14}\xi_3^2$$
$$+ a_{15}\sqrt{r}\, g_2^{\mathrm{I}}(\theta) + a_{16}\sqrt{r}\, g_2^{\mathrm{II}}(\theta) \tag{5.66}$$
$$\tilde{u}_3 = a_{17} + a_{18}\xi_1 + a_{19}\xi_3 + a_{20}\xi_1\xi_3 + a_{21}\xi_1^2 + a_{22}\xi_3^2$$
$$+ a_{23}\sqrt{r}\, \sin\frac{\theta}{2} + a_{24}\sqrt{r}\, \sin\frac{3}{2}\theta \tag{5.67}$$

(c) **surfaces** $\xi_1 = -1$ **and** $\xi_3 = -1$:
These surfaces have one edge with the crack front in common, which is why here only the $\sqrt{r}$-dependence of (3.62) is involved:

$$\tilde{u}_1 = a_1 + a_2\sqrt{r} + a_3\xi_2 + a_4\sqrt{r}\,\xi_2 + a_5r + a_6\xi_2^2 + a_7r\xi_2 + a_8\sqrt{r}\xi_2^2 \tag{5.68}$$

and analogously $\tilde{u}_2$ and $\tilde{u}_3$.

In cases (b) and (c), the 24 coefficients a_i must be substituted by the 24 nodal variables **v** of the surface. For this purpose, we calculate from the shape function $\tilde{\mathbf{u}} = \mathbf{Aa}$ the displacements on the node coordinates $\mathbf{v} = \mathbf{Ma}$ and invert this linear system of equations, from which the interpolation formula for the surface is obtained:

$$\tilde{\mathbf{u}} = \mathbf{AM}^{-1}\mathbf{v}. \tag{5.69}$$

Equation (5.64) is thus composed of such contributions of the six surfaces.

The functions for the stresses σ, tractions **t**, and boundary displacements $\tilde{\mathbf{u}}$ are inserted into the variational principle Π_{CH} (4.28):

$$\Pi_{\mathrm{CH}} = \sum_{e=1}^{n_E}\left[\frac{1}{2}\boldsymbol{\beta}^{\mathrm{T}}\mathbf{H}\boldsymbol{\beta} + \boldsymbol{\beta}^{\mathrm{T}}\mathbf{H}_B\boldsymbol{\beta}_B - \boldsymbol{\beta}^{\mathrm{T}}\mathbf{G}\mathbf{v} - \boldsymbol{\beta}_B^{\mathrm{T}}\mathbf{G}_B\mathbf{v} + \mathbf{v}^{\mathrm{T}}\int_{S_{te}}\mathbf{L}^{\mathrm{T}}\bar{\mathbf{t}}\,dS\right], \tag{5.70}$$

where integration over the element results in the following matrices ($\mathbf{S}$—elastic compliance tensor):

$$\mathbf{H} = \int_{V_e} \mathbf{P}^{\mathrm{T}}\mathbf{SP}\, dV = \mathbf{H}^{\mathrm{T}}, \qquad \mathbf{H_B} = \int_{V_e} \mathbf{P}^{\mathrm{T}}\mathbf{SP_B}\, dV$$

$$\mathbf{G} = \int_{S_e} \mathbf{R}^{\mathrm{T}}\mathbf{L}\, dS, \qquad \mathbf{G_B} = \int_{S_e} \mathbf{R_B}^{\mathrm{T}}\mathbf{L}\, dS. \tag{5.71}$$

Variation of Π_{CH} with respect to $\boldsymbol{\beta}$ provides for each element a relation between the coefficients $\boldsymbol{\beta}$ of the stress function and the nodal displacements $\mathbf{v}$

$$\boldsymbol{\beta} = \mathbf{H}^{-1}\left(\mathbf{Gv} - \mathbf{H_B}\boldsymbol{\beta_B}\right) = \widetilde{\mathbf{B}}\mathbf{v}. \tag{5.72}$$

With this, $\boldsymbol{\beta}$ can be eliminated from (5.70) and subsequent variation with respect to $\delta\mathbf{v}$ yields the sought stiffness relation for a hybrid element:

$$\mathbf{kv} = \mathbf{f} \tag{5.73}$$

with the stiffness matrix $\mathbf{k}$ and the load vector $\mathbf{f}$

$$\mathbf{k} = \mathbf{G}^{\mathrm{T}}\mathbf{H}^{-1}\mathbf{G}, \qquad \mathbf{f} = \mathbf{G}^{\mathrm{T}}\mathbf{H}^{-1}\mathbf{H_B}\boldsymbol{\beta_B} - \mathbf{G_B^T}\boldsymbol{\beta_B} + \int_{S_{te}} \mathbf{L}^{\mathrm{T}}\mathbf{t}\, dS. \tag{5.74}$$

Since (5.73) contains only nodal variables, incorporation of the hybrid elements into the total system is simple. Any combination of displacement and hybrid elements is permitted, provided the displacement functions on the interfaces are identical.

From the nodal displacements of the solution, we obtain via (5.72) the stress intensity factors K_{I}, K_{II} and K_{III} contained in $\boldsymbol{\beta}$ for every crack tip element directly and separately from each other. Since the crack solution in each element was set as constant, we obtain averaged K-factors for the segment of the crack front comprised by the element. Averaging and comparison of the K-factors of all four hybrid elements belonging to a crack front segment provides more exact values and a measure of accuracy.

Numerical Implementation

The essential difficulty in setting up matrices $\mathbf{H}$ and $\mathbf{G}$ according to (5.71) consists in the numerical integration of the singular terms derived from the crack solution $\mathbf{P_S}$, i.e. singularities of type r^{-1} in the volume integral (via strain energy density $\sigma_{ij}S_{ijkl}\sigma_{kl}$) and of type $r^{-1/2}$ (tractions t_i) in all surface integrals that touch the crack front (surface types (b) and (c)). These singularities are removable by switching over to polar coordinates $\int dx_1 dx_2 dx_3 = \int r\, dr\, d\theta\, dz$. In the case of complicated integration domains for arbitrarily shaped elements and due to a multitude of different integrands (about 400 for $\mathbf{H}$), an analytical treatment is futile, so normal Gaussian product formulae are used in conjunction with nested subelements in the direction of the crack front. Inversion of the symmetric positive definite matrix $\mathbf{H}$ is not a problem as long as all zero energy terms have been carefully eliminated in the stress function $\mathbf{P_R}$. Matrices $\mathbf{k}$ and $\widetilde{\mathbf{B}}$ of the hybrid elements were first calculated in the local coordinates of Fig. 5.19 and then transformed into the global Cartesian system according to their real position in space.

Examples

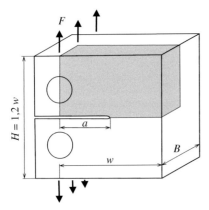

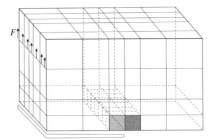

Fig. 5.20 CT-specimen: width w, crack length a, thickness B, height H, force F

Fig. 5.21 FEM model for a quarter of the CT-specimen CT645 (120 elements, 733 nodes)

The 3D hybrid crack tip elements introduced above have been incorporated into a universal FEM program system called FRACTURE [44] and combined with isoparametric hexahedral and pentahedral elements (20 or 15 nodes) of compatible shape functions. As an example, the CT-specimen will be analyzed in three dimensions (see Fig. 5.20). Due to the doubled symmetry, a FEM discretization of the gray-colored quarter is sufficient. Figure 5.21 shows the mesh used ($6 \times 4 \times 5$ elements) for a crack depth of $a = w/2$. In the marked region along the crack front, ten hybrid elements were arranged or, alternately, quarter-point hexahedral elements. The calculated distributions of the stress intensity factor K_I along the crack front are compiled in Fig. 5.22. For comparison, the 2D solution [42] and the 3D numerical solution of Yamamoto and Sumi [45], which is considered to be the most precise, are also shown. The K_I-solutions obtained with the hybrid elements agree well with the 3D reference solution. Use of quarter-point elements and K_I evaluation via displacement extrapolation resulted qualitatively in the same curve, but with a loss of accuracy of about 20 %. If we determine K_I with the method of the equivalent domain

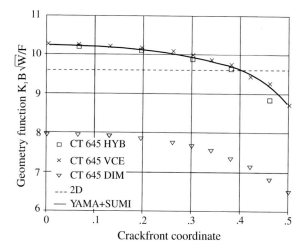

Fig. 5.22 Normalized stress intensity factor K_I versus half the thickness of the CT-specimen using different FEM variants ($a = w/2$, $\nu = 0.3$) mesh CT645

integral using quarter-point elements (see Sect. 6.4) on the other hand, much more exact values are obtainable. It should be noted that the 3D solution in the center of the specimen is about 9 % higher than the 2D-approximation. Strictly speaking, the K_I-value should fall to zero on the specimen's surface, but it is exactly here that the numerical solutions differ the most. Further application examples are published in [34] and [46].

In summary, the hybrid crack tip elements can be appraised as follows: As the examples prove, compared to all other crack elements of comparable mesh quality, they are characterized by a high level of accuracy and optimal user comfort. With hybrid elements, crack configurations can be very coarsely meshed (and partially enclosed). The K-factors are calculated directly without a separate post-process. Nevertheless, hybrid crack tip elements have not become established in commercial FEM codes. This is due to the high level of implementation effort needed and their lack of compatibility with FEM standard algorithms.

5.4 Method of Global Energy Release Rate

5.4.1 Realization Within FEA

In Sect. 3.2.4, G was derived as the released potential energy $-\mathrm{d}\Pi$ of a loaded body during an infinitesimal crack extension $\mathrm{d}a$. In the context of FEM, the potential

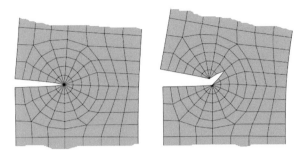

Fig. 5.23 FEM model for a finite crack propagation in the initial and final states

energy $\Pi = \mathcal{W}_{\text{int}} - \mathcal{W}_{\text{ext}}$ can be expressed according to Eq. (4.54) with the help of the nodal variables $\mathbf{V}$, the system stiffness matrix $\mathbf{K}$, and the system load vector $\mathbf{F}$, which are obtained by assembly of the corresponding element components $\mathbf{v}$, $\mathbf{k}$ and $\mathbf{f}$:

$$\Pi(\mathbf{v}) = \sum_{e=1}^{n_E} \left(\frac{1}{2} \mathbf{v}_e^T \mathbf{k}_e \mathbf{v}_e - \mathbf{v}_e^T \mathbf{f}_e \right) = \frac{1}{2} \mathbf{V}^T \mathbf{K} \mathbf{V} - \mathbf{V}^T \mathbf{F}. \tag{5.75}$$

Thus ist is obvious to calculate the energy release rate for a given crack propagation Δa directly with FEM as the difference quotient of two models with crack lengths a and $a + \Delta a$

$$\bar{G} = -\frac{\Delta \Pi}{\Delta a} = -\frac{\Pi(a + \Delta a) - \Pi(a)}{\Delta a}, \tag{5.76}$$

which requires setting up, solving and interpreting two complete FEM models. This process is suitable for any finite crack propagation on a straight, kinked or curved path C (see Fig. 5.23). It provides the total energy difference $\bar{G}$ between the final and initial states per crack extension Δa. The relation with the energy release rate $G = -d\Pi/da$ of infinitesimal crack propagation da is given by the integral along the crack path C

$$\bar{G} = \int_C G(a)\, da. \tag{5.77}$$

If we wish to determine the infinitesimal energy release rate with Eq. (3.78), a very small crack increment $\Delta a \approx 0.001 \ll a$ must be realized numerically, which however may not be too small to avoid accumulation of numerical (rounding) errors.

5.4.2 Method of Virtual Crack Extension

More elegant however is the *method of virtual crack extension* (VCE) as suggested by Parks [47], Hellen [48] and deLorenzi [49]. For this purpose, (5.75) is differentiated with respect to the crack length:

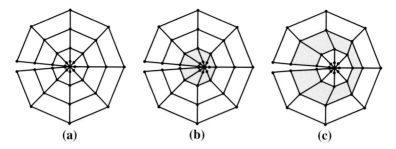

Fig. 5.24 Virtual crack extension with the global energy method

$$G = -\frac{d\Pi}{da} = -\frac{\partial \mathbf{V}^T}{\partial a}\underbrace{(\mathbf{KV} - \mathbf{F})}_{=0} - \frac{1}{2}\mathbf{V}^T\frac{\partial \mathbf{K}}{\partial a}\mathbf{V} + \mathbf{V}^T\frac{\partial \mathbf{F}}{\partial a} \qquad (5.78)$$

The expression in parentheses represents the FEM system of equations and must therefore disappear. If we assume that the external loads $\mathbf{F}$ do not change with the crack length, it follows

$$G = -\frac{1}{2}\mathbf{V}^T\frac{\partial \mathbf{K}}{\partial a}\mathbf{V} \approx -\frac{1}{2}\mathbf{V}^T(a)\frac{\mathbf{K}(a+\Delta a) - \mathbf{K}(a)}{\Delta a}\mathbf{V}(a) = -\frac{1}{2}\mathbf{V}^T\frac{\Delta \mathbf{K}}{\Delta a}\mathbf{V}.$$
$$(5.79)$$

Thus the energy release rate G is calculated from the derivative of the stiffness matrix with respect to the crack length and multiplication from both sides with the displacement solution $\mathbf{V}(a)$, which must only be known for the initial crack length a. This technique is also called the *stiffness derivative method*.

How do we now determine virtual crack extension Δa in the context of FEM? In the original papers [47, 48], the crack tip nodes or a limited domain of elements around the crack tip were shifted by Δa as shown in Fig. 5.24 so that the stiffness matrix is changed de facto only slightly in the crack region V_0 (relevant elements are highlighted in gray). The difference $\Delta \mathbf{K}$ can thus be calculated easily from the few element components, which is simply done in programming if we have direct access the FEM routines to set up the stiffness matrices. This is hardly possible in the case of commercial FEM programs, and the calculation of $\Delta \mathbf{K}/\Delta a$ requires a special post-processor. Nonetheless, computational effort is essentially reduced to a FEM analysis for *one* crack length.

Instead of the difference quotient $\Delta \mathbf{K}/\Delta a$, Lin and Abel [50] have differentiated the stiffness matrix in an analytically exact way in terms of a perturbation analysis $\partial \mathbf{K}/\partial a$ (5.79). For this purpose, the derivative was applied to all terms under the integral for determining $\mathbf{K}$ (4.50). This avoids numerical inaccuracies as a result of rounding errors and the change of FEM discretization for virtual crack extension. However, this method requires additional effort for determining the stiffness derivative, for which we also must have access to the FEM code [51].

While the previously introduced techniques of VCE primarily emanate from the FEM algorithm, a more general continuum-mechanical approach has been pursued by deLorenzi [49, 52]. Thereby, virtual crack extension is considered as a transformation of the initial configuration $x_k(a)$ into the displaced configuration $\bar{x}_k(a+\Delta a)$, whereby the function $\Delta l_k(x)$ describes the virtual displacement of the crack tip and a limited domain V_0 around it (see Fig. 5.24):

$$\bar{x}_k = x_k + \Delta l_k(x) \qquad \text{in } V_0 \tag{5.80}$$

In this way, we investigate the change of potential energy during virtual crack extension (see Sect. 3.2.4), from which the following equation for the global energy release rate G^* is found:

$$G^* \stackrel{\wedge}{=} G = -\int_{V_0}\left[U\delta_{kj} - \sigma_{ij}\frac{\partial u_i}{\partial x_k}\right]\frac{\partial \Delta l_k}{\partial x_j}\,dV + \int_{V_0}\left[\sigma_{ij}\frac{\partial \varepsilon_{ij}^t}{\partial x_k}\Delta l_k - \bar{b}_i\frac{\partial u_i}{\partial x_k}\Delta l_k\right]dV \tag{5.81}$$

This volume integral only extends over that domain altered by the VCE V_0, since outside $\Delta l_k \equiv 0$ holds. It takes volume forces $\bar{b}_i$ and thermal expansions ε_{ij}^t into consideration [53]. The relation (5.81) represents an evaluation formula for the FEM analysis in the initial configuration a, i.e. it can be calculated in the post-processor for the crack extension under consideration Δl_k (as well as any other desired variant). This VCE method is in principle identical to the formulation of the J-integral as an equivalent domain integral, which will be explored in more detail in Sect. 6.4.

The *global energy release method* has a number of *advantages*: Firstly, the FEM as a variational method approximates the most accurately the energy of the structure that is being evaluated. Secondly, for this reason we need not necessarily to furnish the crack region with crack tip elements (which is nonetheless advantageous), since a good level of accuracy can also be achieved with standard elements. In general, we obtain with equal mesh refinement a more exact result for the K-factors with the virtual crack extension method than with displacement interpretation DIM.

The essential *disadvantages* of VCE are the required implementation effort in case the stiffness matrix is directly differentiated and a certain numerical sensitivity with respect to the choice of Δa. To convert the determined energy release rate G into the stress intensity factors, there is only the relation

$$G = G_\mathrm{I} + G_\mathrm{II} + G_\mathrm{III} = \frac{1-\nu^2}{E}(K_\mathrm{I}^2 + K_\mathrm{II}^2) + \frac{1+\nu}{E}K_\mathrm{III}^2. \tag{5.82}$$

But in the case of superimposed crack opening modes I, II, and II, no separation
into the single intensity factors K_I, K_{II} and K_{III} is possible with this equation
alone. This restricts the application area of this method considerably.

5.5 Method of Crack Closure Integral

5.5.1 Basic Equations of Local Energy Method

The *local energy method* was introduced in Sect. 3.2.5 as a comparable approach
to calculating the energy release rate G. It is based on the work ΔW_c that must
be done by the crack face tractions t_i^c on the crack face displacements Δu_i for the
local opening or closing of the crack by Δa (see (3.90)). The basic equations will
be explained with the help of Fig. 5.25 for mode I loading. The top figure shows the
situation for the initial crack length a with the stress curve $t_i^c \triangleq \sigma_{22}(r, \theta = 0; a)$ in
front of the crack. The bottom figure describes the situation after a crack extension
by Δa, leading to an opening displacement of the crack faces to $\Delta u_2 = u_2^+(\Delta a -$
$s, +\pi; a + \Delta a) - u_2^-(\Delta a - s, -\pi; a + \Delta a)$, which is counted from the crack tip
at a distance of $\bar{r} = \Delta a - s$. The work done by the stresses σ_{22} on the displacements
Δu_2 is integrated along Δa:

Fig. 5.25 Local energy
method in the form of the
crack closure integral

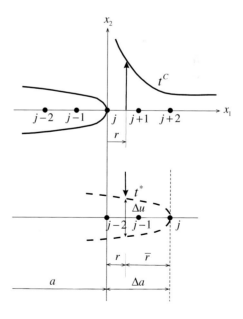

$$G_{\mathrm{I}}(a) = \lim_{\Delta a \to 0} \frac{1}{2\Delta a} \int_0^{\Delta a} \sigma_{22}(r = s, \, \theta = 0; \, a)$$

$$\times \, \Delta u_2(\bar{r} = \Delta a - s, \, \theta = \pm \pi; \, a + \Delta a) \, \mathrm{d}s \, . \tag{5.83}$$

Corresponding relations are obtained for pure mode II with $t_1^c \hateq \tau_{21}$ and Δu_1 as well as $t_3^c \hateq \tau_{23}$ and Δu_3 for mode III:

$$G_{\mathrm{II}}(a) = \lim_{\Delta a \to 0} \frac{1}{2\Delta a} \int_0^{\Delta a} \tau_{21}(s, \, 0; \, a) \, \Delta u_1(\Delta a - s, \, \pm \pi; \, a + \Delta a) \, \mathrm{d}s$$

$$G_{\mathrm{III}}(a) = \lim_{\Delta a \to 0} \frac{1}{2\Delta a} \int_0^{\Delta a} \tau_{23}(s, \, 0; \, a) \, \Delta u_3(\Delta a - s, \, \pm \pi; \, a + \Delta a) \, \mathrm{d}s \tag{5.84}$$

For the general case of mixed crack loading by all modes, these three equations can be summarized:

$$G(a) = G_{\mathrm{I}}(a) + G_{\mathrm{II}}(a) + G_{\mathrm{III}}(a)$$

$$= \lim_{\Delta a \to 0} \frac{1}{2\Delta a} \int_0^{\Delta a} \sum_{i=1}^3 \left[t_i^c(s, \, 0; \, a) - t_i^*(s, \, 0; \, a + \Delta a) \right]$$

$$\times \, \Delta u_i(\Delta a - s, \, \pm \pi; \, a + \Delta a) \, \mathrm{d}s \tag{5.85}$$

In addition, residual stresses t_i^* were introduced that act on the crack faces even after crack extension, such as internal pressure in the crack or cohesive forces between the crack faces. The sign of t_i^* should be set positive if the stresses pull the crack faces together, i.e. in the $-x_i$-direction.

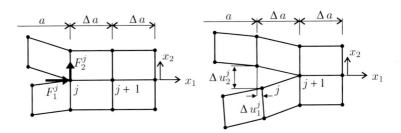

Fig. 5.26 Simple crack closure integral in the FEM-context: **a** forces before and **b** displacements after crack extension

5.5.2 *Numerical Implementation in FEA 2D*

(a) Simple Crack Closure Integral

The simplest numerical implementation of the local energy method consists in executing two FEM calculations, in which the crack is extended on a given path by the increment Δa by separating the mesh along an element edge L. This is illustrated in Fig. 5.26 for two-dimensional 4-noded elements. In the context of FEM, the crack closure work equivalent to (5.83) and (5.84) is calculated directly from the nodal force $F_i^j(a)$ of the crack tip node j in the initial model (Fig. 5.26a) and the displacement of the opening $\Delta u_i^j(a + \Delta a)$ after crack extension (Fig. 5.26b):

$$
\left.
\begin{aligned}
G_{\mathrm{I}}\left(a + \frac{\Delta a}{2}\right) &= \frac{1}{2\Delta a}\left[F_2^j(a)\Delta u_2^j(a + \Delta a)\right] \\
G_{\mathrm{II}}\left(a + \frac{\Delta a}{2}\right) &= \frac{1}{2\Delta a}\left[F_1^j(a)\Delta u_1^j(a + \Delta a)\right]
\end{aligned}
\right\} \text{ plane stress, plane strain}
$$

$$
G_{\mathrm{III}}\left(a + \frac{\Delta a}{2}\right) = \frac{1}{2\Delta a}\left[F_3^j(a)\Delta u_3^j(a + \Delta a)\right] \biggr\} \text{ anti-plane shear} \qquad (5.86)
$$

As for 2D-structures, mode III only occurs in the case of anti-plane shear loading. The result of this difference quotient is to be assigned to the average crack length

Fig. 5.27 Modified crack closure integral for (**a**) linear (*top*) and (**b**) quadratic displacement functions (*bottom*)

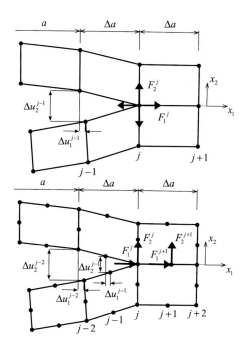

$a + \frac{\Delta a}{2}$. We can immediately see that this method is highly suitable for determining $G_N(a)$ or the K-factors $K_N(a)$ $(N = \text{I, II, III})$ step-by-step for a whole series of crack extensions by one respective element length $L = \Delta a$. With the help of a single FEM mesh, with which the crack path has been discretized by equally large elements, we can thus calculate the fracture parameters in the relevant crack length range via successive node separation, which requires $n + 1$ calculations for n difference quotients (5.86).

(b) Modified Crack Closure Integral MCCI

If we would like to determine the energy release rate, or from it the K-factors only for *one* crack length, it is possible to reduce the cost to *one* FEM calculation according to a suggestion made by Rybicki and Kanninen [54] and Buchholz [55]. It is assumed that the crack extension Δa does not essentially change the loading state at the crack tip. Thus, we can make a good approximation of the crack opening displacement $\Delta u_i^j(a + \Delta a)$ by its value $\Delta u_i^{j-1}(a)$ at node $(j - 1)$ on the crack face of the initial crack length a. This technique is called the *modified crack closure integral* (MCCI) or the *virtual crack closure method* and has become generally established. The process is outlined in Fig. 5.27 for elements with linear (a) or quadratic (b) displacement functions. The index j denotes the crack tip node, so that nodes j, $j + 1$, $j + 2$ lie on the ligament and nodes $j - 2$, $j - 1$, j correspond to the crack faces. For linear elements, the modified crack closure integral results from the work of forces at the crack tip node j with opening displacement at node $j - 1$:

$$
\left.
\begin{aligned}
G_\text{I}(a) &\approx \frac{1}{2\Delta a}\left[F_2^j(a)\Delta u_2^{j-1}(a)\right] \\
G_\text{II}(a) &\approx \frac{1}{2\Delta a}\left[F_1^j(a)\Delta u_1^{j-1}(a)\right]
\end{aligned}
\right\} \text{plane stress, plane strain}
$$

$$
G_\text{III}(a) \approx \frac{1}{2\Delta a}\left[F_3^j(a)\Delta u_3^{j-1}(a)\right] \Big\} \text{ anti-plane shear} \tag{5.87}
$$

In the case of mixed-mode loading and possible residual forces F_i^* on the crack faces $(\hat{=} t_i^*)$, the modified crack closure integral for elements with linear shape functions $(\Delta a \hat{=} \text{element edge length } L)$ is calculated as:

$$
G(a) = G_\text{I}(a)+G_\text{II}(a)+G_\text{III}(a) = \frac{1}{2\Delta a}\sum_{i=1}^{3}\left[(F_i^j(a) - F_i^{*j}(a))\Delta u_i^{j-1}(a)\right].
\tag{5.88}
$$

For elements with quadratic shape functions, which are usually preferred, two nodes must always be separated during crack extension (see Fig. 5.27b, Fig. 5.25).

The crack closure integral is composed of the work terms of the nodal forces j with the displacements at the crack faces nodes $j-2$ (after virtual crack extension $\Delta a = L$) and the forces at the mid-side node $j+1$ with the displacements at $j-1$:

$$
\left.
\begin{aligned}
G_{\mathrm{I}}(a) &\approx \frac{1}{2\Delta a}\left[F_2^j(a)\Delta u_2^{j-2}(a) + F_2^{j+1}(a)\Delta u_2^{j-1}(a) \right] \\
G_{\mathrm{II}}(a) &\approx \frac{1}{2\Delta a}\left[F_1^j(a)\Delta u_1^{j-2}(a) + F_1^{j+1}(a)\Delta u_1^{j-1}(a) \right]
\end{aligned}
\right\} \ \text{plane stress, plane strain}
$$

$$
\left.
G_{\mathrm{III}}(a) \approx \frac{1}{2\Delta a}\left[F_3^j(a)\Delta u_3^{j-2}(a) + F_3^{j+1}(a)\Delta u_3^{j-1}(a) \right]
\right\} \ \text{anti-plane shear}
$$

$$\tag{5.89}$$

Summarized for all crack opening types and residual crack face loads, the modified crack closure integral employed with quadratic element functions thus reads

$$
\begin{aligned}
G(a) &= G_{\mathrm{I}}(a) + G_{\mathrm{II}}(a) + G_{\mathrm{III}}(a) \\
&= \frac{1}{2\Delta a}\sum_{i=1}^{3}\left[\left(F_i^j - F_i^{*j}\right)\Delta u_i^{j-2} + \left(F_i^{j+1} - F_i^{*j+1}\right)\Delta u_i^{j-1} \right].
\end{aligned}
$$

$$\tag{5.90}$$

The formulae (5.88) and (5.90) are generally valid for two-dimensional crack problems (thickness $B=1$, $\Delta A = \Delta a B$) for any anisotropic elastic material behavior. Beyond orthotropy and higher classes of symmetry, the solutions in the plane and longitudinally to it do uncouple. Under these conditions, mode III will not arise in case of loads in the plane (plane stress, plane strain), i.e. $G_{\mathrm{III}} \equiv 0$, and the case of anti-plane shear loading is treated separately, i.e. $G_{\mathrm{III}} \neq 0$, $G_{\mathrm{I}} = G_{\mathrm{II}} = 0$.

(c) Combination of MCCI and Quarter-Point Elements

Using the modified crack closure integral in combination with regular elements at the crack already provides a satisfactory level of accuracy in the calculated fracture parameters G_N and K_N ($N = \mathrm{I}, \mathrm{II}, \mathrm{III}$). The method is thus used mostly in such cases where no crack tip elements are available or applicable. Nevertheless, this technique can also be applied to the quarter-point elements introduced in Sect. 5.2.2. Proceeding from the basic equations (5.83) to (5.85) of the local energy method, the work of crack closure was integrated with the functions of the 2D quarter-point elements [56–58]. This resulted in corresponding work terms for the nodal forces F_i^n ($n = j,\ j+1,\ j+2$) with crack opening displacements Δu_i^n ($n = j-2,\ j-1$) and weighting factors c_l ($l = 1, 2, \ldots, 6$) (see Fig. 5.28).

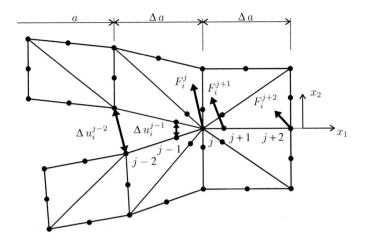

Fig. 5.28 Modified crack closure integral for 2D quarter-point elements

Variant (1)

(The crack face loads F_i^{*n} were omitted for clarity's sake.)

$$G = \frac{1}{2\Delta a} \sum_{i=1}^{3} \left[\left(c_1 F_i^j + c_2 F_i^{j+1} + c_3 F_i^{j+2} \right) \Delta u_i^{j-2} \right.$$

$$\left. + \left(c_4 F_i^j + c_5 F_i^{j+1} + c_6 F_i^{j+2} \right) \Delta u_i^{j-1} \right]$$

$$c_1 = 14 - \frac{33\pi}{8}, \quad c_2 = \frac{21\pi}{16} - \frac{7}{2}, \quad c_3 = 8 - \frac{21\pi}{8}$$

$$c_4 = \frac{33\pi}{2} - 52, \quad c_5 = 17 - \frac{21\pi}{4}, \quad c_6 = \frac{21\pi}{2} - 32 \qquad (5.91)$$

Variant (2)

The nodal force F_i^{j+2} can be eliminated by means of a reduced stress function on the ligament, which leads to a simpler formula [56].

$$G = \frac{1}{2\Delta a} \sum_{i=1}^{3} \left[F_i^j \left(c_1 \Delta u_i^{j-2} + c_2 \Delta u_i^{j-1} \right) + F_i^{j+1} \left(c_3 \Delta u_i^{j-2} + c_4 \Delta u_i^{j-1} \right) \right]$$

$$c_1 = 6 - \frac{3}{2}\pi, \quad c_2 = 6\pi - 20, \quad c_3 = \frac{1}{2}, \quad c_4 = 1 \qquad (5.92)$$

(d) Computation of Nodal Forces

The computation of nodal forces F_i^n ($n = j, j + 1, j + 2$) should be explained in more detail: We are dealing here with sectional forces, since in the context of the FEM model there is equilibrium at every node (resultant force = zero). If we investigate crack problems of the purely symmetrical (mode I) or purely antisymmetrical (modes II and III) type, then corresponding kinematic boundary conditions are prescribed on the ligament. Hence, the forces F_i^n are exactly the associated reaction forces that are available in most FEM codes. In the case of the general mixed mode, the ligament consists of internal nodes whose balanced forces are unavailable. In order to acquire these, the FEM model has to be split into two parts above and below the ligament, and the stiffness matrices $\mathbf{K}^+$ or $\mathbf{K}^-$ of the latter must be determined in a suitable way. From these, with the known displacement solution $\mathbf{V}$, the sought nodal forces $\mathbf{F}^+$ or $\mathbf{F}^-$ of both parts are calculated on the ligament:

$$\mathbf{F}^+ = \mathbf{K}^+\mathbf{V} = -\mathbf{F}^- = -\mathbf{K}^-\mathbf{V}, \quad \text{since} \quad \mathbf{KV} = \mathbf{F} = 0 \tag{5.93}$$

In some commercial FEM codes, the nodal forces of every element can optionally be provided. Subsequently, the respective forces for the ligament nodes in the upper or lower part are summarized from the components of the associated elements, which corresponds de facto to (5.93).

The following trick is simpler and more elegant: The ligament nodes $j, j+1, j+2$ are treated formally as doubled nodes (same coordinates), but their displacements are tied to each other (»multi-point constraint«). Most FEM codes will then provide us with the forces on constraint corresponding exactly to the F_i^n we are looking for. Alternatively, small, very stiff bar elements between the double nodes in all three directions lead to the same result.

5.5.3 Numerical Implementation in FEA 3D

The technique of the crack closure integral can be generalized relatively easily to three-dimensional crack configurations as long as the crack front is a straight line. Essentially, the crack closure integral is carried out locally along one segment Δs of the crack front, whereby one sub-area of the crack ΔA is closed or extended. In the following, only the modified crack closure integral MCCI is treated due to its greater importance. The coordinate s again denotes the arc length along the crack front. F_i^j are the nodal forces on the ligament in front of the crack. The subscript index i indicates the coordinate direction in the local accompanying system, which corresponds to the crack opening types I, II, and III $\widehat{=} i = 2, 1, 3$. The superscript index j numbers the node pairs used for the work integral, so every force F_i^j is multiplied by the crack opening displacement Δu_i^j at the corresponding assigned crack surface node.

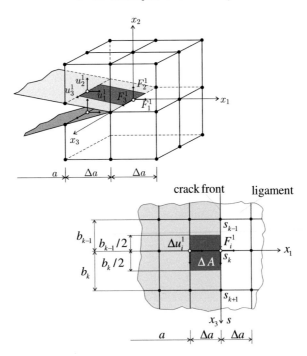

Fig. 5.29 3D crack closure integral for 8-noded hexahedra with a straight crack front

(a) 8-Noded Hexahedron with Straight Crack Front

The geometric relations are sketched out in Fig. 5.29. The elements in front of and behind the crack front must always have the same length $L = \Delta a$ and width b so that the surface areas ΔA are congruent for the virtual crack closure. The position of the nodes along the crack front is numbered consecutively with s_k. The width of the elements is $b_k = s_{k+1} - s_k$ etc. With these elements, the crack closure integral can only be carried out along the element edge between the crack front node F_i^1 and the first crack surface node with the relative displacements Δu_i^1. We must observe the fact that the associated area ΔA of this crack closure consists respectively of half of the contributions of the elements involved (see Fig. 5.29).

$$\Delta A = \frac{1}{2}(b_{k-1} + b_k)\Delta a$$

$$
\begin{aligned}
G(s_k) &= G_\mathrm{I} + G_\mathrm{II} + G_\mathrm{III} \\
&= \frac{1}{2\Delta A}\left(F_2^1 \Delta u_2^1 + F_1^1 \Delta u_1^1 + F_3^1 \Delta u_3^1\right) = \frac{1}{2\Delta A}\sum_{i=1}^{3} F_i^1 \Delta u_i^1
\end{aligned}
\qquad (5.94)
$$

The result is assigned to the position s_k of the corner node. We proceed analogously for all nodes along the crack front and thus obtain the curve $G_N(s)$ or $K_N(s)$ ($N = $ I, II, III).

(b) 20-Noded Hexahedron with Straight Crack Front

In the case of three-dimensional elements with quadratic shape functions (20-noded hexahedra or 15-noded pentahedra), eight nodes lie on the element face. The crack closure integral can be interpreted both for a corner node and for a mid-side node, as illustrated in Fig. 5.30.

The same geometric restrictions and denotations apply as in the previous section. Now the relevant node pairs and representative area ΔA must be determined for the respective MCCI. For a corner node (Fig. 5.30a), half the area of the adjacent

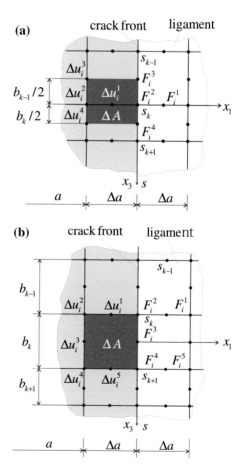

Fig. 5.30 3D crack closure integral for quadratic element functions at (**a**) corner nodes and (**b**) mid-side nodes

elements is regarded again. As opposed to the 2D case, the work terms of the mid-side nodes $j = 3$ and $j = 4$ must also be taken into consideration, whereby only half of the nodal forces may be taken into account.

$$\Delta A = \frac{1}{2}(b_{k-1} + b_k)\Delta a$$

$$G_{\mathrm{I}}(s_k) = \frac{1}{2\Delta A}\left[F_2^1 \Delta u_2^1 + F_2^2 \Delta u_2^2 + \frac{1}{2}F_2^3 \Delta u_2^3 + \frac{1}{2}F_2^4 \Delta u_2^4 \right]$$

$$G(s_k) = \frac{1}{2\Delta A}\sum_{i=1}^{3}\left[F_i^1 \Delta u_i^1 + F_i^2 \Delta u_i^2 + \frac{1}{2}F_i^3 \Delta u_i^3 + \frac{1}{2}F_i^4 \Delta u_i^4 \right] \qquad (5.95)$$

To calculate the MCCI for mid-side nodes, the five nodal forces on the ligament must be correlated to the assigned nodal displacements on the crack surface as shown in Fig. 5.30b. As for the forces F_i^j, only the components of this element may be inserted. Since usually only the total sectional forces are available, a simple weighting B_j was introduced [59] partitioning the forces according to the areas (widths) of the elements involved.

$$\Delta A = b_k \Delta a , \qquad \bar{s} = s_k + \frac{1}{2}b_k$$

$$B_1 = B_2 = b_k/(b_{k-1} + b_k), \qquad B_3 = 1 , \qquad B_4 = B_5 = b_k/(b_k + b_{k+1})$$

$$G_{\mathrm{I}}(\bar{s}) = \frac{1}{2\Delta A}\sum_{j=1}^{5} B_j F_2^j \Delta u_2^j$$

$$G(\bar{s}) = G_{\mathrm{I}} + G_{\mathrm{II}} + G_{\mathrm{III}} = \frac{1}{2\Delta A}\sum_{j=1}^{5}\sum_{i=1}^{3} B_j F_i^j \Delta u_i^j \qquad (5.96)$$

(c) 20-Noded Hexahedron with Curved Crack Front

In the case of a curved crack front, the crack closure integral can only be approximated since the area $\Delta\bar{A}$ in front of the crack front, from which the nodal forces are taken, is not exactly congruent with the crack face element ΔA as illustrated in Fig. 5.31. In order to minimize this misfit, the following geometric conditions should be observed:

- The edges of the element are always perpendicular to the current crack front.
- Both elements in front of and behind the crack front have the same depth L, which corresponds to the crack extension Δa.
- L should be small in comparison to the crack length or the curvature radius of the crack front.
- Then the crack extension area on the crack front segment b_k is calculated $\Delta A \approx \Delta\bar{A} = b_k\Delta a$ (see Fig. 5.31).

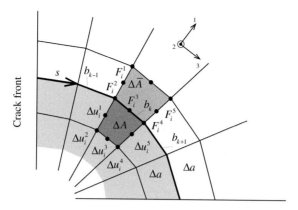

Fig. 5.31 3D crack closure integral for 20-noded hexahedra with a curved crack front

Under these conditions, the formulae (5.95) and (5.96) for corner or mid-side nodes can also be applied to curved crack fronts and provide us with usable levels of accuracy [59, 60].

(d) MCCI for 3D Quarter-Point Elements

To use three dimensional quarter-point elements (see Sect. 5.2.3) with MCCI, adjusted interpretation formulae were developed [61, 62] taking into account the specific displacement functions and the singular stress behavior. Two variants will be described that have proved themselves useful for both straight and curved crack fronts and are equally applicable for distorted 20-noded hexahedral elements, 15-noded pentahedral elements or collapsed distorted hexahedral elements. The geometric assignment of the force and displacement pairs is given in Fig. 5.32.

Variant (1) [62]

$$
\begin{aligned}
G = \frac{1}{2\Delta ab} \sum_{i=1}^{3} \Big[& \left(2c_7 F_i^4 + c_9 F_i^5 + 2c_8 F_i^6 + c_7 F_i^3 + c_8 F_i^7 \right) \Delta u_i^5 \\
& + \left(c_7 F_i^3 + c_8 F_i^7 + 2c_7 F_i^2 + c_9 F_i^1 + 2c_8 F_i^8 \right) \Delta u_i^1 \\
& + \left(c_4 F_i^4 + c_5 F_i^5 + c_6 F_i^6 + c_2 F_i^3 + c_3 F_i^7 + c_1 F_i^2 - c_1 F_i^1/2 + c_1 F_i^8 \right) \Delta u_i^4 \\
& + \left(c_1 F_i^4 - c_1 F_i^5/2 + c_1 F_i^6 + c_2 F_i^3 + c_3 F_i^7 + c_4 F_i^2 + c_5 F_i^1 + c_6 F_i^8 \right) \Delta u_i^2 \\
& + \left(c_1 \left(-2F_i^4 + F_i^5 - 2F_i^6 - 2F_i^2 + F_i^1 - 2F_i^8 \right) + c_{10} F_i^3 + c_{11} F_i^7 \right) \Delta u_i^3 \Big]
\end{aligned}
$$

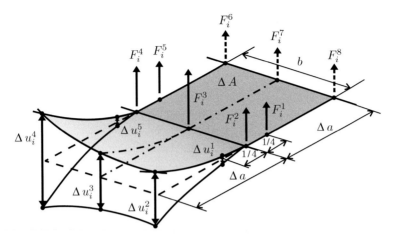

Fig. 5.32 MCCI for 3D quarter-point elements

$$c_1 = (80 - 25\pi)/24, \quad c_2 = (544 - 173\pi)/48, \quad c_3 = (304 - 101\pi)/48,$$
$$c_4 = (104 - 31\pi)/6, \quad c_5 = (11\pi - 31)/6, \quad c_6 = (34 - 11\pi)/3,$$
$$c_7 = (33\pi - 104)/4, \quad c_8 = (21\pi - 64)/4, \quad c_9 = (68 - 21\pi)/4,$$
$$c_{10} = (37\pi - 104)/12, \quad c_{11} = (19\pi - 56)/12 \tag{5.97}$$

Variant (2) [63]

$$
\begin{aligned}
G = \frac{1}{2\Delta ab} \sum_{i=1}^{3} \Bigg[& \left(F_i^1 + c_1 F_i^2 + \frac{c_1}{2} F_i^3 \right) \Delta u_i^1 \\
& + \left(\frac{1}{2} F_i^1 - 2c_2 F_i^2 + c_4 F_i^3 - \frac{c_2}{2} F_i^4 \right) \Delta u_i^2 \\
& + \left(c_2 F_i^2 + c_3 F_i^3 + c_2 F_i^4 \right) \Delta u_i^3 \\
& + \left(\frac{1}{2} F_i^5 - 2c_2 F_i^4 + c_4 F_i^3 - \frac{c_2}{2} F_i^2 \right) \Delta u_i^4 \\
& + \left(F_i^5 + c_1 F_i^4 + \frac{c_1}{2} F_i^3 \right) \Delta u_i^5 \Bigg]
\end{aligned}
$$
$$c_1 = 6\pi - 20, \quad c_2 = \pi - 4, \quad c_3 = \pi - 2, \quad c_4 = (16 - 5\pi)/4 \tag{5.98}$$

5.5.4 Consideration of Crack Face, Volume and Thermal Loading

If tractions t_i^* still affect the crack faces after crack extension, they must be converted into equivalent nodal forces F_i^{*j} with the shape functions (4.51) as is usual in FEM. A symmetric load of both crack faces is assumed, i.e. $t_i^{*+} = -t_i^{*-}$. In case of constant pressure p and rectangular element areas ΔA, the resultant force $F^* = p\Delta A$ is distributed as follows to the n_K nodes:

- 2D and 3D elements with linear shape functions: $F^{*j} = F^*/n_K$
- 2D quadratic shape functions: 2 corner nodes $F^{*j} = \frac{1}{6}F^*$, 1 mid-node $F^{*j} = \frac{2}{3}F^*$
- 3D quadratic shape functions: 4 corner nodes $F^{*j} = -\frac{1}{12}F^*$, 4 mid-nodes $F^{*j} = \frac{1}{3}F^*$
- 2D quarter-point-elements: crack nodes $F^{*1} = 0$, quarter-point $F^{*3} = \frac{2}{3}F^*$, corner point $F^{*2} = \frac{1}{3}F^*$
- 3D quarter-point elements: (see Fig. 5.32)

$$F^{*j} = \left\{ \frac{1}{3}, -\frac{1}{9}, \frac{2}{9}, -\frac{1}{9}, \frac{1}{3}, -\frac{1}{18}, \frac{4}{9}, -\frac{1}{18} \right\} F^*, \quad j = 1, 2, \ldots, 8$$

With cohesive zone models (Sect. 8.5) the crack face loads t_i^* are dependent on the current crack face displacements $\Delta u_i(r, \theta = \pm\pi)$, i.e. the consistent nodal forces have to be integrated anew with (4.51) after every load step.

If there are volume loads (e.g. weight) or forces of inertia in dynamic problems, the modified crack closure integral can be applied unchanged because these loads are contained indirectly in the sectional forces of the ligament nodes (see Sect. 5.5.2). The same is true for thermal loads resulting from inhomogeneous temperature fields.

In conclusion, the modified crack closure integral technique (MCCI) can be appraised in the following way: It is a simple robust and very powerful method to compute energy release rates. We need only the nodal displacements of the crack faces and the sectional forces at the nodes on the ligament in front of the crack, so a simple evaluation in the post-processor of the FEM computation is possible. The interpretation formulae only depend on the type of shape functions along the element edge, i.e. they are independent of whether the crack region was meshed with triangular, quadrilateral or degenerated quadrilateral elements. There are MCCI interpretation formulae for elements with linear, quadratic, or nodal-distorted displacement functions. For dynamic crack analyses, usually linear elements are favored. For static loads, the quadratic shape functions are preferable due to their higher accuracy. The modified crack closure integral generally provides improved accuracy in the stress intensity factors than the displacement interpretation method DIM, since it is based on an energy approach.

Moreover, the method is easily applicable to crack face, volume and thermal loads. One *important advantage* is that the components G_I, G_{II} and G_{III} of the three crack opening modes can be determined separately in the case of mixed-mode loading and thus the stress intensity factors K_I, K_{II} and K_{III} with (3.93) as well. Of course, MCCI also requires a sufficiently fine discretization at the crack, which can capture the near field solution. However, this method is not as sensitive as DIM with respect to element size. The following is recommended:

element edge length L = crack increment Δa < crack length $a/10$.

The modified crack closure integral technique suffers from two *disadvantages*: Firstly, it is limited to linear-elastic material behavior, since the path-independence and reversibility of the crack closure or crack extension process is always assumed. Secondly, some problems arise in the case of three-dimensional crack configurations with curvilinear crack fronts with respect to the geometric compatibility of the crack areas to be closed, which leads to loss of accuracy. Moreover the finite element meshes of the crack region must fulfill specific geometric requirements in MCCI such as equal element sizes in front of and behind the crack and congruent meshing on both crack edges.

5.6 FE-Computation of J-Contour Integrals

At this point, the numerical calculation of the classic J-integral from Sect. 3.2.6 will be explained in the context of FEM for 2D problems. In LEFM, J (3.100) is identical to the elastic energy release rate G, which provides the relation to K_I and K_{II} according to (3.93).

$$J = \int_\Gamma \left(U n_1 - \sigma_{ij} \frac{\partial u_i}{\partial x_1} n_j \right) ds \tag{5.99}$$

The integration path is subdivided into parts Γ_e per element, i.e. $\Gamma = \sum_{e=1}^{n_E} \Gamma_e$ (see Fig. 5.33). The most common method is to place the integration path through the integration points (IP) of the element. This has the advantage that the stresses there are usually known from the FEM analysis and have the highest level of accuracy. Integration over Γ_e should, as in Fig. 5.33, proceed along the natural coordinate $\xi_1 =$ const. with $\xi_2 \in [-1, +1]$ as curve parameter.

The algorithm for parameterized line integrals (4.76) provides the normal unit vector $\boldsymbol{n}$ on Γ_e (see Fig. 5.33)

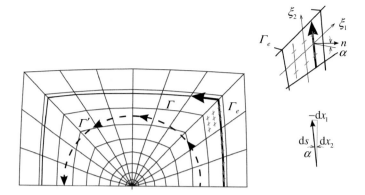

Fig. 5.33 Integration path for the J-integral calculation in the finite element mesh

$$\begin{bmatrix} n_1 \\ n_2 \end{bmatrix} \mathrm{d}s = \begin{bmatrix} \cos\alpha \\ \sin\alpha \end{bmatrix} \mathrm{d}s = \begin{bmatrix} \mathrm{d}x_2 \\ -\mathrm{d}x_1 \end{bmatrix} = \begin{bmatrix} \dfrac{\partial x_2}{\partial \xi_2} \\[2ex] -\dfrac{\partial x_1}{\partial \xi_2} \end{bmatrix} \mathrm{d}\xi_2 \qquad (5.100)$$

and the transformation (4.76) of the line element $\mathrm{d}s = J_\mathrm{L}\mathrm{d}\xi_2$. The 1st term of the integrand (5.99) is the strain energy density U according to (3.71). For two-dimensional elastic problems it reads

$$U = \frac{1}{2}\left(\sigma_{11}\varepsilon_{11} + 2\tau_{12}\varepsilon_{12} + \sigma_{22}\varepsilon_{22}\right). \qquad (5.101)$$

In the 2nd term there are the tractions

$$t_i = \sigma_{ij}n_j, \qquad \begin{bmatrix} t_1 \\ t_2 \end{bmatrix} = \begin{bmatrix} \sigma_{11}n_1 + \tau_{12}n_2 \\ \tau_{12}n_1 + \sigma_{22}n_2 \end{bmatrix} \qquad (5.102)$$

and the derivatives of the displacement vector u_i with respect to x_1:

$$\frac{\partial u_i}{\partial x_1} = \sum_{a=1}^{n_\mathrm{K}} \frac{\partial N_a(\xi_1, \xi_2)}{\partial x_1} u_i^{(a)}. \qquad (5.103)$$

These are expressed by differentiating the shape functions $N_a(\xi_1, \xi_2)$ and displacements $u_i^{(a)}$ of the element nodes a. For the derivation, the inverse Jacobian matrix has to be used (see Sect. 4.4.2):

$$\begin{bmatrix} \dfrac{\partial N_a}{\partial x_1} \\[2ex] \dfrac{\partial N_a}{\partial x_2} \end{bmatrix} = \begin{bmatrix} \dfrac{\partial \xi_1}{\partial x_1} & \dfrac{\partial \xi_2}{\partial x_1} \\[2ex] \dfrac{\partial \xi_1}{\partial x_2} & \dfrac{\partial \xi_2}{\partial x_2} \end{bmatrix} \begin{bmatrix} \dfrac{\partial N_a}{\partial \xi_1} \\[2ex] \dfrac{\partial N_a}{\partial \xi_2} \end{bmatrix} = \begin{bmatrix} J^{-1} \end{bmatrix} \begin{bmatrix} \dfrac{\partial N_a}{\partial \xi_1} \\[2ex] \dfrac{\partial N_a}{\partial \xi_2} \end{bmatrix}. \tag{5.104}$$

Thus the J-integral over one element Γ_e reads:

$$J^{(e)} = \int\limits_{-1}^{+1} \left\{ \frac{1}{2}\left(\sigma_{11}\varepsilon_{11} + 2\tau_{12}\varepsilon_{12} + \sigma_{22}\varepsilon_{22}\right) n_1 \right.$$

$$\left. - (\sigma_{11}n_1 + \tau_{12}n_2)\frac{\partial u_1}{\partial x_1} - (\tau_{12}n_1 + \sigma_{22}n_2)\frac{\partial u_2}{\partial x_1} \right\} J_{\mathrm{L}} \mathrm{d}\xi_2 \tag{5.105}$$

$$= \int\limits_{-1}^{+1} F(\xi_1, \xi_2)\mathrm{d}\xi_2 \approx \sum_{g=1}^{n_{\mathrm{G}}} F(\xi_1 = \mathrm{const.}, \xi_2^g)w_g \,,$$

which is calculated with the last relation by a 1D Gaussian quadrature. For this, it is best to select exactly the integration order n_{G} ($= 3$ in Fig. 5.33) that was already used in the finite element so that we can immediately carry over the σ_{ij}, ε_{ij} and the energy density U (available in many codes) from the FEM result file at the IP.

Finally, we obtain the total value of J by summation of the contributions of all elements in the integration path Γ

$$J = \sum_{e=1}^{n_{\mathrm{E}}} J^{(e)} \,. \tag{5.106}$$

Since the contour Γ consists of paths $\xi_1 = \mathrm{const.}$ through neighboring elements, there are restrictions with respect to the design of the FEM mesh to frame a continuous, closed integration path. Node numbering in the element must be adjusted such that Γ_e always lies on $\xi_1 = \mathrm{const.}$

An alternative calculation variant is possible if the results of the FEM computation are available at every location $\mathbf{x} = [x_1 \, x_2]^{\mathrm{T}}$ independently of the mesh. Some FEM codes offer interpolated and smoothed field quantities in the post-processor. Then we can choose a geometrically simple contour for the J integral such as the semi-circle Γ' drawn dashed in Fig. 5.33. This comfort comes however at the cost of numerical inaccuracies and increased effort, which becomes clear if we imagine preparing all necessary field quantities for (5.105) at a point $\mathbf{x}$ (see Sect. 4.4.4)). This variant of J-calculation can only be recommended if all aforementioned approximation steps are known.

In the meantime there are many extensions and generalizations of the classical J-integral, which is why this topic has been given a separate Chap. 6.

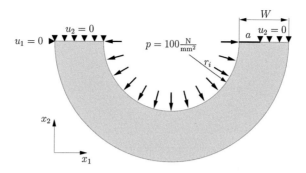

Fig. 5.34 Cylinder with internal crack of length a (half model)

5.7 FE-Calculation of Fracture Mechanics Weight Functions

5.7.1 Determination by Point Forces

In Sect. 3.2.10 we discussed the advantage of fracture-mechanical weight functions, which enable us to determine easily the stress intensity factors for a crack configuration at every desired load. In the following, a few FEM techniques will be introduced to calculate numerically weight functions for 2D crack problems. The cylinder (pipe) with an internal crack shown in Fig. 5.34 will serve as a demonstration example.

One simple but quite practical variant is to determine directly the effect of a single force $F(x) = F_1 e_1 + F_2 e_2$ acting at location x on the K-factors $K_I(a)$ and $K_{II}(a)$ at the tip of a crack of length a. For this purpose, we apply concentrated forces $F^{(l)}$ in sequence upon all nodes $x^{(l)}$ of the crack configuration (surface S_t, volume V or crack face S_c), where later real loads can arise, and calculate the associated K-factors $K_L^{(l)}$ with one of the FEM techniques introduced above. Figure 5.35 illustrates this using the mesh of the cylinder with an internal crack. The direction of each force $F^{(l)}$ corresponds to a coordinate e_1 or e_2. It can also refer to the expected loads (e.g. normal to the surface under pressure). It is practical to set its magnitude to unity, $F^{(l)} = |F^{(l)}| = 1$. The FEM analysis with these $2\,n_L$ »unit loads« $F^{(l)}$ over all $l = 1, 2, \ldots, n_L$ nodes with $i = 1, 2$ components is best carried out simultaneously with $2\,n_L$ right hand sides. The resultant $K_L^{(l)}$-factors ($L = $ I, II) already correspond to the *fracture-mechanical weight functions* (3.147) for node l with force component i:

$$K_L^{(l)}(a) = H_i^L\left(x^{(l)}, a\right) F_i^{(l)}\left(x^{(l)}\right) = H_{il}^L\left(x^{(l)}, a\right). \qquad (5.107)$$

In order to apply the calculated and stored weight functions for a specific load case of this crack configuration, we have to convert the given boundary loads $\bar{t}$, crack face loads $\bar{t}_c$, or volume loads $\bar{b}$ into equivalent nodal forces. To do this, the FEM relations (4.51) and (4.76) are used, which yields, for example, for an element edge

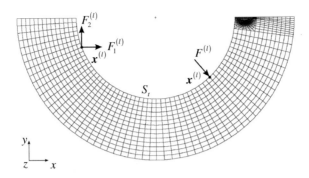

Fig. 5.35 Calculation of weight functions with the help of unit loads. FEM mesh for a cylinder with an internal crack

s on the traction boundary S_{te}:

$$\mathbf{f}_e \triangleq \mathbf{f}_t = \int_{-1}^{+1} \mathbf{N}^{\mathrm{T}}(\xi)\, \bar{\mathbf{t}}(\xi)\, J_L \mathrm{d}\xi \tag{5.108}$$

These nodal forces are integrated and assembled for all loaded element boundaries, from which the equivalent global nodal forces $\hat{\mathbf{F}}^{(l)} = \bigcup_{e=1}^{n_{\mathrm{E}}} \mathbf{f}_e$ result. If we insert their values into (5.107), we obtain both stress intensity factors K_{I} and K_{II} from the weighted sum

$$K_L(a) = \sum_{l=1}^{n_L} \sum_{i=1}^{2} H_{il}^L \, \hat{F}_i^{(l)}. \tag{5.109}$$

The application of weight functions only requires the evaluation of the simple relations (5.108) and (5.109); no further FEM calculation is needed. With a very fine FEM mesh, $\bar{t} \approx$ const. may be assumed along the element edge L so that (5.108) leads to the simplified formula according to which the resulting force $F_R = L\bar{t}$ is subdivided on the edge nodes such as $\begin{bmatrix} \frac{1}{2} & \frac{1}{2} \end{bmatrix}$ with linear and $\begin{bmatrix} \frac{1}{6} & \frac{2}{3} & \frac{1}{6} \end{bmatrix}$ with quadratic shape functions. However, these weight functions are always bound to the geometry of the FEM mesh in use!

For a cylinder with an internal radius $r_i = 40\,\mathrm{mm}$, wall thickness $w = 30\,\mathrm{mm}$, crack length $a = 20\,\mathrm{mm}$ under internal pressure $p = 100\,\mathrm{MPa}$, the stress intensity factor amounts to $K_{\mathrm{I}}^{\mathrm{ref}} = 67.27\,\mathrm{MPa}\sqrt{\mathrm{m}}$ according to the handbook [64, 65]. Application of the unit load method with the FEM mesh in Fig. 5.35 and subsequent summing up of the weight functions for the pressure load at r_i resulted in the value $K_{\mathrm{I}} = 67.01\,\mathrm{MPa}\sqrt{\mathrm{m}}$.

Fig. 5.36 Approximation
of crack loads with power
functions

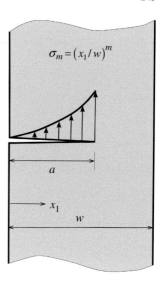

5.7.2 Determination of Parametric Influence Functions

One very practical and well-established method is known under the name of *influence functions*. As opposed to the weight functions for unit loads, these quantify the effect of a parameterized distribution of the traction load on the K_I-factor. Since we can convert every load of the crack configuration with the principle of superposition (Sect. 3.2.10) into an equivalent crack face load $t^c(x)$, these influence functions are developed especially for the crack faces. Often, power functions of order m are used as idealized loads:

$$\sigma_m(x_1/w) = \sigma_m(\bar{x}) = \bar{x}^m \qquad (m = 0, 1, 2, \ldots, n_m) \qquad (5.110)$$

Figure 5.36 shows the conditions for the wall cross-section of the example »cylinder with internal crack«. Now, for every loading function $\sigma_m(\bar{x})$, the stress intensity factors are calculated for a given crack length a with FEM and written in standardized form:

$$K_I^{(m)}(a) = \phi_m(a)\sqrt{\pi a} \qquad (5.111)$$

The functions $\phi_m(a)$ are called *influence functions*.

The stress intensity factor for an actual loading on a structure with a crack is obtained from this as follows: We calculate with FEM or analytically the sectional stresses on the crack line in the crack-free structure, which corresponds to the crack face tractions $t^c(x_1)$. Using a regression analysis, this stress distribution is developed into polynomials such as (5.110), resulting in the coefficients D_m:

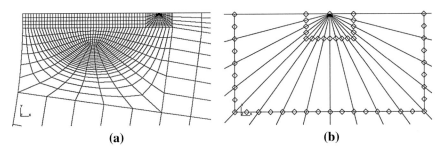

Fig. 5.37 Mesh of the crack tip region of Fig. 5.35 (**a**) with quarter-point triangular elements (**b**)

$$t^c(x_1) = \sum_{m=0}^{n_m} D_m \sigma_m(\bar{x}) = \sum_{m=0}^{n_m} D_m \bar{x}^m \qquad (5.112)$$

The K_I-factor is obtained from the weighted summation of all influence functions:

$$K_I(a) = \sum_{m=0}^{n_m} D_m \phi_m(a) \sqrt{\pi a}. \qquad (5.113)$$

This method is especially effective if we have to analyze the same component with the same crack under varying load cases many times (e.g. thermal shock transients).

For the example under consideration (Fig. 5.34), the influence coefficients were calculated very exactly by Andrasic and Parker [65] with the help of Green's functions and serve as the reference solution ϕ_m^{ref}. Their values for powers $m = 0, 1, \ldots, 4$ are compiled in Table 5.2 ($a = 20\,\text{mm}$). The influence functions determined with FEM are in the 2nd column. To this end, the corresponding stress distributions $\sigma_m(\bar{x})$ were imposed on the crack faces of the mesh (Fig. 5.35). The crack tip was meshed with quarter-point elements CTE (see detail in Fig. 5.37). To calculate the K_I-factors, the displacement interpretation method (DIM) was used. The relative errors of ϕ_m^{CTE} compared to the reference solution are small, but they grow larger with the power m. In order to apply these influence function to the load case »cylinder under internal pressure«, we emanate from the known solution for the circumferential stresses (r_a-outside radius):

$$\sigma_{\theta\theta}(r) = \frac{p r_i^2}{r_a^2 - r_i^2} \left(1 + \frac{r_a^2}{r^2} \right) \qquad (5.114)$$

The stress distribution in the wall $0 \le r_i + x_1 \le r_a$ can be approximated quite accurately as a 4th degree polynomial, from which we determine the concrete coefficients D_m of (5.112) (with $p = 100\,\text{MPa}$):

$$t^c \mathrel{\widehat{=}} \sigma_{\theta\theta}(\bar{x}) = 196.88 - 218.92\bar{x} + 218.07\bar{x}^2 - 140.40\bar{x}^3 + 40.50\bar{x}^4 \qquad (5.115)$$

Table 5.2 Comparison of the influence coefficients for a cylinder with internal crack according to various calculation methods

Order m	ϕ_m^{ref}	ϕ_m^{CTE}	$\Delta\phi_m^{\text{CTE}}$ (%)	ϕ_m^H	$\Delta\phi_m^H$ (%)
0	1.8400	1.8321	−0.4270	1.8341	−0.3186
1	0.6477	0.6452	−0.3774	0.6399	−1.1986
2	0.3090	0.3069	−0.6733	0.3023	−2.1652
3	0.1663	0.1640	−1.3930	0.1599	−3.8102
4	0.0953	0.0930	−2.4465	0.0883	−7.3241

Insertion into (5.110) provides the stress intensity factor $K_I = 66.96$ MPa with an error of -0.46%!

5.7.3 Derivation from Displacement Fields

The actual fracture-mechanical weight functions can be obtained from the derivative of the displacement field $u_i^{(2)}(x, a)$ with respect to the crack length a. According to Sect. 3.2.10, this requires any known solution (in the case of mixed modes, two), index (2), of the crack configuration under consideration. The relationship is given by Eq. (3.157) for mode I and Eq. (3.160) for mixed mode loading. It is obvious how to implement this computational method numerically. For this we determine the required reference solutions (2) with the help of FEM calculations for the crack length of interest a and find out the stress intensity factors $K_I^{(2)}(a)$ and $K_{II}^{(2)}(a)$ with one of the above-described FEM techniques. Derivative of the displacement fields with respect to the crack length must be carried out as a difference quotient $\Delta u_i^{(2)}/\Delta a$. Therefore, the central difference scheme is well suited because of its great precision of $O(\Delta a)^2$. But this means that the displacement fields must be calculated for two neighboring crack lengths $a - \Delta a$ and $a + \Delta a$. The formula (3.157) thus is written at crack length a for mode I:

$$H_i^I(x, a) = \frac{E'}{2K_I(a)} \left(\frac{u_i(a + \Delta a) - u_i(a - \Delta a)}{2\Delta a} \right) \tag{5.116}$$

and (3.160) for mixed loading I and II with $K^2 = K_I^{(2b)} K_{II}^{(2a)} - K_I^{(2a)} K_{II}^{(2b)}$ reads:

$$H_i^I(x, a) = \frac{E'}{2K^2} \left[K_{II}^{(2a)} \frac{u_i^{(2b)}(a + \Delta a) - u_i^{(2b)}(a - \Delta a)}{2\Delta a} \right.$$
$$\left. - K_{II}^{(2b)} \frac{u_i^{(2a)}(a + \Delta a) - u_i^{(2a)}(a - \Delta a)}{2\Delta a} \right]$$

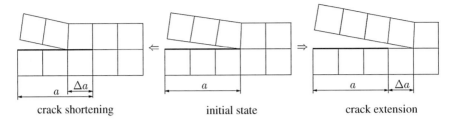

Fig. 5.38 Required crack lengths for the central difference scheme

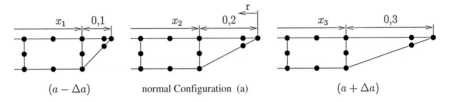

Fig. 5.39 Variation of the crack length by displacement of the crack tip node

Table 5.3 Comparison of the K_I-factors calculated from the weight functions with other calculation methods ($\sigma = 40$ MPa) in MPa$\sqrt{m}$

a (mm)	K_I^{ref} (MPa$\sqrt{m}$)	K_I^{CTE}	K_I^H	ΔK_I^H (%)
6	6.59	6.60	6.50	−1.43
10	9.52	9.52	9.44	−0.83
14	12.73	12.72	12.66	−0.56
18	16.34	16.34	16.30	−0.25
20	18.45	18.37	18.35	−0.54

$$H_i^{\text{II}}(x, a) = \frac{E'}{2K^2}\left[K_I^{(2b)} \frac{u_i^{(2a)}(a + \Delta a) - u_i^{(2a)}(a - \Delta a)}{2\Delta a} \right.$$
$$\left. - K_I^{(2a)} \frac{u_i^{(2b)}(a + \Delta a) - u_i^{(2b)}(a - \Delta a)}{2\Delta a} \right] \tag{5.117}$$

The procedure requires FEM analyses with three different crack lengths as clarified in Fig. 5.38. In the numerical implementation, these three variants can be realized simply and effectively as follows. Since in most cases we set a fan of quarter-point elements around the crack tip anyway (for exact K-calculation) as in Fig. 5.37, we vary the crack length by slight displacement of the crack tip node by $\pm \Delta a < L$. Only the coordinates of the quarter-points are changed slightly in the process–all other nodes of the mesh remain unchanged–so we can evaluate the displacement difference. Figure 5.39 illustrates this technique.

This method was again tested using the mode I example »cylinder with internal crack« for various crack lengths a. A constant crack face load $t^c(x_1) = 40\,\text{MPa}$ on $(0 \leq x_1 \leq a)$ was assumed as a reference solution (2) to calculate the weight functions. This first resulted in the $K_I^{(2)}(a)$-value with the help of DIM. The results are denoted with K_I^{CTE} in Table 5.3 and agree well with the comparison solution K_I^{ref} [65]. Via crack length variation the displacement derivative on the crack faces, and with (5.116) the weight function $H_2^1(x, a)$ were calculated. Finally, this weight function was used to obtain the K_I-factor of the reference load case by numerical integration of (3.147). These K_I^H-values must of course reproduce the K_I^{CTE}-values of the direct FEM-computation and deviate only slightly from the K_I^{ref}-solution (see Table 5.3).

Furthermore, the weight functions of $a = 20\,\text{mm}$ were used to calculate the influence functions of the previous section, i.e. integration of the power functions $\sigma_m(\bar{x})$ with (3.147). The thereby obtained influence coefficients ϕ_m^H are listed in Table 5.2, whereby their error levels increase with higher powers m.

Finally, the load case of »cylinder with internal pressure« was considered. Integration of the sectional stresses (5.114) with these weight functions yielded an intensity factor $K_I = 67.19\,\text{MPa}\sqrt{\text{m}}$ with $-0.1\,\%$ error.

5.7.4 Application of the J-VCE-Technique

The idea behind this method is to use the VCE technique in order to determine the required derivative of the displacement field with respect to crack length. Differentiation of the stiffness relation $\mathbf{KV} = \mathbf{F}$ with respect to $\mathrm{d}a$ at $\mathbf{F} = \text{const.}$ gives

$$\frac{\mathrm{d}\mathbf{K}}{\mathrm{d}a}\mathbf{V} + \mathbf{K}\frac{\mathrm{d}\mathbf{V}}{\mathrm{d}a} = \frac{\mathrm{d}\mathbf{F}}{\mathrm{d}a} = 0 \rightarrow \frac{\mathrm{d}\mathbf{V}}{\mathrm{d}a} = -\mathbf{K}^{-1}\frac{\mathrm{d}\mathbf{K}}{\mathrm{d}a}\mathbf{V}. \qquad (5.118)$$

While Parks and Kamenetzky [66] originally formed the difference quotient $\Delta\mathbf{K}/\Delta a$ using FEM, the more elegant and exact VCE method of deLorenzi [52] should be used. We start from (5.81) for the calculation of the energy release rate in the 2D case $(V_0 \rightarrow A_0)$, where $\Delta l_k(x) = \Delta l_1 \,\hat{=}\, \Delta l$ is true:

$$G = -\int_{A_0} \left[U\frac{\partial\Delta l}{\partial x_1} - \sigma_{ij}\frac{\partial u_i}{\partial x_1}\frac{\partial\Delta l}{\partial x_j} \right] \mathrm{d}A \qquad (5.119)$$

The VCE $\Delta l(x)$ is interpolated with the FEM shape functions in the domain A_0

$$\Delta l(x) = \sum_{a=1}^{n_{\text{K}}} N_a(\boldsymbol{\xi})\Delta l^{(a)}. \qquad (5.120)$$

Applying the shape functions to the integral terms per element

$$U = \frac{1}{2}\sigma_{ij}u_{i,j} = \frac{1}{2}\sigma_{ij}\sum_{a=1}^{n_K}\frac{\partial N_a}{\partial x_j}u_i^{(a)}$$

$$\sigma_{ij}\frac{\partial u_i}{\partial x_1} = \sigma_{ij}\sum_{a=1}^{n_K}\frac{\partial N_a}{\partial x_1}u_i^{(a)}, \qquad \frac{\partial \Delta l}{\partial x_j} = \sum_{b=1}^{n_K}\frac{\partial N_b}{\partial x_j}\Delta l^{(b)}, \qquad (5.121)$$

we obtain

$$G = -\int_{A_0}\left[\frac{1}{2}\sum_a\frac{\partial N_a}{\partial x_j}\sum_b\frac{\partial N_b}{\partial x_1} - \sum_a\frac{\partial N_a}{\partial x_1}\sum_b\frac{\partial N_b}{\partial x_j}\right]\sigma_{ij}u_i^{(a)}\Delta l^{(b)}\,dA.$$

$$(5.122)$$

The nodal variables $u_i^{(a)} \in V$ can be pulled behind the integral. Its evaluation for all elements in A_0 yields a force vector $Q_i^{(a)}$

$$G = -\frac{1}{2}Q_i^{(a)}u_i^{(a)} = -\frac{1}{2}\mathbf{Q}^{\mathrm{T}}\mathbf{V} = -\frac{1}{2}\mathbf{Q}_0^{\mathrm{T}}\mathbf{V}_0 \quad \text{for } i = 1, 2 \text{ and } a = 1, 2, \ldots, N_K.$$

$$(5.123)$$

If we equate (5.123) with (5.79), then follows

$$G = -\frac{1}{2}\mathbf{Q}_0^{\mathrm{T}}\mathbf{V}_0 = -\frac{1}{2}\mathbf{V}_0^{\mathrm{T}}\frac{d\mathbf{K}}{da}\mathbf{V}_0 \quad \Rightarrow \quad \frac{d\mathbf{V}}{da} = \mathbf{K}^{-1}\mathbf{Q}_0. \qquad (5.124)$$

5.7.5 *Calculation by Means of the Bueckner Singularity*

In Sect. 3.2.10 it was shown that the weight functions (3.172) are proportional to a fundamental displacement field, the Bueckner singularity, that appears in the entire crack configuration due to the effect of a force pair B_{I} acting at the crack tip. The direct numerical realization of this approach fails, because the hyper-singular solution of the concentrated forces directly at the crack tip is difficult to model with sufficient accuracy in FEM. In order to get rid of this problem, Paris et al. [67] left open a small hole around the crack tip and applied the fundamental field (3.173) on its boundary. Sham [68, 69] defined a sufficiently small mesh region around the crack tip inside which the hyper-singular field was treated separately.

In an alternative solution approach Busch et al. [70, 71] utilize the superposition principle in order to determine the fundamental displacement field for a finite crack configuration. To this end, the Bueckner singularity known for the infinite domain (3.173), (3.174) is subtracted from the boundary value problem so that only a regular crack problem with stress singularity $1/\sqrt{r}$ remains to be solved via FEM, for which approved solution techniques have already been explained. As we can see in Fig. 5.40, for this purpose the tractions

$$\bar{t}_i^\infty(\boldsymbol{x}) = \sigma_{ij}^\infty(\boldsymbol{x})n_j(\boldsymbol{x}) \qquad (5.125)$$

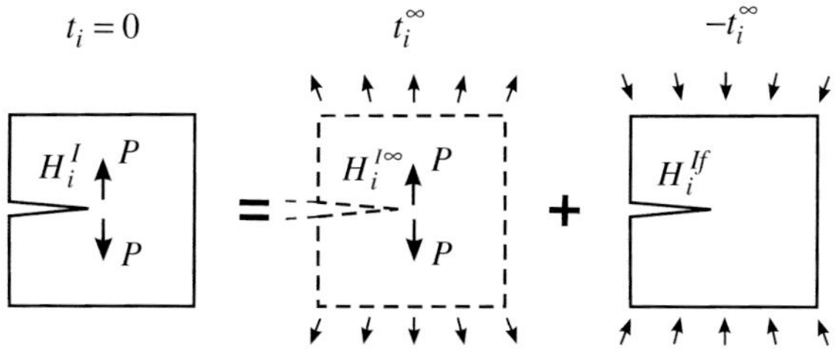

Fig. 5.40 The superposition principle for determining the Bueckner fundamental solution for a finite domain

on the boundary S of the finite crack configuration must be calculated from the hyper-singular Bueckner stress field (3.174) and then applied with reversed sign. The FEM solution thus provides a displacement field $u_i^{(2)f}(\boldsymbol{x})$ which represents the correction of the fundamental solution for the finite domain. The sum with the Bueckner solution (3.173) thus yields the sought fundamental solution for the finite crack configuration under consideration

$$u_i^{(2)}(\boldsymbol{x}) = u_i^{(2)\infty}(\boldsymbol{x}) + u_i^{(2)f}(\boldsymbol{x}), \qquad (5.126)$$

from which the weight functions for mode I with (3.172) follow:

$$H_i^{\mathrm{I}}(\boldsymbol{x}, a) = \frac{2\mu}{\kappa + 1} \frac{1}{\sqrt{2\pi B_{\mathrm{I}}}} u_i^{(2)}(\boldsymbol{x}). \qquad (5.127)$$

5.8 Examples

5.8.1 Tension Sheet with Internal Crack

Figure 5.1 shows the crack of length $2a$ ($a = 10\,\mathrm{mm}$) to be analyzed in a quadratic plate (thickness $B = 1\,\mathrm{mm}$, width $d = 100\,\mathrm{mm}$) subjected to a tensile load of $\sigma = 100\,\mathrm{MPa}$. The material is isotropic elastic with $E = 210,000\,\mathrm{MPa}$ and $\nu = 0.3$. A state of plane strain is assumed. For this simple two-dimensional mode I crack problem, the stress intensity factor is known [64, p. 68]:

$$K_{\mathrm{I}}(a) = \sigma\sqrt{\pi a}\, g\left(\frac{a}{d}\right)$$

$$g = 1.0 \quad \text{infinite plate } d \to \infty \quad \textit{Griffith-crack}$$

$$g\left(\frac{a}{d} = \frac{1}{10}\right) = 1.014 \quad \Rightarrow \quad K_{\mathrm{I}} = 568.35\,\mathrm{MPa}\sqrt{\mathrm{mm}}$$

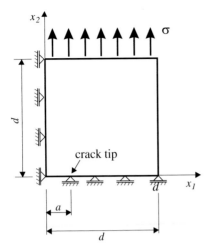

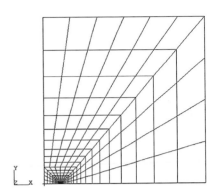

Fig. 5.41 Modeled upper quarter of the plate with internal crack

Fig. 5.42 FEM-discretization with 8-node quadrilateral elements

As explained already in Sect. 5.1, due to symmetry it is sufficient to model one quarter of the plate with corresponding displacement boundary conditions, which are depicted in Fig. 5.41. The finite element mesh used consists of 8-noded quadrilateral elements with quadratic shape functions (see Fig. 5.42). These elements are collapsed to 6-noded triangles at the crack tip and optionally either left in this form or further collapsed to quarter-point elements in accordance with Fig. 5.7. Figure 5.43 shows the details of the mesh at the crack tip. The size of the crack tip elements is $L = a/40 = 0.25$ mm. Alternatively, the crack tip was surrounded by a fan of 14 (left) or 7 (right) elements.

Figure 5.44 gives an impression of the stress concentration at the crack tip and the deformation (shown exaggerated) of the crack faces.

Figures 5.45 and 5.46 show a comparison of crack-opening displacements $u_2(r, \theta = \pi)$ and normal stresses $\sigma_{22}(r, \theta = 0)$ from the FEM solution with the analytical near field solution (see Sect. 3.2.1). In the case of this fine mesh, there is a very high level of agreement.

Calculation of the stress intensity factor $K_I(r^*)$ with the displacement interpretation method according to (5.3) provides the curve illustrated in Fig. 5.47. One can see very clearly that the locally computed K_I-factor declines sharply towards the crack tip in the case of regular elements, whereas quarter-point crack tip elements exhibit no deficits at all. The intensity factor extrapolated to the crack tip has an error margin of +0.08 %.

With quarter-point elements, K_I can also be determined directly with DIM via (5.39). The following parameter study on mesh refinement at the crack is informative. The result is shown in Fig. 5.48. It shows the K_I-factors extrapolated with DIM for various meshes, whereby the size L of the crack tip elements is varied from $a/3.3$

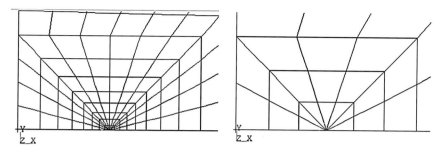

Fig. 5.43 Details of FEM discretization at the crack tip

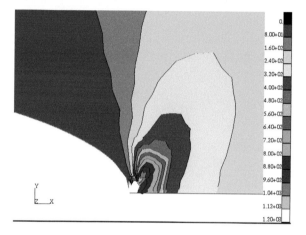

Fig. 5.44 Isoline representation of the v. Mises equivalent stress around the crack

to $a/40$. Only with $L = a/20$ do we reach an accuracy of less than 1 %. Doubling the elements in the circumferential direction does not lead to an improvement.

Finally, K_I or G can also be determined using the modified crack closure integral MCCI. For regular elements, Eq. (5.90) should be used and for quarter-point elements the variants (5.91) or (5.92). With the exception of variant (2), accuracies better than 0.1 % can be realized in K_I.

The computed results for the J-integral are shown in Fig. 5.49. Here, the equivalent domain integral EDI (see Sect. 6.4) was used with various integration domains, i.e. from the first to the sixth element ring around the crack tip with radii r. The J-values are nearly independent of the choice of the integration domain as the theory demands. Only if regular elements are used at the crack, then the evaluation of the first element ring (i.e. only displacement of the crack tip node) is insufficiently accurate. This phenomenon is generally observed, which is why the innermost ring should not be used. Basically, several paths (or element rings in the case of EDI)

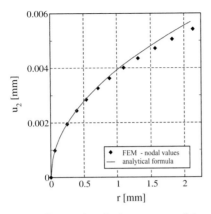

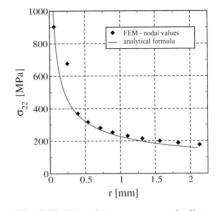

Fig. 5.45 Opening displacement u_2 of the crack face

Fig. 5.46 Normal stresses σ_{22} on the ligament in front of the crack

Fig. 5.47 Stress intensity factors calculated with DIM

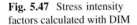

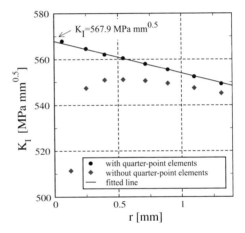

should be evaluated for the J-integral, and it should be tested whether the values are path-independent and converge towards remote rings.

In the case of LEFM holds $J = G$, so we can calculate the respective K-factor from the value of the J-integral with the relation (3.93) if there is pure mode I, II, or III loading. In the above example, we obtain $K_I = \sqrt{JE/(1 - \nu^2)}$. From the third integration path on, accuracy amounts to $\sim$0.05 % for quarter-point elements CTE and $\sim$0.1 % if regular standard elements are used at the crack tip.

The results of all methods and their relative errors compared to the reference solution are again compiled for an overview in Table 5.4.

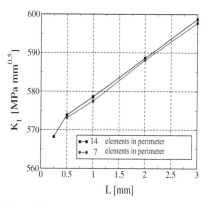
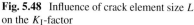
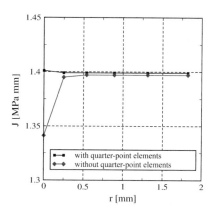

Fig. 5.48 Influence of crack element size L on the K_I-factor

Fig. 5.49 J-integral values for various integration paths

5.8.2 Semi-Elliptical Surface Crack Under Tension

The case being examined is the block-shaped structure with a semi-elliptical surface crack shown in Fig. 5.50. Its dimensions are: $h = 20$ mm, $d = 20$ mm, $b = 30$ mm, $c = 15$ mm and $a = 5$ mm. The axial ratio of the crack $a{:}c = 1{:}3$ corresponds to typical measures of surface errors. The tensile load $\sigma = 30$ MPa is perpendicular to the crack plane, so there is only mode I loading with a varying K_I-factor along the crack front. We again assume isotropic-elastic material with $E = 210{,}000$ MPa and $\nu = 0.3$.

Due to the double symmetry of the problem, only a quarter with corresponding displacement boundary conditions needs to be modeled (see Fig. 5.50). Figure 5.51 shows the chosen FEM meshing of this quarter. When generating meshes around 3D cracks, it is advantageous to start at the crack front and surround it with a »tube« of

Table 5.4 Comparison of the accuracy of various methods for determining K_I and G_I for the tension plate with internal crack (Fig. 5.41)

Method	Crack elements	K_I (MPa$\sqrt{\text{mm}}$)	Error %	G_I (N/mm)	Error %
Reference solution	[64]	568.35		1.3997	
K_I^*-extrapolation	RSE	~555.0	−2.4		
	CTE	567.9	+0.08		
K_I-DIM	CTE	572.5	+0.7		
MCCI	RSE	567.8	−0.1	1.3970	−0.2
	CTE (var. 1)	568.7	+0.07	1.4016	+0.14
	CTE (var. 2)	560.4	−1.4	1.3611	−2.7
J-integral EDI	RSE	567.9	−0.06	1.3979	−0.13
	CTE	568.1	−0.04	1.3985	−0.09

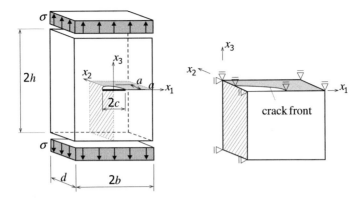

Fig. 5.50 Block-shaped structure with a semi-elliptical surface crack (a:$c = 1$:3) under tension

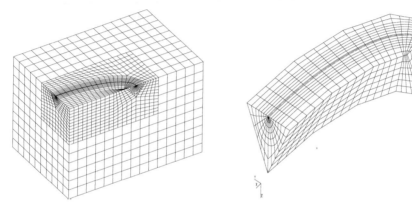

Fig. 5.51 FEM mesh for a semi-elliptical surface crack (1/4-model) with 26,672 nodes and 5,620 hexahedral elements

Fig. 5.52 Tubular meshing of the crack front

concentric element rings. In this way, we have all the crack tip elements collected in one group and can arrange them optimally by size and shape along the curved crack front so that the face surfaces are as perpendicular as possible to it (see Fig. 5.52). The 3D quarter-point hexahedral elements CTE introduced in Sect. 5.2.2 were used at the crack front. The edge length of the crack elements were uniformly set at $L = 0.08$ mm. Proceeding from this tube, we complete and coarsen the FEM mesh up to the outer boundaries of the structure.

Figure 5.53 shows the opened deformed crack profile. To compute the stress intensity factor K_I, all techniques explained above were used. The simplest is the interpretation of the crack face displacements using relation (5.46) (see Fig. 5.11). The result is plotted for all crack front nodes in Fig. 5.53 as a function of the ellipse angle $\varphi = \arcsin(x_2/a)$ (DIM-CTE). For comparison, the reference solution of Raju and Newman [64] is shown in the diagram. Calculation of the 3D J-integral (see

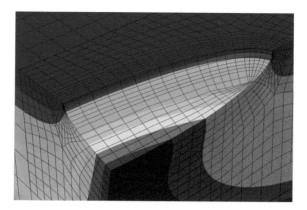

Fig. 5.53 Deformed FEM model with semi-elliptical surface crack

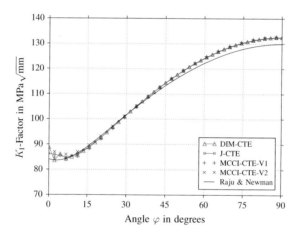

Fig. 5.54 Course of the K_I-factor along the crack front. Comparison of various FEM interpretation methods for quarter-point elements CTE

Sect. 6.3.2) yields results of similar accuracy for $K_I(\varphi)$. Finally, both formulae for calculating the crack closure integral MCCI were tested for the 3D quarter-point elements. The results of the more complex variant 1 of Eq. (5.97) and of the simpler formula (5.98) of variant 2 agree very well with all other FEM results. For simplicity the latter should be favored.

For the sake of comparison, the same problem was calculated with regular standard elements RSE, i.e. the 3D quarter-point elements in the »tube« were replaced by collapsed hexahedral elements with mid-side nodes. The results are compiled in Fig. 5.55 for various interpretation variants. The most precise results for the course of $K_I(\varphi)$ along the crack front is determined with the help of the 3D J-integral, which hardly differs from the result provided by crack tip elements with this high level of mesh refinement at the crack (see Fig. 5.54). On the other hand, the

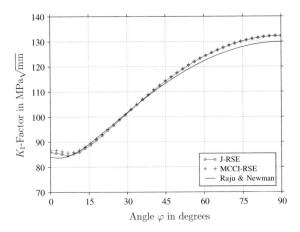

Fig. 5.55 Course of the K_I-factor along the crack front. Comparison of various FEM interpretation methods for regular standard elements RSE

K_I-factors determined by the crack closure integral MCCI-RSE using the interpretation formulae (5.95) and (5.96) are sufficiently accurate as well.

References

1. Byskov E (1970) The calculation of stress intensity factors using the finite element method with cracked elements. Int J Fract 6:159–167
2. Wilson WK (1973) Finite element methods for elastic bodies containing cracks. In: Sih GC (ed) Methods of analysis and solutions of crack problems. Mechanics of fracture, vol 1. Noordhoff, Leyden, pp 484–515
3. Akin JE (1976) The generation of elements with singularities. Int J Numer Methods Eng 10:1249–1259
4. Benzley SF (1974) Representations of singularities with isoparametric finite elements. Int J Numer Methods Eng 8:537–545
5. Blackburn WS (1973) Calculation of stress intensity factors at crack tips using special finite elements. In: Whiteman JR (ed) The mathematics of finite elements and applications. Academic Press, London, pp 327–336
6. Tracey DM (1971) Finite elements for determination of crack tip elastic stress intensity factors. Eng Fract Mech 3:255–256
7. Blackburn WS, Hellen TK (1977) Calculation of stress intensity factors in three-dimensions by finite element methods. Int J Numer Methods Eng 11:211–229
8. Tracey DM (1974) Finite elements for three-dimensional elastic crack analysis. J Nucl Eng Des 26:282–290
9. Henshell RD, Shaw KG (1975) Crack tip finite elements are unnecessary. Int J Numer Methods Eng 9:495–507
10. Barsoum RS (1976) On the use of isoparametric finite elements in linear fracture mechanics. Int J Numer Methods Eng 10:25–37
11. Banks-Sills L, Bortman Y (1984) Reappraisal of the quarter-point quadrilaterial element in linear fracture mechanics. Int J Fract 25:169–180
12. Freese CE, Tracey DM (1976) The natural triangle versus collapsed quadrilateral for elastic crack analysis. Int J Fract 12:767–770

13. Barsoum RS (1977) Triangular quarter point elements as elastic and perfectly-plastic crack tip elements. Int J Numer Methods Eng 11:85–98
14. Hussain MA, Coffin LF, Zaleski KA (1981) Three dimensional singular elements. Comput Struct 13:595–599
15. Manu C (1983) Quarter-point elements for curved crack fronts. Comput Struct 17:227–231
16. Banks-Sills L, Sherman D (1989) On quarter-point three dimensional finite elements in elastic fracture mechanics. Int J Fract 41:177–196
17. Banks-Sills L, Sherman D (1992) On the computation of stress intensity factors for three-dimensional geometries by means of the stiffness derivative and J-integral methods. Int J Fract 53:1–20
18. Hartranft RJ, Sih GC (1968) Effect of plate thickness on the bending stress distribution around through cracks. J Math Phys 47:276–291
19. Zienkiewicz OC, Taylor RL (1991) The finite element method, vol 2, 4th edn. McGraw Hill, London
20. Barsoum RS (1976) A degenerate solid element for linear fracture analysis of plate bending and general shells. Int J Numer Methods Eng 10:551–564
21. Barsoum RS, Loomis RW, Stewart BD (1979) Analysis of through cracks in cylindrical shells by quarter-point elements. Int J Fract Mech 15:259–280
22. Alwar RS, Nambissan KNR (1983) Three-dimensional finite element analysis of cracked thick plates in bending. Int J Numer Methods Eng 19:293–303
23. Zucchini A, Hui CY, Zehnder AT (2000) Crack tip stress fields for thin, cracked plates in bending, shear and twisting: a comparison of plate theory and three-dimensional elasticity theory solutions. Int J Fract 104:387–407
24. Ingraffea AR, Manu C (1980) Stress-intensity factor computations in three dimensions. Int J Numer Methods Eng 15:1427–1445
25. Pian THH, Tong P, Luk CH (1971) Elastic crack analysis by a finite element hybrid method. In: 3. Conference on matrix methods in structural mechanics, Wright Patterson Air Force Base, Ohio, pp 661–682
26. Schnack E, Wolf M (1978) Application of displacement and hybrid stress methods to plane notch and crack problems. Int J Numer Methods Eng 12:963–975
27. Atluri SN, Kobayashi AS, Nakagaki M (1975) An assumed displacement hybrid finite element model for linear fracture mechanics. Int J Fract 11:257–271
28. Tong P, Pian THH, Lasry H (1973) A hybrid element approach to crack problems in plane elasticity. Int J Numer Methods Eng 7:297–308
29. Kuna M, Khanh DQ (1978) Ein spezielles Hybridelement fnr die Spannungsanalyse ebener Körper mit Rissen. Berichte VIII. Int Kongre Mathematik in den Ingenieurwissenschaften, IKM Weimar 2, 71–76
30. Tong P (1977) A hybrid element for rectilinear anisotropic material. Int J Numer Methods Eng 11:377–383
31. Lin KY, Mar JW (1976) Finite element analysis of stress intensity factors for a crack at a bi-material interface. Int J Fract 12:521–531
32. Drumm R (1982) Zur effektiven FEM-Analyse ebener Spannungskonzentrationsprobleme. PhD thesis, Universität Karlsruhe, Karlsruhe, Deutschland
33. Pian THH, Moriya K (1978) Three-dimensional fracture analysis by assumed stress hybrid elements. In: Luxmoore AR (ed) Numerical methods in fracture mechanics. Pineridge Press, Swansea, pp 363–373
34. Kuna M (1982) Konstruktion und Anwendung hybrider Rissspitzenelemente für dreidimensionale Aufgaben. Tech Mech 3:37–43
35. Atluri SN, Kartiresan K (1979) 3D analysis of surface flaws in thick-walled reactor pressure vessels using displacement-hybrid finite element method. Nucl Eng Des 51:163–176
36. Kuna M, Zwicke M (1989) A mixed hybrid finite element for three dimensional elastic crack analysis. Int J Fract 45:65–79
37. Rhee HC, Atluri SN (1982) Hybrid stress finite element analysis of bending of a plate with a through flaw. Int J Numer Methods Eng 18:259–271

38. Moriya K (1982) Finite element analysis of cracked plate subjected to out-of-plane bending, twisting and shear. Bull Jpn Soc Mech Eng 25:1202–1210
39. Atluri SN (1978) Hybrid finite element models for linear and nonlinear fracture mechanics. In: Luxmoore AR (ed) Numerical methods in fracture mechanics. Pineridge Press, Swansea, pp 363–373
40. Tong P, Atluri SN (1977) On hybrid finite element technique for crack analysis. In: Sih GC (ed) Fracture mechanics and technology. Nordhoff, New York, pp 1445–1465
41. Kuna M (1990) Entwicklung und Anwendung effizienter numerischer Verfahren zur bruch-mechanischen Beanspruchungsanalyse am Beispiel hybrider finiter Elemente. Habilitation, Martin Luther Universität Halle
42. Tada H, Paris P, Irwin G (1985) The stress analysis of cracks handbook, 2nd edn. Paris Production Inc., St. Louis
43. Fett T (1998) A compendium of T-stress solutions. Technical report FZKA 6057, Forschungszentrum Karlsruhe, Technik und Umwelt
44. Eisentraut UM, Kuna M (1986) Ein FEM-Programm zur Lösung ebener, axialsymmetrischer und räumlicher Riss-. Festigkeits- und Wärmeleitprobleme. Tech Mech 7:51–58
45. Yamamoto Y, Sumi Y (1978) Stress intensity factors of three-dimensional cracks. Int J Fract 14:17–38
46. Kuna M (1984) Behandlung räumlicher Rissprobleme mit der Methode der finiten Elemente. Tech Mech 5:23–26
47. Parks DM (1974) Stiffness derivative finite element technique for determination of crack-tip stress intensity factors. Int J Fract 10:487–502
48. Hellen TK (1975) On the method of virtual crack extensions. Int J Numer Methods Eng 9:187–207
49. deLorenzi HG (1982) On the energy release rate and the J-integral for 3-D crack configurations. Int J Fract 19:183–193
50. Lin SC, Abel JF (1988) Variational approach for a new direct-integration form of the virtual crack extension method. Int J Fract 38:217–235
51. Wawrzynek PA, Ingraffea AR (1987) Interactive finite-element analysis of fracture processes: an integrated approach. Theor Appl Fract Mech 8:137–150
52. deLorenzi HG (1985) Energy release rate calculations by the finite element method. Eng Fract Mech 21:129–143
53. Bass BR, Bryson JW (1985) Energy release rate techniques for combined thermo-mechanical loading. Int J Fract 22:R3–R7
54. Rybicki EF, Kanninen MF (1977) A finite element calculation of stress intensity factors by a modified crack closure integral. Eng Fract Mech 9:931–938
55. Buchholz FG (1984) Improved formulae for the finite element calculation of the strain energy release rate by modified crack closure integral method. In: Robinson Dorset J (ed) Accuracy, reliability and Training in FEM technology. Robinson and Associates, Dorset, pp 650–659.
56. Raju IS (1987) Calculation of strain-energy release rates with higher order and singular finite elements. Eng Fract Mech 28:251–274
57. Ramamurthy TS, Krishnamurthy T, Narayana KB, Vijayakumar K, Dattaguru B (1986) Modified crack closure integral method with quarter point elements. Mech Res Commun 13:179–186
58. Singh R, Carter B, Wawrzynek P, Ingraffea A (1998) Universal crack closure integral for SIF estimation. Eng Fract Mech 60:133–146
59. Shivakumar KN, Tan PW, Newman JC (1988) A virtual crack-closure technique for calculating stress intensity factors for cracked three-dimensional bodies. Int J Fract 36:R43–50
60. Smith SA, Raju IS (1999) Evaluation of stress intensity factors using general finite element models. In: Panontin TL, Sheppard SL (eds) Fatigue and fracture mechanics. ASTM STP 1332. American Society for Testing and Materials, Philadelphia, pp 176–200
61. Abdel Wahab MM, De Roeck G (1996) A finite element solution for elliptical cracks using the ICCI method. Eng Fract Mech 53:519–526
62. Kemmer G (2000) Berechnung von elektromechanischen Intensitätsparametern bei Rissen in Piezokeramiken. Dissertation, Reihe 18, Nr. 261, TU Dresden. VDI-Verlag Düsseldorf

63. De Roeck G, Abdel Wahab MM (1995) Strain energy release rate formulae for 3D finite element. Eng Fract Mech 50:569–580
64. Murakami Y (1987) Stress intensity factors handbook, vol 1–5. Pergamon Press, Oxford
65. Andrasic CP, Parker AP (1984) Dimensionless stress intensity factors for cracked thick cylinders under polynomial crack face loadings. Eng Fract Mech 19:187–193
66. Parks DM, Kamenetzky EM (1979) Weight functions from virtual crack extensions. Int J Numer Methods Eng 14:1693–1706
67. Paris PC, McMeeking RM, Tada H (1976) The weight function method for determining stress intensity factors. In: Cracks and fracture, STP 601. American Society for Testing of Materials, Philadelphia, pp 471–489
68. Sham TL (1987) A unified finite element method for determining weight functions in two and three dimensions. Int J Solids Struct 23:1357–1372
69. Sham TL, Zhou Y (1989) Computation of three-dimensional weight functions for circular and elliptical cracks. Int J Fract 41:51–75
70. Busch M, Maschke H, Kuna M (1990) A novel BEM-approach to weight functions based on Bueckner's fundamental field. In: Luxmoore A, Owen D (eds) Proceedings of 5th international conference on numerical methods in fracture mechanics, Freiburg, 23–27 April 1990. Pineridge Press, Swansea, pp 5–16
71. Kuna M, Rajiyah H, Atluri SN (1990) A new approach to determine weight functions from Bueckners's fundamental field by the superposition technique. Int J Fract 44(4):R57–R63

Chapter 6
Numerical Calculation of Generalized Energy Balance Integrals

6.1 Generalized Energy Balance Integrals

Based on Eshelby's pioneer work [1, 2], who investigated thermodynamic forces acting on defects in solids by introducing the energy-momentum tensor, a new theory of generalized »material« or »configurational« forces has been developed in the past 15 years (see Maugin [3], Kienzler, Herrmann [4, 5], and Gurtin [6]). In the context of this theory, the invariance properties of mechanical or thermodynamic laws of conservation are investigated with respect to a transformation of the material domain in order to calculate generalized force actions of fields on disturbances in the homogeneous material, i.e. on defects of various forms (such as cracks). These investigations lead far beyond the classic J-integral and permit physical understanding of this integral from a superior point of view.

These ideas will be introduced using the example of a volume defect (an inhomogeneous inclusion, pore or the like) in an isotropic elastic body, which otherwise consists of homogeneous, defect-free material. The body is subject to a certain load, and the solution of the BVP is known. We will consider an arbitrary part V of the body that completely includes the defect (Fig. 6.1).

Let us now consider the change of total potential energy of the system if the defect is shifted into an infinitesimally neighboring position. This virtual displacement of the material defect and its surroundings V relative to the physical space of the field solution is described by varying the coordinates $\delta X_k = \delta l_k$ of the material reference system. (This process must not be confused with the principle of virtual displacements, with which we vary the solution function δu_i!) The energy difference that is removed from the external system and added to the volume V as a result of displacement of the defect can be imagined as the work of a generalized force P_k with δl_k:

$$\delta \Pi = \Pi (X_k + \delta l_k) - \Pi (X_k) = -P_k \, \delta l_k \,. \tag{6.1}$$

To calculate this force, we carry out the following thought experiment: First we cut out the domain V and allow the tractions $t_i = \sigma_{ij} n_j$ to act upon its boundary S, so that

M. Kuna, *Finite Elements in Fracture Mechanics*, Solid Mechanics and Its Applications 201, DOI: 10.1007/978-94-007-6680-8_6,
© Springer Science+Business Media Dordrecht 2013

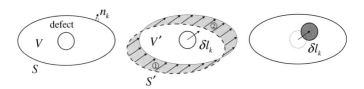

Fig. 6.1 Material force on a defect

there is no deformation due to load removal. Secondly, we define the domain V' that emerges from V in the undeformed state by a rigid body displacement $\delta X_k = -\delta l_k$ (dashed line in Fig. 6.1) but contains the defect in unchanged position. V' too is cut out and locked into place at boundary S' via corresponding tractions. The strain energy of V' differs from that of the original domain V by the addition or subtraction of the energy contributions of the gray-colored areas (1) and (2). During a displacement of $-\delta l_k$ this corresponds exactly to the boundary integral of $U(\mathbf{x})$ over S, whereby the area is measured via projection with the normal vector n_k.

$$\delta \mathcal{W}_{\text{int}} = \mathcal{W}_{\text{int}}^{V'} - \mathcal{W}_{\text{int}}^{V} = -\delta l_k \int_S U(\mathbf{x}) n_k \, \mathrm{d}S \tag{6.2}$$

In the third step, we attempt to fit the displaced domain V' into section S, whose deformation $u_i(X_i + \delta X_i)$ on S' differs however from the original section by the amount

$$\delta u_i = \frac{\partial u_i}{\partial X_k} \delta X_k = u_{i,k}(-\delta l_k). \tag{6.3}$$

This displacement difference is now deformed back, whereby the existing tractions t_i perform the external work (their change δt_i as a result of δX_k may be neglected as a higher order term.):

$$\delta \mathcal{W}_{\text{ext}} = -\int_S \delta u_i t_i \, \mathrm{d}S = \delta l_k \int_S u_{i,k} \sigma_{ij} n_j \, \mathrm{d}S. \tag{6.4}$$

Finally, we can virtually join the displaced domain V' along the boundary S to the total body so that the defect is shifted with respect to the homogeneous material or the elastostatic field solution by δl_k. The difference of total potential energy in the process is calculated with (6.2) and (6.4) as follows:

$$\delta \Pi = \delta \mathcal{W}_{\text{int}} - \delta \mathcal{W}_{\text{ext}} = \left\{ \int_S U n_k \, \mathrm{d}S - \int_S \sigma_{ij} u_{i,k} n_j \, \mathrm{d}S \right\} \delta l_k$$

$$= -\int_S \left[U \delta_{jk} - \sigma_{ij} u_{i,k} \right] n_j \, \mathrm{d}S \, \delta l_k = -\int_S Q_{kj} n_j \, \mathrm{d}S \, \delta l_k = -P_k \, \delta l_k \tag{6.5}$$

The quantity Q_{kj} denotes the *energy–momentum tensor* of elastostatics [1]

$$Q_{kj} = U\delta_{jk} - \sigma_{ij}u_{i,k} \tag{6.6}$$

and the generalized force vector is written

$$P_k = \int_S Q_{kj}n_j \, \mathrm{d}S. \tag{6.7}$$

The P_k-integral thus quantifies the »driving energy« $\delta\Pi = -P_k\delta l_k$, which is provided by the system in the case of a virtual infinitesimal translation δl_k of the defect.

If we assume hyperelastic material behavior, $\partial U/\partial\varepsilon_{mn} = \sigma_{mn}$ is valid, and the tensor Q_{kj} is a unique function of the strains ε_{ij} (or $u_{i,j}$). If the material is inhomogeneous, U also depends explicitly on the spatial coordinates x. Next, we investigate the divergence $Q_{kj,j}$ of the energy–momentum tensor (6.6), whereby the conditions of equilibrium $\sigma_{ij,j} = -\bar{b}_i$ are presupposed, and the chain rule yields:

$$\frac{\partial Q_{kj}(\varepsilon_{mn}, x_l)}{\partial x_j} = \frac{\partial U}{\partial\varepsilon_{mn}}\frac{\partial\varepsilon_{mn}}{\partial x_j}\delta_{jk} - \frac{\partial\sigma_{ij}}{\partial x_j}u_{i,k} - \sigma_{ij}\frac{\partial u_{i,k}}{\partial x_j} + \delta_{jk}\frac{\partial U}{\partial x_j}\bigg|_{\exp}$$

$$= \sigma_{mn}\varepsilon_{mn,j}\delta_{jk} + \bar{b}_i u_{i,k} - \sigma_{ij}u_{i,kj} + U_{,k}|_{\exp} \tag{6.8}$$

Since the 1st and 3rd terms cancel each other, the vector p_k remains, which represents the »material force sources« in the volume:

$$Q_{kj,j} = p_k = \bar{b}_i u_{i,k} + U_{,k}|_{\exp}. \tag{6.9}$$

The divergence $Q_{kj,j}$ of the energy–momentum tensor thus disappears under the following conditions:
- The material is homogeneous, i.e. no *explicit* local dependence of $Q_{kj}(x)$.
- The material is hyperelastic.
- There are no volume forces $\bar{b}_j = 0$.
- The field solution does not contain a singularity in V.

As a consequence, the integral must also become zero over an arbitrary domain V *without defects and force sources* $p_k \sim \bar{b}_j = 0$

$$\int_V Q_{kj,j}\, dV = \int_S Q_{kj}n_j\, dS = P_k = 0. \tag{6.10}$$

A virtual displacement is then not associated with a generalized force. If we interpret (6.10) as a balance equation, it then represents a law of conservation for the energy–momentum tensor.

The energy balance integral introduced above is now applied directly to the defect »crack tip« in the plane, which here is virtually displaced by δl_k. We shrink the considered domain V to the crack tip $r \to 0$, whereby we obtain from S the circular contour Γ_ε shown in Fig. 6.2. Application of (6.7) yields:

$$P_k = J_k = \lim_{r \to 0} \int_{\Gamma_\varepsilon} Q_{kj}n_j\, ds\,, \quad P_1 = J = G. \tag{6.11}$$

By comparing the P_k-integral (6.11) with the J-integral, we can recognize that the x_1-component of P_k is exactly identical to J. This is not surprising since $\delta l_1 = da$ denotes exactly the self-similar crack propagation and yields the energy release rate G. Analogously, the J_2-component describes a parallel displacement of the crack in the x_2-direction and J_3 a translation in the x_3-direction (which of course changes nothing and vanishes in a 2D problem). We have thus found a generalized vectorial form J_k of the J-integral.

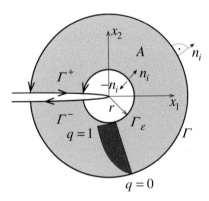

Fig. 6.2 Integration paths around the crack tip and weighting function

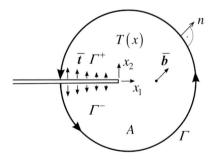

Fig. 6.3 The line-area integral for general loading cases

In the context of LEFM (Chap. 3.2), the K-factor-controlled near field is dominant at the crack tip. The integrals (6.11) can be evaluated along infinitesimal circular contours $r = $ const with the near field solutions (Fig. 6.2).

From this results the relationship between the J_k-integral vector and the stress intensity factors for two-dimensional crack problems of LEFM:

$$J_1 = J = G = \frac{1}{E'}\left(K_I^2 + K_{II}^2\right) + \frac{1+\nu}{E}K_{III}^2$$
$$J_2 = -2K_I K_{II}/E', \quad J_3 = 0. \tag{6.12}$$

6.2 Extension to General Loading Cases

6.2.1 Preconditions for Path-Independence

In order to discuss the path-independence of J_k for two-dimensional crack problems, let us construct a closed integration path $C = \Gamma - \Gamma_\varepsilon + \Gamma^+ + \Gamma^-$ that avoids the crack tip and solely includes the homogeneous, defect-free material region A (Fig. 6.2). Since in domain A no material force sources (6.9) are active, according to (6.10) the domain integral over A and the contour integral over C must be zero. From this identity, we bring the term along Γ_ε to the left hand side:

$$J_k = \lim_{r\to 0}\int_{\Gamma_\varepsilon} Q_{kj}n_j\,\mathrm{d}s = \int_\Gamma Q_{kj}n_j\,\mathrm{d}s + \lim_{r\to 0}\int_{\Gamma^+ + \Gamma^-} Q_{kj}n_j\,\mathrm{d}s - \lim_{r\to 0}\int_A Q_{kj,j}\,\mathrm{d}A\,\overset{0}{\nearrow}.$$
$$\tag{6.13}$$

We thus see that calculating the J_k-integral on any contour Γ yields exactly the same values if the 2nd term over the crack faces would be zero:

$$\lim_{r\to 0}\int_{\Gamma^+ + \Gamma^-} (Un_k - \underbrace{\sigma_{ij}n_j\,u_{i,k}}_{t_i})\,\mathrm{d}s \overset{!}{=} 0. \tag{6.14}$$

This requires firstly that no tractions $t_i = \bar{t}_i$ act on the crack faces Γ^+ and Γ^-:

$$t_i = 0. \tag{6.15}$$

If we assume straight crack faces perpendicular to the x_2-axis, then $n_k = \mp\delta_{2k}$ applies to Γ^+ and Γ^-, whereby the first term of (6.14) is reduced to $(U^- - U^+)\,\delta_{2k}$. From (6.13) follows:

$$J_k = \int_\Gamma \left(U\delta_{kj} - \sigma_{ij}u_{i,k}\right) n_j \, ds + \lim_{r \to 0} \int_{\Gamma^+} \left(U^- - U^+\right) \delta_{2k} \, ds. \tag{6.16}$$

As a result, only the component $k = 1$ of the J_k-integral vector is independent of the integration contour Γ! With regard to component $k = 2$, the difference of strain energy densities on the crack faces must be taken into consideration, which disappears only in the case of pure symmetry or antisymmetric (mode I or II). Another complicating factor is that the energy density U is usually singular in the limiting process $r \to 0$. With curved crack faces, the term $Un_k \, ds$ in (6.14) always exists and creates a path-dependence.

6.2.2 Crack Face, Volume and Thermal Loading

In many practical calculation tasks, loadings on the crack faces (e.g. internal pressure) or inside the volume (e.g. gravity) play a substantial role (Fig. 6.3). Furthermore, it would be worthwhile to apply the efficiency of the J_k-integral also to the analysis of cracks that are loaded by inhomogeneous temperature fields (e.g. thermal shock). It would also be helpful for calculating the J_k-integral using FE-solutions if the integration along an arbitrary path Γ could be carried out at a greater distance from the crack tip, since errors in the numerical solution are reduced there. For these reasons, it is necessary to look for extensions of the J_k- integral.

For this purpose, we will repeat the divergence analysis of Q_{kj} from the previous section according to (6.13), but now abandoning all restrictions. For an elastic material at a given temperature field $T(x)$ the thermal expansions ε^t_{ij} (see appendix A.4.1), are calculated according to (A.85):

$$\varepsilon^t_{ij}(T(x)) = \alpha^t_{ij}(T(x) - T_0) = \alpha^t_{ij}\Delta T(x). \tag{6.17}$$

Expressed with the thermal stress coefficients $\beta_{ij} = C_{ijkl}\alpha^t_{kl}$, Hooke's law is written as:

$$\sigma_{ij} = C_{ijkl}\left(\varepsilon_{kl} - \varepsilon^t_{kl}\right) = C_{ijkl}\varepsilon_{kl} - \beta_{ij}\Delta T(x). \tag{6.18}$$

In the isotropic case, (A.91) holds true with $\beta_{ij} = (3\lambda+2\mu)\alpha_t\delta_{ij}$. The elastic strain energy is generated solely by the elastic strains

$$U^e(\varepsilon^e) = \frac{1}{2}\varepsilon^e_{ij}C_{ijkl}\varepsilon^e_{kl}, \quad U^e(\varepsilon^e) = \mu\varepsilon^e_{ij}\varepsilon^e_{ij} + \frac{\lambda}{2}\left(\varepsilon^e_{kk}\right)^2 \quad \text{(isotropic)}, \tag{6.19}$$

resulting in $\partial U^e/\partial\varepsilon^e_{mn} = \sigma_{mn}$. The divergence of the energy–momentum tensor is calculated with the chain rule

$$\frac{\partial Q_{kj}}{\partial x_j} = \frac{\partial U^e}{\partial \varepsilon^e_{mn}} \frac{\partial \varepsilon^e_{mn}}{\partial x_j} \delta_{jk} - \sigma_{ij,j} u_{i,k} - \sigma_{ij} u_{i,kj}$$

$$= \sigma_{mn} \varepsilon^e_{mn,k} \left(\pm \sigma_{mn} \varepsilon^t_{mn,k} \right) + \bar{b}_i u_{i,k} - \sigma_{ij} (u_{i,j})_{,k} . \tag{6.20}$$

The volume forces come into play via the equilibrium conditions $\sigma_{ij,j} + \bar{b}_i = 0$. In order to compensate the 4th term $\hat{=} \sigma_{ij} \varepsilon_{ij,k}$ with the 1st term, a zero completion is introduced with the thermal strains $(\varepsilon^e_{mn} + \varepsilon^t_{mn} = \varepsilon_{mn})$ so that with (6.17) we obtain:

$$Q_{kj,j} = -\sigma_{mn} \varepsilon^t_{mn,k} + \bar{b}_i u_{i,k} = -\sigma_{mn} \alpha^t_{mn} T_{,k} + \bar{b}_i u_{i,k} = p_k . \tag{6.21}$$

This »source term« should be plugged into the area integral over A of (6.13). In addition, the given loads $\bar{t}_i$ on the crack faces must be taken into consideration via the integral (6.14) over $\Gamma^+ + \Gamma^-$.

The extended J-integral for thermal, volume and crack face loads now has the form of a path-independent line-area integral, which in combination with the strain energy U^e (6.19) is written as:

$$J_k^{te} = \int_{\Gamma} \left[U^e \delta_{kj} - \sigma_{ij} u_{i,k} \right] n_j \, ds + \int_{\Gamma^+ + \Gamma^-} \left[U^e n_k - \bar{t}_i u_{i,k} \right] ds$$

$$+ \int_A \left[\sigma_{mn} \alpha^t_{mn} T_{,k} - \bar{b}_i u_{i,k} \right] dA . \tag{6.22}$$

In addition to this approach made by Nakamura [7] and Aoki [8] ($\widehat{J}$-Integral), another version of the extended J-integral was derived for thermal stresses by Wilson & Yu [9] (J^*) and Atluri [10] (G^*), which emanates from a different thermodynamic definition $\check{U}^{te}$ of internal energy density in the thermoelastic case.

$$\check{U}^{te}(\varepsilon, T) = \int_0^{\varepsilon_{kl}} \sigma_{ij}(\varepsilon, T) \, d\varepsilon_{ij} = \frac{1}{2} \varepsilon_{ij} C_{ijkl} \varepsilon_{kl} - \beta_{ij} \Delta T \varepsilon_{ij}$$

$$\check{U}^{te}(\varepsilon, T) = \mu \varepsilon_{ij} \varepsilon_{ij} + \frac{\lambda}{2} \varepsilon_{kk}^2 - (3\lambda + 2\mu) \alpha_t \Delta T \varepsilon_{kk} \quad \text{(isotropic)} . \tag{6.23}$$

$\check{U}^{te}$ corresponds to a stress work density that no longer has the character of a potential, which in the following will always be symbolized by a check. Nonetheless, the derivative of (6.23) yields the stresses. Considering both variables ε_{ij} and T, the divergence $Q_{kj,j}$ now results in:

$$Q_{kj,j} = \left[\frac{\partial \check{U}^{\text{te}}}{\partial \varepsilon_{mn}} \frac{\partial \varepsilon_{mn}}{\partial x_j} + \frac{\partial \check{U}^{\text{te}}}{\partial T} \frac{\partial T}{\partial x_j} \right] \delta_{jk} - \sigma_{ij,j} u_{i,k} - \sigma_{ij} u_{i,kj}$$

$$= -\beta_{ij} \varepsilon_{ij} T_{,k} + \bar{b}_i u_{i,k} \tag{6.24}$$

The result thus differs from (6.21) in the first term, so J_k^{te} in conjunction with $\check{U}^{\text{te}}$ assumes the form:

$$J_k^{\text{te}} = \int_{\Gamma} \left[\check{U}^{\text{te}} \delta_{kj} - \sigma_{ij} u_{i,k} \right] n_j \, ds + \int_{\Gamma^+ + \Gamma^-} \left[\check{U}^{\text{te}} n_k - \bar{t}_i u_{i,k} \right] ds$$

$$+ \int_A \left[\beta_{ij} \varepsilon_{ij} T_{,k} - \bar{b}_i u_{i,k} \right] dA . \tag{6.25}$$

Physically, the expressions (6.22) and (6.25) have the same meaning and can of course be computationally converted into each other.

Atluri [10] has provided further variants for the case of graded materials, where the thermoelastic constants depend on position. For stationary temperature fields $T_{,kk} = 0$, Gurtin [11] was able to convert the area integral back into a line integral with additional terms.

The relations (6.22) and (6.25) derived here are valid as well for temperature fields of a transient heat conduction analysis. They are highly suitable for the numerical calculation because the temperature fields $T(x)$ are obtained in the nodes. Therefore, their gradients $T_{,k}$ can be determined in the integration points with the same quality as the required strains ε_{mn}.

6.3 Three-Dimensional Variants

According to Sect. 6.1, the energy–momentum tensor Q_{kj} and thermodynamic force P_k are unconditionally true for virtual displacements of a three-dimensional defect in space. Around the defect we now lay an arbitrarily shaped, closed surface S that surrounds the volume V. This can be applied directly to crack problems if the entire crack or the entire crack front is seen as a defect and is displaced as a whole. Normally we are interested however in the local energy release rate G or the K-factors at each point of the crack front. Then, we must carry out a virtual displacement of a single point or segment of the crack front.

We introduce a local system of coordinates $x(s)$ that is carried along an arbitrarily shaped crack front L_c in space, where s denotes the arc length (see Fig. 6.4). The crack surface should be planar and lie in the (x_1, x_3)-plane so that x_2 is always perpendicular to the crack surface.

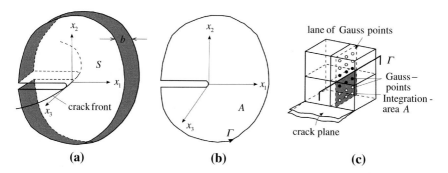

Fig. 6.4 3D-disk integral with **a** closed surface S, **b** area A and integration contour Γ, **c** numerical integration

6.3.1 The 3D-Disk Integral

Aoki et al. [8] defined the volume V around a point on the crack front as a disc of infinitesimal thickness $b = \Delta x_3$ that may have any shape in the (x_1, x_2)-plane. The associated surface $S = A^+ + A^- + b\Gamma$ is composed of the front and rear faces A^+, A^- and the outer edge $b\Gamma$ (see Fig. 6.4). We now analyze the x_1-component of J_k. Inelastic strains ε_{ij}^* of any origin (thermal or plastic) are allowed, but crack face loads are omitted for the sake of clarity. The 3D extension of (6.13) for J_1 is therefore:

$$\widehat{J}(s) = -\lim_{b \to 0} \lim_{\Delta a \to 0} \frac{\Delta \Pi}{b \Delta a} = \lim_{b \to 0} \frac{1}{b} J_1 = \lim_{b \to 0} \frac{1}{b} \left\{ \int_S Q_{1j} n_j \, dS - \int_V Q_{1j,j} \, dV \right\}$$

(6.26)

The tensor Q_{kj} is formed from the pure elastic strain energy density $U^e(\varepsilon^e)$ with $\varepsilon^e = \varepsilon - \varepsilon^*$ according to (6.19). The divergence term in the volume integral is handled exactly as in (6.20), and analogously to (6.21), we find

$$Q_{1j,j} = -\sigma_{mn} \varepsilon_{mn,1}^* + \bar{b}_i u_{i,1} = p_1.$$

(6.27)

If $b \to 0$, the volume integral degenerates into $\int_V (\cdot) dV \to b \int_A (\cdot) dA$. In the case of the surface integral over A^+ and A^-, the normal vectors n_j point in the $\pm x_3$-direction

$$\int_{A^+ + A^-} \left[U^e \delta_{1j} - \sigma_{ij} u_{i,1} \right] n_j \, dA = - \int_{A^+ + A^-} \sigma_{i3} u_{i,1} \, dA,$$

(6.28)

so the term $U^e \delta_{13} = 0$ is cancelled, and the difference between A^+ and A^- arises in the second term. This difference can be converted into $b(\sigma_{i3} u_{i,1})_{,3}$ by means of a Taylor expansion with respect to x_3.

On the whole, (6.26) thus provides us with the *3D-disk integral* $\widehat{J}$:

$$
\begin{aligned}
\widehat{J}(s) &= \lim_{b \to 0} \frac{1}{b} \left\{ b \int_{\Gamma} Q_{1j} n_j \, \mathrm{d}s - b \int_{A^+} \left(\sigma_{i3} u_{i,1} \right)_{,3} \, \mathrm{d}A - b \int_A Q_{1j,j} \, \mathrm{d}A \right\} \\
&= \int_{\Gamma} \left[U^e \delta_{1j} - \sigma_{ij} u_{i,1} \right] n_j \, \mathrm{d}s + \int_A \left[\sigma_{mn} \varepsilon^*_{mn,1} - \bar{b}_i u_{i,1} - \left(\sigma_{i3} u_{i,1} \right)_{,3} \right] \mathrm{d}A \, .
\end{aligned}
$$

$$
(6.29)
$$

Therefore, the energy rate $\widehat{J}$ can be calculated for a virtual displacement of the crack front at point s by a path-independent line-surface integral in the plane perpendicular to the crack front. For two-dimensional problems (plane stress or plane strain state), the last term in the surface integral disappears since there is no dependence on x_3, $\partial(\cdot)/\partial x_3 \equiv 0$.

The numerical calculation for 20-noded hexahedral elements is outlined in Fig. 6.4 c. It is advantageous to arrange the elements along the crack front so that the disk integral coincides exactly with a plane of integration points. For the line integral over Γ, the values of U^e, $u_{i,1}$ and σ_{ij} have to be interpolated between neighboring integration points (• points). For the surface integral over A, we need the derivatives $\varepsilon^*_{mn,1}$, $\sigma_{i3,3}$ and $u_{i,13}$ at the integration points. In FEM, these second order derivatives are already very imprecise and are normally not available as results. Suitable calculation methods for this are provided in Sect. 4.4.4.

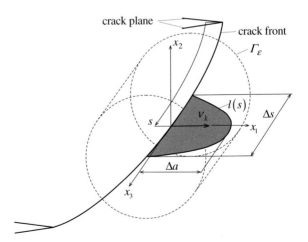

Fig. 6.5 Arbitrary three-dimensional crack configuration with virtual crack front extension

Due to its considerable lack of precision, the geometric requirements on the FE-mesh, and the necessity of laying line and surface integrals through the FE-discretization, the $\widehat{J}$-integral has not become established, although it is theoretically very plausible and seems simple.

6.3.2 Virtual Crack Propagation in 3D

The normal unit vector perpendicular to the crack front at position s is denoted ν_k, which lies in the crack plane as shown in Fig. 6.5. Now let us assume, in a limited segment Δs of the crack front, a virtual displacement $\Delta l_k(s)$ in the crack plane, which should have exactly the normal direction ν_k

$$\Delta l_k(s) = l(s)\Delta a\, \nu_k\,, \qquad \Delta l(s) = |\Delta l_k(s)| = l(s)\Delta a\,. \tag{6.30}$$

In view of the numerical realization, from now on we will use the denotation Δl_k instead of δl_k. In the sectional plane $(x_1, x_2) \perp x_3$, we can evaluate the two-dimensional J_k-integral as a line or a line-surface integral in accordance with Fig. 6.2 or Fig. 6.3. In the system of coordinates selected here, only the J_1- component is relevant to the energy balance during crack extension.

$$J(s) = J_1(s) = J_k(s)\nu_k(s)$$

$$= \left(\lim_{r \to 0} \int_{\Gamma_\varepsilon(s)} \left[U\delta_{kj} - \sigma_{ij}u_{i,k} \right] n_j \, \mathrm{d}\Gamma \right) \nu_k(s) = \left(\lim_{r \to 0} \int_{\Gamma_\varepsilon(s)} Q_{kj}n_j \, \mathrm{d}\Gamma \right) \nu_k(s)\,. \tag{6.31}$$

The energy release per sectional plane amounts to $J(s)\Delta l(s)$. So for the total balance $-\Delta\Pi$ we have to add up over the crack front segment Δs

$$-\Delta\Pi = \int_{\Delta s} J(s)\Delta l(s)\, \mathrm{d}s = \int_{\Delta s} J_k(s)\Delta l_k(s)\, \mathrm{d}s = \bar{J}\Delta A\,. \tag{6.32}$$

Here, $\bar{J}$ means an average value for the total virtual displacement of the segment referred to the area

$$\Delta A = \int_{\Delta s} \Delta l(s)\, \mathrm{d}s = \Delta a \int_{\Delta s} l(s)\, \mathrm{d}s \tag{6.33}$$

of crack extension. Inserting (6.31) into (6.32) leads to:

$$
-\frac{\Delta\Pi}{\Delta A} = \bar{J} = \frac{1}{\Delta A} \int\limits_{\Delta s} \left(\lim_{r \to 0} \int\limits_{\Gamma_{\varepsilon(s)}} Q_{kj} n_j \, \mathrm{d}\Gamma \right) \Delta l_k(s) \, \mathrm{d}s
$$

$$
= \frac{1}{\Delta A} \lim_{r \to 0} \int\limits_{S_\varepsilon} Q_{kj} n_j \Delta l_k \, \mathrm{d}S = \lim_{r \to 0} \int\limits_{S_\varepsilon} Q_{kj} n_j l_k \, \mathrm{d}S \bigg/ \int\limits_{\Delta s} l(s) \, \mathrm{d}s \ . \tag{6.34}
$$

The line integrals Γ_ε along Δs can be combined to make a cylindrical »tube surface« S_ε with the outer normal n_j, which in the limiting case $r \to 0$ is shrunken to the crack front. In this way we have obtained a clear and compact expression for the thermodynamic force (in LEFM, the energy release rate G) associated with the virtual displacement of a crack front segment in space. The result $\bar{J}$ should be assigned to a representative point $\bar{s}$ of the segment Δs.

From the last equation of (6.34), we see that the absolute size Δa of the crack extension is canceled out! It should be stressed that the relations (6.32) and (6.34) are valid for every type of energy-momentum tensor, whereby the meaning of $\bar{J}$ is replaced by the relevant quantity

$$
\bar{P} = \frac{1}{\Delta A} \int\limits_{\Delta s} P_k(s) \Delta l_k(s) \, \mathrm{d}s. \tag{6.35}
$$

$\bar{J}$ was defined for the limiting process $S_\varepsilon \to 0$, because many energy balance integrals have a physical meaning only in this way.

The question of path-independence is posed in space as the independence of an arbitrarily chosen surface S, which is to be placed around the same segment of the crack front. Here we can completely adopt the considerations and divergence investigations made in the two-dimensional case. As Fig. 6.6 shows a closed surface $\bar{S} = S + S^+ + S^- + S_{\mathrm{end}} - S_\varepsilon$ can be formed, from the crack tube S_ε, the arbitrary outer surface S, the crack surfaces S^+, S^-, and the end faces S_{end}.

Inside of $\bar{S}$ there is no defect. In analogy to (6.13), we can now carry out the following adjustment of the integrals using Gauss's theorem:

$$
\bar{J} \Delta A = \lim_{r \to 0} \int\limits_{S_\varepsilon} Q_{kj} n_j \Delta l_k \, \mathrm{d}S = \int\limits_{S} Q_{kj} n_j \Delta l_k \, \mathrm{d}S
$$

$$
+ \lim_{r \to 0} \int\limits_{S^+ + S^- + S_{\mathrm{end}}} Q_{kj} n_j \Delta l_k \, \mathrm{d}S - \lim_{r \to 0} \int\limits_{\bar{V}} \frac{\partial}{\partial x_j} [Q_{kj} \Delta l_k] \, \mathrm{d}V \ . \tag{6.36}
$$

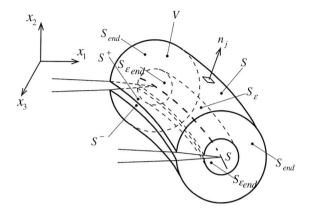

Fig. 6.6 J-Integral Integration domain for the three-dimensional J- integral

6.4 Numerical Calculation as Equivalent Domain Integral

In order to numerically calculate the various energy balance integrals, a line integral or even a combined line-area integral must be calculated in the plane and line, surface and possibly volume integrals in space. In the context of FEM, the geometric and topological determination of integrals in lower order than the dimension of the BVP is quite intricate and their calculation laborious. On the other hand, pure domain integrals (2D or 3D) over a group of elements are among the standard procedures in FEM, for which simple algorithms exist (Sect. 4.4.3). For these reasons, a method will now be introduced which enables to transform any kind of energy balance integral into an equivalent domain integral.

6.4.1 Transformation into an Equivalent Domain Integral 2D

In order to make a transformation into an *equivalent domain integral* (EDI), let us again set up a closed integration path $C = \Gamma + \Gamma^+ + \Gamma^- - \Gamma_\varepsilon$ with the outer normal n_j (see Fig. 6.2). According to the definition (6.13) of J_k, we can write:

$$J_k = - \int_C Q_{kj} n_j \, ds + \lim_{r \to 0} \int_{\Gamma + \Gamma^+ + \Gamma^-} Q_{kj} n_j \, ds. \qquad (6.37)$$

Now a weighting function $q(x)$ is introduced, which must be continuous and differentiable and fulfills the requirements

$$q = \begin{cases} 0 & \text{on} \quad \Gamma \\ 1 & \text{on} \quad \Gamma_\varepsilon \end{cases}, \qquad (6.38)$$

as illustrated in Fig. 6.2. Upon insertion into (6.37), the integral over Γ is dropped

$$J_k = -\int_C Q_{kj} n_j q \, ds + \lim_{r \to 0} \int_{\Gamma^+ + \Gamma^-} Q_{kj} n_j q \, ds. \tag{6.39}$$

By applying the Gaussian divergence theorem, we thus obtain the 2D J_k-integral as a weighted, pure domain integral over the area A enclosed by Γ plus unavoidable crack face integrals (For the sake of simplicity, in the following the limiting process $\lim r \to 0$ will no longer be included.):

$$J_k = -\int_A (Q_{kj} q)_{,j} \, dA = -\int_A (Q_{kj,j} q + Q_{kj} q_{,j}) \, dA + \int_{\Gamma^+ + \Gamma^-} Q_{kj} n_j q \, ds. \tag{6.40}$$

By means of the divergence $Q_{kj,j}$, all previously discussed additional terms such as volume forces $\bar{b}_i$, thermal and inelastic strains $\alpha_{mn} \Delta T$ and ε^*_{mn} or explicit spatial dependence of $U(\mathbf{x})$ is taken into consideration:

$$Q_{kj,j} = p_k = \bar{b}_i u_{i,k} - \sigma_{mn} \alpha_{mn} T_{,k} - \sigma_{mn} \varepsilon^*_{mn,k} + U_{,k}\big|_{\exp}. \tag{6.41}$$

Thus, the generalized J_k-integral for two-dimensional problems can be expressed by the following equivalent domain integral over the area A:

$$J_k = -\int_A (U \delta_{kj} - \sigma_{ij} u_{i,k}) q_{,j} \, dA$$

$$- \int_A \left(U_{,k}\big|_{\exp} + \bar{b}_i u_{i,k} - \sigma_{mn} \alpha_{mn} T_{,k} - \sigma_{mn} \varepsilon^*_{mn,k} \right) q \, dA \tag{6.42}$$

$$+ \int_{\Gamma^+ + \Gamma^-} (U n_k - \bar{t}_i u_{i,k}) q \, ds.$$

If $Q_{kj,j} = 0$ disappears, we actually have a path-independent line integral, and the second integrand in (6.42) is dropped. If the crack face loadings are also neglected, we arrive at the simple J_k-integral according to (6.16), and (6.42) is reduced to:

$$J_k = -\int_A (U \delta_{kj} - \sigma_{ij} u_{i,k}) q_{,j} \, dA + \int_{\Gamma^+ + \Gamma^-} (U n_k) q \, ds. \tag{6.43}$$

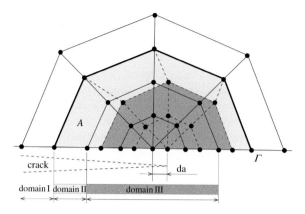

Fig. 6.7 Equivalent domain integral and weighting function q for plane crack problems

How should the weighting function $q(x_1, x_2)$ be chosen in the numerical realization? Mathematically, it has to satisfy the conditions (6.38), but otherwise it is
arbitrary. A diverse range of variants were tested for $q(x_1, x_2)$, among which the
following approach has proven best: The function $q(x_1, x_2)$ is subdivided into three
domains using the FE-mesh as seen in Fig. 6.7:

domain I: $q(x_1, x_2) \equiv 0$
domain II: linear transition from $q = 1$ to $q = 0$
domain III: $q(x_1, x_2) \equiv 1 = \text{const.}$

The function $q(x_1, x_2)$ is represented in the FE-model by nodal point variables $q^{(a)}$
and interpolated with the shape functions of the elements used

$$q(x_1, x_2) = \sum_{a=1}^{n_K} N_a(\xi_1, \xi_2) \, q^{(a)}. \tag{6.44}$$

Domain II usually consists of only one element ring, for which different rings are chosen around the crack tip one after the other. Because in domains I and III $\partial q / \partial x_j = 0$,
only domain II provides with $\partial q / \partial x_j = \text{const}$ a contribution to the 1st domain integral of (6.42). On the other hand, only domains II and III, with $q \neq 0$, contribute
to the 2nd domain integral, which only arises in the case of certain generalizations.
The crack face integral also extends only over II and III.

Yet the weighting function $q(x_1, x_2)$ also has a geometric interpretation as a virtual
displacement $\Delta l = \Delta l_k e_k$ of the crack tip region A. Directly at the crack tip on Γ_ε,
$q = 1$ describes the displacement by $q \Delta l_k$, which falls off to zero towards the contour
Γ with $q \to 0$. Normally, the displacement takes place in the direction of the crack
($k = 1$), since J_1 describes the energy balance for crack propagation $\Delta l_1 = \Delta a$.
However, to determine J_2, a parallel displacement $\Delta l = \Delta l e_2$ of domain III must
be assumed.

6.4.2 Transformation into an Equivalent Domain Integral 3D

We can carry out the transformation of the J-integral (6.34) for three-dimensional crack configurations into an equivalent domain integral in a completely analogous manner. Instead of C, we now consider the closed enveloping surface $\bar{S} = S + S^+ + S^- - S_\varepsilon + S_{end}$ around segment Δs of the crack front (see Fig. 6.6) that surrounds the volume V. Now a continuous differentiable weighting function $q_k(x)$ is introduced again, which becomes zero on the outer surface S and the end faces S_{end} but corresponds to the virtual crack propagation $\Delta l_k(s)$ on the »tube surface« S_ε:

$$q_k = \begin{cases} 0 & \text{on } S, S_{end} \\ \Delta l_k & \text{on } S_\varepsilon, S_{\varepsilon end}. \end{cases} \tag{6.45}$$

In contrast to the 2D case, q_k is now a vector function. The 3D J-integral represents only a scalar quantity, the energy rate during crack propagation. Using the same considerations as were made in the 2D case, the actual definition (6.34) of $\bar{J} = J(\bar{s})$ can now be transformed ($\lim r \to 0$ is again omitted):

$$J(\bar{s}) = \frac{1}{\Delta A} \int_{S_\varepsilon} Q_{kj} n_j \Delta l_k \, dS = \frac{1}{\Delta A} \left[-\int_{\bar{S}} Q_{kj} n_j q_k \, dS + \int_{S+S^++S^-} Q_{kj} n_j q_k \, dS \right]$$

$$= \frac{1}{\Delta A} \left[-\int_V (Q_{kj,j} q_k + Q_{kj} q_{k,j}) \, dV + \int_{S^++S^-} Q_{kj} n_j q_k \, dS \right] \tag{6.46}$$

In the general loading case, the 3D equivalent domain integral is calculated from:

$$J(\bar{s}) = \frac{1}{\Delta A} \left[-\int_V (U \delta_{kj} - \sigma_{ij} u_{i,k}) q_{k,j} \, dV \right.$$

$$- \int_V \left(U_{,k}|_{\exp} + \bar{b}_i u_{i,k} - \sigma_{mn} \alpha_{mn} T_{,k} - \sigma_{mn} \varepsilon^*_{mn,k} \right) q_k \, dV$$

$$\left. + \int_{S^++S^-} (U n_k - \bar{t}_i u_{i,k}) q_k \, dS \right]. \tag{6.47}$$

For the classic J-integral ($Q_{kj,j} = 0$, $\bar{t}_i = 0$), the relation is simplified to:

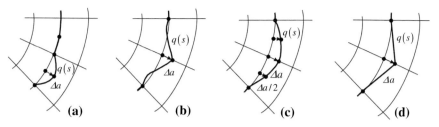

Fig. 6.8 Defining the weighting function q_k along the crack front elements

$$J(\bar{s}) = -\frac{1}{\Delta A} \int\limits_{V} \left(U \delta_{kj} - \sigma_{ij} u_{i,k} \right) q_{k,j} \, dV .$$ (6.48)

Figure 6.8 displays possible variants of virtual crack extension (VCE) of a nodal point on the crack front. With 8-noded hexahedral elements, only the corner node is shifted by Δa and $q(s)$ is linearly interpolated (Fig. 6.8 d). In the case of 20-noded hexahedral elements, either a mid-side node (Fig. 6.8 a) or a corner node (Fig. 6.8 b) can be displaced by Δa, in each case with quadratic interpolation $q(s)$ of the crack front. Variant b) should not be used because of its known low accuracy. Instead of that, variant c) should be favored, whereby the mid-side nodes are shifted along up to 50 %. The VCE applies always to two neighboring element layers in variants (b), (c), and (d). Figure 6.9 illustrates for variant (d) how the planes of the mid-side nodes and corner nodes are displaced. The rings denote precisely the domain V between S_ε and S from Fig. 6.6, in which q_k is reduced from Δl_k to zero in accordance with (6.45). These three-dimensional element rings correspond to domain II of the VCE in the plane (see Fig. 6.7). In practice, one successively displaces rings 1, 2, 3 etc. from elements around the crack front nodes (see Fig. 6.9) so that several equivalent domain integrals can be calculated for one position $\bar{s}$ on the crack front.

6.4.3 Numerical Implementation

J is calculated numerically with the EDI as a post-processor of the FE-analysis. Numerical integration of an arbitrary physical field $f(x)$ over a domain V takes place by means of summation of all finite elements V_e belonging to V. This integration is achieved element by element with the help of the Gaussian integration formulae from Sect. 4.4.3, i.e. the outputs $f^{(g)}(\boldsymbol{\xi}^{(g)})$ at selected integration points IP $\hat{=} g$ are multiplied by the weights $\bar{w}^{(g)}$ and added.

$$\bar{J}_{\text{num}} = \sum_{e} \sum_{g=1}^{m_{G}} f^{(g)} \bar{w}_g \left| \mathbf{J}^{(g)} \right|$$ (6.49)

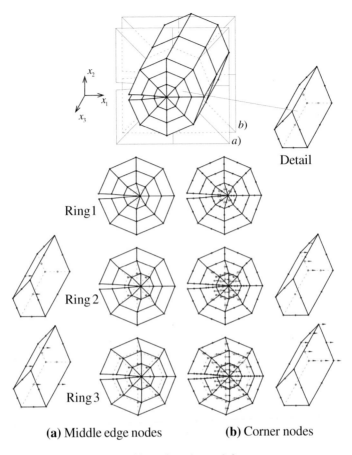

Ring1

Ring 2

Ring 3

(a) Middle edge nodes   (b) Corner nodes

Fig. 6.9 Definition of **q** at various positions along the crack front

In the case of an EDI, the function to be integrated according to (6.42) or (6.47) consists of basic values of type:

$$f^{(g)} = Q^{(g)}_{kj,j} q^{(g)}_k + Q^{(g)}_{kj} q^{(g)}_{k,j} . \tag{6.50}$$

The tensor $Q^{(g)}_{kj} = U^{(g)} \delta_{kj} - \sigma^{(g)}_{ij} u^{(g)}_{i,k}$ is relatively simple to calculate since the stresses are given at the integration points. The stress work density $U^{(g)}$ is usually also available in FE-programs, but in the elastic case it can be calculated from $\sigma^{(g)}_{ij}$ and/or $\varepsilon^{(g)}_{ij}$. If the material laws are non-linear, the integral $U^{(g)} = \int \sigma^{(g)}_{ij} \mathrm{d}\varepsilon^{(g)}_{ij}$ must be evaluated via the loading history, i.e. added incrementally. The derivative of the displacements at the IP $u^{(g)}_{i,k}$ should be carried out in accordance with the method shown in Sect. 4.4.4. The same technique (4.78) is applied for the derivative of the weighting function $q^{(g)}_{k,j}$:

$$\frac{\partial q_k}{\partial x_j} = \frac{\partial \xi_l}{\partial x_j}\frac{\partial q_k}{\partial \xi_l} = J_{lj}^{-1}\frac{\partial q_k}{\partial \xi_l} = J_{lj}^{-1}\sum_{a=1}^{n_K}\frac{\partial N_a(\boldsymbol{\xi})}{\partial \xi_l}q_k^{(a)}. \tag{6.51}$$

Finally, for (6.50) we still need the divergence $Q_{kj,j}^{(g)}$ with the terms shown in (6.41). The gradients of the temperature field (known at the nodes) are determined with the aforementioned technique. The derivative of inelastic strains $\varepsilon_{mn,k}^{*(g)}$ is more difficult as these are only available within a sufficient range of accuracy at the IP. In this case, the interpolation-differentiation method introduced in Sect. 4.4.4 must be employed.

6.5   Consideration of Dynamic Processes

Energy balance integrals for stationary and moving cracks under dynamic loading were already introduced in Sect. 3.5.5. For *stationary cracks*, (3.366) describes the x_1-component $G(t) = J_1$ of a dynamic J-integral vector J_k^*, which corresponds to the virtual displacements Δl_k in all three directions.

Dynamic J-integral 2D with inertia forces:

$$J_k^* = \int_{\Gamma}\left[U\delta_{kj} - \sigma_{ij}\,u_{i,k}\right]n_j\,\mathrm{d}s + \int_A \rho\,\ddot{u}_i\,u_{i,k}\,\mathrm{d}A \tag{6.52}$$

To improve the numerical calculation, it is more advantageous to convert J_k^* again into an equivalent domain integral over the area A as in Sect. 6.4, which yields with the weighting function $q(\boldsymbol{x})$ (6.37):

$$J_k^* = \int_A\left[\left(\sigma_{ij}\,u_{i,k} - U\delta_{kj}\right)q_{,j} + \rho\,\ddot{u}_i\,u_{i,k}\,q\right]\mathrm{d}A. \tag{6.53}$$

In the 3D case, we obtain in exactly the same way the local $J(\bar{s})$-value at position $\bar{s}$ of the crack front using the virtual displacement $\Delta l_k \triangleq q_k$. Instead of (6.43), we have to evaluate a volume integral over the tube-shaped domain V (Figs 6.6, 6.9):

$$J^*(\bar{s}) = \frac{1}{\Delta A}\int_V\left[\left(\sigma_{ij}\,u_{i,k} - U\delta_{kj}\right)q_{k,j} + \rho\,\ddot{u}_i\,u_{i,k}\,q_k\right]\mathrm{d}V. \tag{6.54}$$

The dynamic J-integrals J_1^* (6.53) and (6.54) correspond to the dynamic energy release rate $G(t)$ for stationary cracks Their connection with the stress intensity factors is given by (3.359).

We would have arrived at the same result if the complete divergence of the stress tensor from the equations of motion $\sigma_{ij,j} = -\bar{b}_i + \rho\ddot{u}_i$ had been inserted in the J_k^{te}-integral from Sect. 6.2.2 instead of the volume forces $-\bar{b}_i$. The source terms $p_k = Q_{kj,j}$ of the Eshelby-tensor in (6.21) would thus only have to be extended by the term of inertia $-\rho\ddot{u}_i u_{i,k}$. The expressions (6.22) and (6.34) would then encompass the 2D or 3D J-integral for all conceivable thermal, crack face, volume and inertial loadings of a stationary crack in a thermoelastic material. Analogously, the equivalent domain integrals, 2D (6.42) and 3D (6.47), can be generalized to the combination of all loading types.

The dynamic J_k^{dyn}-integral according to (3.365) for the fast-moving *dynamic crack* (index »dyn«) has a much more complex structure.

$$J_k^{dyn} = \int_\Gamma \left[\left(U + \frac{\rho}{2}\dot{u}_i\dot{u}_i\right)\delta_{kj} - \sigma_{ij}u_{i,k}\right]n_j\,ds + \int_A \left(\rho\ddot{u}_i u_{i,k} - \rho\dot{u}_i\dot{u}_{i,k}\right)dA.$$

(6.55)

The relation between the J_k^{dyn}-components and the K-factors was developed in the dynamic, unsteady case by Nishioka & Atluri [12] for their J_k'-integral ($\equiv J_k^{dyn}$):

$$J_1^{dyn} = G^{dyn} = \frac{1}{2\mu}\left[A_I(\dot{a})K_I^2 + A_{II}(\dot{a})K_{II}^2 + A_{III}(\dot{a})K_{III}^2\right]$$

$$J_2^{dyn} = -\frac{A_{IV}(\dot{a})}{\mu}K_I K_{II}.$$

(6.56)

The functions A_M ($M =$ I, II, III) were already given in (3.367), in addition:

$$A_{IV}(\dot{a}) = \frac{(\alpha_d - \alpha_s)(1 - \alpha_s^2)\bar{D}(\dot{a})}{2[D(\dot{a})]^2}\left[\frac{2 + \alpha_s + \alpha_d}{\sqrt{(1+\alpha_s)(1+\alpha_d)}} - \frac{4(1+\alpha_s^2)}{\bar{D}(\dot{a})}\right]$$

with $\bar{D}(\dot{a}) = 4\alpha_d\alpha_s + (1 + \alpha_s^2)^2$ and $D(\dot{a})$ from (3.332).

(6.57)

With (6.56), two equations are available for determining K_I and K_{II} in the plane mixed-mode case. They are the counterpart to the static relationship (6.12).

Extension of the dynamic J_k^{dyn}-integral to the third dimension and its transformation into equivalent domain integrals is possible in the known manner and yields:

$$2D: \quad J_k^{dyn} = \int_A \left\{\left[\sigma_{ij}u_{i,k} - \left(U + \frac{\rho}{2}\dot{u}_i\dot{u}_i\right)\delta_{kj}\right]q_{,j} + \left(\rho\ddot{u}_i u_{i,k} - \rho\dot{u}_i\dot{u}_{i,k}\right)q\right\}dA$$

(6.58)

$$3\text{D:} \quad J^{\text{dyn}}(\bar{s}) = \frac{1}{\Delta A} \int_V \left\{ \left[\sigma_{ij}\, u_{i,k} - \left(U + \frac{\rho}{2}\dot{u}_i\, \dot{u}_i \right) \delta_{kj} \right] q_{k,j} \right.$$

$$\left. + \left(\rho\, \ddot{u}_i\, u_{i,k} - \rho\, \dot{u}_i\, \dot{u}_{i,k} \right) q_k \right\} \mathrm{d}V \qquad (6.59)$$

Numerical implementation of these domain-independent integrals is done by a FEM post-processor as described in Sect. 4.5.3, whereby the techniques must be extended accordingly to velocities $\dot{u}_i$ and accelerations $\ddot{u}_i$. However, simulation of the crack propagation in the FEM mesh is a completely new aspect. Different techniques for this are introduced in Chap. 8. We can either choose an integration domain for J_k^{dyn} that is sufficiently large to include the crack tips in all phases of the crack propagation, or it can be moved along with the moving crack tip. Due to the path /domain-independence, both possibilities lead to the same result.

6.6 Extension to Inhomogeneous Structures

Frequently, cracks are found in bodies composed of various materials (composites, joints, etc.). First we shall consider the case that each material region has different, but constant mechanical properties, and the crack tip ends in one of these material areas. Fig. 6.10 exemplifies this situation for two material region (α) and (β), which are connected along an *interface* $I_{\alpha\beta}$. As long as the integration contour Γ for a J-integral remains in the homogeneous area of the material (α), the previous equations can be applied. But as soon as Γ or the equivalent domain V include parts of the interface $I_{\alpha\beta}$, we must have recourse to additional terms in order to ensure the path-independence of J.

Fig. 6.10 Integration area for a crack in a heterogeneous structure

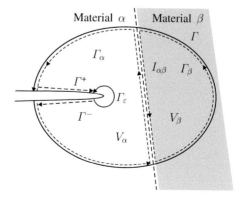

If we approach the interface from both sides, the displacements are continuous, and the tractions must be opposite and equal

$$u_i^{(\alpha)} = u_i^{(\beta)}, \; t_i^{(\alpha)} + t_i^{(\beta)} = \left(\sigma_{ij}^{(\alpha)} - \sigma_{ij}^{(\beta)}\right) n_j^{(\alpha)} = 0 \text{ on } I_{\alpha\beta}. \qquad (6.60)$$

If we now split the integration contour as in Fig. 6.10 into two sub-paths $\Gamma = \Gamma_\alpha + \Gamma_\beta$, which are each restricted to their material region, the interface will be passed through in the opposite direction $n^{(\beta)} = -n^{(\alpha)}$. From this we obtain

$$P_k = \int_{I_{\alpha\beta}} \left[Q_{kj}^{(\beta)} - Q_{kj}^{(\alpha)} \right] n_j^{(\alpha)} \, ds = \int_{I_{\alpha\beta}} \left\{ [\![u_{i,k}]\!]_\alpha^\beta \sigma_{ij}^{(\alpha)} n_j^{(\alpha)} - [\![U]\!]_\alpha^\beta n_k^{(\alpha)} \right\} \, ds, \quad (6.61)$$

where the double brackets $[\![f]\!]_\alpha^\beta = f^{(\beta)} - f^{(\alpha)}$ denote the jump of a variable f on the interface $I_{\alpha\beta}$. This expression can also be interpreted as thermodynamic force associated with the virtual displacement δX_k of the interface. Its amount must be subtracted from the total value of the J_k-integral over Γ in order to obtain exclusively the action of force on the crack tip:

$$J_k = \int_\Gamma \left(U \delta_{kj} - \sigma_{ij} u_{i,k} \right) n_j \, ds - \int_{\Gamma^+ + \Gamma^-} \left(\bar{t}_i u_{i,k} - U n_2 \delta_{2k} \right) \, ds$$

$$- \int_{I_{\alpha\beta}} \left\{ [\![u_{i,k}]\!]_\alpha^\beta \sigma_{ij}^{(\alpha)} n_j^{(\alpha)} - [\![U]\!]_\alpha^\beta n_k^{(\alpha)} \right\} \, ds - \int_{V_\alpha + V_\beta} \left(\bar{b}_i u_{i,k} + U_{,k} \big|_{\exp} \right) \, dV.$$

$$(6.62)$$

As a second example, we will look at a crack in *one* material region, the properties of which however should be a continuous function of the (material) coordinates. Such changes are typical of so-called functionally graded materials. Yet these changes also come about if the mechanical material parameters (elastic modulus, yield stress σ_F, thermal expansion coefficient α_t) are functions of the location indirectly, e.g. via an inhomogeneous temperature field $T(x)$. In these cases, the tensor Q_{kj} also depends *explicitly* on the coordinate x. This is particularly true for the stress work density $\check{U}$, whose thermoelastic variants (6.23) should be considered in more detail:

$$\check{U}^{\text{te}}(\varepsilon, T, x) = \frac{1}{2} \varepsilon_{ij} C_{ijmn}(x) \varepsilon_{mn} - \beta_{ij}(x) \Delta T \, \varepsilon_{ij}$$

$$p_k = \frac{\partial \check{U}^{\text{te}}}{\partial x_k}\bigg|_{\exp} = \frac{1}{2} \varepsilon_{ij} \frac{\partial C_{ijmn}}{\partial x_k} \varepsilon_{mn} - \frac{\partial \beta_{ij}}{\partial x_k} \Delta T \, \varepsilon_{ij}. \qquad (6.63)$$

Physically speaking, this means that a »configurational force« is required for the virtual displacement of a graded material. These relations for the explicit spatial derivative must be inserted in the last term of (6.62).

Thus, Eq. (6.62) is an extension of the 2D J-integral to cracks in heterogeneous bodies, whose material properties either change abruptly or vary continuously with location. This can be straight forward generalized for 3D-crack problems.

It is worth mentioning the special geometric case in which the material properties do *not* change with respect to the crack direction x_1. Then there is no configurational force for the $k = 1$-component of J_k, $p_1 \equiv 0$. This means that material gradients perpendicular to the crack and interfaces parallel to the crack (then $I_{\alpha\beta} = 0$) have *no* effect on the energy balance ($J_1 = G$ (LEFM) or $J_1 = J$ (EPFM))!

6.7 Treatment of Mixed-Mode-Crack Problems

6.7.1 Separation into Crack Opening Modes I and II

Let's assume a pure J line integral for hyperelastic material according to (6.16). As explained already in Chap. 5, the integration path Γ is separated into N single segments for the sake of numerical integration:

$$\Gamma = \sum_{i=1}^{N} \Gamma_i . \tag{6.64}$$

For a better overview, we will limit ourselves to the plane stress state. The x_1-component of the J-integral can then be written in Cartesian coordinates as follows:

$$
\begin{aligned}
J_1 \hat{=} J &= \sum_{i=1}^{N} \int_{\Gamma_i} \left\{ U n_1 - \sigma_{ij} n_j \frac{\partial u_i}{\partial x_1} \right\} \, ds \\
&= \sum_{i=1}^{N} \int_{\Gamma_i} \left\{ \frac{1}{2} [\sigma_{11} \, \sigma_{22} \, \sigma_{12}] \begin{bmatrix} \varepsilon_{11} \\ \varepsilon_{22} \\ 2\varepsilon_{12} \end{bmatrix} n_1 \right. \\
&\qquad\qquad \left. - [\sigma_{11} \, \sigma_{22} \, \sigma_{12}] \begin{bmatrix} n_1 & 0 \\ 0 & n_2 \\ n_2 & n_1 \end{bmatrix} \begin{bmatrix} \partial u_1/\partial x_1 \\ \partial u_2/\partial x_1 \end{bmatrix} \right\} \, ds ,
\end{aligned}
\tag{6.65}
$$

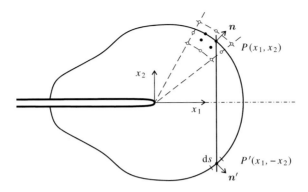

Fig. 6.11 Separation of the J-integral at mixed-mode loading on a symmetrical integration course

where the stresses σ_{ij} and strains ε_{ij} are arranged in column matrices. In order to deal with mixed-mode problems, Ishikawa, Kitagawa & Okamura [13] have proposed to split the J-integral into a mode I and a mode II component. For an arbitrary mixed loading, it is possible to separate the stress, strain, displacement, and traction fields into pure mode I and mode II components provided there is a symmetrical FE-meshing in the crack tip region. Then we consider two points $P(x_1, x_2)$ and $P'(x_1, -x_2)$ arranged mirror-symmetrically to the crack line (see Fig. 6.11). If point $P(x_1, x_2)$ has the field quantities $\sigma_{ij}, t_j, \varepsilon_{ij}$ and u_j and point $P'(x_1, -x_2)$ the field quantities $\sigma'_{ij}, t'_j, \varepsilon'_{ij}$ and u'_j, we can separate into symmetrical and antisymmetric components, whereby the characteristic symmetries and antimetries of the individual quantities are taken into consideration in both modes:

$$\sigma_{ij} = \sigma_{ij}^{I} + \sigma_{ij}^{II} \Rightarrow \begin{bmatrix} \sigma_{11}^{I} \\ \sigma_{22}^{I} \\ \sigma_{12}^{I} \end{bmatrix} = \frac{1}{2} \begin{bmatrix} \sigma_{11} + \sigma'_{11} \\ \sigma_{22} + \sigma'_{22} \\ \sigma_{12} - \sigma'_{12} \end{bmatrix}, \quad \begin{bmatrix} \sigma_{11}^{II} \\ \sigma_{22}^{II} \\ \sigma_{12}^{II} \end{bmatrix} = \frac{1}{2} \begin{bmatrix} \sigma_{11} - \sigma'_{11} \\ \sigma_{22} - \sigma'_{22} \\ \sigma_{12} + \sigma'_{12} \end{bmatrix} \quad (6.66)$$

$$t_j = t_j^{I} + t_j^{II} \Rightarrow \begin{bmatrix} t_1^{I} \\ t_2^{I} \end{bmatrix} = \frac{1}{2} \begin{bmatrix} t_1 + t'_1 \\ t_2 - t'_2 \end{bmatrix}, \quad \begin{bmatrix} t_1^{II} \\ t_2^{II} \end{bmatrix} = \frac{1}{2} \begin{bmatrix} t_1 - t'_1 \\ t_2 + t'_2 \end{bmatrix} \quad (6.67)$$

$$\varepsilon_{ij} = \varepsilon_{ij}^{I} + \varepsilon_{ij}^{II} \Rightarrow \begin{bmatrix} \varepsilon_{11}^{I} \\ \varepsilon_{22}^{I} \\ \varepsilon_{12}^{I} \end{bmatrix} = \frac{1}{2} \begin{bmatrix} \varepsilon_{11} + \varepsilon'_{11} \\ \varepsilon_{22} + \varepsilon'_{22} \\ \varepsilon_{12} - \varepsilon'_{12} \end{bmatrix}, \quad \begin{bmatrix} \varepsilon_{11}^{II} \\ \varepsilon_{22}^{II} \\ \varepsilon_{12}^{II} \end{bmatrix} = \frac{1}{2} \begin{bmatrix} \varepsilon_{11} - \varepsilon'_{11} \\ \varepsilon_{22} - \varepsilon'_{22} \\ \varepsilon_{12} + \varepsilon'_{12} \end{bmatrix} \quad (6.68)$$

$$u_i = u_i^{I} + u_i^{II} \Rightarrow \begin{bmatrix} u_1^{I} \\ u_2^{I} \end{bmatrix} = \frac{1}{2} \begin{bmatrix} u_1 + u'_1 \\ u_2 - u'_2 \end{bmatrix}, \quad \begin{bmatrix} u_1^{II} \\ u_2^{II} \end{bmatrix} = \frac{1}{2} \begin{bmatrix} u_1 - u'_1 \\ u_2 + u'_2 \end{bmatrix} \quad (6.69)$$

Inserting the relations above into (6.65) yields:

$$J = \sum_{i=1}^{N} \int_{\Gamma_i} \left\{ \frac{1}{2} [\sigma^{\mathrm{I}}_{11} \sigma^{\mathrm{I}}_{22} \sigma^{\mathrm{I}}_{12}] \begin{bmatrix} \varepsilon^{\mathrm{I}}_{11} \\ \varepsilon^{\mathrm{I}}_{22} \\ 2\varepsilon^{\mathrm{I}}_{12} \end{bmatrix} n_1 - [\sigma^{\mathrm{I}}_{11} \sigma^{\mathrm{I}}_{22} \sigma^{\mathrm{I}}_{12}] \begin{bmatrix} n_1 & 0 \\ 0 & n_2 \\ n_2 & n_1 \end{bmatrix} \begin{bmatrix} \partial u^{\mathrm{I}}_1/\partial x_1 \\ \partial u^{\mathrm{I}}_2/\partial x_1 \end{bmatrix} \right.$$

$$+ \frac{1}{2} [\sigma^{\mathrm{I}}_{11} \sigma^{\mathrm{I}}_{22} \sigma^{\mathrm{I}}_{12}] \begin{bmatrix} \varepsilon^{\mathrm{II}}_{11} \\ \varepsilon^{\mathrm{II}}_{22} \\ 2\varepsilon^{\mathrm{II}}_{12} \end{bmatrix} n_1 - [\sigma^{\mathrm{I}}_{11} \sigma^{\mathrm{I}}_{22} \sigma^{\mathrm{I}}_{12}] \begin{bmatrix} n_1 & 0 \\ 0 & n_2 \\ n_2 & n_1 \end{bmatrix} \begin{bmatrix} \partial u^{\mathrm{II}}_1/\partial x_1 \\ \partial u^{\mathrm{II}}_2/\partial x_1 \end{bmatrix}$$

$$+ \frac{1}{2} [\sigma^{\mathrm{II}}_{11} \sigma^{\mathrm{II}}_{22} \sigma^{\mathrm{II}}_{12}] \begin{bmatrix} \varepsilon^{\mathrm{I}}_{11} \\ \varepsilon^{\mathrm{I}}_{22} \\ 2\varepsilon^{\mathrm{I}}_{12} \end{bmatrix} n_1 - [\sigma^{\mathrm{II}}_{11} \sigma^{\mathrm{II}}_{22} \sigma^{\mathrm{II}}_{12}] \begin{bmatrix} n_1 & 0 \\ 0 & n_2 \\ n_2 & n_1 \end{bmatrix} \begin{bmatrix} \partial u^{\mathrm{I}}_1/\partial x_1 \\ \partial u^{\mathrm{I}}_2/\partial x_1 \end{bmatrix}$$

$$\left. + \frac{1}{2} [\sigma^{\mathrm{II}}_{11} \sigma^{\mathrm{II}}_{22} \sigma^{\mathrm{II}}_{12}] \begin{bmatrix} \varepsilon^{\mathrm{II}}_{11} \\ \varepsilon^{\mathrm{II}}_{22} \\ 2\varepsilon^{\mathrm{II}}_{12} \end{bmatrix} n_1 - [\sigma^{\mathrm{II}}_{11} \sigma^{\mathrm{II}}_{22} \sigma^{\mathrm{II}}_{12}] \begin{bmatrix} n_1 & 0 \\ 0 & n_2 \\ n_2 & n_1 \end{bmatrix} \begin{bmatrix} \partial u^{\mathrm{II}}_1/\partial x_1 \\ \partial u^{\mathrm{II}}_2/\partial x_1 \end{bmatrix} \right\} ds .$$

$$J = J_{\mathrm{I}} + J_{\mathrm{I,II}} + J_{\mathrm{II,I}} + J_{\mathrm{II}} \tag{6.70}$$

Because we chose an integration path that is symmetrical with respect to the crack (x_1-axis), the components of the normal unit vector (n_1, n_2) at points $P(x_1, x_2)$ and $P'(x_1, -x_2)$ have the following relation to each other (see Fig. 6.11):

$$(n'_1, n'_2) = (n_1, -n_2) . \tag{6.71}$$

With the help of this relation and the Cauchy formula, the tractions of (6.67) at point $P'(x_1, -x_2)$ can be calculated from those from point $P(x_1, x_2)$:

$$\begin{bmatrix} t'^{\mathrm{I}}_1 \\ t'^{\mathrm{I}}_2 \end{bmatrix} = \begin{bmatrix} t^{\mathrm{I}}_1 \\ -t^{\mathrm{I}}_2 \end{bmatrix} \quad \text{and} \quad \begin{bmatrix} t'^{\mathrm{II}}_1 \\ t'^{\mathrm{II}}_2 \end{bmatrix} = \begin{bmatrix} -t^{\mathrm{II}}_1 \\ t^{\mathrm{II}}_2 \end{bmatrix} . \tag{6.72}$$

The other field quantities at P and P' are similarly associated:

$$\begin{bmatrix} \sigma'^{\mathrm{I}}_{11} \\ \sigma'^{\mathrm{I}}_{22} \\ \sigma'^{\mathrm{I}}_{12} \end{bmatrix} = \begin{bmatrix} \sigma^{\mathrm{I}}_{11} \\ \sigma^{\mathrm{I}}_{22} \\ -\sigma^{\mathrm{I}}_{12} \end{bmatrix} , \quad \begin{bmatrix} \sigma'^{\mathrm{II}}_{11} \\ \sigma'^{\mathrm{II}}_{22} \\ \sigma'^{\mathrm{II}}_{12} \end{bmatrix} = \begin{bmatrix} -\sigma^{\mathrm{II}}_{11} \\ -\sigma^{\mathrm{II}}_{22} \\ \sigma^{\mathrm{II}}_{12} \end{bmatrix} ,$$

$$\begin{bmatrix} \varepsilon'^{\mathrm{I}}_{11} \\ \varepsilon'^{\mathrm{I}}_{22} \\ \varepsilon'^{\mathrm{I}}_{12} \end{bmatrix} = \begin{bmatrix} \varepsilon^{\mathrm{I}}_{11} \\ \varepsilon^{\mathrm{I}}_{22} \\ -\varepsilon^{\mathrm{I}}_{12} \end{bmatrix} , \quad \begin{bmatrix} \varepsilon'^{\mathrm{II}}_{11} \\ \varepsilon'^{\mathrm{II}}_{22} \\ \varepsilon'^{\mathrm{II}}_{12} \end{bmatrix} = \begin{bmatrix} -\varepsilon^{\mathrm{II}}_{11} \\ -\varepsilon^{\mathrm{II}}_{22} \\ \varepsilon^{\mathrm{II}}_{12} \end{bmatrix} , \tag{6.73}$$

$$\begin{bmatrix} u'^{\mathrm{I}}_1 \\ u'^{\mathrm{I}}_2 \end{bmatrix} = \begin{bmatrix} u^{\mathrm{I}}_1 \\ -u^{\mathrm{I}}_2 \end{bmatrix} , \quad \begin{bmatrix} u'^{\mathrm{II}}_1 \\ u'^{\mathrm{II}}_2 \end{bmatrix} = \begin{bmatrix} -u^{\mathrm{II}}_1 \\ u^{\mathrm{II}}_2 \end{bmatrix} .$$

Using Eqs. (6.71), (6.72), (6.73) and (6.70), we obtain for the individual four terms of the J-integral:

$$J_I' = J_I, \quad J_{II,I}' = -J_{II,I}, \quad J_{I,II}' = -J_{I,II} \quad \text{und} \quad J_{II}' = J_{II}.\qquad(6.74)$$

Thus the 2nd and 3rd terms cancel out over the entire integration course and (6.70) can be simplified to :

$$J = J_I + J_{II}.\qquad(6.75)$$

With the help of the derived relations, the integral can therefore be decoupled into two separate energy components of a pure mode I and a pure mode II. From these energy release rates $G_I = J_I$ und $G_{II} = J_{II}$, we can then separately calculate both stress intensity factors K_I and K_{II} with (3.93).

$$K_I = \sqrt{E' J_I}, \quad K_{II} = \sqrt{E' J_{II}}.\qquad(6.76)$$

This separation into symmetrical and antimetrical components can be achieved fairly easily in the FEM post-process. But it requires that a symmetrical mesh is generated around the crack tip from the start, which is often impossible in general mixed-mode situations.

6.7.2 Interaction Integral Technique

One disadvantage of energy balance integrals of the J-type is that stress intensity factors cannot be calculated separately in the case of mixed crack loading. Rather, they are contained in a combined form in the calculated energy release rate $J_1 = G$. We could also calculate the x_2-component of J_k with (6.16) for elastic problems, and Eq. (6.12) would provide two equations for determining K_I and K_{II} ($J_3 = K_{III} = 0$). But practically this approach leads to serious inaccuracies in the crack face integrals due to the r^{-1}-singularity of U and the jump-term $(U^+ - U^-)$ and is therefore not useful.

In order to overcome these limitations, the so-called *interaction integral* was developed by Stern et al. [14] and Yau et al. [15] for two-dimensional static elastic problems. The *interaction integral technique* is based on the superposition of two load cases, whereby the reciprocity of the interaction energy of both states is utilized with the help of the Betti theorem. Load case (1) represents the actual loading of the crack configuration, whereas an arbitrary solution with known stress intensity factors is assumed as load case (2). To explain this method, let's first consider a two-dimensional crack problem. Superimposing both load cases, all field quantities add up to:

$$u_i = u_i^{(1)} + u_i^{(2)}, \quad \sigma_{ij} = \sigma_{ij}^{(1)} + \sigma_{ij}^{(2)}, \quad \varepsilon_{ij} = \varepsilon_{ij}^{(1)} + \varepsilon_{ij}^{(2)} \quad \text{etc.}\qquad(6.77)$$

This results in the following energy–momentum tensor:

$$
\begin{aligned}
Q_{kj} &= U(\varepsilon_{pq}^{(1)} + \varepsilon_{pq}^{(2)})\delta_{kj} - \left(\sigma_{ij}^{(1)} + \sigma_{ij}^{(2)}\right)\left(u_i^{(1)} + u_i^{(2)}\right)_{,k} \\
&= \frac{1}{2}\varepsilon_{pq}^{(1)}C_{pqmn}\varepsilon_{mn}^{(1)} + \frac{1}{2}\varepsilon_{pq}^{(2)}C_{pqmn}\varepsilon_{mn}^{(2)} + \frac{1}{2}\varepsilon_{pq}^{(1)}C_{pqmn}\varepsilon_{mn}^{(2)} + \frac{1}{2}\varepsilon_{pq}^{(2)}C_{pqmn}\varepsilon_{mn}^{(1)} \\
&\quad - \sigma_{ij}^{(1)}u_{i,k}^{(1)} - \sigma_{ij}^{(2)}u_{i,k}^{(2)} - \sigma_{ij}^{(1)}u_{i,k}^{(2)} - \sigma_{ij}^{(2)}u_{i,k}^{(1)} \\
&= Q_{kj}^{(1)} + Q_{kj}^{(2)} + Q_{kj}^{(1,2)},
\end{aligned}
\tag{6.78}
$$

where $Q_{kj}^{(l)}$ combines the components of the respective load case $l = \{1, 2\}$, $Q_{kj}^{(1,2)}$ contains the remaining interaction terms. The 3rd and 4th terms of the second line are identical according to the Betti theorem and result in $\varepsilon_{pq}^{(2)}C_{pqmn}\varepsilon_{mn}^{(1)} = \sigma_{mn}^{(2)}\varepsilon_{mn}^{(1)}$.

$$
Q_{kj}^{(1,2)} = \sigma_{mn}^{(2)}\varepsilon_{mn}^{(1)} - \sigma_{ij}^{(1)}u_{i,k}^{(2)} - \sigma_{ij}^{(2)}u_{i,k}^{(1)}
\tag{6.79}
$$

Thus, the J-integral (6.11) is split into three parts:

$$
J_k = \underbrace{\lim_{r\to 0}\int_{\Gamma_\varepsilon} Q_{kj}^{(1)}n_j\,\mathrm{d}s}_{J_k^{(1)}} + \underbrace{\lim_{r\to 0}\int_{\Gamma_\varepsilon} Q_{kj}^{(2)}n_j\,\mathrm{d}s}_{J_k^{(2)}} + \underbrace{\lim_{r\to 0}\int_{\Gamma_\varepsilon} Q_{kj}^{(1,2)}n_j\,\mathrm{d}s}_{J_k^{(1,2)}}.
\tag{6.80}
$$

The last term is called an *interaction integral*:

$$
J_k^{(1,2)} = \lim_{r\to 0}\int_{\Gamma_\varepsilon} Q_{kj}^{(1,2)}n_j\,\mathrm{d}s = \lim_{r\to 0}\int_{\Gamma_\varepsilon}\left[\sigma_{mn}^{(2)}\varepsilon_{mn}^{(1)}\delta_{kj} - \sigma_{ij}^{(1)}u_{i,k}^{(2)} - \sigma_{ij}^{(2)}u_{i,k}^{(1)}\right]n_j\,\mathrm{d}s.
$$

$$
\tag{6.81}
$$

If crack face loads, body forces and inertia forces exist, the integral is converted in analogy to (6.22) into a path-independent integral over an arbitrary contour Γ and the included area A, whereby only $J_1^{(1,2)}$ is of interest:

$$
\begin{aligned}
J_1^{(1,2)} &= \int_\Gamma \left[\sigma_{mn}^{(2)}\varepsilon_{mn}^{(1)}\delta_{1j} - \sigma_{ij}^{(1)}u_{i,1}^{(2)} + \sigma_{ij}^{(2)}u_{i,1}^{(1)}\right]n_j\,\mathrm{d}s \\
&\quad - \int_{\Gamma^+ + \Gamma^-} t_i^{(1)}u_{i,1}^{(2)}\,\mathrm{d}s + \int_A \left(\rho\ddot{u}_i^{(1)} - \bar{b}_i^{(1)}\right)u_{i,1}^{(2)}\,\mathrm{d}A.
\end{aligned}
\tag{6.82}
$$

For the numerical evaluation, it is advantageous to transform this integral again into
an EDI as in Sect. 6.4.1:

$$J_1^{(1,2)} = \int\limits_A \left\{ \left[\left(-\sigma_{mn}^{(2)} \varepsilon_{mn}^{(1)} \right) \delta_{1j} + \sigma_{ij}^{(1)} u_{i,1}^{(2)} + \sigma_{ij}^{(2)} u_{i,1}^{(1)} \right] q_{,j} \, dA \right.$$

$$+ \int\limits_A \left[\left(\rho \ddot{u}_i^{(1)} - \bar{b}_i^{(1)} \right) u_{i,1}^{(2)} \right] q \, dA + \int\limits_{\Gamma^+ + \Gamma^-} \left(-t_i^{(1)} u_{i,1}^{(2)} \right) q \, ds . \quad (6.83)$$

The energy release rate G is equal to the J_1-integral (6.80). According to (3.92),
it has the following relation with the stress intensity factors for the combined load
case $K_N = K_N^{(1)} + K_N^{(2)}$ $(N = \{I, II\})$:

$$G = J_1 = J_1^{(1)} + J_1^{(2)} + J_1^{(1,2)},$$

$$G = \underbrace{\frac{1}{2E'} \left[\left(K_I^{(1)} \right)^2 + \left(K_{II}^{(1)} \right)^2 \right]}_{G^{(1)}} + \underbrace{\frac{1}{2E'} \left[\left(K_I^{(2)} \right)^2 + \left(K_{II}^{(2)} \right)^2 \right]}_{G^{(2)}} + \underbrace{\frac{1}{E'} \left[K_I^{(1)} K_I^{(2)} + K_{II}^{(1)} K_{II}^{(2)} \right]}_{G^{(1,2)}}$$

$$\Longrightarrow J_1^{(1,2)} = G_1^{(1,2)} = \frac{1}{E'} \left[K_I^{(1)} K_I^{(2)} + K_{II}^{(1)} K_{II}^{(2)} \right] \quad (6.84)$$

In order to calculate the two sought quantities $K_I^{(1)}$ and $K_{II}^{(1)}$ from this relation, we
need two known auxiliary load cases, which should be denoted with (2a) and (2b).
The two interaction integrals $J_1^{(1,2a)}$ and $J_1^{(1,2b)}$ provide a linear system of equations
with the solution:

$$K_I^{(1)} = \frac{E'}{K^2} \left[K_{II}^{(2a)} J^{(1,2b)} - K_{II}^{(2b)} J^{(1,2a)} \right]$$

$$K_{II}^{(1)} = \frac{E'}{K^2} \left[K_I^{(2b)} J^{(1,2a)} - K_I^{(2a)} J^{(1,2b)} \right], \quad K^2 = K_I^{(2b)} K_{II}^{(2a)} - K_I^{(2a)} K_{II}^{(2b)} \quad (6.85)$$

Similar to the weight functions (Sect. 3.2.10), it is best to choose a pure mode I
$(K_{II}^{(2a)} = 0)$ for load case (2a) and only mode II $(K_I^{(2b)} = 0)$ for case (2b), whereby
(6.85) is simplified with:

$$K_I^{(1)} = \frac{E'}{K_I^{(2a)}} J^{(1,2a)}, \quad K_{II}^{(1)} = \frac{E'}{K_{II}^{(2b)}} J^{(1,2b)} . \quad (6.86)$$

The auxiliary load cases (2a) and (2b) are arbitrarily selectable as long as they represent adequate solutions of a BVP of this crack configuration (equilibrium, compatibility, identical material law). Thus, simple static solutions may be used without crack face loads even for general loading cases (1). For straight cracks, it is even possible to apply the elastic crack tip fields embodied by (3.12) and (3.23) with given K_I- and K_{II}-factors to the entire body. Since they are in closed form, calculation of the interaction integrals thus becomes especially simple and efficient, which is why we should favor this approach. Be careful with anisotropic materials! The crack's orientation to the material axes in case (2) must be equal to those of the load case (1) being investigated.

In the numerical implementation of the interaction integral as an equivalent domain integral according to (6.83), the following approach should be taken: If an EDI post-processor like (6.42) is available for the ordinary 2D J-integral of the problem class in question (which is presupposed), then the integration algorithm remains unchanged. Only in the integrands of (6.42) the intrinsic energy terms are to be replaced by the interaction energy (6.83) between the numerical solution of load case (1) and the analytic solution of auxiliary load cases (2a, 2b). The required stresses $\sigma_{mn}^{(2)}$ and displacement derivatives $u_{i,1}^{(2)}$ are easily calculated at the IP of the FEM mesh, where the numerical results (1) are given too, thus embedding them into an existing numerical integration routine. Skilled programmers analyze $J_1^{(1,2)}$ simultaneously for both auxiliary load cases (2a) and (2b) in one calculation run.

Compared to the separation technique from Sect. 6.7.1, the interaction integral has several advantages: Firstly, it can be applied to more general loading types, in which case both line and surface integrals arise. Secondly, it also holds true for any crack orientation towards the material axis in anisotropic cases. Thirdly, all numerical advantages of the EDI are exploited. One disadvantage is the necessity of finding solutions for the auxiliary load cases.

6.8 Calculation of T-Stresses

The series expansion of the 2D linear-elastic crack tip solution contains, after the first singular term of the K-factors, a second term which represents a constant normal stress $\sigma_{11} = T_{11}$ acting upon the crack's longitudinal axis (see Sect. 3.2.2). The T_{11}-stresses affect the multi-axiality of the stress state at the crack, which has significant effects on the critical crack initiation value and the slope of the crack resistance curve in EPFM (see Sect. 3.3.6). On the other hand, the T_{11}-stress is responsible for directional stability in the case of fatigue crack propagation (Sect. 3.4.5). Therefore, we should find efficient methods for calculating T_{11} for arbitrary two-dimensional crack configurations as a function of geometry, crack length, and loading. The T_{11}-stress increases proportionally with the load level of a nominal stress σ_n just like the stress intensity factor $K_I = \sigma_n \sqrt{\pi a}\, g(a, w)$. Since T_{11} represents a pure geometrical parameter, a standardized, dimensionless *stress biaxiality ratio* β was introduced:

$$\beta_{\mathrm{T}} = \frac{T_{11}\sqrt{\pi a}}{K_I} \sim \frac{T_{11}}{\sigma_n} \tag{6.87}$$

From the solution of the linear-elastic BVP, we obtain the T_{11}-stress either by interpreting the stress distribution in the ligament ($\theta = 0$) in front of the crack tip

$$T_{11} = \lim_{r \to 0} (\sigma_{11} - \sigma_{22})\Big|_{\theta=0} , \tag{6.88}$$

or via the coefficients a_2 of the second series term according to (3.46)

$$T_{11} = 4a_2 . \tag{6.89}$$

Application of these relations requires a fine FE-discretization at the crack tip, approximately comparable to the mesh quality needed for determining the K-factor. It lacks a certain amount of precision because of the extrapolation or averaging [16]. If in the numerical method the higher eigenfunctions are taken into consideration in the shape function as done in [17], with the boundary collocation method [18] or hybrid crack tip elements (Sect. 5.3), we then obtain T_{11} directly from the coefficient a_2 of the solution. Another possibility is superposition with a single horizontal force at the crack tip, whose energetic interaction with T_{11} according to Eshelby is utilized [19, 20].

One particularly elegant and effective calculation method is based on a path independent integral I_{Γ} developed by Chen [21] that is related to the J-integral. This integral is based on the Betti reciprocity theorem and links together the stresses $\sigma_{ij}^{(1)}$, $\sigma_{ij}^{(2)}$ and displacement fields $u_i^{(1)}$, $u_i^{(2)}$ of two load cases (1) and (2) for the same linear-elastic crack configuration.

$$I_{\Gamma} = \int_{\Gamma} \left(\sigma_{ij}^{(1)} u_i^{(2)} - \sigma_{ij}^{(2)} u_i^{(1)} \right) n_j \, \mathrm{d}s . \tag{6.90}$$

Both load cases must satisfy the equilibrium conditions. As load case (1), let's again take the actual crack problem under consideration. Load case (2) is an auxiliary state that is specifically chosen for the determination of T_{11}. Assuming unloaded crack faces, the path-independence of I_{Γ} can be shown such as in the case of the J-integral. This makes it possible to interpret the FEM results on integration paths chosen outside the numerically less accurate crack tip region. As opposed to the extrapolation method, the total stress and displacement field is included in the calculation of T_{11}. Chen [21] has proved that all unknown coefficients A_n of the Williams eigenfunctions can be determined with the help of I_{Γ}. For this purpose, we arrange the solutions of both load cases as a series expansion in accordance with (3.41) and (3.43) with the complex coefficients A_n and C_m:

load case (1): $\quad \sigma_{ij}^{(1)} = \sum_{n=1}^{\infty} A_n r^{n/2-1} \tilde{\sigma}_{ij}^{(n)}(\theta)\,, \qquad u_i^{(1)} = \sum_{n=1}^{\infty} A_n r^{n/2} \tilde{u}_i^{(n)}(\theta) \quad (6.91)$

load case (2): $\quad \sigma_{ij}^{(2)} = \sum_{m=1}^{\infty} C_m r^{m/2-1} \tilde{\sigma}_{ij}^{(m)}(\theta)\,, \qquad u_i^{(2)} = \sum_{m=1}^{\infty} C_m r^{m/2} \tilde{u}_i^{(m)}(\theta)\,.$

$$(6.92)$$

The eigenfunctions possess the property of orthogonality with regard to the I_Γ-integral, i. e. the following is valid:

$$I_\Gamma(n) = \begin{cases} -\dfrac{\pi(\kappa+1)}{\mu}(-1)^{n+1} n\,\Re(A_n \bar{C}_m) & \text{for } n+m=0 \\[2mm] 0 & \text{for } n+m \neq 0 \end{cases} \qquad (6.93)$$

Thereby, it is possible to filter out the sought components of the nth eigenfunction from the numerical solution of load case (1) by choosing exactly the $(m = -n)$th eigenfunction as auxiliary state (2). If we restrict ourselves to mode I loadings, then both coefficients $A_n = a_n$ and $C_m = a_m$ are real (see (3.41)). If we set $C_m = 1$, so (6.93) yields the sought coefficient

$$A_n = -\frac{\mu}{\kappa+1}\frac{1}{\pi n (-1)^{n+1}} I_\Gamma(n)\,. \qquad (6.94)$$

To determine the T_{11}-stress, the coefficient of the $(n = 2)$th term $A_2 = a_2$ is sought, which is why the $(m = -2)$th eigenfunction should be set as auxiliary state (2). The associated field quantities are calculated from (3.41) and (3.43)

$$\sigma_{11}^* \equiv \sigma_{11}^{(m=-2)} = -\frac{2}{r^2}(\cos 2\theta + \cos 4\theta)$$

$$\sigma_{22}^* \equiv \sigma_{22}^{(m=-2)} = -\frac{2}{r^2}(\cos 2\theta - \cos 4\theta) \qquad (6.95)$$

$$\sigma_{12}^* \equiv \sigma_{12}^{(m=-2)} = -\frac{2}{r^2}\sin 4\theta$$

$$u_1^* \equiv u_1^{(m=-2)} = \frac{1}{2\mu r}(\kappa \cos\theta + \cos 3\theta)$$

$$u_2^* \equiv u_2^{(m=-2)} = \frac{1}{2\mu r}(-\kappa \sin\theta + \sin 3\theta)$$

From (6.94) then the relation between T_{11} and the I_Γ-integral follows:

$$T_{11} = 4A_2 = \frac{2\mu}{\pi(\kappa+1)} I_\Gamma(n=2) = \frac{E'}{4\pi} I_\Gamma(n=2)\,. \qquad (6.96)$$

We thus arrive at the following procedure: We calculate the crack problem of interest with FEM for load case (1) and obtain the solutions $\sigma_{ij}^{\mathrm{FEM}}$, u_i^{FEM}. In the post-process, one or several integration paths Γ are defined. With (6.95), the auxiliary state (2) is calculated on Γ and then the paths-independent I_Γ-integral is numerically calculated:

$$I_\Gamma(n=2) = \int_\Gamma \left(\sigma_{ij}^{\mathrm{FEM}} u_i^* - \sigma_{ij}^* u_i^{\mathrm{FEM}} \right) n_j \, \mathrm{d}s . \tag{6.97}$$

Finally, (6.96) yields the sought T_{11}-stress or, if the K_{I}-factor is known, the biaxiality parameter β via (6.87).

In case the user has already implemented the more favorable EDI-technique to calculate the J-integral, the line integral (6.97) for I_Γ can be transformed into an equivalent domain integral in the same manner as in Sect. 6.4.1 ($\sigma_{ij,j}^{\mathrm{FEM}} = \sigma_{ij,j}^* = 0$) [22].

$$I_\Gamma(n=2) = \int_A \left(\sigma_{ij}^{\mathrm{FEM}} u_i^* - \sigma_{ij}^* u_i^{\mathrm{FEM}} \right) q_{,j} \, \mathrm{d}A . \tag{6.98}$$

It should be noted that the intensity factor $K_{\mathrm{I}} = \sqrt{2\pi} A_1$ of the $(n=1)$th eigen-function can also be determined with the I_Γ-integral if the corresponding orthogonal eigenfunction $m = -n = -1$ is used as an auxiliary state (2) [23].

6.9 Examples

6.9.1 Internal Crack Under Crack Face Loading

As an example of the J-integral in the case of crack face loading, the central crack in a sheet $b = 0.1$ m under constant internal pressure $\sigma_0 = 100$ MPa will be analyzed (Fig. 6.12). For the marked quarter of the geometry, the quite coarse FE-mesh of Fig. 6.13 was used while considering the symmetries. It consists of 8-noded quadri-lateral elements that have been collapsed around the crack tip into triangular quarter-point elements. The hatched element rings around the crack tip are used to calculate the equivalent domain integral (6.42). To determine $J_1 \cong G$, besides the area integral over the element rings A (1st line of (6.42)) the crack face integral with the tractions $\bar{t}_2 = -\sigma_0$ must also be evaluated (3rd line of (6.42)).

Fig. 6.14 shows both portions denoted as »J-contour« and »J-crack face« as well as their sum »J-total« for all four rings of integration. Although both portions change considerably if the included crack faces become larger, their sum remains constant. Among the rings of integration, deviations from the average value $\bar{J} = 15.279$ MN/m amount to less than 0.8 %. Converting into the stress intensity factor via $K_{\mathrm{I}} = \sqrt{\bar{J} E}$ (plane stress) yields

$$K_{\mathrm{I}}^{\mathrm{FEM}} = 39.088 \, \mathrm{MPa} \sqrt{\mathrm{m}} \quad \leftrightarrow \quad K_{\mathrm{I}}^{\mathrm{ref}} = 39.327 = 1.109 \sigma_0 \sqrt{\pi a} ,$$

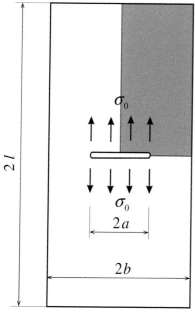

Fig. 6.12 Crack with stress σ_0 on the faces
$a : b : l = 0.4 : 1 : 2.5$

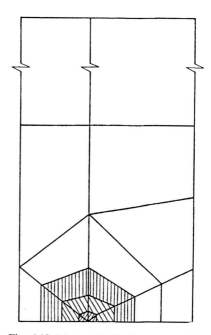

Fig. 6.13 Mesh with 22 elements and 81 nodes

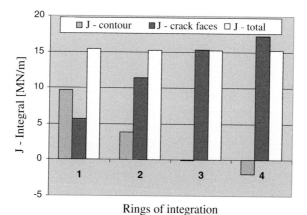

Fig. 6.14 Portions of the J-integral for crack face loading

which deviates from the reference solution [24] by -0.6%. According to the super-position principle of Fig. 3.24, this result is completely identical with the case of load-free crack faces but applied external tension σ_0.

6.9.2 Edge Crack Under Thermal Shock

For a component with a crack, thermal shock represents an extreme loading case because the temperature gradients lead usually to high local stresses. In order to calculate fracture-mechanical parameters, first the transient heat conduction problem must be solved, which provides as a result the temperature distribution in the component for every instant of time. This is followed by the thermomechanical BVP for determining the transient course of the fracture parameters.

In the concrete example, a sheet with an external crack (Fig. 6.15) of length a/b is loaded by an abrupt cooling ΔT (thermal shock) of the medium on the crack side. The thermally induced, time-varying stresses cause an opening of the crack under mode I. Assuming linear-elastic material behavior, the function $K_I(t)$ is sought. A state of plane strain is chosen, i. e. $E' = E/(1 - \nu^2)$ and the thermal expansion coefficient $\alpha' = \alpha(1 + \nu)$. For transient heat conduction, a heat transition coefficient ϑ is set on the edge and a heat conduction coefficient k in the volume. Provided that there is no heat flow on all other edges and the crack faces (isolation), we arrive at a one-dimensional heat conduction with respect to x_1. To better present the results, dimensionless quantities will be introduced (c_v – specific heat capacity):

coordinate	$\hat{x} = x/b$,	temperature	$\Delta \hat{T} = T(\hat{x}, \hat{t})/\Delta T$
heat transition coefficient	$\hat{\vartheta} = \vartheta b/k$,	time	$\hat{t} = \sqrt{\dfrac{k}{\rho c_v}}\, t \Big/ b$

$$(6.99)$$

The intensity of the thermal shock is characterized by the ratio between heat transition and heat conduction, i. e. by the size of $\hat{\vartheta}$, for which a quite severe value of 10 was defined.

Figure 6.16 shows the utilized FE-discretization of the upper half with 8-noded quadrilateral elements and 12 quarter-point elements CTE at the crack. This very fine mesh is necessary in order to determine accurately the steep temperature gradients. The results of the heat conduction analysis are exhibited in Fig. 6.17 as a dimensionless temperature profile over specimen width as a function of time $\hat{t}$. Starting with a uniform initial temperature, the specimen is cooled down by the temperature jump with increasing time.

Subsequent to the heat conduction analyzes, the same mesh was used to solve the crack problem for selected times. To determine the $K_I(t)$-factors, the extended J-integral in the EDI-form according to (6.42) was used and for comparison the displacements were interpreted (DIM) with (5.39). For the x_1-component $J_1 = G$, the thermal part in the 2nd integral of (6.42) with $\sigma_{mn}\alpha_{mn}T_{,1} = \sigma_{mm}\alpha'T_{,1}$ becomes important. From the results obtained (see [25]), the stress intensity factor K_I is shown as an example in Fig. 6.18 normalized with the »thermal stress« $E'\alpha'|\Delta T|$ as a function of time $\hat{t}$ during the cooling process. The K_I-factor increases rapidly as a result of thermally induced stresses, reaches a maximum at $\hat{t} = 0.26$, and then

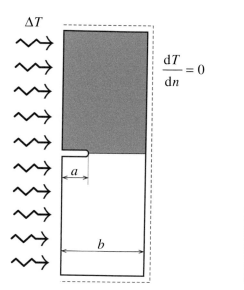

Fig. 6.15 Thermal shock of a sheet with an edge crack

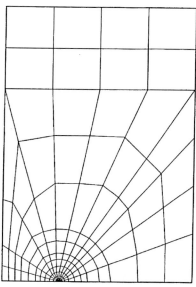

Fig. 6.16 FE-mesh with 128 elements and 475 nodes

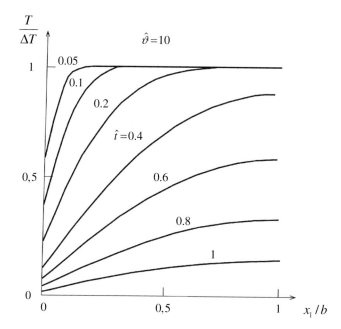

Fig. 6.17 Temporal temperature profile over specimen width during thermal shock

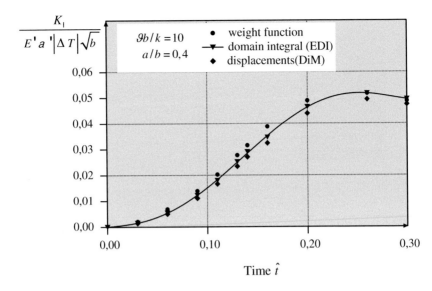

Fig. 6.18 Temporal course of the K_I-factor during thermal shock for crack depth $a/b = 0.4$

declines to zero for $\hat{t} \to \infty$ with increasing cooling of the sheet. The results from the J-integral agree very well with the solution gained via weight functions [25]. As the example shows, DIM realizes usable K_I-values also in the case of thermal loadings.

6.9.3 Dynamically Loaded Internal Crack

This example deals with the two-dimensional dynamic problem (plane strain) of a central internal crack in a sheet (Fig. 6.19). A sudden tension of magnitude σ_0 is imposed on the upper and lower edge at time $t = 0$, which is expressed by the Heaviside jump function $H(t)$. The material constants amount to: $\nu = 0.3$, $E = 200\,\text{GPa}$, density $\rho = 5.000\,\text{kg/m}^3$, from which a dilatational wave velocity of $c_d = 7.338\,\text{m/s}$ is calculated. Exploiting the symmetries, one quarter of the crack configuration was discretized with isoparametric 4-noded elements (see Fig. 6.20). The use of these linear elements is preferred in the case of transient dynamic FE-analyses with explicit time integration [26] (Sect. 4.6). However, this precludes the possibility of using special quarter-point elements at the crack tip, so the dynamic J^*-integral is particularly advantageous in this situation. Figure 6.20 shows coarse and fine mesh variants of the crack region, whereby the smallest elements L at the crack tip are about 1/12 or 1/60 of the crack length a.

For this mode I problem, $J_1^*(t) = G(t)$ was computed as an EDI according to Eq. (6.53) in the post-processor for every time t. With (3.359), this results in the stress intensity factor $K_I(t) = \sqrt{E'J_1^*}$. Its temporal course is shown in Fig. 6.21 normalized to the static value $K_0 = \sigma_0\sqrt{\pi a}$ for the infinite sheet. The plane wave

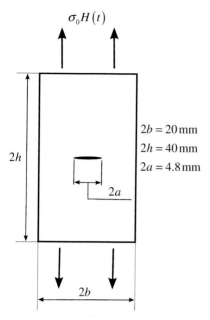

$\sigma_0 H(t)$

$2h$

$2b = 20\,\text{mm}$
$2h = 40\,\text{mm}$
$2a = 4.8\,\text{mm}$

$2a$

$2b$

Fig. 6.19 Sheet with internal crack $a = 0.24b$ under a jump loading

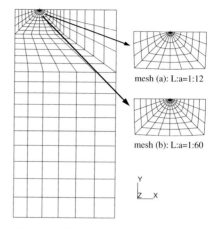

mesh (a): L:a=1:12

mesh (b): L:a=1:60

Y

Z   X

Fig. 6.20 FE-mesh of the *lower right* quarter of the sheet with an internal crack

triggered at the edges requires the time $\tau = h/c_d = 2.73\,\mu\text{s}$ until it reaches the crack faces, whereupon the K_I-factor increases steeply and attains almost three times the reference value at the maximum. After this, the stress waves are reflected and scattered in the body without energy absorption so that further crack loading proceeds in an oscillating manner. To check the results, $K_I(t)$ was determined additionally with two alternative calculation methods from the FE-solution. On the one hand, the displacement interpretation method (Sect. 5.1) according to formula (5.3) was employed. Interpretation of the crack opening displacement u_2^{FEM} directly at the crack tip element as well as the extrapolation of several nodal values on the crack face as in Fig. 5.3 yielded consistent K_I-factors. On the other hand, the technique of the modified crack closure integral (MCCI) (Sect. 5.5 Fig. 5.27) was used with the formula (5.87) for linear elements. All the MCCI interpretation formulae given in Sect. 5.5 for the static case are also valid for stationary cracks under transient loading, since the relation (3.359) between the energy release rate $G(t)$ and the $K(t)$-factors remains unchanged! $G(t)$ is calculated at every time using the dynamic field solutions via virtual crack closure. The results of both alternative methods DIM and MCCI are plotted as well in Fig. 6.21 and show an agreement of $\pm 5\,\%$ with the J^*-results. The coarse mesh deviates the most in the peak area for obvious reasons. It has been numerically confirmed that the EDI-form (6.58) of the 2D dynamic J-integral is actually independent of the choice of the integration area A. The values converge

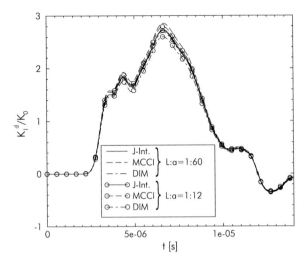

Fig. 6.21 Transient course of the stress intensity factor. Comparison of various methods and meshes

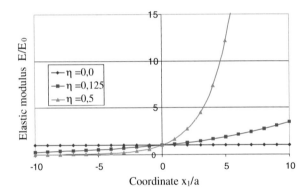

Fig. 6.22 Location-dependence of the elastic modulus in the functionally graded material

with an accuracy of $<1\,\%$ already from the 3rd element ring around the crack tip (Fig. 6.20).

Reference [27] contains further information and 3D dynamic benchmark problems.

6.9.4 Crack in a Functionally Graded Material

Functionally graded materials (FGM) are materials with specially adjusted location-dependent mechanical properties. By means of smooth transitions of the elastic

Fig. 6.23 FE-mesh with 736 elements (8-node quadrilateral) and 64 quarter-point elements at the crack tips

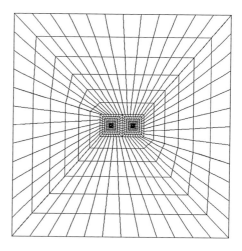

modulus or the thermal expansion coefficient, for example jumps in stiffness or stresses between different material regions can be avoided. Let's consider the simple example of a Griffith crack of length $2a$ in a sheet of width $2b$ under tension σ^∞, whereby the elastic modulus should be varied exponentially with the coordinate x_1:

$$E(x_1) = E_0 \exp\left(\eta \frac{x_1}{a}\right). \tag{6.100}$$

The parameter η defines the »strength« of the material gradient which is illustrated in Fig. 6.22 for the values $\eta = 0$ (homogeneous), 0.125 and 0.5. It could be proven [28, 29], that for cracks in elastic, continuously changing functionally graded materials, the same near field solution exists as those in homogeneous materials (Sect. 3.2.1). The crack loading is thus characterized by the stress intensity factors K_I and K_{II}, yet the strain and displacement fields are calculated from the local elastic constants at the location of the crack tip, e.g. $E(x_1 = a)$. The values of the K-factors depend however on the global material gradient and differ from those of the homogeneous case for the same BVP!

The FE-analysis was carried out in the standard way with the mesh shown in Fig. 6.23. The stress intensity factor K_I was calculated with the 2D J-integral in accordance with Eq. (6.42). The graded property must be taken into consideration in the 2nd integral via the explicit spatial derivative of the strain energy density, which means in the present isotropic case with (A.76) and (A.91)

$$\left.\frac{\partial U}{\partial x_1}\right|_{\exp} = \frac{1}{2} \varepsilon_{ij} \frac{\partial C_{ijkl}}{\partial x_1} \varepsilon_{kl} = \frac{1}{2}\left(\varepsilon_{ij}\varepsilon_{ij} + \frac{\nu}{1-2\nu}\varepsilon_{kk}^2\right)\frac{\partial E(x_1)}{\partial x_1}, \tag{6.101}$$

whereby we still have to insert the concrete derivative of (6.100). The J-integral was implemented as an EDI over various integration domains with radii $\{0.4\ 0.8\ 1.2\ 1.6\ 2.0\}a$. From here, the stress intensity factor is obtained by means of $K_I = \sqrt{E(a)J/(1-\nu^2)}$ with $\nu = 0.3 = \text{const}$ in the state of plane strain. The

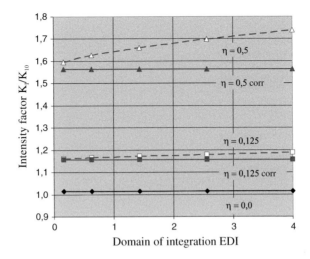

Fig. 6.24 Dependence of the J-integral on the domain and the correction terms

results at the right crack tip are shown in Fig. 6.24 for three different gradients η and normalized to the classical Griffith crack $K_{I0} = \sigma^\infty \sqrt{\pi a}$. Firstly, we see that the simple J-integral would be path-dependent (open symbols) and the necessary correction was only made by the 2nd integral of (6.42) with (6.101) (full symbols). Secondly, the material gradient η causes a considerable increase of K_I at the same loading σ^∞ (and a reduction at the left crack tip).

6.10 Concluding Assessment of Methods

This chapter has shown that energy balance integrals of the J-integral type can be developed and applied to almost all problems in fracture mechanics. This was shown in detail for cracks in heterogeneous, functionally graded, and anisotropic elastic structures under thermal, crack face, weight and inertia loadings. In addition, the verification examples in Sect. 5.8 demonstrate the efficiency of J-integrals in comparison to other FE-techniques. Moreover, it has been proven and recognized that the generalizations of the J-integral represent highly universal and meaningful loading parameters for cracks, for which still further applications will follow in Chaps. 7 and 8. The essential reason for this is to be found in its consistent physical interpretation as the energy release rate in elastic conservative systems or as energy flux into the process zone in dissipative non-conservative systems.

Especially with respect to numerical calculation, energy balance integrals have considerable advantages compared to other methods for determining fracture parameters. Since they can always be formulated as path-independent or at least domain-independent integrals, we can avoid in the interpretation the direct near field of the crack, where the numerical solution is the least accurate. In addition, the accuracy of the numerical solution can be evaluated exactly using different integration paths, since theoretically the determined J-values should all be equal. A deviation thus results either from the inaccuracy of the underlying FE-solution or from an imprecise integration method for J.

Another advantage of J-integrals is that their calculation can be carried out downstream the FE-analysis as a post-processor. They thus do not require changing the FE-program like other techniques. The advantages of the energy balance integrals have been confirmed by numerous verifications and comparative calculations. As a result, J-integrals of various specifications have been offered in some commercial FE-codes (ABAQUS, ANSYS, among others).

References

1. Eshelby JD (1970) Energy relations and the energy momentum tensor in continuum mechanics. In: Kanninen MF (ed) Inelastic behavior of solids. McGraw Hill, New York, pp 77–114
2. Eshelby JD (1975) The elastic energy-momentum tensor. J Elast 5:321–335
3. Maugin GA (1993) Material inhomogeneities in elasticity. Chapman & Hall, London
4. Kienzler R (1993) Konzepte der Bruchmechanik. Vieweg, Wiesbaden
5. Kienzler R, Herrmann G (2000) Mechanics in material space. With application to defects and fracture mechanics. Springer, Berlin u.a
6. Gurtin ME (2000) Configurational forces as basic concept of continuum physics. Springer, Berlin u. a
7. Shih CF, Moran B, Nakamura T (1986) Energy release rate along a three-dimensional crack front in a thermally stressed body. Int J Fract 30:79–102
8. Aoki S, Kishimoto K, Sakata S (1982) Elastic-plastic analysis of crack in thermally-loaded structures. Eng Fract Mech 16:405–413
9. Wilson WK, Yu IW (1979) The use of the J-integral in thermal stress crack problems. Int J Fract 15:377–387
10. Atluri SN (1986) Computational methods in the mechanics of fracture. Elsevier Science Publisher, Noorth-Holland
11. Gurtin ME (1979) On a path-independent integral for thermoelasticity. Int J Fract 15:R169–R170
12. Nishioka T, Atluri SN (1983) Path-independent integrals, energy release rates, and general solutions of near-tip fields in mixed-mode dynamic fracture mechanics. Eng Fract Mech 18:1–22
13. Ishikawa H, Kitagawa H, Okamura H (1980) J-integral of mixed mode crack and its application. In: Proceedings of 3rd international conference on mechanical behaviour of materials, vol 3. Pergamon Press, pp 447–455
14. Stern M, Becker EB, Dunham RS (1976) Contour integral computation of mixed-mode stress intensity factors. Int J Fract 12(3):359–368
15. Yau J, Wang S, Corton H (1980) A mixed-mode crack analysis of isotropic solids using conservation laws of elastics. J Appl Mech 47:335–341
16. Larsson SG, Carlsson AJ (1973) Influence of non-singular stress terms and specimen geometry on small-scale yielding at track tips in elastic-plastic materials. J Mech Phys Solids 21:263–277
17. Leevers PS, Radon JC (1982) Inherent stress biaxiality in various fracture specimen geometries. Int J Fract 19:311–325
18. Fett T (1998) A compendium of T-stress solutions. Technical report. Report FZKA 6057, Forschungszentrum Karlsruhe, Technik und Umwelt

19. Kfouri AP (1986) Some evaluations of the elastic T-term using Eshelby's method. Int J Fract 30:301–315
20. Nakamura T, Parks DM (1992) Determination of elastic T-stress along threedimensional crack fronts using an interaction integral. Int J Solids Struct 29:1597–1611
21. Chen YZ (1985) New path independent integrals in linear elastic fracture mechanics. Eng Fract Mech 22:673–686
22. Chen CS, Krause R, Pettit RG, Banks-Sills L, Ingraffea AR (2001) Numerical assessment of T-stress computation using a P-version finite element method. Fatigue Fract Eng Mater Struct 107:177–199
23. Peters B, Barth FJ, Hahn HG (1995) Klassifizierung von angerissenen Bauteilen mit Hilfe der T-Spannung. In: Tagungsband zur 27. Vortragsveranstaltung des DVM-Arbeitskreises Bruchvorgänge
24. Tada H, Paris P, Irwin G (1985) The stress analysis of cracks handbook, 2nd edn. Paris Production Inc., St.Louis
25. Bahr A, Balke H, Kuna M, Liesk H (1987) Fracture analysis of a single edge cracked strip under thermal shock. Theor Appl Fract Mech 8:33–39
26. Abaqus: ABAQUS theory and user manual. Dassault SystFmes Simulia Corp., Pawtucket (2010)
27. Enderlein M, Ricoeur A, Kuna M (2003) Comparison of finite element techniques for 2D and 3D crack analysis under impact loading. Int J Solids Struct 40:3425–3437
28. Eischen JW (1987) Fracture of nonhomogeneous materials. Int J Fract 34:3–22
29. Kim JH, Paulino GH (2002) Finite element evaluation of mixed mode stress intensity factors in functionally graded materials. Int J Numer Methods Eng 52:1903–1935

Chapter 7
FE-Techniques for Crack Analysis in Elastic-Plastic Structures

FEM has become an indispensable tool for the stress analysis of crack configurations in elastic-plastic materials, as the physically and sometimes geometrically non-linear IBVPs in finite structures are not solvable with analytical methods. To model the material behavior, predominantly the incremental laws of plasticity with various hardening types introduced in Sect. A.4.2 come in to question. Here too, the goal of the computations is to determine the fracture-mechanical loading parameters for ductile crack initiation and crack propagation. For this purpose, we got acquainted with the crack opening displacement δ_t, the crack opening angle γ_t, the J-integral and the multi-axiality parameters T and Q in Sect. 3.3. In the EPFM, a variety of model parameters (geometry, loading level, material behavior) influence the result in different, complex ways, so we must proceed carefully. In particular the fracture-mechanical interpretation must also be considered with caution.

Despite enormous advances in computer technology, non-linear FE-analyses for cracks need an intense amount of numerical operations and storage due to the great effort of mesh discretization and the incremental solution algorithm. Thus, the basic principle, »As simple as possible, as complicated as necessary !« is true in this context as well, i.e. we should start with 2D-models, geometrically linear and with simple hardening laws, in order to understand the important effects before carrying out the necessary model extensions.

7.1 Elastic-Plastic Crack Tip Elements

At stationary crack tips (at rest) in elastic-plastic materials under monotonic loading, the asymptotic near field is known (see Sect. 3.3.6):

- ideal plastic (3.223): $\qquad\qquad \varepsilon_{ij} \sim 1/r, \quad \sigma_{ij} \sim \text{const.},$ (7.1)

- power-law hardening (3.238): $\varepsilon_{ij} \sim r^{-\frac{n}{n+1}}, \quad \sigma_{ij} \sim r^{-\frac{1}{n+1}}$ (7.2)

M. Kuna, *Finite Elements in Fracture Mechanics*, Solid Mechanics and Its
Applications 201, DOI: 10.1007/978-94-007-6680-8_7,
© Springer Science+Business Media Dordrecht 2013

It has been attempted to develop elastic-plastic crack tip elements for these asymptotics as well. Again, a modification of the isoparametric elements has proved successful. In Sect. 5.2.2, the collapsed isoparametric quadrilateral element with quadratic shape functions was already introduced (see Fig. 5.7a). By combining the coordinates of nodes 1, 4 and 8 as well as the quarter-point shift of nodes 5 and 7, we obtain the strain singularities (5.28)–(5.30) with respect to the distance r to the crack tip:

$$\varepsilon_{ij}(r,\,\theta) = \frac{A_{0ij}(\theta)}{r} + \frac{A_{1ij}(\theta)}{\sqrt{r}} + A_{2ij}(\theta). \qquad (7.3)$$

For application in LEFM, the $1/r$-behavior is undesirable and was eliminated by binding the displacements of these three nodes together with (5.31). For EPFM, we make use of this property, i. e. the three nodes may now move independently of each other, whereby in addition the ideal-plastic $1/r$-singularity (7.1) is activated. Figure 7.1 shows this procedure. By combining quarter-point shift and free crack tip nodes, initially elastic and subsequently ideal-plastic behavior can be simulated this way.

In order to analyze the properties of the collapsed elements with *unchanged mid-side nodes*, let's return to Sect. 5.2.2. If we evaluate the Eqs. (5.17)–(5.18) for the value $\varkappa = 1/2$ we obtain in the place of (5.19):

$$x_1 = \frac{L}{2}(1 + \xi_1), \quad x_2 = \frac{H}{2}\xi_2\,(1 + \xi_1)$$

$$r = \frac{1}{2}\sqrt{L^2 + H^2\xi_2^2}\,(1 + \xi_1) \quad \Rightarrow (1 + \xi_1) = \frac{r}{\frac{1}{2}\sqrt{L^2 + H^2\xi_2^2}} \qquad (7.4)$$

The mapping of $(1+\xi_1)$ on r is now linear. For the JACOBIan matrix (5.20, 5.21) and its inverse (5.22) now follows:

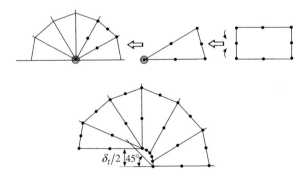

Fig. 7.1 Collapsed 8-noded quarter-point elements at the crack tip with elastic-plastic material behavior

$$J_{11} = L, \quad J_{21} = 0, \quad J_{12} = H\xi_2, \quad J_{22} = \frac{H}{2}(1 + \xi_1) \sim r,$$

$$J^{-1} = \frac{1}{J}\begin{bmatrix} J_{22} & -J_{12} \\ 0 & J_{11} \end{bmatrix} \sim \begin{bmatrix} 1 & \frac{\xi_2}{r} \\ 0 & \frac{1}{r} \end{bmatrix} \quad \text{with } J = J_{11}J_{22} = \frac{1}{2}HLr. \tag{7.5}$$

In contrast to $\varkappa = 1/4$, the $\sqrt{r}$-terms have disappeared in (7.4) and (7.5). The derivatives of the shape functions (5.24)–(5.27) remain unchanged, so we get for the strains according to (5.28)–(5.30) now the functions

$$\varepsilon_{11} = a_0 + a_1(1 + \xi_1) + \frac{b_0 + b_1(1 + \xi_1) + b_2(1 + \xi_1)^2}{r}$$

$$= \frac{b_0}{r} + e_1' + e_2'r \quad \text{and also} \tag{7.6}$$

$$\varepsilon_{22} = \frac{d_0}{r} + d_1 + d_2 r$$

$$\varepsilon_{12} = \frac{b_0 + d_0}{r} + f_1' + f_2'r$$

with modified constants e_i' and f_i'. Thus, the collapsed quadrilateral elements with mid-side nodes have no $1/\sqrt{r}$- singularity, but rather a linear term $\sim r$. They are generally recommended for elastic-plastic crack problems and preferred to the quarter-point variant (7.3).

> If we collapse one element edge of the isoparametric 8-noded quadrilateral element to one point but allow free nodal displacements, a 2D crack tip element is generated, which possesses a $1/r$-singularity in the strains on all radial rays.

With these elements, a fan is placed around the crack tip. Under loading, all crack tip nodes can move individually so the blunting is simulated by a »pearl necklace« (see Fig. 7.1). The ideal-plastic crack tip elements are compatible with each other and with the standard elements at the outer edges. To generalize this idea to the 3D-case, we adapt this technique to collapsed 20-noded hexahedral elements arranged in a »tube« around the crack front.

Special crack tip elements for modeling the HRR-singularity (7.2) have not become established. For hardening exponents $n > 10$, the asymptotic behavior approaches anyway Eq. (7.1). Moreover, real yield curves usually reach a saturation $\sigma_F \to$ const. For this reason, these collapsed elements are used with success in most elastic-plastic crack analyses.

For elastic-plastic crack problems it is more difficult to capture the near field solution than in LEFM since it (1) occurs deep inside the plastic zone, (2) its area of existence enlarges with growing plastification and (3) its character can change with the loading history under certain circumstances. On the other hand, the correct modeling of the asymptotic near field plays a minor role for the global behavior of structures with cracks in EPFM. Fracture parameters such as the J-integral can be calculated accurately enough even at greater distances from the crack tip. Yet crack tip elements are indispensable if we want to investigate the details of plastic strain and the state of stress at the very crack tip.

7.2 Determination of Crack Tip Opening Displacements

Permanent plastic deformations of the crack faces are utilized as relevant parameters for different concepts of EPFM. According to the *CTOD concept*, the opening displacement of the blunted crack tip δ_t represents a parameter for *stationary cracks* (see Sect. 3.3.4). In order to calculate the *crack opening displacement* δ_t accurately with FEM, a concentric fan-shaped mesh around the crack tip is needed, whereby the innermost ring should consist of the aforementioned collapsed elements (see Fig. 7.1). The simulation must be carried out as geometrically non-linear so that the large deformations at the crack tip are reproduced correctly. We then obtain a nicely stretched node chain at the blunted crack as exemplified in Fig. 7.2. To determine δ_t it is best to use the $\pm 45°$ secant method shown in Fig. 3.35a. To this end, the intersection with the element edges on the crack faces must be found. An easier possibility is to choose the u_2-displacement of an FE-node on the crack face close behind the blunting, e. g. the nodes of the last crack tip element ($\theta = 180°$) in Fig. 7.1. However, this variant depends considerably on the used FE-discretization at the crack tip. Following the experimental methods for determining the CTOD, occasionally a linear extrapolation is carried out of the opening displacements $u_2(x_1)$ from areas further away from the crack tip where the crack faces are almost linear. According to this definition, a relatively coarse mesh with a geometrically linear option may be used, since the far field displacements to be interpreted are largely independent of it. Frequently, the opening displacement of the crack notch at the specimen's front edge—the *crack opening displacement V* (COD) is also sought for comparison with test results.

In the case of *growing cracks*, a linear opening profile is formed at the crack tip as discussed in Sect. 3.3.8, which is characterized quite well by a *crack opening angle* $\gamma_t \cong$ CTOA (see Fig. 3.35). The CTOA is utilized successfully as a criterion for ductile crack growth in thin-walled structures under plane stress (metal sheets, aircraft skin) [2].

In order to simulate crack propagation, a regular FE-mesh made of equally large elements along the crack path is necessary. Figure 7.3 provides an example. Suitable FE-techniques for simulating crack propagation are addressed in Chap. 8. From the numerical results of the deformed crack faces, a crack tip opening angle can be determined, for example, in the manner shown in Fig. 7.3 [3]. Unfortunately, there is no agreed definition for CTOA. Its assessment depends in a sensitive manner on the chosen mesh and interpolation technique.

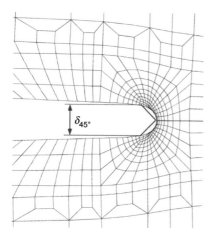

Fig. 7.2 Determining the crack tip opening displacement CTOD from the FE- analysis of the crack at rest [1]

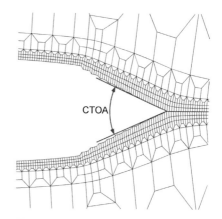

Fig. 7.3 Determining the crack tip opening angle CTOA from the FE-analysis of the crack in motion [1]

For three-dimensional crack configurations, the above-introduced techniques can be applied analogously to every position of the crack front and the plane perpendicular to it. Schwalbe [4] proposed another pragmatic technical measure of deformation near the crack tip, the δ_5-concept, which has become established.

7.3 Calculation of the J-Integral and its Meaning

7.3.1 Elastic-Plastic Extensions of J

Since the validity of the classic J-integral is restricted to (non-) linear elasticity theory or plastic deformation theory (see Sect. 3.3.6), various extensions to incremental plasticity theory and large deformations have been suggested. For further explanation we resort directly to Chap. 6.

(a) Elastic-plastic contour integral

For two-dimensional crack problems, one obvious method is simply to evaluate the RICE line integral (3.100) with the results of the elastic-plastic FE-analysis.

$$J(\Gamma) = \int_{\Gamma} \left[\check{U}^{\mathrm{ep}} \delta_{1j} - \sigma_{ij} u_{i,1} \right] n_j \, \mathrm{d}s = \int_{\Gamma} \check{Q}_{1j} n_j \, \mathrm{d}s \qquad (7.7)$$

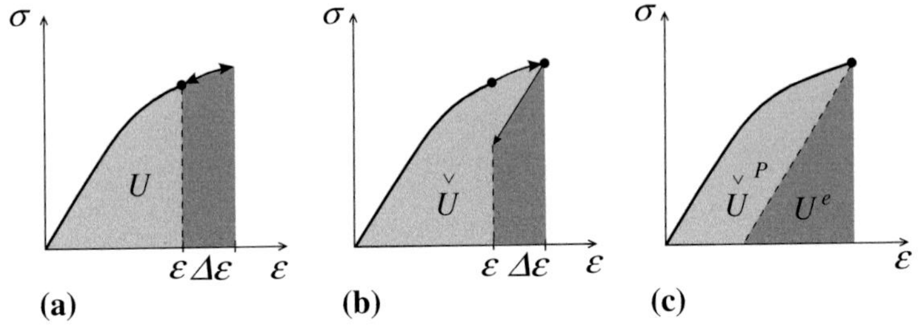

Fig. 7.4 Difference between the total **a** and incremental **b** plasticity theories in loading and load removal **c**

Unlike elasticity theory, now the total elastic-plastic stress work density $\check{U}^{\mathrm{ep}}$ per volume should be used instead of the elastic specific strain energy $U(\varepsilon_{ij})$:

$$\check{U}^{\mathrm{ep}}(\varepsilon_{ij},\, x_k) = \int\limits_0^{\varepsilon_{ij}} \sigma_{kl}(\bar{\varepsilon}_{mn})\,\mathrm{d}\bar{\varepsilon}_{kl} = U^{\mathrm{e}}(\varepsilon_{ij}^{\mathrm{e}}) + \check{U}^{\mathrm{p}}(x_k) \tag{7.8}$$

It is formed with the elastic and irreversible plastic strains $\mathrm{d}\varepsilon_{kl} = \mathrm{d}\varepsilon_{kl}^{\mathrm{e}} + \mathrm{d}\varepsilon_{kl}^{\mathrm{p}}$, i. e. the potential character is lost, which is why the quantity is marked with a check ˇ. This becomes understandable if we consider a loading and unloading process of $\pm\Delta\varepsilon$ at the material point as shown in Fig. 7.4a, b. Neither the stresses nor the stress work density $\check{U}^{\mathrm{ep}}$ are a unique function of the strains! In particular, the plastic component $\check{U}^{\mathrm{p}}(x_k)$ is a function of the total loading history in the material point and must therefore be conceived as an explicit function of the coordinate x_k.

Many numerical results have shown that in the case of *monotonously increasing loading* for *stationary cracks* and *infinitesimal deformations* J according to (7.7) is independent of the integration path in very good approximation (The strict proof only holds true in the case of proportional stresses). Thus, we can also shrink the path Γ around the crack tip $\Gamma_\varepsilon \to 0$, where (with sufficient mesh refinement) the HRR-solution is prevalent. The relevance of J as a characteristic parameter of the HRR-field is thus retained. In the case of load removal or crack growth, J loses these properties and is worthless.

(b) Path-independent formulation $\tilde{J}$

Generally, in order to eliminate path-dependence, a domain integral over the enclosed area A must be added to the line integral along Γ. We shall apply the same formalism as in Sect. 6.2 to the integrands of (7.7) (see Fig. 6.2):

$$J(\Gamma_\varepsilon) \mathrel{\widehat{=}} \int\limits_{\Gamma_\varepsilon} \check{Q}_{1j} n_j\,\mathrm{d}s = \int\limits_{\Gamma} \check{Q}_{1j} n_j\,\mathrm{d}s - \int\limits_{A} \check{Q}_{1j,j}\,\mathrm{d}A \mathrel{\widehat{=}} \tilde{J} \tag{7.9}$$

If we subdivide the strains and stress work density into their elastic and plastic components, the divergence leads to:

$$\check{Q}_{1j,j} = \frac{\partial \check{U}^{ep}}{\partial x_1} - \sigma_{ij,j} u_{i,1} - \sigma_{ij} u_{i,j1} = \frac{\partial U^e}{\partial x_1} + \frac{\partial \check{U}^p}{\partial x_1} - \sigma_{ij} \varepsilon_{ij,1}$$

$$= \frac{\partial U^e}{\partial \varepsilon^e_{ij}} \varepsilon^e_{ij,1} + \check{U}^p_{,1} - \sigma_{ij} (\varepsilon^e_{ij,1} + \varepsilon^p_{ij,1}) = \check{U}^p_{,1} - \sigma_{ij} \varepsilon^p_{ij,1} \qquad (7.10)$$

Here, the relation $\partial U^e / \partial \varepsilon^e_{ij} = \sigma_{ij}$ was used and homogeneous equilibrium conditions $\sigma_{ij,j} = 0$ were assumed. Insertion into (7.9) yields the equivalent algorithm for the sought near field value $J(\Gamma_\varepsilon)$:

$$\tilde{J} := J(\Gamma_\varepsilon) = \int_\Gamma \left[\check{U}^{ep} \delta_{1j} - \sigma_{ij} u_{i,1} \right] n_j \, ds - \int_A \left[\check{U}^p_{,1} - \sigma_{ij} \varepsilon^p_{ij,1} \right] dA \qquad (7.11)$$

The area integral thus quantifies exactly the difference between the two line integrals, which are calculated around the crack tip Γ_ε in the near field and along an arbitrary outer path Γ through the far field. This correction term effects the path-independence of the extended $\tilde{J}$-integral for any load path in the context of the plastic flow theory.

The formal mathematical transformation of $J(\Gamma_\varepsilon)$ into a line-area integral $\tilde{J}$ changes nothing in its physical property! But what is the fracture-mechanical meaning of $\tilde{J}$? The interpretation as a potential energy release rate must be abandoned due to the dissipation component in $\check{U}^{ep}$. However, we can conceive $\tilde{J}$ as an elastic-plastic work rate during virtual crack propagation or as an energy flux that is supplied across the contour Γ_ε from outside to the crack tip. According to ESHELBY, $J(\Gamma_\varepsilon)$ is the configurational force associated with a virtual displacement of the crack tip, while in the line integral over Γ also the displacement of the included plastic zone ($\hat{=}$ defects) is contained. Whether the integral $\tilde{J}$ for $\Gamma_\varepsilon \to 0$ converges towards a finite value depends on the asymptotics of the crack tip solution. As proved already in Sect. 3.3.6, this necessitates a $1/r$-singularity in the stress work density $\check{U}^{ep}$. This condition is pretty satisfied in the context of the incremental plasticity theory for stationary cracks assuming infinitesimal deformations. Considering finite deformations, the stress fields at the blunted crack tip-that now acts de facto like a notch-take on finite values. Therefore, $\tilde{J}$ drops to zero in the near field of about $r < 4\delta_t$, and becomes path-dependent (see example in Sect. 7.4.1).

The elastic-plastic formulation (7.11) can be readily generalized to the J-integral vector J_k, if we replace x_1 by x_k:

$$\tilde{J}_k = \int_{\Gamma_\varepsilon} \left[\check{U}^{ep} \delta_{kj} - \sigma_{ij} u_{i,k} \right] n_j \, ds$$

$$= \int_\Gamma \left[\check{U}^{ep} \delta_{kj} - \sigma_{ij} u_{i,k} \right] n_j \, ds - \int_A \left[\check{U}^p_{,k} - \sigma_{ij} \varepsilon^p_{ij,k} \right] dA \qquad (7.12)$$

Likewise, the transformation into an equivalent domain integral (6.42) is possible as shown in Sect. 6.4.1. In the 2D case, this leads with the weighting function $q(x)$ to:

$$\tilde{J}_k = -\int_A \left[\check{U}^{\mathrm{ep}} \delta_{kj} - \sigma_{ij} u_{i,k} \right] q_{,j} \, \mathrm{d}A - \int_A \left[\check{U}^{\mathrm{p}}_{,k} - \sigma_{ij} \varepsilon^{\mathrm{p}}_{ij,k} \right] q \, \mathrm{d}A. \quad (7.13)$$

For three-dimensional crack configurations, we obtain in analogy to (6.47) for a virtual crack propagation $q_k \,\widehat{=}\, \Delta l_k$ the expression:

$$\tilde{J}(\bar{s}) = -\frac{1}{\Delta A} \left\{ \int_V \left[\check{U}^{\mathrm{ep}} \delta_{kj} - \sigma_{ij} u_{i,k} \right] q_{k,j} \, \mathrm{d}V + \int_V \left[\check{U}^{\mathrm{p}}_{,k} - \sigma_{ij} \varepsilon^{\mathrm{p}}_{ij,k} \right] q_k \, \mathrm{d}V \right\}$$
$$(7.14)$$

If there are volume, crack face or thermal loadings, the relations (7.13) and (7.14) must be extended by the corresponding additional terms from (6.42) or (6.47). These extensions of J were introduced by Moran and Shih [5] as well as Carpenter et al. [6]. In the numerical realization, Eqs. (7.13) and (7.14) are favored (see Sect. 6.4).

(c) The incremental ΔT^*-Integral

Atluri, Nishioka et al. [7, 8] have developed a path-independent integral T^*, which should be valid for static and dynamic crack problems with any inelastic material law. The vectorial integral represents the sum of the increments ΔT^*_k over the loading history:

$$T^*_k = \sum \Delta T^*_k, \qquad \Delta T^*_k = \int_{\Gamma_\varepsilon} \Delta \check{Q}_{kj} n_j \, \mathrm{d}s \qquad (7.15)$$

The increment of the energy–momentum tensor $\Delta \check{Q}_{kj} = \check{Q}_{kj}(t + \Delta t) - \check{Q}_{kj}(t)$ takes on the concrete form:

$$\Delta \check{Q}_{kj} n_j = \left[\Delta \check{U}^{\mathrm{ep}} \delta_{jk} - (\sigma_{ij} + \Delta \sigma_{ij}) \Delta u_{i,k} - \Delta \sigma_{ij} u_{i,k} \right] n_j$$
$$= \Delta \check{U}^{\mathrm{ep}} n_k - (t_i + \Delta t_i) \Delta u_{i,k} - \Delta t_i u_{i,k} \qquad (7.16)$$
$$\Delta \check{U}^{\mathrm{ep}} = (\sigma_{ij} + \frac{1}{2}\sigma_{ij}) \Delta \varepsilon_{ij}$$

We pursue the same idea as in Eq. (7.9) for the correction of path-dependence with the domain integral A, but this time applied to the increment $\Delta \check{Q}_{kj}$ (7.16):

$$\oint_{\Gamma - \Gamma_\varepsilon} \Delta \check{Q}_{kj} n_j \, \mathrm{d}s = \int_A \Delta \check{Q}_{kj,j} \, \mathrm{d}A$$

$$= \int_A \left\{ \Delta \check{U}^{\text{ep}}_{,k} - \left[(\sigma_{ij} + \Delta\sigma_{ij})\Delta u_{i,k}\right]_{,j} - \left[\Delta\sigma_{ij} u_{i,k}\right]_{,j} \right\} \, \mathrm{d}A$$

$$= \int_A \left[(\sigma_{ij,k} + \frac{1}{2}\Delta\sigma_{ij,k})\Delta\varepsilon_{ij} - (\varepsilon_{ij,k} + \frac{1}{2}\Delta\varepsilon_{ij,k})\Delta\sigma_{ij} \right] \, \mathrm{d}A ,$$

$$(7.17)$$

whereby in the last step $\Delta \check{U}^{\text{ep}}$ from (7.16) and $(\sigma_{ij} + \Delta\sigma_{ij})_{,j} = 0$ were used. After inserting Hooke's law $\sigma_{ij} = C_{ijkl}\varepsilon^{\text{e}}_{kl}$ only the plastic strains $\varepsilon^{\text{p}}_{ij} = \varepsilon_{ij} - \varepsilon^{\text{e}}_{ij}$ remain.

We thus obtain a path-independent incremental line-area integral:

$$\Delta T^*_k = \int_\Gamma \Delta \check{Q}_{kj} n_j \, \mathrm{d}s - \int_A \left[(\sigma_{ij,k} + \frac{1}{2}\Delta\sigma_{ij,k})\Delta\varepsilon^{\text{p}}_{ij} - (\varepsilon^{\text{p}}_{ij,k} + \frac{1}{2}\Delta\varepsilon^{\text{p}}_{ij,k})\Delta\sigma_{ij} \right] \, \mathrm{d}A$$

$$(7.18)$$

The ΔT^*_k-integral is de facto nothing else than the incremental form of the $\tilde{J}$-integral from (7.11), which is easily proved by integration over the loading history. Conversely, in the numerical realization (7.12) of $\tilde{J}$ per load step, it is precisely the operations (7.18) that are to be carried out and added. Thus, ΔT^*_k does not differ from $\tilde{J}$ with reference to the energetic interpretation as well. Corresponding three-dimensional generalizations with volume and inertia forces are found in [7].

(d) Energy flux integral $\widehat{J}$ by Kishimoto

Kishimoto, Aoki and Sakata [9, 10] have put forwards so-called $\widehat{J}$-integrals, which are based on the model of a fictitious fracture process zone A_B as discussed already in Sect. 3.3.8 (Fig. 3.48). Energy input into the process zone A_B is calculated by means of the flux integral $\mathcal{F}$ according to (3.280) over the boundary $\Gamma_B \cong \Gamma_\varepsilon$ of the process zone or by means of the equivalent line-area integral (3.282) on an external contour Γ. The authors [9] purposely employed only the elastic part U^{e} for the strain energy density in order to quantify the potentially available energy release rate. The $\mathcal{F}$- integral (3.282) leads, with $U^{\text{e}}_{,1} = \sigma_{ij}\varepsilon^{\text{e}}_{ij,1}$ and the separation of the plastic strain $\varepsilon^{\text{p}}_{ij}$, directly to the elastic-plastic $\widehat{J}$-integral

$$\widehat{J} = \int_\Gamma \left[U^{\text{e}} n_1 - \sigma_{ij} u_{i,1} n_j \right] \, \mathrm{d}s + \int_A \sigma_{ij}\varepsilon^{\text{p}}_{ij,1} \, \mathrm{d}A.$$

$$(7.19)$$

In contrast to $\widetilde{J}$ (7.11), the plastic work $\check{U}^{\mathrm{p}}$ is lacking in both the line and area integral. The 3D extension is the disk integral indicated in Sect. 6.3.1 (Fig. 6.4), whereby now instead of the thermal strains ε_{mn}^{*} the plastic strains $\varepsilon_{mn}^{\mathrm{p}}$ would have to be inserted or additionally included. With the volume forces $\bar{b}_i$, we obtain from (6.29) at position s of the crack front:

$$\widehat{J}(s) = \int_{\Gamma} \left[U^e \delta_{1j} - \sigma_{ij} u_{i,1} \right] n_j \, \mathrm{d}s + \int_{A} \left[\sigma_{mn} \varepsilon_{mn,1}^{\mathrm{p}} - \bar{b}_i u_{i,1} - (\sigma_{i3} u_{i,1})_{,3} \right] \mathrm{d}A$$

$$(7.20)$$

For numerical implementation, the term $(\sigma_{i3}u_{i,1})_{,3}$ is for reasons already mentioned complex and disadvantageous compared to $\widetilde{J}$ and T^*.

In all elastic-plastic generalizations $\widetilde{J}$, ΔT^* and $\widehat{J}$ terms appear in the surface or volume integrals, where the stress work densities $\check{U}^{\mathrm{ep}}$, $\check{U}^{\mathrm{p}}$, plastic strains $\varepsilon_{ij}^{\mathrm{p}}$, or stresses σ_{ij} must be differentiated with respect to the coordinate x_k. In FEM these quantities themselves result from the derivatives of the primary displacement variables and are generally only available at the integration points. Therefore, their derivatives are flawed with a high level of inaccuracy. Suitable interpolation and differentiation techniques were described in Sect. 4.4.4. Another difficulty comes from the fact that the mentioned fields exhibit in the plastic zone at the crack high gradients, which must be differentiated again! Therefore, we need to take the utmost care in the numerical calculation of these integral terms. Positioning the integration paths Γ far from the crack does not help because the domain integrals nonetheless comprise the entire included area-i. e. the inaccurate crack region as well.

It should also be mentioned that the above-introduced extensions of J have so far not been implemented in any commercial FEM programs.

7.3.2 Application to Stationary Cracks

The elastic-plastic variants of J will now be investigated using the simple 2D example of a tension sheet with an internal crack, paying special attention to the fracture-mechanical interpretation. Figure 7.5 shows the FE-mesh and boundary conditions for a quarter of the specimen from Fig. 5.1 with crack length $a = 0.4\,d$. For the material behavior, incremental plasticity with isotropic hardening was assumed. The FE-mesh consists of 8-noded quadrilateral elements each with 3×3 integration points. The crack tip is surrounded by the collapsed special elements with the $1/r$-singularity described in Sect. 7.1. The external tension σ is applied as a sequence of four load and load removal steps as shown in Fig. 7.5. The figure also shows the plastic zones in the crack region at load levels 1–4. We can clearly see that the crack blunting CTOD is based on plastic deformations, since it hardly changes during the load removal steps $1 \to 2$ and $3 \to 4$.

With the help of the FEM results, the RICE line integral J (7.7) and its path-independent extension $\widetilde{J}$ or T^* were calculated with (7.11), whereby for both

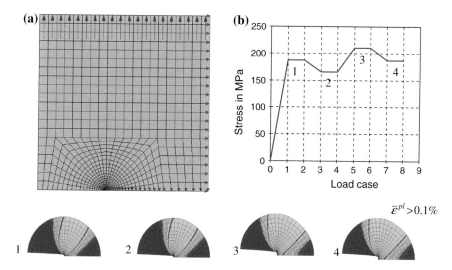

Fig. 7.5 **a** FE-model of a sheet with internal crack and **b** sequence of tensile loading

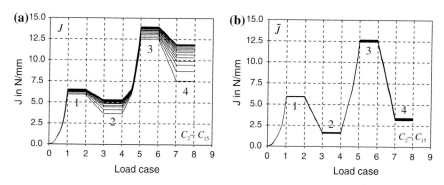

Fig. 7.6 **a** J-integral and **b** $\tilde{J}$-integral versus loading for the integration paths C2-C15

integrals the equivalent domain version was used in the form (7.13). As integration domains A, 15 semicircles around the crack tip were chosen, whose outer radii correspond to the integration paths Γ. Figure 7.6a proves quite clearly that J is nearly path-independent in the 1st load step, which rises monotonously. In the further course, especially during load removal, drastic differences arise between the J-values of the various paths. In contrast to this, the $\tilde{J}$-integral yields identical values on all paths, even during unloading (see Fig. 7.6b). The difference between J and $\tilde{J}$ consists precisely in the second integral of (7.13), with which the path-independence of $\tilde{J}$ or T^* was achieved . This means, deviations between the paths are of basic nature for J, whereas for $\tilde{J}$ they must be ascribed to numerical inaccuracy!

7.3.3 Application to Moving Cracks

Formally, all previously introduced elastic-plastic integrals J, $\tilde{J}$, T^* and $\hat{J}$ can also be applied to cracks that propagate quasi-statically in ductile materials. However, their physical meaning must be scrutinized carefully. To elucidate this further, let's return to the model of a fracture process zone A_B discussed in Sect. 3.3.8, which moves with the crack tip (Fig. 3.48). A fixed material point, over which the crack moves away with the process zone, necessarily experiences a non-proportional plastic loading and subsequent elastic unloading phase. For this reason, the classical J-integral is unusable. The extended line-area integrals $\tilde{J}$, T^* and $\hat{J}$ are indeed path-independent, but their actual value must become zero in the context of continuum-mechanical modeling, since with an asymptotic approach $\Gamma_\varepsilon \to 0$ towards the crack tip, we encounter the weak logarithmic singularity (3.261) at the moving crack. Thus, these integrals are worthless as fracture-mechanical intensity parameters in the case of crack propagation.

We will therefore investigate the energetic interpretation. Energy is supplied to the process zone A_B from the surrounding continuum across its boundary Γ_B. This prevailing irreversible energy flow is characterized by the flow integral $\mathcal{F}$ (3.280) across Γ_B, which can be converted into a path-independent integral (3.282) over an external contour Γ and the enclosed area A. Moran and Shih [5] have shown that $\tilde{J}$, T^* and $\hat{J}$ are more or less special cases of the flow integral $\mathcal{F}$ derived by them for crack propagation. Now, the choice of model for the process zone is decisive. If we stay with the continuum-mechanical / plasticity theory, the energy flux $\mathcal{F} \equiv \tilde{J} \equiv T^* \equiv \hat{J}$ will inevitably become zero because the process zone $\Gamma_B = \Gamma_\varepsilon \to 0$ is only a singular point-the crack tip. Various authors [11, 12] have tried to introduce a process zone A_B of *finite* size, which either moves with the crack tip unchanged as a whole or elongates with the crack as a strip (see Fig. 7.7). Then the T^*- or $\hat{J}$-integral is calculated along its fixed boundary $\Gamma_B \cong \Gamma_\varepsilon \neq 0$, which actually provides a finite, path-independent J-integral value during ductile crack propagation. Yet this value is a function of the arbitrarily set contour size Γ_B and thus is not a true fracture parameter but rather only an artifact of the model. Consequently, it was shown in [13] that with a shrinking process zone size $\Gamma_B \to 0$, the T^*-integral (as well as J, $\tilde{J}$ and $\hat{J}$) vanishes

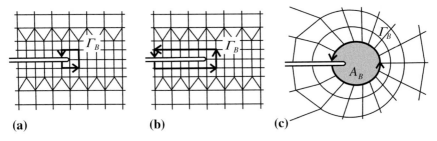

(a)   **(b)**   **(c)**

Fig. 7.7 Fictitious process zones as **a** co-moving or **b** strip-shaped contour Γ_B, **c** true physical model of a process zone

towards zero. The true cause of this calamity is again the continuum model itself, i.e. the process zone A_B may not be defined merely fictitiously as an integration contour, but must be discretely represented with other material-mechanical models! For example, a simple linear form of the process zone is the cohesive zone model, which is addressed more closely in Chap. 8. Modern damage-mechanical models are well-founded on material's microstructure and physics. They allow to describe failure mechanisms in a realistic process zone A_B of finite extension. Then, a genuine physical meaning is attributed to Γ_B, and the flux integrals $\mathcal{F}$ or $\tilde{J}$, T^* and $\widehat{J}$ become meaningful.

> *Note:* In case the classical J-integral (7.7) is still used for ductile crack growth, because the FE-code has no other options for example, the following must be taken into consideration: As soon as the crack has left the restricted J-controlled initial phase, J becomes extremely path-dependent and does not arrive at a finite saturation value with paths placed more tightly around the crack tip $\Gamma \to 0$. Instead, this »crack tip value« $J_{\text{tip}} = J(\Gamma \to 0)$ becomes zero! If we wish to establish a relationship with testing standards for J-Δa-curves as presented in Sect. 3.3.8, then the outermost converging integration paths Γ must be used [13, 14].

7.4 Examples

7.4.1 Compact-Tension Specimen

The FE-techniques introduced above will be first clarified using the 2D example of the compact tension specimen (see Fig. 3.12). The test body investigated has the following dimensions: width $w = 50$, height $h = 60$, thickness $B = 25$, crack length $a = 30$ (everything in mm). A state of plane strain is assumed. Due to symmetry, only the upper half is meshed, and on the ligament in front of the crack the displacement $u_2 = 0$ is set. Figure 7.8 shows the utilized FE-mesh, which was designed according to the recommended rules for elastic-plastic analyses of cracks at rest. Thus, 16 collapsed quadrilateral elements were placed around the crack tip (see Fig. 7.9). The size of the elements was $L = 0.05\,\text{mm} = 0.00167\,a$. This was followed by a very fine discretization of the entire specimen.

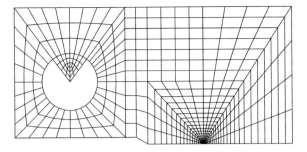

Fig. 7.8 FE-mesh of the CT-specimen (561 quadrilateral 8-noded elements, 1821)nodes

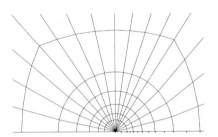

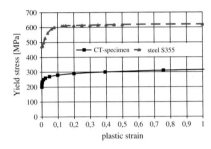

Fig. 7.9 Mesh at crack tip with 16 collapsed crack tip elements

Fig. 7.10 Yield curves for the CT-specimen and the plate tension test

Note: In elastic-plastic analyses, we have to mesh with sufficient fineness not only the crack area but also all those areas in which the plastic zone is formed. Otherwise the global plastic deformations and fracture parameters such as the J-integral will be underestimated.

The CT-specimen is loaded through bolts in pin holes by the force F, which actually represents a contact problem. For the sake of simplicity, the bolt is modeled as a firmly fixed quadrant at whose tip the loading is imposed. In order to avoid unrealistic, exaggerated plastic strain at the force application point, purely elastic material properties with a high elastic modulus are assigned to the quadrant.

Note: In elastic-plastic problems, individual forces or bearings should never be set at only one node but should rather be applied by means of contact models or elastically stiffened sub-domains. Also, it is more advantageous to impose a monotonously increasing displacement instead of the force because in this way the convergence of the solution algorithm above the limit load is simplified. The magnitude of force is obtained from the bearing reaction at the point of action.

In the example of the CT-specimen, a x_2-displacement q is prescribed from 0 to 1 mm. The material is a ductile steel with the yield curve shown in Fig. 7.10 (yield stress−plastic strain) as well as $E = 210000$ MPa and $\nu = 0.3$. Figure 7.11 a shows the formation of the plastic zones with increasing loading drawn on the deformed model. First, a limited plastic zone is formed at the crack tip, which then propagates completely across the ligament to a plastic hinge. The plastic limit load F_L is hereby reached, which leads to a flattening of the force-displacement curve (Fig. 7.13). The sequence of Fig. 7.11b illustrates the shape of the plastic core zone ($\varepsilon_v^p > 0.5\% \,\widehat{=}\,$ grey) directly at the crack.

Deformations at the crack tip are depicted in Fig. 7.12a, b, where we distinguish between a geometrically linear and a non-linear analysis. Considering large deformations, the blunting of the crack tip as a result of plastically highly strained elements is modeled much better than with infinitesimal strains. For this reason, a geometrically non-linear analysis is absolutely necessary to determine the crack tip opening precisely, whereas the global deformation behavior is hardly affected, as Fig. 7.13 proves. Numerical effort increases considerably (in this example, by a factor of four) in the case of large deformations due to the finer load-step incrementation required.

The J-integral is calculated according to Eq. (7.13) (without correction terms) on 15 element rings around the crack tip having a distance of $R = 0.0016 - 0.2\,a$. The results are again shown for a geometrically linear and non-linear analysis in Figs. 7.15a, b. In the SSY range, J grows quadratically with loading q, but for LSY a linear relation is observed. Assuming small deformations, J is independent of the integration path. If large deformations are being considered on the other hand, we observe a strong path-dependence. In particular, the values of the paths lying very close to the crack tip clearly decline because the $1/r$-singularity in the specific stress work density $\check{U}$ is lost due to crack blunting. As we found in Sect. 7.3, under these conditions only the J_{ff} calculated from the far field provides a meaningful fracture-mechanical parameter in the context of EPFM. It is thus necessary to determine the J-values from contours as far as possible from the crack tip or from the associated domain integrals. There the integrals also converge towards a common value (Fig. 7.15b) J_{ff}, which agrees with J from the geometrically linear analysis (Fig. 7.15a).

(a)   **(b)**

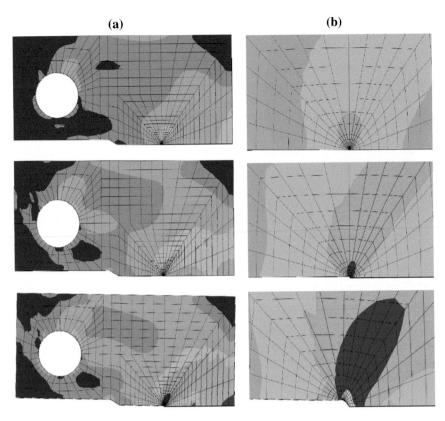

Fig. 7.11 Shape of the plastic zones in the CT-specimen at loading stages $q = 0.1, 0.22$ and 1.0 mm. **a** general view (*left*), **b** crack region (*right*)

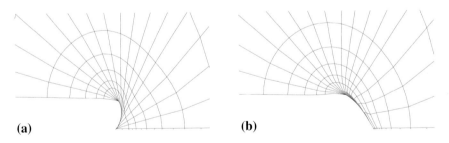

(a)   **(b)**

Fig. 7.12 Blunting of the crack tip in the FE-model assuming **a** small and **b** large strains (scale 1 : 1)

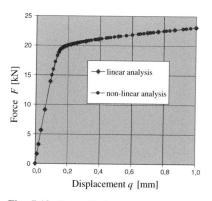

Fig. 7.13 Force-displacement diagram of the CT-specimen

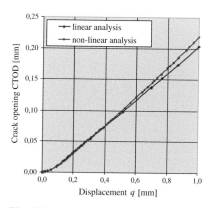

Fig. 7.14 Crack tip opening displacement CTOD as a function of displacement q of force application point

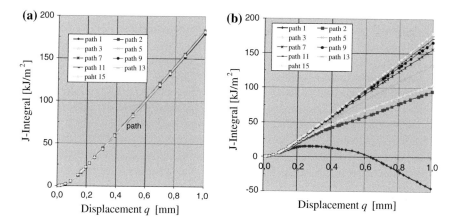

Fig. 7.15 The J-integral as a function of loading with **a** small and **b** large deformations

7.4.2 Tensile Plate with Surface Crack

In the context of a testing program [15], strip-shaped segments were separated from a pipe DN800 and tested with the tensile test (see Fig. 7.16). Semi-elliptical surface cracks with the axial ratio $a : c = 1 : 3$ were placed in the center of the inner surface of these plate tension specimens. Three different crack sizes $a : c = 3 : 9$, $6 : 18$ and $9 : 27$ mm were prepared, which amount to 0.25, 0.5 and 0.75 of wall thickness h.

Since the tension specimen, its loading and bearing conditions have two planes of symmetry, the center plane and the longitudinal plane, only the quarter hatched in Fig. 7.16 needs to be modeled. This area was meshed with 20-node hexahedral elements, whereby an extreme refinement in the crack region was carried out in

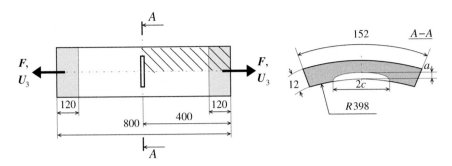

Fig. 7.16 Tension plate specimen with semi-elliptical surface crack (in mm)

order to calculate the inhomogeneous stress and deformation state arising there with sufficient precision. In total, the FE-model consisted of 6585 hexahedral elements and 31954 nodal points. Figure 7.17b provides a general view of the FE-model. The boundary conditions of the tension specimen correspond to the standard bearing and symmetry conditions.

The curved pipe segments were clamped with clamping jaws and stretched. We therefore assume in the FE-model that the contact areas between the specimen and the clamping jaws (gray fields in Fig. 7.16) move rigidly in the longitudinal direction, i. e. monotonously growing, identical node displacements U_3 are applied here. In the FE-calculation, the tensions for all load increments result from the reaction forces at the clamping jaws. The isotropic elastic-plastic material behavior of the steel S355 was determined with the help of standard tension specimens taken from this pipe section. The obtained yield curves (yield stress–plastic strain) are shown in Fig. 7.10. Further material parameters included: $R_{p0,2} = 472\,\text{MPa}$, $R_m = 610\,\text{MPa}$, $E = 210000\,\text{MPa}$ and $\nu = 0.3$. In the FE-calculation, the influence of large deformations on the geometry and loading was taken into consideration.

Figure 7.17a shows a detailed view of the mesh of the half crack configuration for crack depth $a{:}c = 6{:}18$. Similar discretizations were generated for the two other crack depths $a{:}c = 3{:}9$ and $a{:}c = 9{:}27$. Along the crack front, the elements were concentrated in a fan-shape in a tube of 20 segments. The smallest elements directly

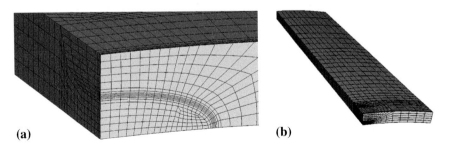

(a)   **(b)**

Fig. 7.17 **a** FE-mesh at the crack tip 6:18, **b** general view of the meshed model

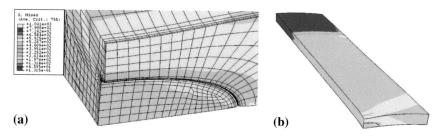

(a)   **(b)**

Fig. 7.18 Elastic-plastic deformation and V. MISES-stress of the tension specimen with crack 6:18.
a Detailed view of the crack **b** general view

at the crack tip have a length of $L = a/50$. Along the entire crack front, special crack
tip elements with a $1/r$-strain singularity are arranged, for which 20-node hexahedral
elements collapsed to pentahedral elements are used. To calculate the J-integral, four
integration paths were defined around each segment of the crack front. In this way,
$J(s)$ can be determined along the crack front s for the entire loading course. The
calculation results show that the J-vales are nearly independent of the integration
path. Only the narrowest path declines somewhat for known numerical reasons, which
is why the average value of the three outer paths was taken as the result for J.

Figure 7.18b represents a general view of the deformed tension specimen with
crack $a{:}c = 6{:}18$ and Fig. 7.18 a a detail near the crack front. The colors correspond
to isoareas of the MISES equivalent stress. All stresses above the yield strength $R_{p0,2}$
indicate plastified domains. As a whole, the calculation results lead to the following
conclusions:

- All specimens reach the fully plastic state. Proceeding from the cracks, pronounced
  plastic zones arise that extend diagonally over the entire cross-section, as is visible
  in Fig. 7.18b.
- Excess of the plastic limit load F_L is characterized by a distinctive kinking of
  all force-elongation curves of the tension specimens. As long as plastification is
  restricted by the surrounding elastic domain, the curves progress almost linearly.
  This is especially noticeable in the case of the calculated crack opening displacements (COD) of the cracks on the specimen surface (see Fig. 7.19). Notch opening
  (compliance) understandably increases with crack size. At the same time, the transition from elastic to fully plastic state becomes softer. The crack opening (COD)
  is an important measured variable, from which often the crack tip opening (CTOD)
  is extrapolated.
- The cracks are heavily blunted via plastic deformations, as we can see in Fig. 7.18a.
  The crack tip opening δ_t was evaluated using the secant method ($\pm45°$) at all
  positions s of the crack front on perpendicular sectional planes of the FE-mesh.
  The maxima occur on the apex of the semi-elliptical cracks and are plotted in
  Fig. 7.20 versus tensile force. Also included is the fracture toughness CTODi $=$
  $\delta_i^{SZH} = 142\,\mu\text{m}$. At this value ductile crack growth should initiate according to
  the CTOD-criterion.

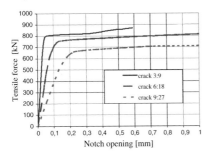

Fig. 7.19 Diagram: tensile force—max. crack opening (COD) on the specimen edge (crack center)

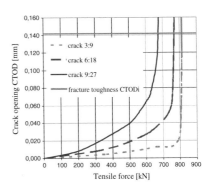

Fig. 7.20 Max. crack opening displacements (CTOD) as a function of loading for all three crack depths

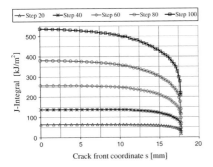

Fig. 7.21 Course of the J-integral along the crack front with increasing loading for crack 6:18

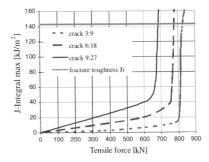

Fig. 7.22 Max. values of the J-integral as a function of loading for all three crack depths

According to the J-integral concept of ductile fracture mechanics, stable ductile crack growth starts, when the value of J exceeds the physical initiation parameter J_i ($J > J_i$). Figure 7.21 shows the calculated J-integral values along the crack front as a function of load (steps). The maximum J-values arise in each case at the apex of the semi-elliptical crack front. Figure 7.22 summarizes the maximum values J_{max} for all three crack configurations. A moderate increase of J can be observed in the range of limited plastic deformation (SSY). As soon as the tension exceeds the corresponding plastic limit load, the values of J increase sharply in all three analyzed cases. For comparison, the fracture toughness parameter $J_i = 143\,\text{kJ/m}^2$ is plotted in Fig. 7.22. Thus, the fracture criterion provides the critical forces F_i or elongation U_3 of the specimens for a crack initiation: crack 3:9: $F_i = 814\,\text{kN}$, $U_3 = 6.8\,\text{mm}$, crack 6:18: $F_i = 779\,\text{kN}$, $U_3 = 1.53\,\text{mm}$, crack 9:27: $F_i = 686\,\text{kN}$, $U_3 = 1.31\,\text{mm}$. With increasing crack length, a smaller force F_i or a larger elongation U_3 is necessary to initiate crack growth.

The predictions of elastic-plastic FE-analyses regarding crack initiation were confirmed by the experimental results of the plate tension tests [15].

References

1. Brocks W, Cornec A, Scheider I (2003) Computational aspects of nonlinear fracture mechanics. In: Milne I, Ritchie RO, Karihaloo B (eds) Comprehensive structural integrity—numerical and computational methods, vol 3. Elsevier, Oxford, pp 127–209
2. Newman JC, James MA, Zerbst U (2003) A review of the CTOA/CTOD fracture criterion. Eng Fract Mech 70:371–385
3. Scheider I, Schödel M, Brocks W, Schönfeld W (2006) Crack propagation analyses with ctoa and cohesive model: comparison and experimental validation. Eng Fract Mech 73:252–263
4. Schwalbe KH (1995) Introduction of δ_5 as an operational definition of the CTOD and its practical use. In: ASTM STP 1256, American Society of Testing and Materials, pp 763–778
5. Moran B, Shih CF (1987) Crack tip and associated domain integrals from momentum and energy balance. Eng Fract Mech 27:615–642
6. Carpenter WC, Read DT, Dodds RH (1986) Comparison of several path independent integrals including plasticity effects. Int J Fract 31:303–323
7. Atluri SN, Nishioka T, Nakagaki M (1984) Incremental path-independent integrals in inelastic and dynamic fracture mechanics. Eng Fract Mech 20(2):209–244
8. Atluri SN (1982) Path-independent integrals in finite elasticity and inelasticity, with body forces, inertia, and arbitrary crack-face conditions. Eng Fract Mech 16:341–364
9. Kishimoto K, Aoki S, Sakata M (1980) Dynamic stress intensity factors using $\hat{J}$-integral and finite element method. Eng Fract Mech 13:387–394
10. Aoki S, Kishimoto K, Sakata M (1984) Energy flux into the process region in elastic-plastic fracture problems. Eng Fract Mech 20:827–836
11. Nishioka T, Fujimoto T, Atluri SN (1989) On the path independent t^* integral in nonlinear and dynamic fracture mechanics. Nucl Eng Des 111:109–121
12. Brust FW, Nishioka T, Atluri SN, Nakagaki M (1985) Further studies on elastic-plastic stable fracture utilizing the T^* integral. Eng Fract Mech 22:1079–1103
13. Brocks W, Yuan H (1989) Numerical investigations on the significance of J for large stable crack growth. Eng Fract Mech 32:459–468
14. Brocks W, Scheider I (2003) Reliable J-values, numerical aspects of the path-dependence of the J-integral in incremental plasticity. Materialprufung 45:264–275
15. Kuna M, Wulf H, Rusakov A, Pusch G, Hübner P (2003) Entwicklung und Verifikation eines bruchmechanischen Bewertungssystems fnr Hochdruck-Ferngasleitungen. In: 35.Tagung Arbeitskeis Bruchvorgänge, DVM, Freiburg, pp 153–162

Chapter 8
Numerical Simulation of Crack Propagation

Prediction of the crack propagation process is of great importance for many fracture mechanical issues. Numerical simulation offers outstanding possibilities and has become an indispensable tool in performing this task. In particular, there is high technical interest in modeling subcritical growth of fatigue cracks, stable crack growth in ductile materials and unstable dynamic fracture processes. As explained in Sects. 3.2–3.5, fracture mechanics provides criteria and principles for these cases that specify:

- at what load level crack propagation starts,
- in what direction θ_c crack propagation occurs,
- how large the amount Δa of crack propagation is.

Thus the object of numerical simulation is to implement these laws in the context of the finite element method by appropriate solution algorithms. Within the framework of continuum mechanics, crack propagation means a change of the BVP, because thereby new boundaries (crack faces) are generated with altered conditions. Consequently, in the FEM analysis we are faced with the problem that the spatial discretization has to be cut and adapted consecutively for the propagating crack. For this purpose various techniques have been developed, some of which are presented and discussed below.

In this procedure, crack propagation is usually modeled as a temporal sequence of BVP with *discrete*, growing crack lengths a_i, and a material separation along the crack increment Δa_i is assumed as *discontinuous* (sharp discontinuity). Of course, in reality the crack propagation occurs *continuously*, and the failure of the material in the process zone is a *continuous* process, too. Some numerical approaches try to simulate these phenomena as well. To avoid any misinterpretation, one has to distinguish clearly between the material model and the numerical technique, although both are closely intertwined in the simulation.

M. Kuna, *Finite Elements in Fracture Mechanics*, Solid Mechanics and Its Applications 201, DOI: 10.1007/978-94-007-6680-8_8,
© Springer Science+Business Media Dordrecht 2013

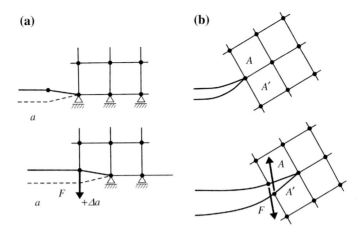

Fig. 8.1 Schematic representation of node separation for **a** symmetric and **b** general loading

8.1 Nodal Release Technique

The simplest method to simulate crack propagation in a FEM-mesh consists in a *disconnection of a node*, so that the crack is enlarged by an increment Δa along the element edge up to the next node. This is shown in Fig. 8.1 both for the symmetrical mode I case and the general mixed-mode loading case. Under mode I (Fig. 8.1a) crack propagation proceeds always along a symmetry line, so that only the bond of the normal displacement has to be released at the crack tip node. In the general case (Fig. 8.1b) the crack tip node and the element edge must at first be doubled on Δa and afterwards separated, i. e. an additional node appears in the FEM-model and the nodal correlation of crack elements A and A' has to be modified. At the highly loaded crack tip node there exists in terms of FEM an equilibrium of forces, which after separation has to be split into a couple of equal and opposite directed forces F (case b) or one reaction force (case a). An abrupt canceling of these binding forces would lead to a spontaneous unloading in the crack tip region, which is physically not correct and might cause numerical problems. Therefore it is advisable to run the unloading process gradually. In Fig. 8.2 well-tried functions are depicted that lower the force F from its maximum initial value F_0 down to zero, which implies a shifting of the true crack tip position x/d.

Let us assume that in the course of a FEM analysis the fracture criterion is reached at the current crack tip. For the required direction θ_c the best possible orientation is selected in the FEM mesh. The amount of crack propagation Δa is linked to the element size L. Then the algorithm of *nodal release technique* follows:

(a) Freeze (keep constant) the external loads
(b) Determine the sectional force F_0 from the bearing reaction (case a) or with the methods described in Sect. 5.5.2 d) (case b)
(c) Release the nodal bonding (mode I case a) or introduce a double node (mixed-mode case b) at the crack tip

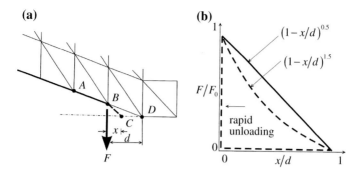

Fig. 8.2 Continuous unloading of the reaction force at the crack tip node

(d) Replace the kinematic bonds at the crack tip node by an equivalent sectional force F_0

(e) Release stepwise the force from F_0 to zero, in this way making the crack faces stress-free

(f) Check, whether the fracture criterion is fulfilled at the new crack length:

$$\text{yes} \rightarrow \text{unstable crack propagation, goto point 2}$$
$$\text{no} \rightarrow \text{continue the FEM-analysis with next load step}$$

Actually, the technique of nodal release can then be applied only if the path of crack propagation is known in the structure. In this case the elements of the mesh can be arranged in an appropriate manner with respect to the required size and orientation. Otherwise the solution becomes strongly mesh-dependent. The technique is particularly suitable in combination with standard finite elements, which is why the fracture parameters need to be determined with robust methods such as the J-integrals. Nevertheless, the technique is useful for a known fracture path, whereby the kinematics of crack propagation is retraced in the simulation to calculate the stress state at the crack. This approach is often applied to analyze fracture mechanics specimens in order to determine the corresponding fracture mechanics parameters (J-integral, K-factors, CTOD et al.) from the measured relationship between crack length (usually mode I case) and load (force or displacement) by means of simulation. It is important to note that the energy dissipation by the fracture process itself is not taken into account by the nodal release technique introduced here !

8.2 Techniques of Element Modification

8.2.1 Element Splitting

A more powerful but elaborate technique for the FEM simulation of crack propagation is the splitting of finite elements. Depending on element type and problem statement there exist many variations. The methodology will be exemplified at a two-dimensional discretization with six-node triangular elements. The starting point is again a critical crack of length a at the position P in the current FEM mesh. The direction θ_c and amount Δa of crack propagation are given by fracture mechanics. Therefore, the new position P_{new} of the crack tip in the FEM mesh is known and has to be realized precisely by element divisions. This goal can be achieved by various *subdivision algorithms* that are illustrated in Fig. 8.3. Hereby, from a »father element« two or four »child elements« are created of the same type, whereas the surrounding elements are kept unchanged. The combination of algorithms is illustrated in Fig. 8.4 as an example with a typical mesh. The newly added elements are drawn as dashed lines.

The *element splitting technique* allows the exact numerical simulation of crack paths to be well-founded by fracture mechanics, which is a great *advantage*. As a *disadvantage* the inevitable interference in the data structure of the FEM model has to be mentioned, which is relatively complicated and requires availability of the source code. Moreover, for inelastic material behavior the stresses and state variables must be transferred from the solution of the »father elements« to the newly generated »child elements«. For 3D crack configurations this technique turns out to be quite difficult due to the spatial geometry and topology [1].

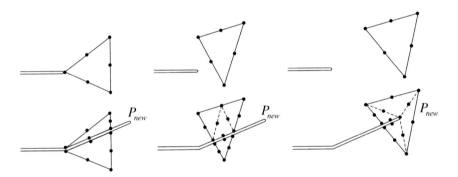

Fig. 8.3 Various algorithms to subdivide triangular elements

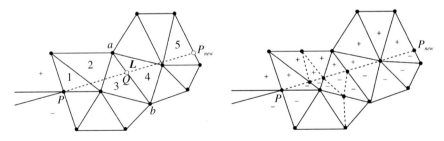

Fig. 8.4 Sample application for the technique of element splitting

8.2.2 *Element Elimination Technique*

In this technique, the propagation of a crack is simply realized in such a way that the most highly stressed finite element at the crack tip is removed from the FEM model. Many commercial FEM-programs offer a so-called *element elimination technique* or *element death option*. As failure criterion any strength hypothesis can be selected out of classical engineering mechanics (v. Mises, maximum principal stress, …), fracture mechanics (K, J, δ, …) or damage mechanics (microcrack density, porosity, …). The impressive simplicity of this technique is opposed to various disadvantages. Obviously the result of simulation depends on the size of the finite elements and on the shape of the FEM-mesh. The smaller the elements at the tip of the main crack, the higher is their stress and the sooner their failure. What is more crucial, both the balance equation of energy (no dissipation) and the mass conservation are violated.

The method is therefore less suited for simulating the propagation of macrocracks, where dominant singularities prevail, but is better used for modeling of various damage mechanisms in the microstructure of a materials such as e. g. the formation of microcracks or the growth of micropores. Figure 8.5 shows a typical application of the element elimination technique to a tool steel with brittle carbide particles [2]. The considered cut-out of the microstructure $100 \times 100\,\mu$m was reproduced in fine detail with finite elements. With increasing load, microcracks appear at the carbides and coalesce later to form a macrocrack.

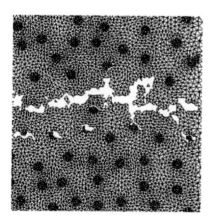

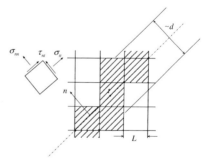

Fig. 8.5 Simulation of crack propagation in a tool steel using the element elimination technique [2]

Fig. 8.6 Simulation of facture by means of adjusted orthotropic material stiffness

8.2.3 Adapting Element Stiffness

This technique has been developed to model fracture processes in brittle materials such as concrete, stone or ceramic. For it the name *smeared crack model* was coined, which is best interpreted as a homogenized microcrack model. If in those materials the maximum principal stress exceeds a critical tensile strength then a narrow band of microcracks is formed perpendicular to the principal stress direction as outlined in Fig. 8.6. Since the material can carry only low tensions σ_{nn} in the normal direction, an orthotropic anisotropy of the elastic properties appears, characterized by the basis vectors $(\boldsymbol{n}, \boldsymbol{t})$ in the normal and tangential directions with respect to the microcrack band. The reduced material stiffness values are described in this coordinate system using a modified elasticity matrix (see (4.46)), which has the following form for a plane stress state:

$$\boldsymbol{\sigma} = \tilde{\boldsymbol{C}}\boldsymbol{\varepsilon}, \qquad \begin{bmatrix} \sigma_{nn} \\ \sigma_{tt} \\ \tau_{nt} \end{bmatrix} = \frac{1}{1-\nu^2} \begin{bmatrix} (1-w)E & 0 & 0 \\ 0 & E & 0 \\ 0 & 0 & \mu\frac{(1-\nu)}{2}E \end{bmatrix} \begin{bmatrix} \varepsilon_{nn} \\ \varepsilon_{tt} \\ \gamma_{nt} \end{bmatrix} \qquad (8.1)$$

E and ν denote the elastic constants of the intact material. The density of the microcracks is specified by a *damage variable* $w(\varepsilon)$ that evolves with deformation ε from the value $w = 0$ (undamaged initial state) up to the final value $w = 1$ (total failure). Thus the modulus of elasticity in the normal direction diminishes with growing damage from the initial value E down to zero, whereas in the direction tangentially to the microcrack band the stiffness sustains unaltered, see (8.1). Due to surface roughness and material bridges between the crack faces, the shear stiffness of the microcrack band does not disappear completely under mode-II loading, but is reduced in (8.1) by a shear correction factor $\mu \approx 0.2$. Further details can be found in [3, 4].

In order to implement this technique in the FEM-algorithm, only the elasticity matrix $\boldsymbol{C}$ in the integration points of the involved elements must be exchanged by the modified matrix $\tilde{\boldsymbol{C}}(\boldsymbol{\sigma}, w)$ and be transformed into the global $\boldsymbol{x}$ coordinate system. The FEM mesh does not need to be changed, which is the great *advantage* of this approach. On the other hand, a *disadvantage* consists in a certain mesh dependency, in particular the width d of the microcrack band is determined arbitrarily by the choice of element size L. In contrast to the technique of element splitting, where the crack is represented by a sharp geometric discontinuity, material separation is modeled in this method as a geometrically continuous transition inside the microcrack band, which corresponds to an extreme localization of strain ε_{nn} on the macroscopic level. The technique of adjusted element stiffness was successfully applied mainly in fracture mechanics of concrete [5].

8.3 Moving Crack Tip Elements

A disadvantage of the previously described techniques is that only regular finite elements are employed. Of course, the use of special *crack tip elements* (see Chap. 5) would be far more advantageous and precise. This requires however a different technique, because these special elements always surround the crack tip and thus must be moved with it. Various variants of local entrainment of crack tip elements in a FEM-mesh are known that have been developed by Nishioka [6] in particular for dynamic crack problems. Figure 8.7 illustrates the procedure in the example of a mode-I crack propagation along a line of symmetry. The region moving with the crack tip elements is denoted by A, region B comprises the adjacent regular elements to be modified and C describes the remote unchanged mesh. As the sequence of images shows, the local adaptation of the mesh is exclusively made in region B, which then jumps forward by an »element grid«. As a crack tip element, either a hybrid element (Sect. 5.3) or a quarter-point element (Sect. 5.2.2) can be used, which enables an exact calculation of the stress intensity factors despite very coarse meshes, because of their built-in crack singularities.

Besides this essential *advantage*, the technique of *moving crack tip elements* distinguishes itself from other methods by the fact that the length of the crack increments Δa is not linked with the FEM discretization but can be adjusted continuously on the desired size. With some hybrid elements, the increment Δa can even be varied in the element itself, see Fig. 5.15. These benefits are paid for by a higher effort of continuous remeshing, i. e. at least in region B the new element stiffnesses have to be built and assembled. Certain *limitations* result from the fact that powerful crack tip elements are only available for static and dynamic crack problems in LEFM. In other cases (e. g. EPFM) the region A also has to be discretized with standard elements.

With negligible additional effort this technique can also be extended to curvilinear crack propagation under mixed-mode loading. Figure 8.8 shows an example of a circular core region A enclosing the moving crack tip by quarter-point elements, which is connected to the outer network by a transition region (B) of concentric ring

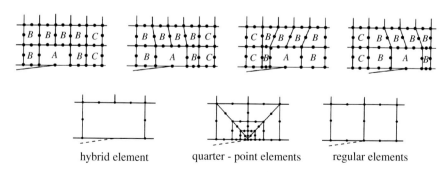

hybrid element   quarter - point elements   regular elements

Fig. 8.7 Simulation of crack propagation under mode-I by shifting special elements with the crack tip

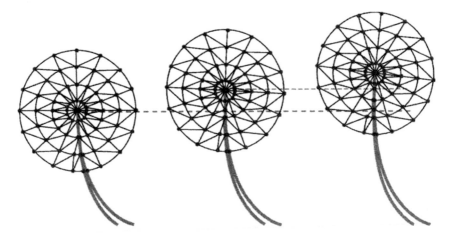

Fig. 8.8 Simulation of crack propagation under mixed-mode loading using moving quarter-point elements in a circular region [7, 8]

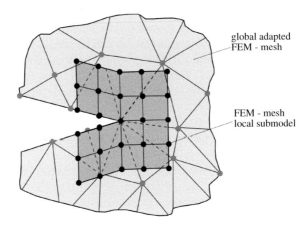

Fig. 8.9 Combined technique of global crack discretization and local submodel with a special element arrangement for crack analysis [10]

elements [7, 8]. For the mesh adaptation one can take advantage of automatic mesh generators such as DeLaunay-algorithm [9] for triangles.

An interesting technique for mixing of standard and crack tip elements was proposed in [10] for plane and spatial crack propagation. A simple remeshing strategy for regular elements is combined with the better accuracy of crack tip elements. The algorithm is explained in the following with reference to Fig. 8.9:

(a) Remeshing of the crack region with standard elements (2D triangles or 3D tetrahedrons) for the current crack increment
(b) Perform a FEM-analysis with this global mesh

(c) Define a submodel around the crack tip that consists of optimal arranged elements (regular 2D quadrilaterals or 3D hexahedrons or corresponding quarter-point elements, respectively)
(d) Transfer the displacement field from the global analysis to the boundary nodes of the submodel
(e) Perform a FEM-analysis with the local submodel and determine the fracture parameters either by applying the crack closure integral (Sect. 5.5) or the quarter-point rule (Sect. 5.2.3).

8.4 Adaptive Remeshing Strategies

8.4.1 Error-Controlled Adaptive Meshing

In the previous Sect. 8.3, we used an *automatic meshing* of the whole crack configuration or selected regions performed by means of conventional mesh generators. In contrast to that we call it an *adaptive meshing*, when the algorithm performs an adaptation of the discretization on the basis of the FEM-solution itself. This requires a criterion to decide on a local refinement or coarsening of the mesh based on a measure of the numerical quality at every location. For this purpose, there exist various a posteriori error estimators, see e. g. [9, 11], which either use an energy error norm between the local FEM-result and a global improved FEM-based approximation or which evaluate the residuals of field quantities on the element boundaries $\tilde{S}_e$. Since in the displacement-based FEM the tractions on the edges between neighboring elements are not exactly reciprocal (4.6) but show a discontinuity (see Fig. 4.2), the quadratic norm of the traction jump $\Delta t_i = t_i^+ + t_i^- = (\sigma_{ij}^+ - \sigma_{ij}^-) n_j$ is employed as error indicator for each element e:

$$\eta_e^2 = \sum_{k=1}^{N_K} l_k \| \Delta t_i \|^2 . \tag{8.2}$$

Hereby, the sum is taken over all N_K edges $\tilde{S}_e$ of an element weighted by their length l_k. The total error of the numerical solution obtained with the given discretization is estimated by the mean value of all elements

$$\eta = \sqrt{\sum_{e=1}^{n_e} \eta_e^2} . \tag{8.3}$$

Now, the adaptive modification of the mesh is controlled by comparing the local element error with the mean value. Elements with errors above average $\eta_e^2 > \alpha_{\text{fein}} \eta^2$ will be refined ($\alpha_{\text{fein}} \approx 0.8$), while elements with low error $\eta_e^2 < \alpha_{\text{grob}} \eta^2$ will be coarsened again ($\alpha_{grob} \approx 0.001$). This way, an error-controlled gradual improvement of the discretization is achieved in a series of FEM analyzes, until the total error indicator η is below a tolerance limit.

The adaptive meshing algorithm can be realized very effectively in combination with an iterative FEM-solver (e. g. conjugate gradient method), if the data structures and results of the previous refinement step are used for preconditioning [12].

8.4.2 Simulation of Crack Propagation

If the adaptive meshing algorithm is applied to crack problems, then the FEM-mesh in the crack tip region will be automatically refined in a very comfortable manner, since here the local error of the numerical solution is highest due to the stress singularity. To calculate the stress intensity factors, in principle all FEM-techniques explained in Chap. 5 can be used. Since the automatically generated mesh mostly consists of triangular elements with a rather irregular arrangement around the crack tip, the calculation of J as an equivalent domain integral is best suited. In the case of mixed-mode loading it is recommended to employ the interaction integral presented in Sect. 6.7.2 in order to separate K_I and K_{II}. When should the successive adaptive refinement at the crack be terminated? Experience has shown, that monitoring of the convergence behavior of the J-integral or K-factors is advisable to control the quality of the mesh in terms of fracture parameters. If the relative improvement of the results is below a prescribed empirical limit, the solution is accurate enough.

The thus obtained FEM-solution forms the basis to go forward to simulate crack propagation, i.e. to decide on the size and the direction of the crack increment. Thereupon, a mesh modification follows, which is best done by element division according to Sect. 8.2.1. Starting with the modified mesh of the extended crack, thereafter a new FEM-solution is calculated using the adaptive remeshing technique. The complete algorithm for the crack propagation analysis in combination with the adaptive refinement technique is listed in Fig. 8.10.

As an example the crack propagation in a symmetric tensile specimen is examined, which is loaded in mode I by a prescribed opening displacement on its left side, see Fig. 8.11. The computation starts with a very coarse mesh, cf. Fig. 8.11a. The error-controlled algorithm generates an extremely refined mesh at the crack, so that a sufficiently accurate calculation of the K_I-factor is ensured. Please note that the discretization is also improved at the highly stressed left corners on top and bottom, cf. Fig. 8.11b. The Figure 8.11c and d represent the mesh in various stages of crack propagation. The mesh refinement during the crack propagation can be seen very clearly. After the crack has left its earlier position behind, the discretization at the old crack tip is even coarsened again.

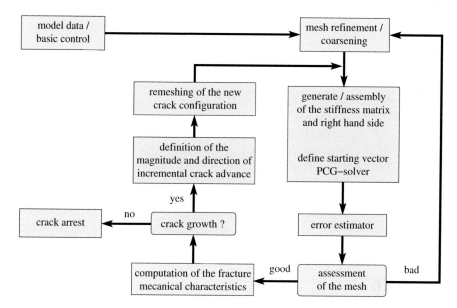

Fig. 8.10 Schematic of crack propagation simulation in combination with an adaptive, error-controlled remeshing strategy

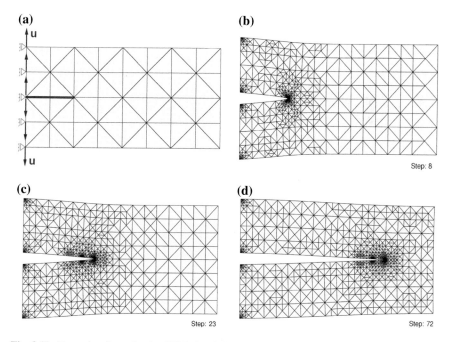

Fig. 8.11 Example of an adaptive FEM-simulation of crack propagation in a tensile specimen: **a** Initial mesh, **b** refinement at initial crack length, **c–d** propagation

8.5 Cohesive Zone Models

8.5.1 Physical Background

The idea of a *cohesive zone model* is based on the assumption that the material's failure process during fracture occurs only in a narrow strip-shaped zone in front of the main crack. According to this approach the damage of the material until its final separation takes place primarily in this limited region, while the rest of the body obeys the common laws of deformation and remains free of damage.

The first model of this kind came from Barenblatt [13], who considered the atomic interaction forces across the faces of an opening crack and referred to this region as *cohesive zone*. In this physically motivated approach the failure is modeled *continuously*. As a consequence the unrealistic stress singularities at the crack tip disappear, which is an essential property of all cohesive zone models. A similar model was developed by Dugdale [14] to simulate a strip-shaped plastic zone ahead of the crack in ductile metal sheets, see Sect. 3.3.3.

Meanwhile, the cohesive zone model has gained widespread applications, which are mainly motivated by typical phenomena of material failure in a narrow band. Thus, in ceramic materials or concrete [5] material bridges can be observed transmitting some forces across the crack faces (Fig. 8.12). Characteristic phenomena of this kind are also present in fiber-reinforced materials or polymers, where interaction forces are formed by fiber pullout or by stretched molecular chains (*crazes*). Also the process zone in ductile fracture can approximately be reduced to a narrow band, where as the result of formation, growth and coalescence of microvoids, a geometric

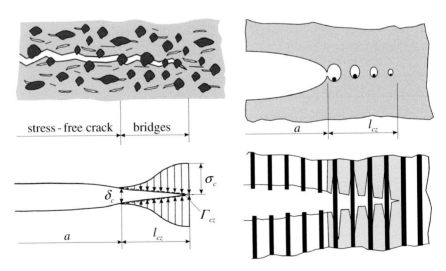

Fig. 8.12 Examples of using cohesive zone models: Brittle heterogeneous materials, ductile dimple fracture, fiber reinforced composites (l_{cz}—length of the cohesive zone)

softening of the remaining ligaments occurs, see Fig. 8.12. Further applications of cohesive zone models are found in adhesive compounds or welded joints. Especially in conjunction with numerical FEM-calculations, cohesive zone models have achieved great importance, since they allow us to easily simulate crack propagation processes. For the first time Needleman [15] has in this way modeled crack propagation in ductile materials by FEM.

The central point of all cohesive zone models is the function that describes the interaction force between the two interfaces (crack faces). This law represents a real local material property that is independent of the external load. The so-called *cohesive law* or *separation law* is usually a relation between the boundary tractions σ and the separation $\delta_n = u_n^+ - u_n^-$ of the interfaces, i. e. the distance between the crack faces. Meanwhile, many proposals for cohesive laws exist in the literature, which differ according to various materials and failure mechanism, see e. g. the overviews of Brocks and Cornec [16, 17]. Some typical shapes are shown in Fig. 8.13. Initially, the stress increases with growing distance up to a maximum that is called the *cohesive strength* σ_c of the material. If the separation has reached a critical *decohesion length* δ_c, then the material is completely separated and no stress can be transmitted.

Integrating the separation law up to failure δ_c yields the area under the curve that corresponds to the dissipated work during a material's separation—the specific fracture energy per surface area $G_c = 2\gamma$ as introduced by Griffith.

$$G_c = \int_0^{\delta_c} \sigma(\delta_n) d\delta_n \quad \text{energy of separation} \tag{8.4}$$

According to Sect. 3.2.5 the separation energy has to be supplied by the local energy release rate $\Delta \mathcal{W}_c$ of the system as expressed by the relationship (3.84). On the other hand, one can evaluate the J-integral $\widehat{=} J_{\text{tip}}$ via (3.100) directly at the crack tip, for which the integration path is placed along the boundary Γ_{cz} of the cohesive zone, see Fig. 8.12. Thus, the first term in (3.100) with U vanishes. For pure mode-I loading the second term contains only the product of normal stresses $-\sigma_{22} = \sigma(\delta_n)$ and opening displacements u_2^+ of the upper crack face, which is repeated with opposite sign ($u_2^- = -u_2^+, +\sigma_{22}$) at the lower crack face. With $\delta_n = u_2^+ - u_2^-$ we get:

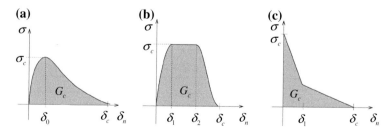

Fig. 8.13 Typical forms of separation laws

$$J_{\text{tip}} = \oint_{\Gamma_{cz}} \left(-\sigma_{ij} \frac{\partial u_i}{\partial x_1} \right) n_j \, ds = \oint_{\Gamma_{cz}} \left(-\sigma_{22} \frac{\partial u_2}{\partial x_1} \right) ds$$

$$= \int_0^{l_{cz}} \sigma(\delta_n) \frac{\partial \delta_n}{\partial x_1} \, dx_1 = \int_0^{\delta_c} \sigma(\delta_n) \, d\delta_n = G_c . \tag{8.5}$$

The crack initiates when $\delta = \delta_c$ is attained at the actual crack tip ($x_1 = 0$), which is equivalent to the CTOD value (Fig. 8.12). Within the frame of LEFM, the correlation with the fracture toughness characteristics $G_c \triangleq G_{Ic} = K_{Ic}^2 / E'$ is immediately given. In EPFM, the separation energy correlates with the physical initiation value $G_c \approx J_i$, provided that the J-integral is sufficiently path independent. Thus, the relationship between cohesive zone model and classical fracture mechanics is established.

For brittle metals the exponential law (8.6) depicted in Fig. 8.13a is suited, which is based on an energy potential of atomic bonds by Rose et al. [18] and has been introduced by Needleman [19] in modified form for cohesive zone models.

$$\sigma(\delta_n) = \frac{G_c}{\delta_0} \frac{\delta_n}{\delta_0} \exp\left(-\frac{\delta_n}{\delta_0} \right), \qquad G_c = e \, \sigma_c \, \delta_0, \qquad (e \approx 2.718) \tag{8.6}$$

In the beginning the linear part dominates until the maximum $\sigma_c = G_c/(e\delta_0)$ of the function is reached at δ_0, after which the curve decays exponentially.

A trapezoidal shape of the separation law (cf. Fig. 8.13b) was proposed by Tvergaard and Hutchinson [20] and Scheider [21] for ductile crack propagation. The parameters for initial stiffness, the region of constant maximum tension and the softening curve, respectively, can be freely chosen. The smooth, differentiable curve shape facilitates the numerical realization.

$$\sigma(\delta_n) = \begin{cases} \sigma_c \left[2 \left(\dfrac{\delta_n}{\delta_1} \right) - \left(\dfrac{\delta_n}{\delta_1} \right)^2 \right] & \text{for } \delta_n < \delta_1 \\[2ex] \sigma_c & \text{for } \delta_1 \leq \delta_n \leq \delta_2 \\[2ex] \sigma_c \left[2 \left(\dfrac{\delta_n - \delta_2}{\delta_c - \delta_2} \right)^3 - 3 \left(\dfrac{\delta_n - \delta_2}{\delta_c - \delta_2} \right)^2 + 1 \right] & \text{for } \delta_2 < \delta_n < \delta_c \end{cases} \tag{8.7}$$

$$G_c = \sigma_c \left(\frac{1}{2} \delta_c - \frac{1}{3} \delta_1 + \frac{1}{2} \delta_2 \right) \tag{8.8}$$

For brittle materials such as concrete, functions with infinite initial stiffness are preferred, because otherwise an artificial numerical softening of the structure would be induced. The decreasing slope is assumed linearly Hillerborg [22] or bi-linearly Bazant [23], see Fig. 8.13c.

$$\sigma(\delta_n) = \sigma_c \left(1 - \frac{\delta_n}{\delta_c} \right) \tag{8.9}$$

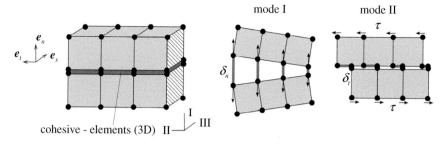

Fig. 8.14 Finite element realization of cohesive zone models

The essential three parameters σ_c, G_c and δ_c of every cohesive law can be relatively well obtained from experiments:

- σ_c from the ultimate strength of smooth or notched tensile specimens,
- G_c from fracture mechanics experiments via K_{Ic} or J_i and
- δ_c from measurements of the fracture process zone.

In the general case, the interfaces do not only move in the perpendicular direction δ_n (mode I), but also shift to each other in the tangential δ_t (mode II) and transversal δ_s (mode III) directions. Then, a vector of separation $\boldsymbol{\delta} = \begin{bmatrix} \delta_n & \delta_t & \delta_s \end{bmatrix}^T$ is defined in the local coordinate system $(\boldsymbol{e}_n, \boldsymbol{e}_t, \boldsymbol{e}_s)$, see Fig. 8.14. In the two-dimensional case two cohesive laws are required for the normal and shear separations, which may be coupled to each other [21, 24, 25]. They link the separation vector $\boldsymbol{\delta}$ with the vector of cohesive stresses $\boldsymbol{t}$:

$$\boldsymbol{t} = \sigma \boldsymbol{e}_n + \tau \boldsymbol{e}_t, \qquad \boldsymbol{\delta} = \delta_n \boldsymbol{e}_n + \delta_t \boldsymbol{e}_t \tag{8.10}$$

$$\boldsymbol{t} = \boldsymbol{f}(\boldsymbol{\delta}) \quad \text{or} \quad \sigma = f_n(\delta_n, \delta_t) \quad \text{and} \quad \tau = f_t(\delta_n, \delta_t) \tag{8.11}$$

Figure 8.15 shows typical cohesive laws for both modes of separation. The shear stresses at sliding change their sign, if the direction of separation δ_t changes. The law for the normal stresses σ is confined to $\delta_n \geq 0$, since otherwise a contact of crack faces occurs that generates reaction forces. In case of compressive loads assumptions have to be made for the friction between both interfaces.

In addition, a method for extending the cohesive laws to local mixed-mode conditions is presented, which traces back to Ortiz and Pandolfi [25]. An *effective separation* δ is introduced, whereby the factor $0 \leq \eta \leq 1$ determines the ratio between shear and tensile stiffness in the cohesive law.

$$\delta = \sqrt{\delta_n^2 + \eta^2 \delta_t^2} \tag{8.12}$$

Every cohesive law can be derived from an associated potential of internal energy ψ_e. For the exponential law in (8.6) the function

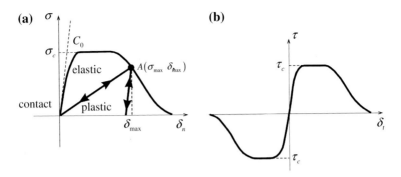

Fig. 8.15 Cohesive laws for separation in the **a** normal and **b** tangential direction

$$\psi_e(\delta) = \int_0^\delta \sigma(\bar{\delta})\,d\bar{\delta} = G_c\left[1 - \left(1 + \frac{\delta}{\delta_0}\right)\exp\left(-\frac{\delta}{\delta_0}\right)\right] \qquad (8.13)$$

exists, from where $\sigma = \partial\psi_e/\partial\delta$ is obtained. Using (8.12) and the chain rule, the two cohesive laws are found as

$$\sigma = \frac{\partial\psi_e}{\partial\delta_n} = \frac{\partial\psi_e}{\partial\delta}\frac{\partial\delta}{\partial\delta_n} = \frac{t}{\delta}\delta_n \quad \text{and} \quad \tau = \frac{\partial\psi_e}{\partial\delta_t} = \frac{\partial\psi_e}{\partial\delta}\frac{\partial\delta}{\partial\delta_t} = \frac{t}{\delta}\eta^2\delta_t, \qquad (8.14)$$

whereby from (8.12) and (8.14) follows:

$$t = \sqrt{\sigma^2 + \frac{1}{\eta^2}\tau^2}. \qquad (8.15)$$

The model can be generalized also to the 3-D case, if the two equitable (at isotropy) shear separations in the interface are added vectorially and if the relationship (8.14) is understood as cohesive law in this effective T-direction:

$$\left.\begin{array}{l}\boldsymbol{\delta}_T = \delta_t\boldsymbol{e}_t + \delta_s\boldsymbol{e}_s, \quad \delta_T = \sqrt{\delta_t^2 + \delta_s^2} \\[2mm] \boldsymbol{\tau}_T = \tau_t\boldsymbol{e}_t + \tau_s\boldsymbol{e}_s, \quad \tau_T = \sqrt{\tau_t^2 + \tau_s^2}\end{array}\right\} \Rightarrow \tau_T = \frac{t}{\delta}\eta^2\delta_T. \qquad (8.16)$$

Finally, we want to note the differences between the loading and unloading regimes of cohesive laws, compare Fig. 8.15. Up to stress values of the cohesive strength σ_c, it is generally assumed that unloading runs backward on the same curve to the origin. After exceeding the maximum this is not correct anymore. From the curve point $A(\sigma_{max}, \delta_{max})$ attained until now, the unloading and eventual re-loading occur along a different path, from which the cohesive law is continued. Thus, δ_{max} has the meaning of an internal variable. The unloading runs either parallel to the slope in the origin C_0, if the cohesive law describes plastic deformation, or it goes straight back to the origin, if the failure is elastic in nature and merely decreases the stiffness.

Loading: $\delta = \delta_{\text{max}}$ and $\dot{\delta} \geq 0$
Unloading: $\delta < \delta_{\text{max}}$ or $\dot{\delta} < 0$

$$-\text{elastic: } t = C_0 \delta, \qquad C_0 = \frac{\partial t(0)}{\partial \delta} = \frac{\partial^2 \psi_e(0)}{\partial \delta^2} \qquad (8.17)$$

$$-\text{plastic: } t = \frac{t_{\text{max}}}{\delta_{\text{max}}} \delta$$

Detailed information on cohesive laws can be found in [16, 17].

8.5.2 Numerical Realization

The FEM model consists of damage-free continuum elements obeying an arbitrary material law, and interface elements that capture the material separation by means of a cohesive zone model, see Fig. 8.14. These interface elements or *cohesive elements* are opening in accordance with the separation law and lose their stiffness, when the normal or tangential separation reach their critical values δ_{nc} and δ_{tc}, respectively. Thereafter, the formerly joined continuum elements will be separated, i.e. the material has failed at this point. The crack is only allowed to propagate along the cohesive elements. If the crack path is not known a priori, then different paths must be provided in the FEM-mesh and in the extreme case, cohesive elements need to be prepared between all continuum elements. Cohesive elements constitute the mechanical interaction between two interfaces. For this, they don't need to have an extension in vertical direction (e_{n}-coordinate). The elements connect pairs of nodes on opposite surfaces of the adjacent continuum elements. Figure 8.16 shows typical linear and planar cohesive elements as they are used for two- and three-dimensional crack problems. In the undeformed state the pairs of nodes lie on top of each other.

The separation of the cohesive interfaces is calculated from the jump of the displacement vector $\boldsymbol{\delta} = [\![\mathbf{u}]\!] = \mathbf{u}^+ - \mathbf{u}^- = [\delta_{\text{n}} \; \delta_{\text{t}} \; \delta_{\text{s}}]^{\text{T}}$. The displacements on the faces of the cohesive elements and thus their difference $[\![\mathbf{u}]\!]$ are set using the same

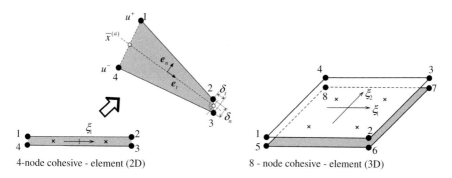

4-node cohesive - element (2D) 8 - node cohesive - element (3D)

Fig. 8.16 Cohesive elements for interfaces in planar and spatial structural models

shape functions as in the adjacent continuum elements, see Sect. 4.3.1. The cohesive elements in Fig. 8.16 have e. g. linear functions, only corner nodes and two integration points per coordinate. The separations $[\![\mathbf{u}]\!]$ are interpolated analogously to Eq. (4.40) by means of the nodal variables $[\![\mathbf{v}]\!]$:

$$[\![\mathbf{u}(x)]\!] = \sum_{a=1}^{n_K} N_a(\xi)[\![\mathbf{u}^{(a)}]\!] = \mathbf{N}[\![\mathbf{v}]\!]. \tag{8.18}$$

In all integration points, the material behavior of the cohesive elements is calculated at each load step, i. e. the cohesive stresses $t([\![\boldsymbol{u}]\!]) = \begin{bmatrix} \sigma_n & \tau_t & \tau_s \end{bmatrix}^T$ between the interfaces are determined from the separations $[\![\boldsymbol{u}]\!]$ via the cohesive law. In order to derive the stiffness relationship for these elements, the principle of virtual work is used. The inner work of a cohesive element reads:

$$\delta \mathcal{W}_{\text{int}}^{(e)} = \int_{S_e} \delta[\![\boldsymbol{u}]\!] \cdot t \, dS. \tag{8.19}$$

Because we are dealing with a non-linear analysis, a linearization of the increment is required at the current load level:

$$\delta \Delta \mathcal{W}_{\text{int}}^{(e)} = \int_{S_e} \delta[\![\boldsymbol{u}]\!] \cdot \frac{\partial t}{\partial[\![\boldsymbol{u}]\!]} \cdot \Delta[\![\boldsymbol{u}]\!] \, dS. \tag{8.20}$$

With (8.18) the separation and its variation $\delta[\![\boldsymbol{u}]\!]$ are replaced by the nodal variables $[\![\mathbf{v}]\!]$, providing the matrix equation:

$$\delta \Delta \mathcal{W}_{\text{int}}^{(e)} = \delta[\![\mathbf{v}]\!]^T \underbrace{\int_{S_e} \mathbf{N}^T \frac{\partial t}{\partial[\![\boldsymbol{u}]\!]} \mathbf{N} \, dS}_{\mathbf{K}([\![\boldsymbol{u}]\!])} \; \Delta[\![\mathbf{v}]\!]$$

$$= \delta[\![\mathbf{v}]\!]^T \qquad \mathbf{K}([\![\boldsymbol{u}]\!]) \qquad \Delta[\![\mathbf{v}]\!]. \tag{8.21}$$

$\mathbf{K}$ is the sought stiffness matrix of the cohesive element, which depends on the current material tangent $\partial t/\partial[\![\boldsymbol{u}]\!]$ of the separation law. The numerical integration over the interface S_e is carried out by means of FEM-standard procedure (Sect. 4.4.3).

In case of large deformations and rigid body rotations of the cohesive elements, it is necessary to determine the separation in a co-rotating Lagrangian coordinate system $(\boldsymbol{e}_n, \boldsymbol{e}_t)$, see Fig. 8.16. If the coordinates of a nodal pair a are denoted by $\mathbf{x}^{(a)}$ in the initial state, then the deformed positions are calculate from the displacements of the upper $\mathbf{u}^+(\mathbf{x}^{(a)})$ and lower $\mathbf{u}^-(\mathbf{x}^{(a)})$ faces of the cohesive element. The coordinates of an average reference point $\bar{\mathbf{x}}^{(a)}$ are therefore obtained as

$$\bar{\mathbf{x}}^{(a)} = \mathbf{x}^{(a)} + \frac{1}{2} \left(\mathbf{u}^+(\mathbf{x}^{(a)}) + \mathbf{u}^-(\mathbf{x}^{(a)}) \right), \tag{8.22}$$

from which the reference system of the element is determined (dashed line in Fig. 8.16). The separations δ_n and δ_t in the normal and tangential directions e_n and e_t are calculated as depicted.

Because of their versatile feasibility, cohesive zone models are also applied for the simulation of crack propagation in fatigue, in dynamic fracture processes or at viscoplastic material behavior. They are best suited to treat interfacial cracks and delaminations in composites or welds. More examples, numerical guidelines and further reading are given in the review articles [16, 17, 26].

8.6 Damage Mechanical Models

At this point damage mechanics models have to be mentioned, since they were developed very intensively and successfully for the simulation of ductile crack propagation in metallic materials in the last two decades. As has been already explained in Chap. 2, the ductile failure originates from micromechanical damage processes in the material. At the beginning of loading, micropores arise that enlarge and grow together during subsequent plastic deformation, which finally leads to the local failure of material at the micro level. In order to describe these processes by continuum mechanics methods, so-called *damage mechanics models* have been created. The formulation of the material laws is done similarly as in the theory of plasticity using phenomenological approaches and thermodynamic principles. Some models are inspired by specific micro-mechanical processes and try to capture them in homogenized form. To quantify the material damage, internal state variables-*damage variables*- are introduced into the constitutive material laws.

As a result of local stresses and plastic deformation, the state of damage in the material increases. This is expressed by an evolution law that quantifies the development of the damage variables. In this way, ductile damage models allow us to simulate, in addition to plastic deformation and hardening, also deterioration and softening in constitutive laws. The main advantage of damage mechanics compared to traditional or fracture mechanical strength hypotheses is that the deformation and failure behavior are now coupled at a local level, i. e. a criterion is provided capturing the current stress state as well as the history of loading and deformation. Local failure is postulated to occur, if the damage variable has attained a critical value.

Therefore, damage mechanical models are qualified exceptionally to simulate the ductile failure in the process zone at the tip of a macrocrack, because the local stress and strain state (e. g. multi-axiality h) can be taken into account. The price for it is a complex structure of the material laws, an increased number of parameters, and a sensitivity to numerical instabilities. The best known ductile damage models stem from Rousselier [27] and Gurson [28]. As an example, we will introduce and apply the GTN-model, which represents a generalization of the work of Gurson by Tvergaard and Needleman [29].

An elastic-plastic continuum is modeled in which spherical cavities (micropores, voids) can develop and grow. The volume fraction of voids f is considered as a measure of material damage and used as damage variable (or the modified quantity f^*). Figure 8.17 shows the underlying conceptual model of a representative volume element, the mechanical response of which has to be formulated at any given macroscopic stress state σ_{ij}. The ductile matrix material deforms according to the laws of v. Mises plasticity. For isotropic hardening, the yield stress of the matrix material is given by $\sigma_M = R(\varepsilon_M^P)$ as a function of equivalent plastic strain ε_M^P. The main item of this model is the extended yield condition

$$\Phi = \left[\frac{\sigma_v}{\sigma_M}\right]^2 + 2q_1 f^* \cosh\left[\frac{3}{2}q_2\frac{\sigma^H}{\sigma_M}\right] - (1 + q_3 f^{*2}) = 0 \qquad (8.23)$$

with v. Mises equivalent stress $\sigma_v = \sqrt{\frac{3}{2}\sigma_{ij}^D\sigma_{ij}^D}$ and hydrostatic stress $\sigma^H = \sigma_{kk}/3$, expressed by the macroscopic stress tensor σ_{ij}. The yield surface has the shape of an ellipsoid in the principal stress space, see Fig. 8.17. Without damage ($f^* = 0$) Eq. (8.23) corresponds to the v. Mises cylinder. With increasing damage this limit surface shrinks, so that the load-carrying capacity of the material diminishes. The individual terms in the yield condition (8.23) can be weighted by means of the empirical parameters q_1, q_2 and q_3, resp.

The modified damage variable f^* in Eq. (8.23) is a function of the void volume fraction f:

$$f^* = \begin{cases} f & \text{for } f \le f_c \\ f_c + \dfrac{f_f^* - f_c}{f_f - f_c}(f - f_c) & \text{for } f_c < f < f_f \\ f_f^* & \text{for } f \ge f_f \end{cases} \qquad (8.24)$$

The quantity f_c in (8.24) denotes that specific void volume fraction, at which due to the coalescence of voids an accelerated damage evolution begins, which is modeled as a bi-linear curve. Complete local failure of the material will occur, if the critical void volume fractions f_f resp. $f_f^* = 1/q_1$ are reached.

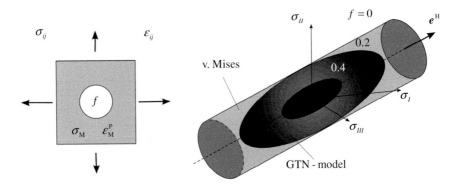

Fig. 8.17 Representative Volume Element RVE and yield surface of the Gurson-Model

The macroscopic plastic strain rate $\dot{\varepsilon}_{ij}^{\mathrm{P}}$ is directed normally to the yield surface, whereby $\dot{\Lambda}$ denotes the plastic multiplier

$$\dot{\varepsilon}_{ij}^{\mathrm{P}} = \dot{\Lambda}\frac{\partial \Phi}{\partial \sigma_{ij}}. \tag{8.25}$$

The evolution equation for the equivalent plastic strain rate of the matrix material $\varepsilon_{\mathrm{M}}^{\mathrm{P}}$ is deduced by the equivalence of the macroscopic and microscopic plastic work rates in the damaged volume element:

$$(1-f)\sigma_{\mathrm{M}}\,\dot{\varepsilon}_{\mathrm{M}}^{\mathrm{P}} = \sigma_{ij}\,\dot{\varepsilon}_{ij}^{\mathrm{P}}, \quad \varepsilon_{\mathrm{M}}^{\mathrm{P}} = \left.\varepsilon_{\mathrm{M}}^{\mathrm{P}}\right|_{0} + \int \frac{\sigma_{ij}\,\dot{\varepsilon}_{ij}^{\mathrm{P}}}{(1-f)\sigma_{\mathrm{M}}}. \tag{8.26}$$

The change of void volume fraction consists of two additive terms

$$\dot{f} = \dot{f}_{\mathrm{grow}} + \dot{f}_{\mathrm{nucl}}, \tag{8.27}$$

whereby $\dot{f}_{\mathrm{grow}}$ describes the increase by growth of voids, and $\dot{f}_{\mathrm{nucl}}$ accounts for the nucleation of new voids. The growth term is based on the conservation law of mass in the representative volume element

$$\dot{f}_{\mathrm{grow}} = (1-f)\,\dot{\varepsilon}_{kk}^{\mathrm{P}}. \tag{8.28}$$

For the formation of new voids a statistical, strain-controlled process is assumed, which obeys a Gaussian normal distribution function with the mean value ε_{N} and the standard deviation s_{N}:

$$\dot{f}_{\mathrm{nucl}} = A\,\dot{\varepsilon}_{\mathrm{M}}^{\mathrm{P}}, \quad A = \frac{f_{\mathrm{N}}}{s_{\mathrm{N}}\sqrt{2\pi}}\exp\left[-\frac{1}{2}\left(\frac{\varepsilon_{\mathrm{M}}^{\mathrm{P}} - \varepsilon_{\mathrm{N}}}{s_{\mathrm{N}}}\right)^{2}\right]. \tag{8.29}$$

The nucleation of new voids is proportional to the equivalent plastic strain $\varepsilon_{\mathrm{M}}^{\mathrm{P}}$ in the matrix and to the density of overall available nuclei f_{N}.

8.7 Examples of Fatigue Crack Propagation

8.7.1 Shear Force Bending Specimen

As an example of fatigue crack propagation under mixed-mode loading, the shear force bending test of Fig. 8.18 is chosen, because corresponding experimental results are available. A starting crack of length 6 mm is located on the top edge of the specimen at a distance of $b = 30$ mm to the drill-hole. The specimen was clamped

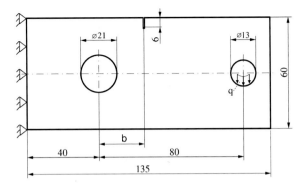

Fig. 8.18 Shear force bending specimen with starting crack under cyclic loading

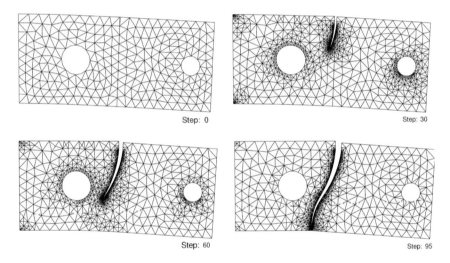

Fig. 8.19 Adaptive simulation of crack propagation in the shear force bending specimen

on the left side and subjected to a cyclic (zero-tension) force at the hole with a magnitude, lying above the fatigue threshold value ΔK_{th}, so that the crack will propagate. The simulation was performed using the adaptive technique of Sect. 8.4. Figure 8.19 represents the coarse initial mesh for the FEM-simulation and three stages of crack propagation including automatic adaptive mesh refinement. In this example, the crack extension was prescribed to go in constant increments of size $\Delta a = 3$ mm. The result of the FEM-simulation shows good agreement with the experimentally observed path of the fatigue crack (cf. Fig. 8.20). The small deviations are due to the fact that the crack propagation is done piecewise linear in the numerical simulation. Thus, the real curved crack path can only be followed approximately, but this could be improved by using smaller increments Δa.

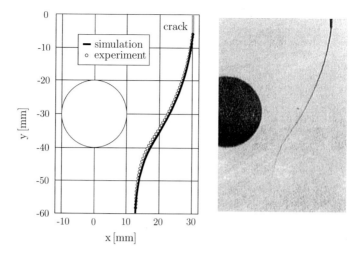

Fig. 8.20 Comparison of FEM-simulation with the experimental path of cracking

8.7.2 ICE-Wheel Failure

In 1998, an accident happened with the ICE high-speed train W.-C. Roentgen at Eschede. This fatal disaster claimed 101 human lives and catastrophic property damage, see Fig. 1.3. The derailment was caused by the breakage of a rim of the rubber damped wheels of the ICE. In this design of railway wheels, an annular rubber body is clamped between the rim and the wheel disk to reduce noise, see Fig. 8.21. During a straight ahead drive, the wheel is substantially exposed to a vertical wheel force of $Q = 98\,\mathrm{kN}$, that acts on the contact area between wheel and rail. It causes pressure and bending loads of the wheel rim, which are repeated during every wheel revolution and may push a fatigue crack. The failure analysis of the broken wheel rim revealed that fracture originated from a metallurgical defect on the inner side of the rim in the vicinity of the »roof ridge«. Under in-service loads the fatigue crack spread into the cross-section and finally lead to the complete fracture of the remaining ligament, see the Figs. 8.21 and 8.24.

In the context of evaluating this failure case, Richard et al. [30] performed fracture mechanical analyses of the loading situation and the crack propagations by means of the automatic remeshing technique reported in Sect. 8.3, Fig. 8.9.

In the first step, the state of stress in the uncracked wheel was calculated which results from mounting pre-loads and the load at straight-ahead driving. Figure 8.23 represents the elaborate 3D FEM-discretization of the half wheel. The situation in the contact region wheel–rail is shown in the used FEM-submodel of the rim, see Fig. 8.22. As a result, it turned out, that the highest hoop stresses occur in the wheel rim on the inner side at the »roof ridge«. In the course of one revolution of the wheel, they change between $\sigma_{max} = 220\,\mathrm{MPa}$ and $\sigma_{min} = 6\,\mathrm{MPa}$. Thus, the largest alternating load range exists in the vicinity of the origin of the observed crack.

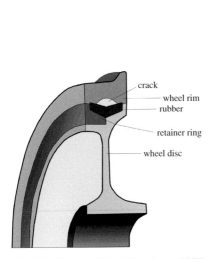

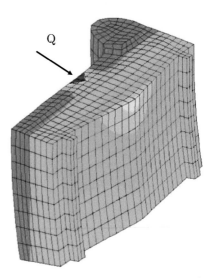

Fig. 8.21 Design of the rubber-damped ICE railway wheel

Fig. 8.22 Distribution of maximum principal stresses in the wheel rim [30]

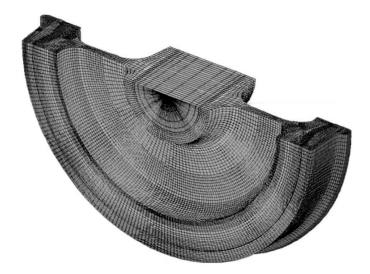

Fig. 8.23 FEM-discretization of the rubber-damped railway wheel (diameter 862 mm) with about 130000 hexahedral elements [30]

Therefore, in the second step, a submodel with 52000 finite elements was generated to simulate the crack propagation in this cut-out of the wheel. Since the fatigue crack propagation in the real rim didn't start directly at the »roof ridge« but 13 mm aside, a semi-circular crack of 1.5 mm depth was placed at this position in

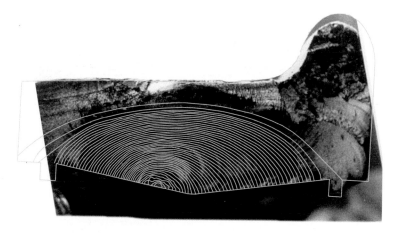

Fig. 8.24 Fracture surface of the ICE railway wheel with results of FEM-simulations of crack propagation [30]

the simulation model. The stress intensities along the entire crack front exceeded the threshold value $\Delta K_{\text{th}} = 8.2\,\text{MPa}\,\text{m}^{1/2}$ of the wheel steel. From the FEM-analysis the range of the K-factor ΔK along the current crack front is obtained for each revolution of the wheel, whereof the local crack growth increment was calculated via the Paris-relationship (3.287). Figure 8.24 illustrates the crack fronts obtained this way at each simulation step. Initially, the crack grows roughly semicircular, whereas later it spreads much faster in the width. The complete crack growth analysis included 26 simulation steps and ended only when the fracture toughness of $K_{\text{Ic}} = 86.8\,\text{MPa}\,\text{m}^{1/2}$ was achieved. The critical crack at onset of catastrophic breakage had a depth of 31.7 mm and a maximum length of 71.1 mm on the inner side of the rim. Figure 8.24 shows for comparison the simulated crack front positions together with the actual crack propagation on the fracture surface of the wheel rim.

Supposing a constant cyclic loading of the wheel, one can calculate on the basis of the crack growth curve da/dN of the wheel steel the critical number of load cycles $N_B \approx 1.4$ million that correspond to about 3791 km driven distance. This estimate is based on a linear damage accumulation hypothesis, which does not account for any other load cases or sequence effects. In fact, the fracture pattern exhibits lines of rest due to overloads and others. Further details can be found in [30].

8.8 Examples of Ductile Crack Propagation

8.8.1 Cohesive Zone Model for CT-Specimen

Scheider et al. [17, 21] applied the cohesive zone model to simulate the ductile crack propagation in fracture mechanics specimens. One example is the 3D analysis of

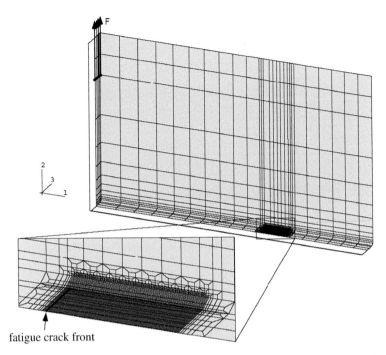

F

fatigue crack front

Fig. 8.25 Three-dimensional FEM-model of the CT-specimen with 6732 hexahedral elements (8-nodes) and 910 cohesive-elements [21]

a CT-specimen shown in Fig. 3.12 (width $w = 100$ mm, thickness $B = 10$ mm with 20 % side-grooves, initial crack length $a_0 = 60$ mm). The investigations were performed on the ferritic reactor pressure vessel steel 20MnMoNi55 that is distinguished by high ductility and large fracture toughness. At first all the necessary input-parameter for the model had to be determined. The true stress-strain-curve of the steel was measured by experiments on round tensile specimens and approximated by a power-law $\sigma = 925\, \varepsilon^{0.14}$. The trapezoidal function (8.7) was chosen as cohesive law with $\delta_1 = 0.01\, \delta_c$ and $\delta_2 = 0.75\, \delta_c$, so that still the three parameters G_c, σ_c and δ_c had to be determined. The energy of separation G_c was identified with the physical initiation value of the fracture toughness $J_i = 120$ N/mm and not with the engineering value $J_{0.2}$, since therein already energy terms of plastic deformation are included. In order to determine the maximum cohesive stress σ_c, notched tensile specimens have proved to be best suited. To this end notched tensile specimens with various notch radii were tested and subsequently simulated with axial-symmetrical FEM-models using the known yield curve. The value of σ_c is obtained from the maximum normal stress in the narrowest cross section of the specimen at that force, where the specimen fails in the experiment (usually unstable fracture). In the present case, for all notch radii approximately the same value $\sigma_c = 1460$ MPa of the cohesive stress was obtained. The last missing parameter δ_c was calculated from G_c and σ_c using the relationship (8.8).

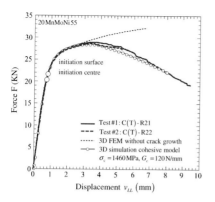

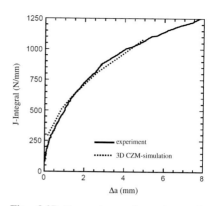

Fig. 8.26 Force-displacement-diagram of the CT-specimen. Comparison of FEM-simulation with experiment for steel 20MnMoNi55

Fig. 8.27 Comparison of crack-growth resistance curve for steel 20MnMoNi55, measured at CT-specimen, with the numerical simulation

After these preliminary studies, the CT-specimen was simulated using the cohesive zone model and the identified material parameters. Fig. 8.25 shows the generated FEM-model of one quarter (due to double symmetry) of the three-dimensional specimen. The side grooves are inserted in order to increase the stress triaxiality in the crack region. As can be seen in the detail screen of Fig. 8.25, the region of the ligament ahead of the fatigue crack is meshed very fine (element edge length $L = 0.075$mm) to resolve the crack propagation in the calculation sufficiently. This high-resolution area along the symmetry plane is uniformly discretized by 8-node cohesive elements (cf. Fig. 8.16) to enable the simulation of crack extension.

The results of several fracture mechanics tests with CT-specimens were available: The force-displacement curves of the load application point F-v_{LL}, the crack length measured with the potential method, and the corresponding J-integral values determined according to the ESIS standard [31], resp. In Fig. 8.26 the experimental force-displacement curves of two specimens are contrasted with the results of the numerical simulations. A FEM-computation carried out without crack propagation results in a too stiff behavior, whereas the simulation with the cohesive zone model shows a very good agreement. Likewise, the point of crack initiation, which occurs well before the maximum load, is correctly reproduced by the simulation. The comparison of experimental and numerical J-Δa crack resistance curves (Fig. 8.27) verifies the accuracy of the cohesive zone model and demonstrates its ability for predicting fracture mechanical material properties. Furthermore, the numerical crack propagation simulation is validated by comparing the crack front reached in the CT-specimen at a deformation of $v_{LL} = 5.35$ mm. Figure 8.28 displays a SEM-image of the fracture surface. The area of ductile dimples distinguishes significantly from the preceding and subsequent fine pattern of fatigue cracking. Also, the shape of the

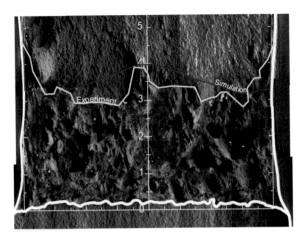

Fig. 8.28 Electron microscopic picture of the fracture surface of the CT-specimen. Comparison of crack front with FEM-simulation [21]

crack front is well reproduced by the simulation. It decreases a little bit from the middle, but hastens ahead towards the edges due to the side grooves.

8.8.2 Damage Mechanics for SENB-Specimen

As an example of the applicability of damage mechanical models to simulate ductile crack propagation, the investigations of Abendroth and Kuna [32, 33] will be presented. The tested material is steel S690Q, for which the yield curve and all parameters of the GTN-damage model had been determined by means of a miniaturized deep drawing test (small punch test) [32]. Here, only the results are reported. The hardening of the matrix material was best fitted by the power law

$$\sigma_M(\varepsilon_M^p) = \sigma_{F0}\left(\frac{\varepsilon_M^p - \varepsilon_L^p}{\varepsilon^*} + 1\right)^{1/n}, \tag{8.30}$$

which contains as parameters the initial yield strength $\sigma_{F0} = 690$ MPa, the hardening exponents $n = 11$, a reference strain $\varepsilon^* = 0.589\%$ and the Lüders-strain $\varepsilon_L^p = 1\%$, resp. The exact determination of all nine parameters of the GTN-damage model (Sect. 8.7) is generally quite complex. Usually, the most important parameters are identified from notched tensile specimens and fracture mechanics tests. For all other parameters plausible values are taken from the literature. In the present case the set of parameters listed in Table 8.1 was determined from small punch tests and validated at tensile specimens. In order to characterize the fracture properties of the steel S690Q, crack resistance curves J-Δa were measured by means of three-point-bend

Table 8.1 GTN-damage parameters for steel S690Q at room temperature

f_0	f_c	f_f	q_1	q_2	q_3	f_N	ε_N	s_N
0.002	0.1357	0.2	1.419	1.213	q_1^2	0.0273	0.5352	0.1

specimens (SENB-specimens, Fig. 3.12) at BAM Berlin [34]. The specimens had the dimensions: Width $w = 26$ mm, thickness $B = 13$ mm with 20 % side grooves, initial crack length $a_0 = 0.51w$, resp.

These fracture experiments were numerically recomputed using the GTN-damage model. Figure 8.29 depicts the FEM-discretization of one half of the specimen as a two-dimensional model assuming a plane strain state. To simulate the ductile crack propagation with the help of damage mechanics, a rather fine, uniform mesh of elements (here 0.10×0.10 mm^2) is required in the whole region, into which the crack is expected to move. In the simulation, the crack propagation takes place in the following way: In the extremely stressed elements in front of the crack tip, damage reaches its critical value f_f, whereby the load carrying capacity of the material (stress response and yield condition (8.23)) is reduced to zero. Thus crack propagation proceeds as a successive sequence of material failure in the integration points of elements along the ligament. Formally, these elements are still involved in the FEM-system, but their stiffness got lost. The extreme deformations of these elements represent the crack opening. The detail in Fig. 8.30 (left) shows the values of the damage variables f^* and those elements which have already failed.

The measured force-deflection curves $F(u)$ of the SENB-specimen are compared in Fig. 8.31 with different versions of the FEM-analysis. In the elastic range, the results of all 2D (plane strain) and 3D FEM-calculations are in good agreement with experiment. After complete plastification of the specimen's cross section (cf. Fig. 8.30), the FEM-results *without* damage mechanics lie considerably above the experimental curves. However, by using the GTN-damage model, a large reduction of the F-u-curves is effected, because here the crack propagation is considered. The 2D (plane strain) analysis exhibits a too stiff specimen behavior, since with increasing plastification the side grooves lose their global impact and the stresses in

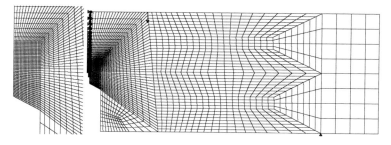

Fig. 8.29 FEM-mesh of the SENB-specimen: Detail of the crack tip region (*left*), whole specimen with boundary conditions (*right*)

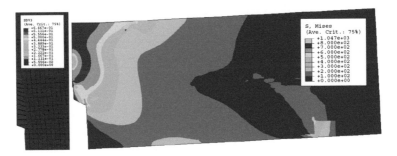

Fig. 8.30 Damage (crack propagation) and v. MISES equivalent stress in the SENB-specimen

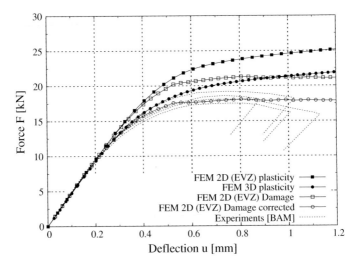

Fig. 8.31 Force-deflection-curve of a SENB-specimen of steel S690Q. Comparison of FEM and experiment

thickness direction are reduced. This effect can be taken into account by introducing an effective specimen thickness B_{eff}, which decreases from B to the value $0.8B$ in the course of deformation u. This correction function was calibrated from the ratio of 2D to 3D-curves without damage, and the modified 2D F-u-curve was derived out of it, which agrees very well with the experiment (Fig. 8.31).

From the FEM-results, the elastic-plastic J-integral (see Sect. 7.3) was evaluated as a function of the force F. Since in case of crack propagation, J is path-dependent in regions of local stress relief, only paths far-away from the crack tip can be used, where J reaches a stabilized value. This value is equivalent to the evaluation formula used in the test standard [35]. In addition, the amount of crack extension Δa is determined on the basis of failed finite elements, from which the numerically simulated crack growth resistance curve in Fig. 8.32 is derived.

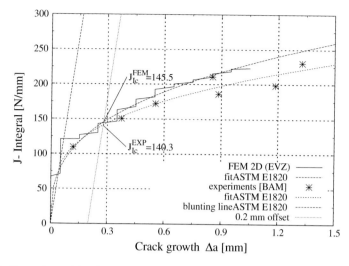

Fig. 8.32 Comparison of simulated and experimental J-Δa-curves of the SENB-specimen of steel S690 Q

Its stepwise course results from the successive failure in the integration points of elements. The outcome of simulation shows a good agreement with the experiment. If the engineering initiation values J_{Ic} are investigated at an offset of $\Delta a = 0.2$ mm according to the ASTM-standard [34], then simulation $J_{Ic}^{FEM} = 145.5$ N/mm and experiment $J_{Ic}^{Exp} = 140.3$ N/mm yield nearly identical results.

This example and similar experience from the literature show that with the help of damage mechanics a fairly good prediction of ductile crack propagation is possible in specimens and components. In contrast to the criteria of EPBM, in this connection the influence of the local stress state (triaxiality $\hbar$) and its changes (loading and unloading) is taken into account at any point of the crack front. An essential *advantage* follows from the fact that now real material parameters are concerned, so that the transferability between different specimen geometries (even without crack) as well as from small samples to large components is ensured. A prerequisite, however, is knowledge of the damage mechanics parameters of the material, the determination of which is complex and not always clear. This makes the application of damage mechanics difficult in practice. Another *disadvantage* is the dependency of the numerical solution on the size (especially the height) of the finite elements in the crack tip region, because damage and failure (at equal external load) occur the sooner the smaller the distance to the crack tip is made. The reason for that is the conjunction of a softening material model and an inappropriate numerical solution algorithm. As long as the damage mechanical law does not include an inherent material-specific length parameter, the size of the FEM-discretization causes quasi a »numerical homogenization«. At present, several methods such as non-local integral and differential approaches are investigated for regularizing these numerical difficulties. Very often as a pragmatic way out, the size of the elements at the crack is empirically adjusted and then interpreted as a characteristic length (e. g. void distance) of the material, which of course must be set constant in all simulations.

References

1. Dhondt G (1998) Cutting of a 3-D finite element mesh for automatic mode I crack propagation calculations. Int J Numer Meth Eng 42:749–772
2. Mishnaevsky L, Weber U, Schmauder S (2004) Numerical analysis of the effect of microstructures of particle-reinforced metallic materials on the crack growth and fracture resistance. Int J Fract 125:33–50
3. Bazant ZP, Oh BH (1983) Crack band theory for fracture and concrete. Mater Struct 16:155–177
4. deBorst R (2002) Fracture in quasi-brittle materials: a review of continuum damage-based approaches. Eng Fract Mech 69:95–112
5. Hillerborg A, Rots JG (1989) Crack concepts and numerical modelling. In: Elfgren L (ed) Fracture mechanics of concrete structures. Chapman & Hall, London, New York, pp 128–146
6. Nishioka T, Atluri SN (1980) Numerical modeling of dynamic crack propagation in finite bodies by moving singular elements. J Appl Mech 47:570–582
7. Murthy KSRK, Mukhopadhyay M (2000) Adaptive finite element analysis of mixed-mode crack problems with automatic mesh generator. Int J Numer Meth Eng 49:1087–1100
8. Nishioka T, Furutsuka J, Tchouikov S, Fujimoto T (2002) Generation-phase simulation of dynamic crack bifurcation phenomenon using moving finite element method based on Delaunay automatic triangulation. Comput Model Eng Simul 3:129–145
9. Zienkiewicz OC, Taylor RL, Zhu JZ (2005) The finite element method: its basis and fundamentals, vol 1. Elsevier, Amsterdam
10. Schöllmann M, Fulland M, Richard HA (2003) Development of a new software for adaptive crack growth simulations in 3D structures. Eng Fract Mech 70:249–268
11. Wriggers P (2001) Nichtlineare Finite-Elemente-Methoden. Springer, Berlin
12. Meyer A, Rabold F, Scherzer M (2006) Efficient finite element simulation of crack propagation using adaptive iterative solvers. Commun Numer Methods Eng 22:93–108
13. Barenblatt GI (1962) The mathematical theory of equilibrium cracks in brittle fracture. Adv Appl Mech 7:55–129
14. Dugdale D (1960) Yielding of steel sheets containing slits. J Mech Phys Solids 8:100–104
15. Needleman A (1987) A continuum model for void nucleation by inclusion debonding. J Appl Mech 54:525–531
16. Brocks W, Cornec A (2003) Cohesive models—special issue. Eng Fract Mech 70(14):1741–1986
17. Brocks W, Cornec A, Scheider I (2003) Computational aspects of nonlinear fracture mechanics. In: Milne I, Ritchie RO, Karihaloo B (eds) Comprehensive structural integrity—numerical and computational methods, vol 3. Elsevier, Oxford, pp 127–209
18. Rose JH, Ferrante J, Smith JR (1981) Universal binding energy curves for metals and bimetallic interfaces. Phys Rev Lett 47:675–678
19. Needleman A (1990) An analysis of tensile decohesion along an imperfect interface. Int J Fract 42:21–40
20. Tvergaard V, Hutchinson JW (1992) The relation between crack growth resistance and fracture process parameters in elastic-plastic solids. J Mech Phys Solids 40:1377–1397
21. Scheider, I (2001) Bruchmechanische Bewertung von Laserschweißverbindungen durch numerische Rissfortschrittsimulation mit dem Kohäsivzonenmodell. Ph.D. thesis, Technische Universität Hamburg
22. Hillerborg A, Modeer M, Petersson PE (1976) Analysis of crack formation and crack growth in concrete by means of fracture mechanics and finite elements. Cem Concr Res 6:773–782
23. Bazant ZP (1993) Current status and advances in the theory of creep and interaction with fracture. In: Bazant ZP, Ignacio CC (eds) Proceedings of the 5th international RILEM symposium on creep and shrinkage of concrete, Chapman & Hall, Barcelona, pp 291–307
24. Xu X, Needleman A (1994) Numerical simulations of fast crack growth in brittle solids. J Mech Phys Solids 42:1397–1434
25. Ortis M, Pandolfi A (1999) Finite-deformation irreversible cohesive elements for three-dimensional crack-propagation analysis. Int J Numer Methods Eng 44:1267–1282

26. Schwalbe KH, Scheider I, Cornec A (2013) Guidelines for applying cohesive models to the damage behaviour of engineering materials. Springer, Heidelberg
27. Rousselier G (1987) Ductile fracture models and their potential in local approach of fracture. Nucl Eng Des 105:97–111
28. Gurson AL (1977) Continuum theory of ductile rupture by void nucleation and growth: Part I—Yield criteria and flow rules for porous ductile materials. J Eng Mater Technol 99:2–15
29. Tvergaard V, Needleman A (1984) Analysis of the cup-cone fracture in a round tensile bar. Acta Metall 32(1):157–169
30. Richard HA, Sander M, Kullmer G, Fulland M (2004) Finite-Elemente-Simulation im Vergleich zur Realität—Spannungsanalytische und bruchmechanische Untersuchungen zum ICE-Radreifenbruch. Materialprüfung 46:441–448
31. ESIS P2 (1992) Procedure for determining the fracture behaviour of materials. Technical report, European Structural Integrity Society
32. Abendroth M (2005) Identifikation elastoplastischer und schädigungsmechanischer Materialparameter aus dem Small Punch Test. Ph.D. thesis, TU Bergakademie Freiberg
33. Abendroth M, Kuna M (2006) Identification of ductile damage and fracture parameters from the small punch test using neural networks. Eng Fract Mech 73:710–725
34. Klingbeil D, Brocks W, Fricke S, Arndt S, Reusch F, Kiyak Y (1998) Verifikation von Schädigungsmodellen zur Vorhersage von Rißwiderstandskurven für verschiedene Probengeometrien und Materialien im Rißinitiierungsbereich und bei großem Rißwachstum. Technical report, BAM-V.31 98/2, Bundesanstalt für Materialforschung und -prüfung
35. ASTM-E 1820 (2007) Standard test method for measurement of fracture toughness. Technical report, American Society for Testing and Materials, West Conshohocken

Chapter 9
Practical Applications

9.1 Fatigue Crack Growth in a Railway Wheel

Bainitic cast iron with nodular graphite (Austempered Ductile Iron ADI) shows a good ductility, a superior wear resistance and a high fatigue strength, which makes it an interesting alternative to steel for producing railway wheels. However, ADI possesses a lower fracture toughness and is, due to the casting process, more prone to defects. A railway wheel is exposed to high static and cyclic loading. Therefore, besides the classical service strength analysis also fracture mechanics concepts have to be applied to ensure sufficient safety against fatigue crack growth and brittle fracture. In order to assess fracture safety and lifetime, it is necessary to calculate numerically the stress state in an ADI-wheel under static and cyclic loads presuming hypothetical cracks [1, 2]. The aim of the investigation is to derive critical crack sizes or admissible limit values of loading as early as during the design stage of the wheel, and to specify suitable inspection strategies.

9.1.1 Material Data of Austempered Ductile Iron ADI

The conventional mechanical properties and the fracture mechanics data of the ADI material are compiled in Table 9.1. In laboratory test specimens ADI material fails in a ductile manner. The fracture toughness J_i^{BL} is determined from the crack resistance curve (see Fig. 3.46) at the moment of physical crack initiation. For the fracture mechanical stress analysis of the wheel, linear-elastic material behavior is assumed (SSY at the crack), and the failure is supposed to be macroscopically brittle. One can expect that the transferability between specimen and component is given. The transformation of the fracture toughness J_i^{BL} into the corresponding critical K-value is carried out by

$$K_{Ji} = \sqrt{\frac{E J_i^{BL}}{1 - \nu^2}} .$$ (9.1)

M. Kuna, *Finite Elements in Fracture Mechanics*, Solid Mechanics and Its
Applications 201, DOI: 10.1007/978-94-007-6680-8_9,
© Springer Science+Business Media Dordrecht 2013

Table 9.1 Conventional and fracture mechanical material parameters of ADI

Parameter	Value
Elasticity modulus E	170 GPa
POISSON's ratio ν	0.3
0.2 %-yield strain $R_{p0.2}$	637 MPa
Ultimate strength R_{m}	893 MPa
Fracture toughness J_i^{BL} or K_{Ji}	11.0 kJ/m^2 or 45.3 MPa $\sqrt{m}$
Fatigue crack growth at $R = 0.1$:	
Threshold value ΔK_{th}	5.4 MPa $\sqrt{m}$
C	$0.94 \cdot 10^{-08}$
m	2.9
Fatigue crack growth at $R = 0.5$:	
Threshold value ΔK_{th}	4.3 MPa $\sqrt{m}$
C	$1.0 \cdot 10^{-08}$
m	3.2

Fig. 9.1 Cyclic crack growth curve of ADI at $R = 0.1$ and 0.5

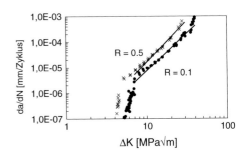

This way, the fracture mechanical assessment using K_{Ji} lies on the conservative side and the ductility of the ADI implies an additional safety reserve. The calculation of the *fatigue strength* and the *residual life time* at cyclic loads is carried out via the threshold value ΔK_{th} and the parameters C and m of the PARIS-ERDOGAN equation (3.287) of Sect. 3.4.1. The parameters are determined with the help of the cyclic crack growth curve depicted in Fig. 9.1 for the stress ratios $R = 0.1$ and 0.5, see Table 9.1.

9.1.2 Finite Element Calculation of the Wheel

Geometry

The investigated railway solid wheel with a bent disk has a diameter of $\oslash = 920$ mm and a standardized profile of the wheel rim as shown in Fig. 9.2.

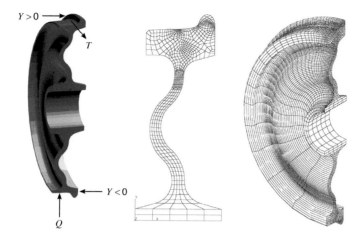

Fig. 9.2 Geometry, loading and FEM-discretization of the railway wheel (the forces $Y > 0$ and T are plotted for clarity above.)

Load Cases

For the whole wheel set, an axle load of 180 kN is assumed, which forms the basic value for all other load cases. In accordance with regulatory rules for experimental service strength proof of railway solid wheels, the loads of the wheel are specified as follows: In the radial direction the vertical wheel force Q acts, whereas perpendicular to the wheel, lateral forces apply due to cornering (wheel flange position $+Y$) or switch course (guide rail position $-Y$). These forces are illustrated in Fig. 9.2. In the following particular example, two extreme load cases are considered, which are enhanced by a safety factor of $f = 1.8$ to take higher dynamic wheel loads (impact, oscillations) into account, see Table 9.2. At braking events, an additional friction force T occurs between wheel and rail, acting in the tangential direction at the running tread. The magnitude of the braking force was given by $T = 0.2\,Q$. It is modeled as a uniformly distributed load on the contact area.

Table 9.2 Forces for specific load cases

Load case	Vertical wheel force Q in kN	Lateral wheel force Y in kN	Braking force T in kN
1	Extreme load -159	Wheel flange position $+62$	Tangential force 31.8
2	Extreme load -159	Guide rail position -62	Tangential force 31.8

Modeling

Due to the symmetry properties of geometry and loading conditions, only one half of the wheel needs to be modeled, cf. Fig. 9.2. It can be assumed that the wheel hub is rigidly fixed on the axle. Therefore in the FEM-model, the inner surface of the hub is treated as clamped support. The vertical wheel force is applied by a pressure load on an area of total $1.58\,\text{cm}^2$ on the tread. The positive lateral force $+Y$ is placed horizontally on the inner side of the wheel flange. The negative lateral force $-Y$ operates on the wheel flange at the height of the tread. These assumptions agree with the application points of forces in experimental strength tests. Linear-elastic isotropic material behavior is assumed in the FEM-calculations. The material parameters are given in Table 9.1.

Stresses

During one revolution of the wheel, the load application point moves by $360°$ on a circular orbit. Since the geometry of the wheel is rotationally symmetric, it is sufficient to perform the calculation only for one force application point. Each body-fixed point at a distance r from the axis then passes, during one revolution, all stress states that are on the same circumference of the wheel. Hereby, the principal stress directions change at a specific point of the wheel. Crack propagation is aligned perpendicular to the maximum principal stress direction. Subcritical crack growth is controlled by the range of alternating stresses during one revolution of the wheel. In order to find the largest stress range, the maximum principal stress is determined on a circumferential line and compared with all other stresses that lie on the same line but have been transformed before into that direction of the maximum principal stress.

Extreme Stresses in the Wheel Disk

According to common experience the highest stresses in railway wheels appear in the transition zone from hub to disk. The FEM-computations done for the considered ADI-wheel show as well very high stresses in the radial direction in this zone (distance r of about 162 mm from the wheel center), compare Fig. 9.3. For the considered case, extreme tensile and compressive stresses in the wheel disk are not confined to this position alone but were also observed at larger radii r, which are listed in Table 9.3 together with the respective stresses. They are mainly caused by bending of the wheel disk due to lateral forces.

Table 9.3 Maximum and minimum principal stresses

Load case	Inner wheel side		Outer wheel side	
1	$-241\,\text{MPa}$,	$r = 160\,\text{mm}$	$198\,\text{MPa}$,	$r = 264\,\text{mm}$
2	$196\,\text{MPa}$,	$r = 157\,\text{mm}$	$-189\,\text{MPa}$,	$r = 355\,\text{mm}$

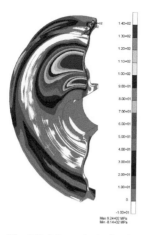

Fig. 9.3 Max. principal stress at load case 1

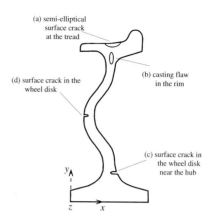

Fig. 9.4 Postulated crack configurations

9.1.3 Specification of Crack Postulates

Next the crack configurations are specified that need to be considered in the ADI-wheel. Hereby, both the results of the strength analysis with respect to the maximum appearing stresses are taken into account as well as possible defect positions resulting from the casting process of the wheel. The positions of the selected cracks are drawn in Fig. 9.4.

Crack Configuration (a): Semi-Elliptical Surface Crack on the Tread

Because of the rolling contact between wheel and rail, initial cracks may be formed by friction and slipping effects on the surface of the wheel tread lying transversely to the circumferential direction. These cracks experience at every rolling-over process a tension-pressure-tension load cycle. Therefore it has to be clarified above what size a_{th} those cracks are able to propagate by fatigue and after how many cycles they reach their critical crack length a_c to initiate brittle fracture. As an intensifying load assumption, also braking loads T were taken into account leading to tangential crack opening stresses.

Crack Configuration (b): Casting Defect in the Wheel Rim

Due to the manufacturing of the wheels by casting, the occurrence of casting defects like voids or pores is possible particularly in the region from wheel–rim to wheel–disk where the cross-section changes. These casting defects are modeled conservatively as circular cracks, orientated perpendicular to the maximum normal stress. The critical crack size is sought as a margin for the resolution sensitivity of non-destructive testing methods.

Crack Configuration (c): Surface Crack at the Transition Disk—Hub

Since the transition region disk—hub is exposed to the highest stresses, it has to be verified experimentally by means of cyclic service-strength testing that no fatigue pre-crack is formed at this place. This implicates the fracture mechanical safety proof as well. Hence, no further considerations are needed for this region.

Crack Configuration (d): Surface Crack at the Disk's Bend

In the area of the wheel disk's bend, the FEM-computations yield bending stresses of almost comparable magnitude like those at the transition between hub and wheel disk. Since this area is not covered by the cyclic service-strength testing, a surface crack is postulated and assessed at this position in the circumferential direction.

9.1.4 Fracture Mechanical Analysis

To determine the stress intensity factors, usually the so-called *decoupling method* is tried as first approach in fracture mechanical stress analysis. To do this, at first a simplified *substitute model* is selected as a cutout of that region of the component where the postulated crack is located. Second, we look for K-factor solutions and geometry functions $g(a)$, which are available for this crack configuration from handbooks such as e. g. Murakami [3].

$$K_I(a) = g(a)\,\sigma(\mathbf{x})\sqrt{\pi a} \qquad (9.2)$$

Thereby, the stress intensity factors are simply calculated by inserting the stress distribution $\sigma(\mathbf{x})$, which results from the FEM computation at the crack position. In this way, the fracture mechanics evaluation can be appended downstream to a normal strength calculation without modeling the crack. This decoupling method represents a good approximation as long as the assumed cracks are sufficiently small compared to the load-bearing cross section, making any feedback on the global stress state in the component negligible. This prerequisite is fulfilled for the crack configurations (a) and (b) of the wheel. At large crack depth, it no longer applies to the crack configuration (d), which is why for this case the K-factors must be calculated by an explicit FEM-modeling of the crack.

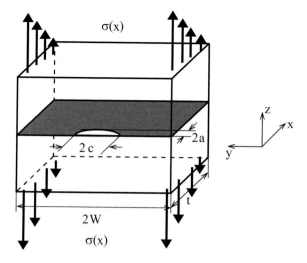

Fig. 9.5 Prismatic substitute model with semi-elliptical surface crack under given stress distribution $\sigma(x)$

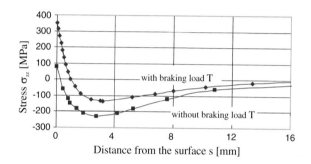

Fig. 9.6 FEM-result of crack-opening stresses for crack configuration (**a**)

Crack Configuration (a)

This semi-elliptical surface crack is assumed at the middle of the tread crossways to the rolling direction. The ratio of crack depth a to half crack length c is chosen as $a : c = 1 : 3$, which is typical for surface cracks. As a suitable substitute model for this crack configuration an accordant surface crack in a rectangular plate is selected, the dimensions of which correspond to the width $2W = 135$ mm and height $t=17$ mm of the wheel rim, cf. Fig. 9.5. To control the influence of the crack shape, an edge crack of the same depth a across the complete width of the wheel rim ($c = \infty$) is examined as an extreme case.

Position and orientation of the crack correspond to the location and direction of the maximum normal stresses σ_{zz} in circumferential direction on the wheel-rim. Here the stresses reach their peak values at the very surface shortly before and

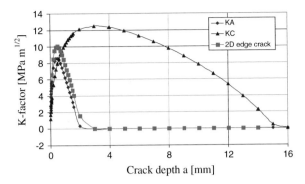

Fig. 9.7 Stress intensity faktors as function of crack depth (3D surface crack)

behind the contact point due to the contact pressure wheel-rail. As can be seen from Fig. 9.6, the high tensile stresses at the surface fade away with increasing depth, where compressive stresses occur due to the pressing contact. The circumferential stresses at the surface will increase approximately to a threefold value, if the braking load T is taken into account. These maximum loads have to be supposed for the assessment of both brittle fracture and fatigue crack growth.

Based on this stress distribution $\sigma_{zz}(x)$ and by means of the substitute model, the stress intensity factors are calculated. Hereby, K_A refers to the deepest point A (apex) of the crack front and K_C to the two surface points C, where the crack front intersects the tread, respectively. Figure 9.7 shows K_A and K_C as a function of the crack depth at a constant aspect ratio $a : c$. It can be recognized that, with growing crack depth, both K-factors attain at first a maximum, then drop almost down to zero, if the crack reaches the compressive region. By virtue of the high stress gradient the K_C-factor at the surface is greater than the apex value K_A. This means the crack would grow first into its lateral direction and later into the depth. The substitute model of a 2D edge crack yielded a somewhat higher K_A-factor.

$$K_{A_{max}} = 8.53\,\mathrm{MPa}\,\sqrt{m} \qquad \text{at } a_{max} = 0.5\,\mathrm{mm}, \quad a_{th} = 0.05\,\mathrm{mm}$$
$$K_{C_{max}} = 12.49\,\mathrm{MPa}\,\sqrt{m} \qquad \text{at } a_{max} = 3.0\,\mathrm{mm}, \quad c_{th} = 0.38\,\mathrm{mm} \qquad (9.3)$$

All maximum stress intensity factors lie far below the fracture toughness of $K_{Ji} = 45.3\,\mathrm{MPa}\,\sqrt{m}$, so that brittle fracture can be excluded for a surface crack in the tread under service loads.

In the next step the question of a possible crack propagation by fatigue has to be answered. For this purpose it is assumed that at every wheel revolution (load cycle) the stress distribution at the surface crack $\Delta\sigma(s)$ oscillates from zero up to the maximum state, which resulted from the above cyclic stress analysis. To determine above what size a_{th} a crack would be able to grow even under this stress range, the threshold value $\Delta K_{th} = 4.3\,\mathrm{MPa}\,\sqrt{m}$ is inserted into (9.2) and rearranging gives

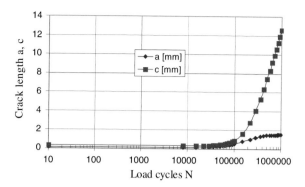

Fig. 9.8 Crack growth due to cyclic loading

$$a_{th} = \frac{1}{\pi} \left(\frac{\Delta K_{th}}{\Delta \sigma g(a_{th})} \right)^2 . \tag{9.4}$$

By using the geometry functions of these crack configurations the threshold values of crack sizes a_{th} and c_{th} are calculated and given in (9.3).

According to these preliminary studies, rather small cracks are able to grow under alternating loads. By integrating the Paris-Erdogan equation (3.291) (Sect. 3.4.1) we obtain the number of load cycles N required to propagate a crack from its initial length a_0 up to the size a. An initial crack length of $a_0 = 0.1$ mm is chosen that corresponds approximately to the depth of surface flaws, which are formed by rolling friction fatigue with the rail. With the parameters C and m listed in Table 9.1, the crack growth was calculated for $R = 0.5$ and depicted in Fig. 9.8. The results show that the crack grows faster in a lateral direction (points C, length c) than into the depth (point A, crack depth a) because of the higher ΔK-value, which is however no critical issue. After about 500,000 cycles the crack growth stagnates at the depth of $a = 1.5$mm, since the crack comes into the compression region. Thus it is guaranteed that, in the case of a fatigue crack growth, the crack will come to rest.

Crack Configuration (b)

The FEM computations for the most severe load case give maximum principal normal stresses of 60 MPa inside the wheel rim. This crack configuration can be approximately treated by a substitute model of a circular crack in an infinite domain, according to(3.59). If in addition to the primary loading positive residual stresses of $R_{p0.2}/2$ are presumed, K-factors will be obtained, all of which lie below K_{Ji}, so that even for such extreme load assumptions a brittle fracture can be excluded. According to (9.4) the crack would need to have a size of $a_{th} = 4$ mm, if it were even able to overcome the threshold load necessary for fatigue crack growth.

Crack Configuration (d)

In the bend region of the wheel disk a semi-elliptical surface crack is presumed. In accordance with experimental results and the extent of the stress maximum in a circumferential direction, an aspect ratio of $a : c = 1 : 5$ is chosen. The substitute model shown in Fig. 9.5 corresponds to a plate with this crack under combined bending and tensile stress distribution.

To check the applicability and accuracy of the substitute model used for this crack configuration, a detailed three-dimensional FEM-analysis is carried out for the wheel cutout depicted in Fig. 9.9 (left) including a surface crack of $a = 12$ mm. Because of symmetry in a circumferential direction the model is reduced to one half. Figure 9.9 (right) represents the employed FEM-mesh for this *submodel* with crack, on the surface of which the displacement fields of the preceding *global analysis* are imposed. The calculated stress distribution is illustrated by Fig. 9.10. By means of the J-integral technique the distribution of the K-factor is determined for the 3D elasticity problem, see Fig. 9.11.

The FEM-results reveal that the substitute model overestimates the true stress intensity factors by a factor of about 5! The main reasons for this are seen in the rotational symmetry and the higher bending stiffness of the ADI wheel. Further FEM-calculations with varying crack lengths would be necessary to determine these geometry functions completely. To gain anyhow an improved geometry function for K_A, the function $K_A(a)$ received by the substitute model »semi-elliptical surface crack«, is first fitted by a power law function and then scaled down over the whole range of $0 \leq a \leq 15$ mm so that it agrees at $a = 12$ mm with the FEM solution, see

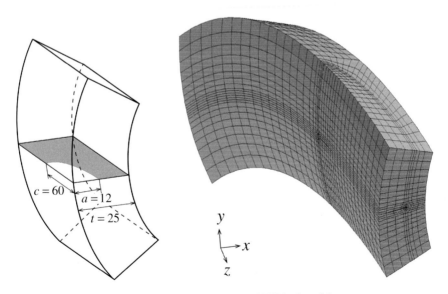

Fig. 9.9 Detail of the wheel disk with surface crack and FEM submodel

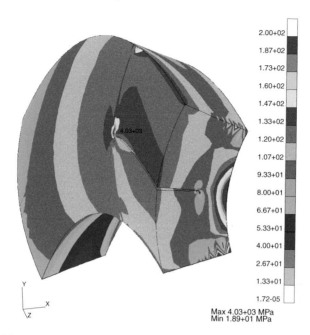

Fig. 9.10 V. MISES stress distribution in submodel for crack configuration (**d**)

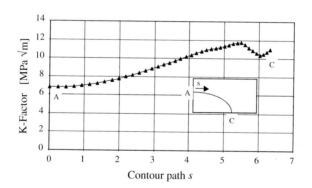

Fig. 9.11 Course of stress intensity factor along the semi-elliptical crack front (**d**)

Fig. 9.12. The corrected K-factor function $K_{A_{\text{new}}}(a) = 2.079 \cdot a^{0.478}$ is represented in Fig. 9.12 as well.

On the basis of the very conservative substitute model the fracture toughness would be attained at a critical crack depth of $a_c = 15\,\text{mm}$. The more precise FEM-computation leads to the result that brittle fracture will *not* occur for cracks up to this size.

Also with respect to the crack depth a_{th}, at which fatigue crack growth starts, the corrected K-solution $K_{A_{\text{new}}}$ gives a much more favorable value 4.6 mm than

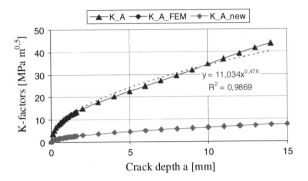

Fig. 9.12 FEM-corrected geometry function for the stress intensity factor K_A

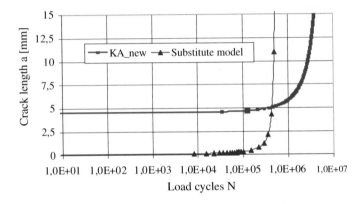

Fig. 9.13 Simulation of fatigue crack growth for crack (**d**)

the substitute model $K_{A\text{subst}}$ with 0.14 mm. Figure 9.13 represents the crack depth as a function of load cycles, starting at the according threshold values a_{th}. In the very conservative substitute model, about $5.3 \cdot 10^5$ cycles lead to the crack depth $a_c = 15$ mm, this means to brittle fracture, while with the corrected solution this value will be reached after $3.7 \cdot 10^6$ cycles, without leading to brittle fracture. The great difference between both models effects substantially the prediction of lifetime.

This practical example proves consequently the benefit that a more precise FEM stress analysis of cracks may have to quantify the safety margins.

9.2 Brittle Fracture Assessment of a Container Under Impact Loading

In the development of transport and storage casks for nuclear fuel elements the integer confinement of the inventory has to be ensured even in the case of extreme accidents. Since the failure of the structure must absolutely be prevented, safety precautions against fracture and inadmissible plastic deformation have to be verified under various loading conditions. According to guidelines of the International Atomic Energy Agency IAEA [4], especially impulsive, dynamic load cases, such as the drop of a container from a 9 m level onto a rigid foundation, is an example of such a severe loading situation requiring verification. In fact, one has to presume the lower limits of the material properties (fracture toughness), low temperatures and unfavorable combinations of all emergency scenarios. The determination of safety reserves against brittle fracture demands the application of computational methods in fracture mechanics. Under dynamic loading of a stationary crack the stress situation is characterized by means of the dynamic stress intensity factors K_I^d, K_{II}^d, K_{III}^d, which comprise a function of time t. In order to solve such complicated three-dimensional initial boundary value problems with crack, the use of FEM is inevitable. In the following example a practical application of the dynamic J^*-integral (Sect. 6.5) is demonstrated for the computation of K-factors by simulating the drop test of a transportation cask.

9.2.1 FE-Model of the Drop Test

The numerical analyses concern a real 9 m free drop test of a transportation cask performed in the CRIEPI study [5], which was especially addressed to fracture mechanical investigations. The test container was made from ductile cast iron. It was equipped with shock absorbers at both front walls and had been cooled down to $-40°C$ before the experiment. Moreover, an artificial semi-elliptical surface crack was prepared at the bottom side of the container, where due to deflection the highest tensile stresses are expected. The most important geometrical dimensions of container and crack are drawn in Fig. 9.14. The simulations of the drop test were carried out by means of the FEM-program Abaqus/Explicit [6] that works with an explicit time integration algorithm. For the fracture mechanical evaluation a post-processor was developed in [7] to compute numerically the dynamic J^*-integral by means of an equivalent domain integral EDI as derived in (6.54). Using relation (3.359), the $J^* = G$ value is converted into the dynamic stress intensity factor K_I^d.

Figure 9.15 shows the FEM-model used for the container drop test [7]. It mainly consists of the following components: the cask body, the secondary lit, the fuel rod carrier and the two shock absorbers. Due to symmetry of the container one half-model is sufficient for the analysis, discretized by first-order hexahedral elements. All geometric data and material properties are taken from the CRIEPI report [5]. Instead

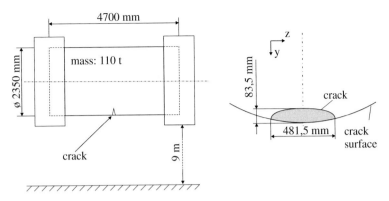

Fig. 9.14 Geometrical dimensions of a container with shock absorber and crack

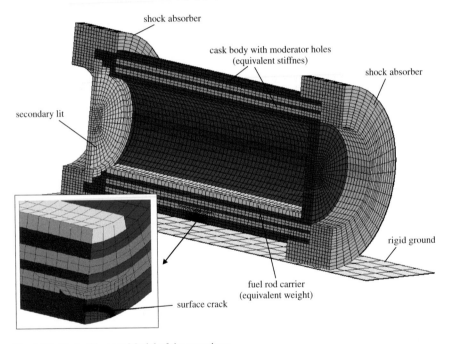

Fig. 9.15 Finite-Element-Model of the container

of using the real load of the fuel rods and their carrier, the interior of the container was replaced in the experiment by a substitute construction. This dummy is realized by one additional layer of elements with equivalent mass but without stiffness. On both ends of the container, shock absorbers were mounted made of plywood with a steel liner. The difficult material behavior of such absorbers was simulated in a homogenized manner by an elastic-plastic material model, the parameters of which were given in [5]. In order to prevent mutual penetration of the different parts of the

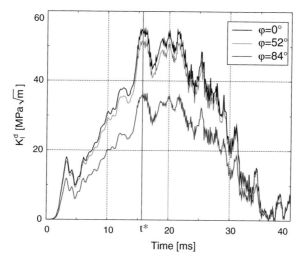

Fig. 9.16 Dynamic stress intensity factor as function of time

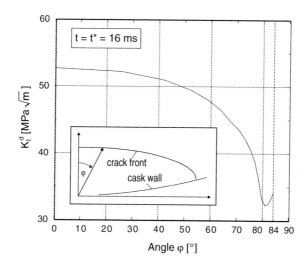

Fig. 9.17 Dynamic stress intensity factor along the crack front

model, the concerned surfaces of the cask, the shock absorbers and the ground were treated as contact pairs.

9.2.2 Fracture Mechanical Results of the Simulation

For the considered semi-elliptical crack ($a : c \approx 1 : 3$) the FEM-calculations yield the dynamic stress intensity factor K_I^d as a function of time t and position along the crack front, which is indicated by the angle φ as shown in Fig. 9.17. At $\varphi = 84°$ the crack front intersects the outer surface of the cask. In Fig. 9.16, K-factors $K_I^d(t)$ are plotted in dependence on time for three selected angles φ. All three curves have an absolute maximum at $t^* = 16$ ms. The highest values of K_I^d are reached at the apex point A of the semi-elliptical crack at $\varphi = 0°$. This becomes clear again in Fig. 9.17 that shows the $K_I^d(\varphi)$ distribution along the crack front for t^*. As can be seen, K_I^d has a maximum of 53 MPa m$^{1/2}$ at the apex point and decreases in the direction of the container's surface to a minimum of 33 MPa m$^{1/2}$ (the slight increase at the edge is a numerical artifact). The calculated maximum value of K_I^d is compared with the dynamic fracture toughness given in the CRIEPI-report as $K_{Id} = 69$ MPa m$^{1/2}$ for $T = -40°C$. Thus the loading of the crack lies about 30 % lower and the fracture safety of the container against this impact load is proved. This result agrees with the experimental observations, where no crack initiation was found after the drop test.

9.2.3 Application of Submodel Technique

In order to reduce the effort for discretizing and computing, the applicability of the *submodel technique* to dynamic load cases was investigated. Using this method, the displacement-time-courses obtained in the *global analysis without* crack are imposed onto the boundary nodes of the submodel, which now includes the crack. Figure 9.18

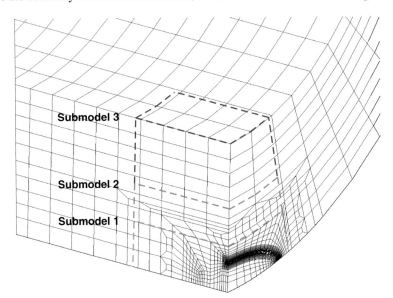

Fig. 9.18 Size of three investigated submodels

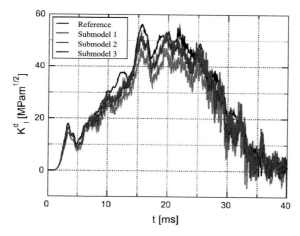

Fig. 9.19 Influence of submodel size on K_I^d—courses

shows a detail of the FEM-model with three crack-submodels, which differ in their radial extension. The results computed this way are depicted in Fig. 9.19. With increasing size of the submodel the results converge to the reference solution given in Fig. 9.16 of the previous section, which was based on a FEM-model of the complete cask with crack. A further enlargement of the submodel in the circumferential direction leads to a nearly perfect agreement with the reference solution. These findings prove on the one hand that the J-integral method is a very expedient technique to determine stress intensity factors under dynamic loading as well. On the other hand, they point out that always non-conservative (too low) K_I^d-values are computed if the submodel technique is used, which requires certain attention in practice.

9.3 Ductile Fracture of a Weldment in a Gas Pipeline

9.3.1 Introduction

Gas pipelines are subject to regular monitoring and maintenance procedures. The piping is tested by modern pig systems using non-destructive inspection methods (magnetic flux measurements, ultrasound, X-ray examination et al.). Thus mounting seams (girth welds) are especially in our focus, since they are the most likely to exhibit weld defects due to their manufacturing conditions on construction sites.

At TU Bergakademie Freiberg [8], a fracture mechanics approach to safety analysis of welds was developed and implemented as a computer-based *assessment code*. This code is intended to assist the test engineer during operational non-destructive inspections of gas pipelines. In particular, if weld defects have been detected, they must be assessed by fracture mechanics under the given operating pressure and

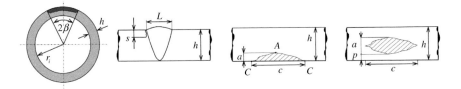

Fig. 9.20 Considered defect and weld seam geometries in piping

possible additional loads (laying, earth movement, subsidence phenomena, residual stresses etc.). From this, conclusions can be drawn regarding technical safety and the need for further inspection or maintenance activities. To verify the assessment concept it was necessary to check its transferability to real conditions by means of a component test. Likewise, the validity of the simplifying assumptions contained in the concept had to be proven. In any case it must be ensured that the assessment system yields conservative, i. e. safe, statements. For these reasons, the component test was analyzed by finite element calculations considering in detail the specific defect geometry, the elastic-plastic material behavior and the actual course of loading [8, 9].

9.3.2 Fracture Mechanics Assessment Concept FAD

Defect Postulates

The defects in the seam of the girth welds are safety-related, assumed conservatively as cracks, and classified into interior and surface defects as illustrated in Fig. 9.20. The assessment of the cracks is carried out by applying the fracture mechanics concept of FAD (*Failure Assessment Diagram*) already introduced in Sect.3.3.5. Typical geometries and load cases relevant for weld joints were taken into consideration in accordance with the ÖSTV-directive [10], the CEGB-R6-Routine [11] and the SINTAP procedure [12]. The FAD method evaluates the stress situation at the crack according to two criteria:

(a) The parameter $K_r = K_I/K_{Ic}$ relates the crack tip loading to a critical material value (fracture toughness) and is a measure of the threat by brittle fracture.
(b) The parameter $L_r = \sigma_{n\,\text{Rohr}}/\sigma_F$ relates a representative stress in the net-section to the yield strength of the material. It characterizes the amount of plastification and is a measure for the threat of plastic collapse.

In the Failure Assessment Diagram, structural failure is described by a limit curve, which interpolates between the two extreme states of brittle fracture and plastic collapse. The area within the limit curve indicates the safe region. For a specific component with crack an associated point $P(K_r, L_r)$ is plotted in the diagram. Its relative distance $S = \overline{0F}/\overline{0P}$ to the limit curve is a measure of safety. The limit

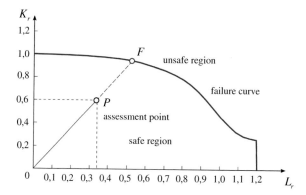

Fig. 9.21 Failure Assessment Diagram (FAD) with failure limit curve

curve employed in the assessment code of [8] is rather closely oriented towards the actual material behavior. It is drawn in Fig. 9.21 and has the form:

$$K_r = f(L_r) = (1 - 0.14\,L_r^2)[0.3 + 0.7\exp(-0.65\,L_r^6)]. \qquad (9.5)$$

The stress state $P(K_r, L_r)$ of a component with defect has to be calculated for the specific crack configuration and material from primary and secondary loads. The approach will be exemplified at a part-through surface crack in the weld seam, see Fig. 9.20.

Calculation of $K_r = K_I/K_{Ic}$

The value K_r is calculated from the stress intensity factor K_I of the crack and the fracture toughness K_{Ic}. For the assessment always the lowest material parameter K_{Ic} of base metal, heat affected zone and weld metal should be chosen, because the crack could run into any of these material areas. The loading parameter K_I for a semi-elliptical surface crack is obtained from [3] as

$$K_I = \frac{1}{\Phi}\sqrt{\pi a}\,(\sigma_m M_m M_{Km} + \sigma_E M_m + \sigma_b M_b M_{Kb})\,g_K\,, \qquad (9.6)$$

with $\Phi = \sqrt{1.464\left(\frac{2a}{c}\right)^{1.65} + 1}$ and a factor $g_K = 1.2$ accounting for the curvature effect. In this equation, σ_m means the membrane stress acting in the defect-free component, σ_E is the residual stress, σ_b is the pure bending stress across the wall thickness h and a is the depth of the surface crack.

The factors M_m and M_b describe the influence of geometrical parameters of the crack configuration on the stress intensity factor resulting from the acting stress portions. The stress concentration at weld notches is accounted for by the factors

M_{Km} and M_{Kb} (> 1) [9]. They depend on the ratio of weld width L to weld thickness h and on the relative depth coordinate s/h, see Fig. 9.20. In case of a surface crack, either the deepest point A or the near-surface points C can become critical. Therefore, the assessment has to be carried out for both types of points. Corresponding calculation formulas and geometry factors are used for other crack and weld geometries as well [8].

Calculation of $L_r = \sigma_{n\,pipe}/\sigma_F$

The value L_r denotes the stress state in the relevant crack configuration with respect to the *plastic limit load*. Here, we have to distinguish between a global plastic collapse and a local plastic failure. For the evaluation of gas piping having comparatively small defects (relative to the total cross section) and deep cracks, one must always assume a local plastic failure. The employed limit load solution [10] was specifically designed for surface flaws in pressurized components with curved walls. Thereby, σ_{npipe} represents the effective net-section stress, at which local plastic failure is reached in the remaining cross-section around the crack. This stress value can be calculated from the membrane and bending stresses as well as the internal pressure p as follows:

$$\sigma_{n\,pipe} = g_s\bar{\sigma} + p\,, \qquad \bar{\sigma} = \frac{\sigma_b + \sqrt{\sigma_b^2 + 9\sigma_m^2\left(1 - \frac{a}{h}\right)^2}}{3\left(1 - 2\frac{a}{h}\right)^2}$$

$$g_s = \frac{1}{1 - \left[\beta(1 - \vartheta) + 2\arcsin\left((1 - \vartheta)\sin\frac{\beta}{2}\right)\right]/\pi}\,, \qquad \vartheta = \frac{h - a}{h} \qquad (9.7)$$

Depending on the assessment mode, for σ_F either the yield strength $R_{p0.2}$ is applied or hardening is partially taken into account by $\sigma_F = \frac{1}{2}(R_{p0.2} + R_m)$. To get exact evaluations of the various crack configurations, the relations (9.7) were complemented by corresponding local limit load solutions [8].

Computer-Based Assessment Code

Based on the above-described FAD-concept, a PC program was developed for the fracture mechanical assessment of weld defects. An interactive dialog box allows us to input all geometric parameters of the crack configuration, to provide the necessary material parameters and to specify the stress state (pressure, bending, residual stresses). Next, in an evaluation module the parameters (K_r, L_r) are calculated by the relations (9.6)–(9.7) and finally displayed graphically together with the failure limit curve (9.5) in the assessment diagram, see Fig. 9.21. If the point P lies inside the limit curve, then the stress state is rated as admissible and the relative distance S to the limit curve is displayed as »safety against failure«. Moreover, by means of variant calculations, the user can estimate the effects of various input data and their range of tolerance (due to lack of knowledge or measurement uncertainty) on the safety. This option allows the safety-related assessment of a piping depending on the concrete situation and obeys the requirement for strict conservatism.

Material Data of Piping and Weldment

To determine the conventional and fracture mechanics material properties of the pipe material, samples were removed from the base material (BM) S355 J2, the weld metal (WM) and the heat affected zone (HAZ) and tested. The stress-strain curves of the tensile test were described by the Ramberg-Osgood power law.

$$\varepsilon = \varepsilon_e + \varepsilon_p = \frac{\sigma}{E} + \left(\frac{\sigma}{D}\right)^{1/N}. \tag{9.8}$$

The yield strength $R_{p0.2}$, the tensile strength R_m, the hardening exponent $N \mathrel{\widehat{=}} 1/n$ and the coefficient D are listed in Table 9.4 for all three materials.

The fracture-mechanical parameters were determined for all three material regions BM, WM and HAZ with the help of three-point bending (SENB) specimens ($10 \times 20 \times 100$ mm) having 20 % side grooves. All material regions show at room temperature a ductile, stable crack growth. The fracture toughness at the onset of crack growth (initiation) can be characterized by the J-integral and the crack opening displacement $\delta =$ CTOD.

$$J_i = 2\delta_i \sigma_F, \qquad \sigma_F = 0.5(R_{p0.2} + R_m), \qquad K_{Ji} = \sqrt{\frac{E J_i}{1 - \nu^2}} \tag{9.9}$$

The characteristic values for all three material regions are summarized in Table 9.4. The weld metal exhibits a much higher initiation fracture toughness than the base material and the heat affected zone.

Table 9.4 Data of tensile test and crack initiation values for J-integral and CTOD-concept

	$R_{p0.2}$ (MPa)	R_m (MPa)	N	D	J_i^{BL} (kJ/m^2)	δ_i^{SZW} (μm)	K_{Ji} (MPa m$^{\frac{1}{2}}$)
Base material	403	575	0.15	801	63	69	119.4
Weld metal	432	583	0.13	773	143	142	179.9
Heat affected zone	477	620	0.12	779	56	58	112.6

9.3.3 Large Scale Test of a Piping with Pre-cracked Weldments

Experimental Procedure

In collaboration with the Welding Institute in Halle, a component test of a gas piping DN 920 was carried out, see [8]. The test specimen and the experimental setup are shown schematically in Fig. 9.22. The specimen has been made of two existing pipes in such a way that, in the central region of highest stresses, two original girth weld seams (1 and 2) are installed at a distance of 2 m. The aim of this large-scale experiment was to test the strength of the welds with artificially induced crack-like defects. Manufacturing defects in welds are typically aligned in the circumferential direction. Therefore, in both welds a roughly semi-elliptical surface notch was machined from outside at the 6-o'clock position by means of a special sawing fixture. Afterwards a fatigue crack was created from the notch by cyclic internal pressure load.

Dimensions of test specimen and cracks:

External radius	$r_a =$	460 mm
Internal radius	$r_i =$	447 mm
Wall thickness	$h =$	13 mm
Distance between force application points	$2l_1 =$	4000 mm
Distance between bearings	$2l_2 =$	8800 mm

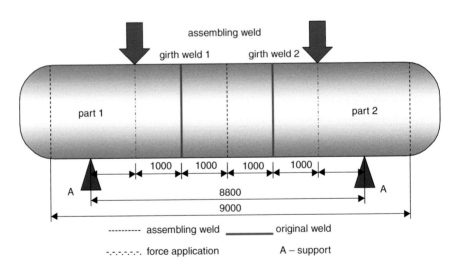

Fig. 9.22 Component test piping with two weld seams under internal pressure and four-point-bending

Fig. 9.23 Experimental setup for pressure and bending loading of the test piping

Long surface crack 1 in girth weld 1: Depth $a = 8.5$ mm, Length $c = 160$ mm
Short surface crack 2 in girth weld 2: Depth $a = 8.5$ mm, Length $c = 16$ mm

The experimental setup is illustrated in Fig. 9.23. It was so designed that the test piping could be loaded both by internal pressure p (water) and in four-point bending by two hydraulic actuators, each having a maximum force of $F = 1000$ kN. The purpose of the bending load, which causes the highest stresses at the place of the cracks, was to simulate additional stresses in the longitudinal direction of the piping that occur in practice. For weld cracks in the circumferential direction, these axial tensile stresses have far more impact on the fracture behavior and the plastification of the remaining wall thickness than the hoop stresses generated by pressure. The aim was to stress the test piping until failure or rupture (leakage) happens.

The loading program consisted of five levels:

(1) no bending load, applying the pressure up to $p = 5.5$ MPa (operating pressure)
(2) applying bending load up to a actuator forces of $F = 600$ kN, hold pressure at $p = 5.5$ MPa = constant
(3) hold the bending load at $F = 600$ kN, increasing the pressure to $p = 7.0$ MPa
(4) increasing the bending load to maximum $F = 1000$ kN, hold pressure $p = 7.0$ MPa = constant. Since up to this load level no failure of the specimen occurred, a last step followed:
(5) bending load $F = 1000$ kN = constant, increasing the pressure to bursting

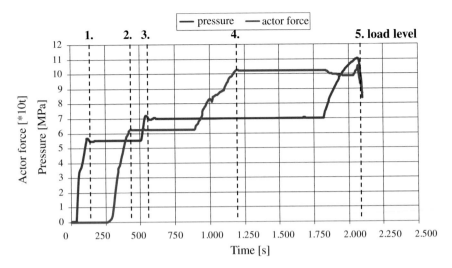

Fig. 9.24 Loading program of the test piping

The temporal course of internal pressure p and actuator forces F is shown in Fig.9.24.

Experimental Results

The test piping failed by stable growth of the artificially induced long surface crack 1 in the girth weld 1, resulting in an about 50 mm large leakage and a total pressure drop. In terms of safety, this situation »leak before break« is more favorable than the reverse case »unstable break before leak«.

The load values at failure amounted to:

- Pressure $p_c = 11$ MPa (burst pressure)
- Bending load $F_c = 1,000$ kN (i. e. bending stresses of $\sigma_b = 289$ MPa)

The failure occurred well before reaching the theoretical burst pressure of $p_{th} = 16.7$ MPa calculated for a defect-free pipe. The perforated pipe wall shows a clear division into notch area, fatigue crack and leak area, see Fig. 9.25. Fractographic investigations in the SEM clearly reveal the transition region from fatigue cracking to ductile fracture. Thus, the fatigue pre-crack in the weld metal served as a starter crack and the leak was caused by ductile crack growth in the burst test. This means, an assessment by ductile fracture mechanics is appropriate.

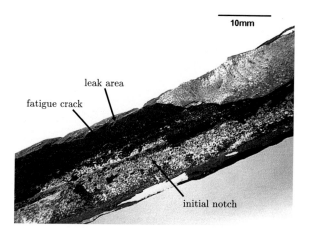

Fig. 9.25 Overall view of the fracture surface of crack 1

Application of Assessment Code

A major goal of the component test was to verify the developed assessment code
for applications to defective real welds in piping. For this purpose the assessment
code was applied to those parameters of the component test, which existed at actual
failure, i. e. burst pressure $p = 11$ MPa, additional bending stress of 289 MPa and
geometry of the critical crack 1. In terms of material behavior, the evaluation was
not only done using the properties of the weld metal WM, where the crack initiated,
but also with the much less favorable characteristics of the heat-affected zone HAZ,
because in engineering practice a defect in the HAZ can not be completely excluded.
Furthermore, variants were calculated by the program system supposing more severe,
fracture favoring conditions. In all cases the »safety against failure« calculated by
the assessment code was $S < 1$, i. e. failure had to occur and the evaluation point
P lay beyond the failure limit curve in the diagram 9.21. It was also investigated, at
which crack depth a the assessment code would have predicted initiation. Depending
on the material, values of $a = 4$ mm (HAZ) and $a = 5$ mm (WM) were obtained. In
conclusion, the component test has shown that with the help of the computer-aided
assessment code in all cases a conservative safety-related proof of the integrity of
gas pipelines with defective girth welds can be made. The assessment system evaluates only crack initiation, i. e. crack growth and strain hardening are considered as
additional safety reserves.

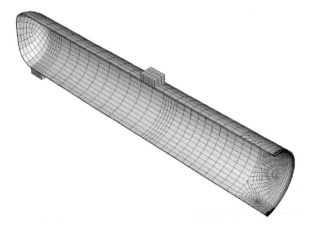

Fig. 9.26 Finite-element-mesh of one quarter of the test piping

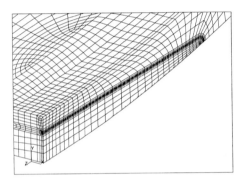

Fig. 9.27 Mesh detail in the crack region

9.3.4 FE-Analysis of Large Scale Piping Test

FE-Model of the Component

Figure 9.26 displays the FEM-discretization of one quarter of the test piping, whereby advantage was taken of twofold symmetry (about $14,000$ hexahedron elements and $70,000$ nodes). The pressure punches and bearing shells were modeled by volume elements of appropriate stiffness. The initial shape of crack 1 after fatigue pre-cracking was measured at the fracture surface (depth $a = 8.5$ mm and length $2c = 160$ mm) and approximated in the FEM-model as an elongated semi-ellipse. A very fine FEM mesh is required to discretize the crack region in order to capture the deformation and stress concentration correctly, see Fig. 9.27. Along the crack front the elements were concentrated fan-shaped in a tube of 20 segments. The smallest elements directly at the crack tip have a size of about $a/50$. Just around the crack tip,

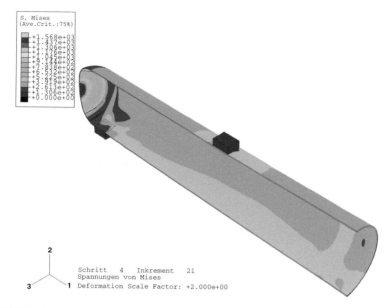

S, Mises
(Ave.Crit.:75%)

Schritt   4   Inkrement   21
Spannungen von Mises
Deformation Scale Factor: +2.000e+00

Fig. 9.28 Deformations and equivalent stress in the test piping at load level 5

special elastic-plastic crack tip elements were arranged along the entire crack front. These are collapsed 20-node hexahedral elements explained in Chap. 7. In order to calculate the J-integral, four integration paths were defined around each segment of the crack front. The calculated results show that the J values are almost independent of the integration path. Only the narrowest path yields a drop-off from known numerical reasons and was therefore not used.

In the FEM analysis, the entire course of loading in the component test was simulated comprising the described five levels. The actuator force F was applied as corresponding surface load on the top of the punch. The internal water pressure p was imposed on the entire inner wall. The material behavior was modeled as elastic-plastic with the yield curve of (9.8). The influence of large deformations was taken into account.

The calculated results show good agreement with strain gauge measurements on the test piping. The stress and deformation state is characterized as follows:

- From load level 4 on, the entire ligament of the pipe wall ahead of the crack is plastified, i. e. the plastic limit load is exceeded here locally.
- At load level 5, the test pipe is plastified completely in the cross section (tensile side), see Fig. 9.28.
- A pronounced plastic zone develops at the crack. The crack tip is blunting due to intense plastic deformations (CTOD), see Fig. 9.29.

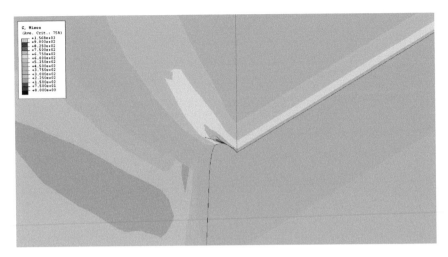

Fig. 9.29 Detail of the FEM-solution at crack 1 of the test piping at load level 5

Fracture Mechanical Evaluation

From the FEM analysis of the test piping both fracture mechanical parameters J-integral and crack opening displacement CTOD were calculated along the crack front. Figure 9.30 illustrates the behavior of J at the end of each load level. The calculations show that the J-values at apex A are always much larger than at the surface points C. This behavior explains why the crack growth has started in the apex region and led to a wall-penetrating fracture. Also on the fracture surface no crack growth could be observed in a lateral direction (in C).

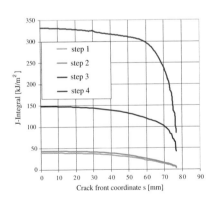

Fig. 9.30 Course of the J-integral along the semi-elliptical crack front at load levels 1–4

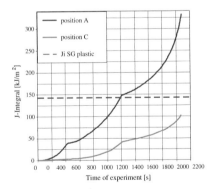

Fig. 9.31 J-integrals at the apex point A and the surface points C of the crack versus time

The temporal course of the calculated J-values at both vertex points A and C during the experiment is reported in Fig. 9.31. A monotonic increase of J-values can be observed during the test until failure. According to the J-integral concept a ductile stable crack growth is initiated, when J exceeds the physical initiation parameter J_i. Since the crack started in the weld metal (WM), its critical material parameter ($J_i = 143\,\text{kJ/m}^2$) is taken from Table 9.4 for evaluation. This value is also drawn in Fig. 9.31 as a horizontal line. The fracture criterion thus provides the statement that at the end of load level 4 ($p = 7\,\text{MPa}$, $F = 1{,}000\,\text{kN}$) crack initiation would have occurred in the piping. Obviously during load level 5 the ductile crack has spread through the wall until leakage, which was not yet modeled in the FEM-simulation. An evaluation based on the CTOD-concept led to comparable results.

Hence it was shown that the ductile fracture behavior in gas pipelines can be predicted quite accurately according to the J-integral concept by means of numerical FEM stress analysis, if detailed information about material parameters, defect geometry and loading are provided. Such elastic-plastic fracture mechanical analyses are much more expensive, but confirm on the other hand the results of the assessment code.

References

1. Kuna M, Springmann M, Mädler K, Hübner P, Pusch G (2002) Anwendung bruchmechanischer Bewertungskonzepte bei der Entwicklung von Eisenbahnrädern aus bainitischem Gusseisen. Konstruieren & Gießen pp 27–32
2. Kuna M, Springmann M, Mädler K, Hübner P, Pusch G (2005) Fracture mechanics based design of a railway wheel made of austempered ductile iron. Eng Fract Mech 72:241–253
3. Murakami Y (1987) Stress intensity factors handbook. vol. 1–5. Pergamon Press, Oxford
4. IAEA (2002) Advisory material for the IAEA regulations for the safe transport of radioactive material. International Atomic Energy Agency Safety Standards Series, TS-G-1, Appendix VI
5. CRIEPI (1990) Integrity of cast-iron cask against free drop test, part iii: Verification of brittle failure, design criterium. Technical Report, Central Research Institute for Electric Power Japan
6. Abaqus (1998) ABAQUS Theory und User Manual. Pawtucket, USA
7. Enderlein M, Klein K, Kuna M, Ricoeur A (2003) Numerical fracture analysis for the structural design of castor casks. In: 17th International conference on structural mechanics in reactor technology (SMIRT 17). Prague, Czech Republic
8. Kuna M, Wulf H, Rusakov A, Pusch G, Hübner P (2002) Ein computergestütztes bruch-mechanisches bewertungssystem für hochdruck-ferngasleitungen. In: MPA-Seminar, vol. 28. Stuttgart, pp 4/1–4/20
9. Kuna M, Wulf H, Rusakov A, Pusch G, Hübner P (2003) Entwicklung und verifikation eines bruchmechanischen bewertungssystems für hochdruck-ferngasleitungen. In: 35.Tagung Arbeitskeis Bruchvorgänge, DVM, Freiburg, pp 153–162
10. ÖSTV (1992) Empfehlungen zur bruchmechanischen bewertung von fehlern in konstruktio-nen aus metallischen werkstoffen. Technical Report, Österreichischer Stahlbauverband, AG Bruchmechanik
11. Milne I, Ainsworth RA, Dowling AR, Stewart AT (1991) Assessments of the integrity of structures containing defects. British Energy-Report, R6-Revision 3
12. Zerbst U, Wiesner C, Kocak M, Hodulak L (1999) Sintap: Entwurf einer verein-heitlichten europäischen fehlerbewertungsprozedur - eine einführung. Technical Report, GKSS-Forschungszentrum, Geesthacht

Appendix
Fundamentals of Strength of Materials

In this chapter some basics of higher Theory of Strength of Materials are presented. It is a compressed compilation of the essential concepts and relationships that are necessary for understanding the book without giving derivations. For the reader interested in detailed studies, advanced textbooks on strength of materials, continuum mechanics and material modeling are recommended.

A.1 Mathematical Representation and Notation

The mathematical representation of vectors and tensors as well as algebraic and analytic calculation rules are given both in symbolic form (bold italics) and in index notation. We follow the generally accepted rules and notation such as Einstein's summation convention and comma form of differentiation. The index notation is basically restricted to a Cartesian, space-fixed coordinate system. Upper-case letters for variables and indices generally refer to the reference configuration, whereas lower-case letters mark the current configuration. Matrices are written bold but upright.

Vector: $\quad \vec{a} \mathrel{\hat=} \boldsymbol{a} = a_1\boldsymbol{e}_1 + a_2\boldsymbol{e}_2 + a_3\boldsymbol{e}_3 = \sum_{i=1}^{3} a_i\boldsymbol{e}_i = a_i\boldsymbol{e}_i$

Tensor second order: $\quad \overset{\Rightarrow}{A} \mathrel{\hat=} \boldsymbol{A} = \sum_{i=1}^{3}\sum_{j=1}^{3} A_{ij}\boldsymbol{e}_i\boldsymbol{e}_j = A_{ij}\boldsymbol{e}_i\boldsymbol{e}_j$

Tensor 4th order: $\quad \mathbb{C} \mathrel{\hat=} C_{ijkl}\boldsymbol{e}_i\boldsymbol{e}_j\boldsymbol{e}_k\boldsymbol{e}_l$

Invariants of a tensor $\boldsymbol{A}$: $\quad I_1^A,\ I_2^A,\ I_3^A$

Unit tensor second order: $\boldsymbol{I} = \delta_{ij}\boldsymbol{e}_i\boldsymbol{e}_j$ with Kronecker-symbol: δ_{ij}

Scalar product: $\quad \boldsymbol{a} \cdot \boldsymbol{b} = a_i b_i$

Double scalar product: $\quad \boldsymbol{A} : \boldsymbol{B} = A_{ij}B_{ij}$

Column matrix $(m \times 1)$: $\mathbf{a} = [a_1 a_2 \cdots a_m]^{\mathrm{T}}$

M. Kuna, *Finite Elements in Fracture Mechanics*, Solid Mechanics and Its
Applications 201, DOI: 10.1007/978-94-007-6680-8,
© Springer Science+Business Media Dordrecht 2013

Matrix $(m \times n)$:

$$\mathbf{A} = \begin{bmatrix} A_{11} & A_{12} \cdots A_{1n} \\ A_{21} & A_{22} \cdots A_{2n} \\ \vdots & \vdots \\ A_{m1} & A_{m2} \cdots A_{mn} \end{bmatrix}$$

Transpose/inverse matrix: $\mathbf{A}^{\mathrm{T}}$, $\mathbf{A}^{-1}$, $\left(\mathbf{A}^{-1}\right)^{\mathrm{T}} = \mathbf{A}^{-\mathrm{T}}$

Variational symbol: δ

Partial derivative: $\partial(\cdot)/\partial x_i = (\cdot)_{,i}$

Nabla-operator: $\nabla(\cdot) = \dfrac{\partial(\cdot)}{\partial x_1}\boldsymbol{e}_1 + \dfrac{\partial(\cdot)}{\partial x_2}\boldsymbol{e}_2 + \dfrac{\partial(\cdot)}{\partial x_3}\boldsymbol{e}_3 = (\cdot)_{,i}\,\boldsymbol{e}_i$

Laplace-operator: $\Delta(\cdot) = \dfrac{\partial^2(\cdot)}{\partial x_1^2} + \dfrac{\partial^2(\cdot)}{\partial x_2^2} + \dfrac{\partial^2(\cdot)}{\partial x_3^2} = (\cdot)_{,ii}$

A.2 State of Deformation

A.2.1 Kinematics of Deformation

As a result of loading, every deformable body experiences a motion in space and time as illustrated in Fig. A.1. The kinematics of continua deals with the geometrical aspects of the motion. The entire motion is composed of translation and rotation of the body as a whole (*rigid body motions*) as well as the relative displacement of its particles (*deformation*). For this purpose it is necessary to assign at any time to each particle (material point) of the body a corresponding location in physical space (spatial point). Such a one-to-one mapping defines a *configuration* of the body. The undeformed state of the body at time $t = 0$ is denoted as *reference configuration* comprising the volume V and the surface A. The location of each particle P is

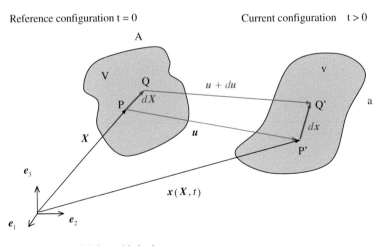

Fig. A.1 Kinematics of deformable body

identified by its position vector $X = X_M \, e_M$ in a Cartesian coordinate system with basis vectors e_M. These *material coordinates* do not change during the motion, but mark uniquely each particle and are also called Lagrangian coordinates. At a later time $t > 0$ of motion, the deformed body occupies the *current configuration* with the changed volume v and surface a. The particle P runs along the trajectory to P', its current position vector $x = x_m \, e_m$ is described by the *spatial coordinates* x_m (Eulerian coordinates). The motion of a body is thus the temporal sequence of such configurations. The function

$$x = x(X, t) \tag{A.1}$$

describes the position of the particles P' in space, whereby the time t is the curve parameter and the initial position X is the group parameter.

During the motion each material point P is shifted by the vector u to the position P'. This *displacement vector* marks the difference between the current position x of a particle and its initial one X.

$$u(X, t) = x - X \quad \text{or} \quad u_m(X, t) = x_m - X_m$$
$$x(X, t) = X + u \quad \text{or} \quad x_m(X, t) = X_m + u_m \tag{A.2}$$

Hence, the deformation state of a body is uniquely characterized in space and time by the displacement field $u(X, t)$.

The physical properties of the particles (density, temperature, material state, etc.) and the field quantities yet to be determined by the initial boundary value problem (displacements, strains, stresses, etc.) change during the motion. To describe the temporal changes of these field quantities χ, there are two fundamentally different approaches. In the Lagrang*ian (material) description* the changes of χ are traced for each particle and are characterized by the functional dependence on the material coordinates X.

$$\chi = \chi(X, t) \tag{A.3}$$

The observer is quasi attached to the moving particle and measures the changes of the field variables. In the Euler*ian (spatial) description* the changes of a field quantity χ are observed at a fixed spatial point and are therefore expressed as a function $\varXi$ of the spatial coordinates x.

$$\chi = \varXi(x, t) \tag{A.4}$$

An observer fixed at the position x measures the changes resulting from the fact that unlike particles with different properties are passing by. In principle both formulations are equivalent and can be converted into each other, if the motion is known. The Eulerian description is preferred in fluid mechanics, because there changes of field quantities (pressure, velocity) at fixed positions are commonly of interest. The Lagrangian description has advantages in solid mechanics, because here the initial state is usually known and the individual properties of the particles (deformation, stress or state variables) need to be traced during the loading history.

A.2.2 Deformation Gradient and Strain Tensors

To characterize the stress in the interior of a solid body, the relative displacements between two infinitesimally neighboring particles P and Q are of primary importance. In Fig. A.1 a material line element $\overline{PQ} = \mathrm{d}X = \mathrm{d}X_M \, e_M$ is considered in the initial configuration, which is deformed into $\overline{P'Q'} = \mathrm{d}x = \mathrm{d}x_m \, e_m$ during its motion in the current configuration.

$$\mathrm{d}x_m = \frac{\partial x_m(X, t)}{\partial X_M} \mathrm{d}X_M = F_{mM} \, \mathrm{d}X_M \quad \text{or} \quad \mathrm{d}x = F \cdot \mathrm{d}X \qquad (A.5)$$

The partial derivatives of x_m with respect to X_M form a tensor of second order that is called a *deformation gradient* F. It can be expressed by the displacement field and the unit tensor of second order I using (A.2).

$$F_{mM} = \frac{\partial x_m}{\partial X_M} = \delta_{mM} + \frac{\partial u_m}{\partial X_M}, \quad I = \delta_{mM} e_m e_M \qquad (A.6)$$

The mathematical inversion of (A.5) represents the deformation of the reference configuration as viewed from the current configuration.

$$\mathrm{d}X_M = \frac{\partial X_M(x, t)}{\partial x_m} \mathrm{d}x_m = F^{-1}_{Mm} \, \mathrm{d}x_m \quad \text{or} \quad \mathrm{d}X = F^{-1} \cdot \mathrm{d}x \qquad (A.7)$$

$$F^{-1}_{Mm} = \frac{\partial X_M}{\partial x_m} = \delta_{Mm} - \frac{\partial u_M}{\partial x_m} \qquad (A.8)$$

The *deformation gradient* provides the relationship between the material line elements in the initial and the current configuration. It is defined on the basis vectors of both configurations (two-point tensor) and in general not symmetric. To make the affine mapping $X \leftrightarrow x$ unique and reversible, the Jacobian functional determinant J must be definite.

$$J = \det \left[\frac{\partial x_m}{\partial X_M} \right] = \det[F_{mM}] \neq 0 \qquad (A.9)$$

The deformation gradient involves both the elongation and the local rigid body rotation in the vicinity of a material point X. By $\overline{PQ} = \mathrm{d}X$ an arbitrary direction in the neighborhood of P is defined. The elongation and the local rotation of the line element can be separated from each other by means of the *polar decomposition*:

$$F = R \cdot U = V \cdot R \quad \text{or} \quad F_{kN} = R_{kM} \, U_{MN} = V_{km} \, R_{mN}. \qquad (A.10)$$

The rotation is described by the orthogonal tensor R,

$$R^{\mathrm{T}} = R^{-1} \quad \text{or} \quad R_{Lk} = R^{-1}_{kL} \quad \text{and} \quad \det[R_{kL}] = 1, \qquad (A.11)$$

whereas both *stretch tensors* U and V quantify solely the stretch (final length / initial length) of the line elements. The mapping F can be imagined as a sequence of a rotation R and an elongation V or as an elongation U followed by a rotation R. The tensors U and V are symmetric and positive definite.

$$U = U_{MN}\, e_M\, e_N \quad \text{right stretch tensor (reference configuration)}$$
$$V = V_{mn}\, e_m\, e_n \quad \text{left stretch tensor (current configuration)}$$
$$V = R \cdot U \cdot R^{\mathrm{T}} \quad \text{and} \quad U = R^{\mathrm{T}} \cdot V \cdot R \tag{A.12}$$

To split off the less interesting rotation term, the so-called right and left Cauchy-Green *deformation tensors* are introduced, which are obtained from the stretch tensors or straight from the deformation gradient F as follows:

$$C = F^{\mathrm{T}} \cdot F = U^{\mathrm{T}} \cdot U \quad \text{or} \quad C_{MN} = F_{kM}\, F_{kN} = U_{LM}\, U_{LN} \quad \text{(right)}$$
$$b = F \cdot F^{\mathrm{T}} = V \cdot V^{\mathrm{T}} \quad \text{or} \quad b_{mn} = F_{mL}\, F_{nL} = V_{ml}\, V_{nl} \quad \text{(left)} \tag{A.13}$$

All measures of deformation U, V, b and C migrate into the unit tensor I if no stretching occurs.

The squares of the arc lengths of material line elements are calculated in the reference configuration as

$$(\mathrm{d}L)^2 = \mathrm{d}X \cdot \mathrm{d}X = (F^{-1} \cdot \mathrm{d}x) \cdot (F^{-1} \cdot \mathrm{d}x)$$
$$= \mathrm{d}x \cdot (F^{-\mathrm{T}} \cdot F^{-1}) \cdot \mathrm{d}x = \mathrm{d}x \cdot b^{-1} \cdot \mathrm{d}x = \mathrm{d}x_k\, b_{kl}^{-1}\, \mathrm{d}x_l \tag{A.14}$$

and in the current configuration as

$$(\mathrm{d}l)^2 = \mathrm{d}x \cdot \mathrm{d}x = (F \cdot \mathrm{d}X) \cdot (F \cdot \mathrm{d}X)$$
$$= \mathrm{d}X \cdot (F^{\mathrm{T}} \cdot F) \cdot \mathrm{d}X = \mathrm{d}X \cdot C \cdot \mathrm{d}X = \mathrm{d}X_K\, C_{KL}\, \mathrm{d}X_L . \tag{A.15}$$

The elongation of the line element amounts to

$$(\mathrm{d}l)^2 - (\mathrm{d}L)^2 = 2\,\mathrm{d}X \cdot E \cdot \mathrm{d}X = 2\,\mathrm{d}x \cdot \eta \cdot \mathrm{d}x . \tag{A.16}$$

This way the following strain measures are defined:
Green-Lagrang*ian strain tensor* (related to the reference configuration):

$$E = \frac{1}{2}(C - I) \quad \text{or} \quad E_{KL} = \frac{1}{2}(C_{KL} - \delta_{KL}) \tag{A.17}$$

Euler-Almansi *strain tensor* (related to the current configuration):

$$\boldsymbol{\eta} = \frac{1}{2}(\boldsymbol{I} - \boldsymbol{b}^{-1}) \quad \text{or} \quad \eta_{kl} = \frac{1}{2}(\delta_{kl} - b_{kl}^{-1}) \qquad (A.18)$$

Both strain tensors yield the information about the relative changes of lengths and angles of material line elements in the vicinity of a point P due to deformation. They are symmetric and reduce in the undeformed state to zero. By means of the deformation gradient the strain tensors of both configurations can be converted into each other.

$$\boldsymbol{E} = \boldsymbol{F}^{\mathrm{T}} \cdot \boldsymbol{\eta} \cdot \boldsymbol{F} \quad \text{and} \quad \boldsymbol{\eta} = \boldsymbol{F}^{-\mathrm{T}} \cdot \boldsymbol{E} \cdot \boldsymbol{F}^{-1} \qquad (A.19)$$

By applying relation (A.6) we get the non-linear relationship between displacements and strains in the form:

$$E_{KL} = \frac{1}{2}\left(\frac{\partial u_K}{\partial X_L} + \frac{\partial u_L}{\partial X_K} + \frac{\partial u_M}{\partial X_K}\frac{\partial u_M}{\partial X_L}\right) \quad \text{and}$$

$$\eta_{kl} = \frac{1}{2}\left(\frac{\partial u_k}{\partial x_l} + \frac{\partial u_l}{\partial x_k} - \frac{\partial u_m}{\partial x_k}\frac{\partial u_m}{\partial x_l}\right). \qquad (A.20)$$

A.2.3 Rate of Deformation

In the following the temporal changes of motion and deformation of a material particle and its vicinity will be examined in more detail. Velocity $\boldsymbol{v}$ and acceleration $\boldsymbol{a}$ of a particle are obtained from (A.2) by the first and second material time derivative of the displacement vector $\boldsymbol{u}$.

$$\boldsymbol{v} = \dot{\boldsymbol{u}} \quad \text{or} \quad v_i(\boldsymbol{X}, t) = \frac{\partial u_i(\boldsymbol{X}, t)}{\partial t} = \dot{u}_i$$

$$\boldsymbol{a} = \dot{\boldsymbol{v}} = \ddot{\boldsymbol{u}} \quad \text{or} \quad a_i(\boldsymbol{X}, t) = \frac{\partial v_i(\boldsymbol{X}, t)}{\partial t} = \dot{v} = \ddot{u}_i \qquad (A.21)$$

The velocity (time-rate) of the deformation process is important for the treatment of inelastic material laws, which are formulated as constitutive relationship between stress-rates and strain-rates. To this end we examine the relative velocity $d\boldsymbol{v}$ of two neighboring particles of a line element $d\boldsymbol{x} = \overline{P'Q'}$ in the current configuration at time $t > 0$, see Fig. A.2. A Taylor-expansion of the velocity at point $P(\boldsymbol{x})$ gives $d\boldsymbol{v}$, whereby $\boldsymbol{l}$ denotes the spatial *velocity gradient*.

Fig. A.2 Velocity field of neighboring particles

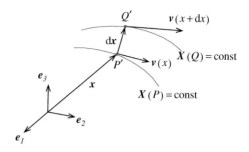

$$dv = d\dot{x} = l \cdot dx \quad \text{or} \quad dv_i = l_{ij}\,dx_j \tag{A.22}$$

$$l = \frac{\partial v}{\partial x} = \nabla \cdot v = l_{ij}e_ie_j \quad \text{with} \quad l_{ij} = \frac{\partial v_i}{\partial x_j} = v_{i,j} \tag{A.23}$$

$$l = \dot{F} \cdot F^{-1} \quad \text{or} \quad l_{ij} = \dot{F}_{iM} F_{Mj}^{-1} \tag{A.24}$$

The velocity gradient can be decomposed into a symmetric and skew-symmetric part:

$$l = d + w$$

$$d = \frac{1}{2}(l + l^{\mathrm{T}}) \quad \text{or} \quad d_{ij} = \frac{1}{2}(v_{i,j} + v_{j,i}) \tag{A.25}$$

$$w = \frac{1}{2}(l - l^{\mathrm{T}}) \quad \text{or} \quad w_{ij} = \frac{1}{2}(v_{i,j} - v_{j,i})$$

The tensor d denotes the rate at which lengths and angles of a material line element change in time and is called *rate of deformation* tensor. The *spin tensor* w indicates the angular velocity by which a line element rotates. There are the following relationships with the time derivatives of the strain tensors:
In the current configuration:

$$d = \dot{\eta} + l^{\mathrm{T}} \cdot \eta + \eta \cdot l \quad \text{or} \quad d_{ij} = \dot{\eta}_{ij} + v_{k,i}\,\eta_{kj} + \eta_{ik}\,v_{k,j} \tag{A.26}$$

and in the reference configuration:

$$D = F^{\mathrm{T}} \cdot d \cdot F = \dot{E} \quad \text{with} \quad \dot{E} = \frac{1}{2}(v + v^{\mathrm{T}} + v^{\mathrm{T}} \cdot u + u^{\mathrm{T}} \cdot v)$$

$$D_{MN} = x_{m,M}\,d_{mn}\,x_{n,N} = \dot{E}_{MN} \quad \text{with} \tag{A.27}$$

$$\dot{E}_{MN} = \frac{1}{2}(v_{M,N} + v_{N,M} + v_{K,M}\,u_{K,N} + u_{K,M}\,v_{K,N}).$$

A.2.4 Linearization for Small Deformations

For many tasks in strength of materials the strains can be considered as infinitesimally small. In this case, the above presented theory of large finite deformations simplifies significantly, because the non-linear quadratic terms of the displacement gradient in the Green-Lagrange (A.17) and Euler-Almansi strain tensors (A.18) may then be neglected in (A.20). Likewise, it can be shown that the derivatives of the displacement vector with respect to material and spatial coordinates coincide

$$\frac{\partial u_M}{\partial X_N} \approx \frac{\partial u_m}{\partial x_n}.$$ (A.28)

Hence we obtain the well-known infinitesimal strain tensor ε

$$\varepsilon \approx \eta \approx E = \frac{1}{2}(\nabla u + (\nabla u)^{\mathrm{T}}) \quad \text{or} \quad \varepsilon_{ij} \approx \eta_{ij} \approx E_{ij} = \frac{1}{2}(u_{i,j} + u_{j,i}).$$
(A.29)

$$\varepsilon = \varepsilon_{ij}\, e_i\, e_j = \varepsilon^{\mathrm{T}} \quad \text{or} \quad [\varepsilon_{ij}] = \begin{bmatrix} \varepsilon_{11} & \varepsilon_{12} & \varepsilon_{13} \\ \varepsilon_{21} & \varepsilon_{22} & \varepsilon_{23} \\ \varepsilon_{31} & \varepsilon_{32} & \varepsilon_{33} \end{bmatrix} = [\varepsilon_{ij}]^{\mathrm{T}}$$ (A.30)

In the same way the rate of deformation is reduced to

$$\dot{\varepsilon} \approx \dot{\eta} \approx \dot{E} = \frac{1}{2}(\nabla v + (\nabla v)^{\mathrm{T}}) \quad \text{or} \quad \dot{\varepsilon}_{ij} \approx \dot{\eta}_{ij} \approx \dot{E}_{ij} = \frac{1}{2}(v_{i,j} + v_{j,i}).$$ (A.31)

The geometrical meaning of the individual components of the strain tensor is illustrated in Fig. A.3 for the (x_1, x_2)-plane. We consider three points $P(x_1, x_2)$, $Q(x_1 + dx_1, x_2)$ and $R(x_1, x_2 + dx_2)$, the positions of which are marked on the undeformed body. As a result of deformation, they move into the positions P', Q' and R'. The strain tensor (A.29) contains derivatives of the displacements that correspond to the Taylor-expansion of $u(x)$ at point P as displayed in Fig. A.3. The relative elongations of the line elements $\overline{PQ}$ and $\overline{PR}$ in their axial direction x_1 and x_2 are called *normal strains* (analogously for the x_3-coordinate):

$$\varepsilon_{11} = \frac{\overline{P'Q'} - \overline{PQ}}{\overline{PQ}} = \frac{\partial u_1}{\partial x_1}, \quad \varepsilon_{22} = \frac{\partial u_2}{\partial x_2}, \quad \varepsilon_{33} = \frac{\partial u_3}{\partial x_3}.$$ (A.32)

Shear strains are a measure for the change of angles at a deformed volume element with respect to the rectangular initial state, see Fig. A.3:

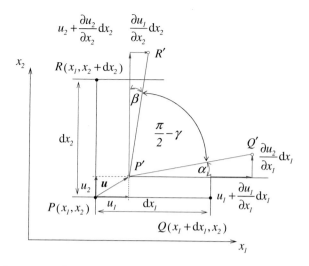

Fig. A.3 Displacements and strains in the (x_1, x_2)-plane

$$\gamma_{12} = \alpha + \beta \approx \tan \alpha + \tan \beta = \frac{\partial u_2}{\partial x_1} + \frac{\partial u_1}{\partial x_2}. \tag{A.33}$$

The associated component ε_{12} of the strain tensor has half the value of the engineering glide angle γ_{12}. Transferring these considerations to the other two coordinate planes, we can express all shear strains as follows:

$$\begin{aligned}
\varepsilon_{12} &= \frac{1}{2}\left(\frac{\partial u_2}{\partial x_1} + \frac{\partial u_1}{\partial x_2}\right) = \varepsilon_{21} = \frac{1}{2}\gamma_{12} \\
\varepsilon_{23} &= \frac{1}{2}\left(\frac{\partial u_3}{\partial x_2} + \frac{\partial u_2}{\partial x_3}\right) = \varepsilon_{32} = \frac{1}{2}\gamma_{23} \\
\varepsilon_{31} &= \frac{1}{2}\left(\frac{\partial u_1}{\partial x_3} + \frac{\partial u_3}{\partial x_1}\right) = \varepsilon_{13} = \frac{1}{2}\gamma_{31}.
\end{aligned} \tag{A.34}$$

The normal and shear strains summarized in the strain tensor ε of (A.30) characterize the state of deformation at a point P of the body. The terms on the main diagonal (index $i = j$) correspond to the normal strains, whereas the shear strains form the off-diagonal terms (index $i \neq j$).

For the strain tensor one can find so-called *principal axes*, i.e. such a coordinate system, where all shear strains vanish and the normal strains assume extreme values (*principal strains*). The associated coordinate transform into the principal axes system will be explained in detail for the stress tensor in Sect. A.3.3. The three principal strains ε_α (with $\alpha = \{$I, II, III$\}$) belong to spatial directions orientated orthonormal to each other. They are sorted by magnitude.

$$\text{Convention:} \quad \varepsilon_I \geq \varepsilon_{II} \geq \varepsilon_{III} \tag{A.35}$$

The shear strains attain extreme values in those coordinate systems rotated each by 45 degrees with respect to the principal axes. These three extreme values are called *principal shear strains* and are calculated as follows:

$$\gamma_I = (\varepsilon_{II} - \varepsilon_{III}), \quad \gamma_{II} = (\varepsilon_I - \varepsilon_{III}), \quad \gamma_{III} = (\varepsilon_I - \varepsilon_{II}) \tag{A.36}$$

The following decomposition of the strain tensor is important in material theory: ε_{ij} can be split into a portion ε_{ij}^D describing a pure shape change of the volume element (distortion), and a portion ε^H representing only the change in volume (dilatation).

$$\varepsilon_{ij} = \varepsilon_{ij}^D + \varepsilon^H \delta_{ij} \tag{A.37}$$

The relative volume change is the sum of all normal strains. It is expressed as an average volumetric strain on all sides by the so-called *spherical tensor*. The remaining shape changing portion is called a *deviator*.

Volume change: $\quad \dfrac{\Delta V}{V_0} = \varepsilon_{11} + \varepsilon_{22} + \varepsilon_{33} = \varepsilon_{kk} = 3\,\varepsilon^H$

Spherical tensor: $\quad \varepsilon^H \delta_{ij}\,, \quad \varepsilon^H = \dfrac{\varepsilon_{kk}}{3}\,, \quad$ Deviator: $\varepsilon_{ij}^D = \varepsilon_{ij} - \varepsilon^H \delta_{ij}$

$$\tag{A.38}$$

A.3 State of Stress

A.3.1 Stress Vector and Stress Tensor

A deformable body is exposed to external forces that may, depending on their physical origin, either act on the surface a or in the volume v. This will be discussed first in the current configuration shown in Fig. A.4. *Surface loads* $\bar{t}$ are forces ds per unit area, imposed on specific areas on the surface of the body, such as e.g. an external pressure. Under *body forces* $\bar{b}$ we mean external forces per unit volume acting at particles in the interior of the body, such as e.g. gravity or electromagnetic fields. Distributed line loads and concentrated forces known from engineering mechanics are special degenerated cases of these surface and volume forces.

$$\text{Surface loads:} \ \ \bar{t} = \frac{ds}{da}\,, \quad \text{Volume forces:} \ \ \bar{b} = \frac{ds}{dv} \tag{A.39}$$

Fig. A.4 Body with surface
tractions and volume forces

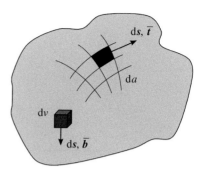

Due to external loading *internal* forces arise in the body, which can be made visible by the method of sections on imaginary cutting planes. They have the character of surface loads and are denoted by the terms *traction vector* and *stress tensor*. At any point P, we define a differentially small section da with an arbitrary orientation, fixed by its unit normal vector n as depicted in Fig. A.5. From the acting differential section force ds per area da we obtain in the limit process the *traction vector* t

$$t(x, n, t) = \frac{ds}{da} = \frac{\text{current section force}}{\text{current sectional area}}.$$ (A.40)

Since we are dealing with the actual section force on the deformed surface element in the current configuration, this is called the *true* or *Cauchy stress vector*. The Cauchy stress vector can be decomposed into a component perpendicular to the surface, the *normal stress* σ, and a *shear stress* τ acting tangentially to the surface.

The stress or traction vector t depends on the location $P(x)$, the orientation n of the sectional area and possibly on time t. This means *one* sectional orientation n alone is not sufficient to describe the stress state at P uniquely with respect to any other sectional areas. Therefore, we investigate the stress state at P with respect to three cutting planes aligned perpendicular to the coordinate axes $n_1 = e_1$, $n_2 = e_2$ and $n_3 = e_3$, which delivers in each case a traction vector $t_1(n_1)$, $t_2(n_2)$ and $t_3(n_3)$, respectively. Now, the traction vectors t_i on each section are split into their three Cartesian components σ_{ij} as illustrated in Fig. A.6.

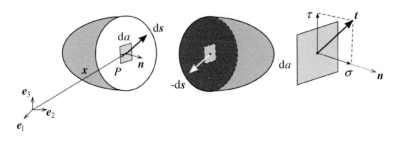

Fig. A.5 Traction vector t on a cutting plane da with orientation n

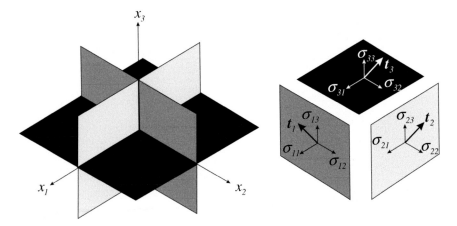

Fig. A.6 On the definition of the stress tensor

$$t_i = \sigma_{i1}e_1 + \sigma_{i2}e_2 + \sigma_{i3}e_3 = \sigma_{ij}e_j \qquad (A.41)$$

Hereby, the 1st index i denotes the orientation of the section, whereas the 2nd index j indicates the direction of the stress component σ_{ij}. Logically, components owing the same indices $i = j$ are normal stresses $\sigma_{11}, \sigma_{22}, \sigma_{33}$, which act perpendicular on the respective cutting plane. If the stress components have different indices $i \neq j$, it concerns the six shear stresses $\sigma_{12}, \sigma_{21}, \sigma_{23}, \sigma_{32}, \sigma_{31}, \sigma_{13}$, which we denote also by τ_{ij}. With regard to the sign of the stresses the same rules hold as common for all sectional variables, i.e. stress components are defined as positive, if they point on the right-hand (resp. left-hand) section in the positive (resp. negative) coordinate direction.

If the components of the three tractions t_i are arranged as rows in a 3×3 matrix, they form the nine elements of a second-order tensor.

This tensor $\boldsymbol{\sigma}$ is called the *Cauchy stress tensor*.

$$\begin{bmatrix} t_1 \ t_2 \ t_3 \end{bmatrix}^{\mathrm{T}} = \begin{bmatrix} \sigma_{11} & \sigma_{12} & \sigma_{13} \\ \sigma_{21} & \sigma_{22} & \sigma_{23} \\ \sigma_{31} & \sigma_{32} & \sigma_{33} \end{bmatrix} = [\sigma_{ij}], \qquad \boldsymbol{\sigma} = \sigma_{ij}e_i e_j \qquad (A.42)$$

By applying the equilibrium of torque to the volume element it can be proved that shear stresses, assigned to each other on orthonormal sections, are equal, i.e. $\tau_{21} = \tau_{12}, \tau_{32} = \tau_{23}$ and $\tau_{13} = \tau_{31}$. This makes the stress tensor symmetric

$$\boldsymbol{\sigma} = \boldsymbol{\sigma}^{\mathrm{T}} \quad \text{or} \quad \sigma_{ij} = \sigma_{ji} \qquad (A.43)$$

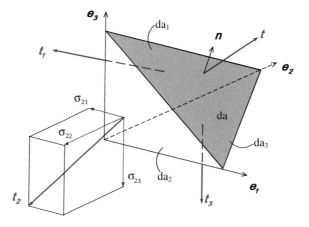

Fig. A.7 Stress state at a tetrahedral element

and simplifies it to six independent scalar elements $\sigma_{11}, \sigma_{22}, \sigma_{33}, \tau_{12}, \tau_{23}, \tau_{31}$.

> The Cauchy stress tensor σ characterizes completely the stress state in an infinitesimal volume element at point $P(x)$. This means, the entity of all traction vectors t at P for all possible cut-orientations n is uniquely determined by σ.

In order to prove this, we need to find a relationship between σ and t. To this end, we investigate at P in the current configuration a differentially small tetrahedron, delimited by the already declared sectional area da with its normal vector n and another three triangular faces da_i, aligned perpendicular to the coordinate axes e_i, see Fig. A.7. The areas of these triangles are calculated by projecting the sectional area da onto the respective coordinate axes using the cosines of the included angles:

$$da_i = n \cdot e_i \, da = \cos(n, e_i) \, da = n_i \, da \,.$$

If the equilibrium of forces is analyzed now for the tetrahedral element together with (A.41), we obtain

$$t \, da - t_i \, da_i = 0$$
$$t_j \, e_j \, da - \sigma_{ij} \, n_i \, e_j \, da = (t_j - \sigma_{ij} n_i) \, e_j \, da = 0 \,.$$

Because the expression in parenthesis must vanish, the sought relationship between traction vector and stress tensor (employing its symmetry) is found:

$$t_j = \sigma_{ij} \, n_i = \sigma_{ji} \, n_i \quad \text{or} \quad t(x, n, t) = \sigma^{\mathrm{T}}(x, t) \cdot n = \sigma \cdot n \,. \tag{A.44}$$

or in matrix notation:

$$
\begin{bmatrix} t_1 \\ t_2 \\ t_3 \end{bmatrix} = \begin{bmatrix} \sigma_{11} & \tau_{12} & \tau_{13} \\ \tau_{21} & \sigma_{22} & \tau_{23} \\ \tau_{31} & \tau_{32} & \sigma_{33} \end{bmatrix} \begin{bmatrix} n_1 \\ n_2 \\ n_3 \end{bmatrix} , \quad [t_i] = [\sigma_{ij}][n_j]. \tag{A.45}
$$

Thus, the traction vector t can be calculated for an arbitrarily oriented cutting plane by a scalar product between stress tensor σ and the normal unit vector n. This relation is called *Cauchy's stress formula*.

This holds also for the limiting case if the point P lies on the surface of the body and the area element da becomes a part of the surface with outward directed normal vector n. Then the local stress state (tensor) has to be in accordance with the resulting traction vector $t = \sigma \cdot n$ to equal the external surface load $\bar{t}$.

A.3.2 Stresses in the Reference Configuration

The Cauchy stress tensor describes the true stresses in the Eulerian description in the current configuration. If stress quantities are needed in the Lagrangian description, then the force and area variables at location x have to be converted into the reference configuration as sketched in Fig. A.8. By relating the surface load $d\bar{s}$ or the section force ds to the initial surface dA, the vector of *nominal stresses* is obtained:

$$
p = \frac{ds}{dA} = \frac{\text{current section force}}{\text{initial sectional area}} \tag{A.46}
$$

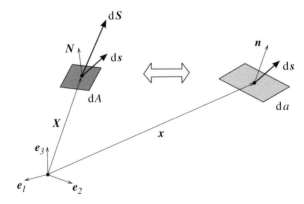

Fig. A.8 On the definition of stress tensors in the reference configuration

The associated stress tensor P is called *1. Piola-Kirchhoff stress tensor*. The Cauchy formula (A.44) applies to it in analogous form, whereby now the normal vector N is used on dA in the reference configuration.

$$p(X, N, t) = N \cdot P \quad \text{or} \quad p_l = N_M P_{Ml} \tag{A.47}$$

The conversion into the Cauchy stress tensor is read as follows:

$$P = \det(F) F^{-1} \cdot \sigma \quad \text{or} \quad P_{Ml} = J F_{Mk}^{-1} \sigma_{kl}. \tag{A.48}$$

The 1. Piola-Kirchhoff stress tensor is not symmetric, which is very inappropriate for the formulation of constitutive equations. Therefore, the 2. *Piola-Kirchhoff stress tensor* was introduced as a pseudo-stress tensor to achieve symmetry. To this end an »fictitious« section force vector dS is defined, which is calculated formally by a pull-back transformation of ds into the reference configuration.

$$dS = F^{-1} \cdot ds \quad \text{analogous to} \quad dX = F^{-1} \cdot dx \tag{A.49}$$

This yields the stress vector $\hat{T}$,

$$\hat{T} = \frac{dS}{dA} = \frac{\text{»initial« section force}}{\text{initial sectional area}} = F^{-1} \cdot p \tag{A.50}$$

and the associated 2. Piola-Kirchhoff stress tensor T,

$$\hat{T} = N \cdot T(X, t) \quad \text{or} \quad \hat{T}_L = N_M T_{ML}. \tag{A.51}$$

By means of the deformation gradient F the following relationships can be established to the two other stress definitions:

$$\begin{aligned} T &= P \cdot F^{-T} & \text{or} \quad T_{ML} &= P_{Ml} F_{Ll}^{-1} \\ T &= \det(F) F^{-1} \cdot \sigma \cdot F^{-T} & \text{or} \quad T_{ML} &= J F_{Mm}^{-1} \sigma_{ml} F_{Ll}^{-1}. \end{aligned} \tag{A.52}$$

The symmetry of σ_{ml} implies the symmetry of $T_{ML} = T_{LM}$.

A.3.3 Transformation into Principal Axes

In all previous definitions and considerations of mechanical field quantities, a consistent global Cartesian coordinate system (x_1, x_2, x_3) with basis vectors e_i was used. Frequently, it turns out to be necessary or reasonable to change the reference coordinate system. During a parallel shift of the coordinate system the strain and stress measures do not change, since they represent already differentiated quantities. Therefore, we focus on the investigation of a reference coordinate system (x'_1, x'_2, x'_2), whose basis e'_i is rotated with respect to the original e_j system. This rotation is described by an orthogonal transformation matrix $\mathbf{r} = [r_{ij}]$, relating the basis vectors of both reference systems to each other,

$$e'_i = r_{ij}\, e_j .$$

(A.53)

The elements of the transformation matrix are simply obtained as the components of the new basis vectors expressed by those of the old one, $r_{ij} = e'_i \cdot e_j = \cos(e'_i, e_j)$.

According to the laws of tensor algebra any vector $\mathbf{a}$ or any second order tensor $\mathbf{A}$ of the old coordinate system e_j can be converted into the rotated coordinate system e'_i by applying the following transformation rules to their components:

$$
\begin{aligned}
a'_i &= r_{ij}\, a_j , & \mathbf{a}' &= a'_i e'_i \stackrel{!}{=} \mathbf{a} = a_i e_i \\
A'_{ij} &= r_{ik}\, A_{kl}\, r_{jl} , & \mathbf{A}' &= A'_{ij} e'_i e'_j \stackrel{!}{=} \mathbf{A} = A_{ij} e_i e_j .
\end{aligned}
$$

(A.54)

By using this »conversion rule« all previously introduced quantities as displacement vector, force vectors, stretch, strain and stress tensors can be transferred into a rotated coordinate system, of course without altering their physical meaning. Symmetric tensors of second order have special properties that will be exemplified though the Cauchy stress tensor in more detail.

Of all possible rotated reference systems, the so-called *principal axes system* is distinguished by a transformed tensor that assumes the form of a diagonal matrix. In case of the stress tensor σ' this means that in the new coordinate directions there are only three normal stresses and all shear stresses disappear.

$$
[\sigma'_{ij}] = \begin{bmatrix} \sigma'_{11} & 0 & 0 \\ 0 & \sigma'_{22} & 0 \\ 0 & 0 & \sigma'_{33} \end{bmatrix}
$$

(A.55)

This special reference system e'_i is called a *principal axes system* and the corresponding directions are denoted as *principal axes*. The three normal stresses associated with the principal axes system represent extreme values irrespective of the former reference systems and are named *principal normal stresses*.

A principal axis has a descriptive picture: The traction vector t is colinear with the direction of the surface normal vector $\mathbf{n} = e'_i$, whereby the normal stress has a yet unknown magnitude σ. By using the Cauchy formula (A.45) and the unit tensor δ_{ij}, we find

$$
\begin{aligned}
t_i &= \sigma_{ij}\, n_j \stackrel{!}{=} \sigma\, n_i \\
\sigma_{ij}\, n_j - \sigma\, \delta_{ij}\, n_j &= (\sigma_{ij} - \sigma\, \delta_{ij})\, n_j = 0
\end{aligned}
$$

(A.56)

or in matrix notation

$$\begin{bmatrix} \sigma_{11} - \sigma & \tau_{12} & \tau_{13} \\ \tau_{21} & \sigma_{22} - \sigma & \tau_{23} \\ \tau_{31} & \tau_{32} & \sigma_{33} - \sigma \end{bmatrix} \begin{bmatrix} n_1 \\ n_2 \\ n_3 \end{bmatrix} = \begin{bmatrix} 0 \\ 0 \\ 0 \end{bmatrix}. \tag{A.57}$$

From the mathematical point of view the determination of principal axes represents an eigenvalue problem. Hereby, σ means the sought eigenvalue, and the corresponding eigenvector $\boldsymbol{n}$ gives the principal direction. The homogeneous linear system of equations for the sought principal directions has *non-trivial solutions*, if and only if the determinant of coefficient matrix is zero.

$$\det(\sigma_{ij} - \sigma \, \delta_{ij}) = \begin{vmatrix} \sigma_{11} - \sigma & \tau_{12} & \tau_{13} \\ \tau_{21} & \sigma_{22} - \sigma & \tau_{23} \\ \tau_{31} & \tau_{32} & \sigma_{33} - \sigma \end{vmatrix} = 0 \tag{A.58}$$

The resolution of the determinant leads to an equation of 3rd degree

$$\sigma^3 - I_1^\sigma \, \sigma^2 + I_2^\sigma \, \sigma - I_3^\sigma = 0. \tag{A.59}$$

The three real roots give the principal stresses σ_α (with $\alpha = \{I, II, III\}$). They are sorted by magnitude.

$$\text{Convention:} \quad \sigma_I \geq \sigma_{II} \geq \sigma_{III} \tag{A.60}$$

Solving the homogeneous system of equations for each principal stress σ_α provides the three orthonormal principal axes, which have yet to be normalized to unit length.

$$\boldsymbol{n}_\alpha = n_{\alpha 1} \, \boldsymbol{e}_1 + n_{\alpha 2} \, \boldsymbol{e}_2 + n_{\alpha 3} \, \boldsymbol{e}_3 = n_{\alpha i} \, \boldsymbol{e}_i, \quad \alpha = \{I, II, III\} \tag{A.61}$$

The quantities $I_k^\sigma (\sigma_{ij})$ (with $k = \{1, 2, 3\}$) in (A.59) denote the three *invariants* of the stress tensor. As the name implies, these are characteristic figures of a second-order tensor that do not depend on the coordinate system.

$$I_1^\sigma(\sigma_{ij}) = \sigma_{11} + \sigma_{22} + \sigma_{33} = \sigma_{kk}$$

$$I_2^\sigma(\sigma_{ij}) = \begin{vmatrix} \sigma_{11} & \tau_{12} \\ \tau_{12} & \sigma_{22} \end{vmatrix} + \begin{vmatrix} \sigma_{11} & \tau_{13} \\ \tau_{13} & \sigma_{33} \end{vmatrix} + \begin{vmatrix} \sigma_{22} & \tau_{23} \\ \tau_{23} & \sigma_{33} \end{vmatrix}$$

$$= \sigma_{11} \sigma_{22} + \sigma_{11} \sigma_{33} + \sigma_{22} \sigma_{33} - \tau_{12}^2 - \tau_{23}^2 - \tau_{13}^2 \tag{A.62}$$

$$= \frac{1}{2} (\sigma_{kk} \sigma_{ll} - \sigma_{kl} \sigma_{lk})$$

$$I_3^\sigma(\sigma_{ij}) = \det[\sigma_{kl}]$$

$$= \sigma_{11} \sigma_{22} \sigma_{33} + 2 \tau_{12} \tau_{23} \tau_{13} - \sigma_{11} \tau_{23}^2 - \sigma_{22} \tau_{13}^2 - \sigma_{33} \tau_{12}^2$$

The invariants are particularly easily written in the principal axes system:

$$I_1^\sigma = \sigma_I + \sigma_{II} + \sigma_{III}$$
$$I_2^\sigma = \sigma_I \sigma_{II} + \sigma_{II} \sigma_{III} + \sigma_{III} \sigma_I \qquad (A.63)$$
$$I_3^\sigma = \sigma_I \sigma_{II} \sigma_{III}$$

The shear stresses attain extreme values on such planes rotated by $45°$ to the principal axes and are called *principal shear stresses*.

$$\tau_I = \frac{1}{2}(\sigma_{II} - \sigma_{III}), \quad \tau_{II} = \frac{1}{2}(\sigma_I - \sigma_{III}), \quad \tau_{III} = \frac{1}{2}(\sigma_I - \sigma_{II}) \qquad (A.64)$$

Similar to the strain tensor, also the stress tensor can be decomposed into a *deviator* and a *spherical tensor* (hydrostatic stress):

$$\sigma_{ij} = \sigma_{ij}^D + \sigma^H \delta_{ij} \qquad (A.65)$$

Hydrostatic stress: $\sigma^H \delta_{ij}$, $\quad \sigma^H = \frac{1}{3}\sigma_{kk}$, $\quad$ Deviator: $\sigma_{ij}^D = \sigma_{ij} - \sigma^H \delta_{ij}$
$$(A.66)$$

A.3.4 Equilibrium Conditions

The basic laws of mechanics are valid for a deformable body, too. This requires in statics that the resultant force and resultant torque of all occurring forces must be zero. For dynamic processes, the inertia forces of the accelerated particles have additionally to be involved in the equilibrium considerations. Based on the principles of linear momentum by Newton and angular momentum by Euler, the corresponding equations of motion are derived for the deformable body. These considerations are explained in the following using the current configuration as an example, whereby we refer to Fig. A.4. In an analogous way the derivation can be done for the initial configuration as well.

The resultant force F_R of the applied external loads on the body is calculated by the integral over all body forces $\bar{b}$ in the volume v and all surface loads $\bar{t}$ acting on its boundary a. The total linear momentum I_P of the body is composed of the masses of all moving particles $dm = \rho\,dv$ multiplied by their velocities $v(x,t) = \dot{x}(x,t) = \dot{u}(x,t)$, whereby ρ denotes the mass density in the current configuration.

$$F_R = \int_v \bar{b}\,dv + \int_a \bar{t}\,da, \quad I_P = \int_v \rho\,v(x,t)dv = \int_v \rho\,\dot{u}(x,t)dv \qquad (A.67)$$

According to Newton's law of motion, the time derivative of the momentum equals the resultant force. Taking the consistency of mass $\Delta m = $ const. of each particle into account, the time derivative of I_p affects only $\dot{v} = a = \ddot{u}$. Employing Eq. (A.67), we find the *global principle of momentum* for the whole body,

$$F_R = \dot{I}_P \quad \Rightarrow \quad \int_v \bar{b}\,dv + \int_a \bar{t}\,da = \int_v \rho\,\ddot{u}(x,\,t)\,dv\,. \tag{A.68}$$

The *local form of the principle of momentum* for a material volume element Δv is obtained from (A.68) by using the Cauchy formula (A.44) $t = \sigma \cdot n$ and the Gaussian integral theorem to convert the surface integral (A.68) into a volume integral. Finally, this yields the following relationship:

$$\int_{\Delta v} [\nabla \cdot \sigma + \bar{b} - \rho\,\ddot{u}]\,dv = 0\,, \tag{A.69}$$

which must be valid for an arbitrary subregion $\Delta v \to 0$. This demands the expression in brackets to vanish:

$$\nabla \cdot \sigma + \bar{b} = \rho\ddot{u} \quad \text{or} \quad \sigma_{ij,j} + \bar{b}_i = \rho\,\ddot{u}_i\,, \tag{A.70}$$

or in elaborate notation:

$$\frac{\partial \sigma_{11}}{\partial x_1} + \frac{\partial \tau_{12}}{\partial x_2} + \frac{\partial \tau_{13}}{\partial x_3} + \bar{b}_1 = \rho\,\ddot{u}_1$$

$$\frac{\partial \tau_{21}}{\partial x_1} + \frac{\partial \sigma_{22}}{\partial x_2} + \frac{\partial \tau_{23}}{\partial x_3} + \bar{b}_2 = \rho\,\ddot{u}_2 \tag{A.71}$$

$$\frac{\partial \tau_{31}}{\partial x_1} + \frac{\partial \tau_{32}}{\partial x_2} + \frac{\partial \sigma_{33}}{\partial x_3} + \bar{b}_3 = \rho\,\ddot{u}_3$$

In the case of statics ($\ddot{u} = 0$), these equations represent the local *equilibrium conditions* of Cauchy. They form a system of three partial differential equations PDE, which the stress tensor σ has to comply with at each position. In kinetics these expressions are known as *equations of motion*. Their twofold integration with respect to time provides the deformations $u(x, t)$ of the body.

A.4 Material Laws

The relationship between the stress state and the appearing strains depends on the mechanical properties of the materials—specifically on their deformation behavior. In continuum mechanics this correlation is commonly modeled phenomenologically and mathematically formulated by so-called *material laws*. The deformation and failure behavior of materials can be quite clearly classified according to the influence of location, direction and time:

(a) **Spatial dependency**

If the material behavior depends on the location (coordinates x) of testing, this will be called *inhomogeneous* (e.g. forging, weldments). On the contrary, »*homogeneous*« means the same properties exist everywhere.

(b) **Direction dependency**

If the material behavior distinguishes itself at the same location in different directions of loading, we will call this property *anisotropy* (e.g. elasticity modulus of composites). Otherwise (no directional dependency) *isotropy* is present.

(c) **Time dependency**

If the temporal course of loading does not matter to the material behavior, this will be called a *skleronom* deformation behavior. The constitutive laws are then independent of time t and rate of deformation. In many engineering materials, the response to loading depends essentially on the velocity of the process, so that the time enters as a variable in the constitutive law. This kind of deformation behavior is called *rheonom*. The time dependence is observed in viscoelastic or viscoplastic properties.

If the material properties change exclusively in time without loading, this will be called *aging*. In addition to the above mentioned quantities, material laws and parameters may depend indirectly on other physical factors such as e.g. temperature, moisture content, chemical reactions or radioactive radiation.

A.4.1 Elastic Material Laws

Elastic material behavior is characterized by two features:

• The deformations are *reversible*, i.e. during unloading the body deforms back into its original shape. This is illustrated in Fig. A.9 by means of the uniaxial stress-strain curve $\sigma - \varepsilon$. There is a one-to-one correlation between the current stress σ and the instantaneous elastic strain ε. The ultimate stress state is independent of the deformation history.

• The deformations depend neither on time nor loading rate (skleronom).

The elastic deformation behavior of almost all materials is linear for small strains, which is known as *Hooke's law* $\sigma = E\varepsilon$. The *modulus of elasticity* (*Young's modulus*) E is given by the slope of the stress-strain curve $E(\varepsilon) = \mathrm{d}\sigma/\mathrm{d}\varepsilon$. There are however also materials that can accommodate large, purely elastic deformations, which show considerable nonlinearities, such as e.g. rubber or plastics.

Hyperelastic Material Laws

In the general case of multiaxial loading and large deformations, the elastic constitutive law must be formulated as a function expressing the stress tensor by the strain

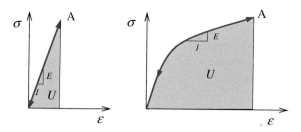

Fig. A.9 Linear and nonlinear elastic material behavior at uniaxial loading

tensor. Hereby, the associated stress and strain measures of the current configuration or the initial configuration have to be employed, see Sects. A.2 and A.3.

$$\sigma = \sigma(\eta) \quad \text{or} \quad T = T(E) \tag{A.72}$$

To this end we consider in Eulerian description the work in a volume element, which is performed by the true stresses σ_{kl} on the strains η_{kl} during a deformation process from the undeformed initial state up to the final state η_{ij}:

$$\mathrm{d}U = \sigma_{kl}\,\mathrm{d}\eta_{kl} \quad \Rightarrow \quad U(\eta_{ij}) = \int_0^{\eta_{ij}} \sigma_{kl}\,\mathrm{d}\eta_{kl}\,. \tag{A.73}$$

Hereby, U denotes the *strain energy density* per volume for the general multiaxial case. Because for elastic material behavior a unique relationship exists between current stresses σ_{ij} and elastic strains η_{ij}, the integral U must be independent of the deformation path. This means physically that the strain energy density is stored as a specific potential energy in the volume, which can be converted with the density ρ into a free energy per unit mass $\psi_e = U/\rho$. In mathematical terms, U or ψ_e represent a path-independent integral of the state variables η_{ij}. Vice versa we get:

$$\sigma_{ij} = \frac{\partial U(\eta_{ij})}{\partial \eta_{ij}} = \rho \frac{\partial \psi_e(\eta_{ij})}{\partial \eta_{ij}} \quad \text{or} \quad \sigma = \frac{\partial U(\eta)}{\partial \eta}\,. \tag{A.74}$$

This general form of the elasticity law is named *hyperelasticity* and is valid for large deformations, anisotropy and arbitrary nonlinearities.

The specific formulation depends on the choice of the elastic potential $U(\eta)$. A more detailed description of hyperelastic material models for large deformations can be found in Ogden [1] and Haupt [2].

Generalized Hooke's Law

In most engineering applications we can restrict ourselves to small deformations. This fact allows us to formulate the elasticity law by the true stress tensor σ_{ij} and infinitesimal strain tensor ε_{ij}:

$$\boldsymbol{\sigma} = \boldsymbol{\sigma}(\varepsilon) \quad \text{or} \quad \sigma_{ij} = \sigma_{ij}(\varepsilon_{ij}) \,. \tag{A.75}$$

For isothermal states the function $U(\varepsilon_{ij}) \geq 0$ has a quadratic form in ε_{ij},

$$U(\varepsilon_{ij}) = \frac{1}{2}\varepsilon_{ij}C_{ijkl}\varepsilon_{kl} \,, \tag{A.76}$$

yielding the linear relationship

$$\sigma_{ij} = \frac{\partial U}{\partial \varepsilon_{ij}} = C_{ijkl}\,\varepsilon_{kl} \quad \text{and} \quad \frac{\partial^2 U}{\partial \varepsilon_{ij}\partial \varepsilon_{kl}} = C_{ijkl} \,. \tag{A.77}$$

This is Hooke's law in its most general form including the *elasticity tensor* C_{ijkl} of 4th order. Because σ_{ij} and ε_{kl} are symmetric and the partial differentiations are permutable, it possesses the symmetry properties $C_{ijkl} = C_{klij} = C_{jikl} = C_{ijlk} = C_{jilk}$, which reduces the number of essential elastic constants to 21.

For a better illustration Hooke's law is transferred from tensor calculus into matrix notation. In accordance with the Voigt rule each index pair $(ij) \equiv (ji) \rightarrow (\alpha)$ is mapped to one single Greek index (α):

$(11) \rightarrow (1)\,,\ (22) \rightarrow (2)\,,\ (33) \rightarrow (3)\,,\ (23) \rightarrow (4)\,,\ (31) \rightarrow (5)$ and $(12) \rightarrow (6)\,.$
The summation of Greek indices covers $\alpha, \beta, \ldots = \{1, 2, \ldots, 6\}\,.$ $\tag{A.78}$

$$\begin{bmatrix} \sigma_1 \\ \sigma_2 \\ \sigma_3 \\ \sigma_4 \\ \sigma_5 \\ \sigma_6 \end{bmatrix} \cong \begin{bmatrix} \sigma_{11} \\ \sigma_{22} \\ \sigma_{33} \\ \tau_{23} \\ \tau_{31} \\ \tau_{12} \end{bmatrix} = \begin{bmatrix} C_{11} & C_{12} & C_{13} & C_{14} & C_{15} & C_{16} \\ & C_{22} & C_{23} & C_{24} & C_{25} & C_{26} \\ & & C_{33} & C_{34} & C_{35} & C_{36} \\ & & & C_{44} & C_{45} & C_{46} \\ & \text{sym} & & & C_{55} & C_{56} \\ & & & & & C_{66} \end{bmatrix} \begin{bmatrix} \varepsilon_{11} \\ \varepsilon_{22} \\ \varepsilon_{33} \\ \gamma_{23} \\ \gamma_{31} \\ \gamma_{12} \end{bmatrix} \cong [C_{\alpha\beta}] \begin{bmatrix} \varepsilon_1 \\ \varepsilon_2 \\ \varepsilon_3 \\ \varepsilon_4 \\ \varepsilon_5 \\ \varepsilon_6 \end{bmatrix}$$

$$[\sigma_\alpha] = [C_{\alpha\beta}]\,[\varepsilon_\beta]\,, \quad \text{symmetry: } [C_{\beta\alpha}] = [C_{\alpha\beta}] \tag{A.79}$$

Thus, in the general anisotropic case every given strain component induces a complete multiaxial stress state!

The inversion of Hooke's law for the strains is done by inverting the *elasticity matrix* $[C_{\alpha\beta}]$, which yields the *compliance matrix* $[S_{\alpha\beta}]$.

$$\left[\varepsilon_\alpha\right] = \left[C_{\alpha\beta}^{-1}\right]\left[\sigma_\beta\right] = \left[S_{\alpha\beta}\right]\left[\sigma_\beta\right]$$

$$
\begin{bmatrix}
\varepsilon_{11} \\
\varepsilon_{22} \\
\varepsilon_{33} \\
\gamma_{23} \\
\gamma_{31} \\
\gamma_{12}
\end{bmatrix}
=
\begin{bmatrix}
S_{11} & S_{12} & S_{13} & S_{14} & S_{15} & S_{16} \\
 & S_{22} & S_{23} & S_{24} & S_{25} & S_{26} \\
 & & S_{33} & S_{34} & S_{35} & S_{36} \\
 & & & S_{44} & S_{45} & S_{46} \\
 & \text{sym} & & & S_{55} & S_{56} \\
 & & & & & S_{66}
\end{bmatrix}
\begin{bmatrix}
\sigma_{11} \\
\sigma_{22} \\
\sigma_{33} \\
\tau_{23} \\
\tau_{31} \\
\tau_{12}
\end{bmatrix}
\tag{A.80}
$$

Many materials and crystal structures have symmetry properties by which the number of elastic constants is reduced. The most important classes are mentioned in the following.

(a) Orthotropic material behavior

If the material has three preferred directions perpendicular to each other x_1, x_2 and x_3 with different elastic properties, we will call it orthotropy. Examples for this are orthorhombic crystals or fiber reinforced composites. The shear and elongation strains become decoupled. In total nine independent elastic constants remain.

$$
[C_{\alpha\beta}] =
\begin{bmatrix}
C_{11} & C_{12} & C_{13} & 0 & 0 & 0 \\
C_{12} & C_{22} & C_{23} & 0 & 0 & 0 \\
C_{13} & C_{23} & C_{33} & 0 & 0 & 0 \\
0 & 0 & 0 & C_{44} & 0 & 0 \\
0 & 0 & 0 & 0 & C_{55} & 0 \\
0 & 0 & 0 & 0 & 0 & C_{66}
\end{bmatrix}
\tag{A.81}
$$

More descriptive is the representation of the anisotropic Hooke's law in engineering constants E_i (elasticity modulus in x_i-direction), μ_{ij} (shear modulus in the (x_i, x_j)-plane) and the Poisson's ratios ν_{ij} (necking in x_i-direction at tension in x_j). The application to orthotropic materials (A.81) reads:

$$
\begin{bmatrix}
\varepsilon_{11} \\
\varepsilon_{22} \\
\varepsilon_{33} \\
\gamma_{23} \\
\gamma_{31} \\
\gamma_{12}
\end{bmatrix}
=
\begin{bmatrix}
1/E_1 & -\nu_{21}/E_2 & -\nu_{31}/E_3 & 0 & 0 & 0 \\
-\nu_{12}/E_1 & 1/E_2 & -\nu_{32}/E_3 & 0 & 0 & 0 \\
-\nu_{13}/E_1 & -\nu_{23}/E_2 & 1/E_3 & 0 & 0 & 0 \\
0 & 0 & 0 & 1/\mu_{23} & 0 & 0 \\
0 & 0 & 0 & 0 & 1/\mu_{31} & 0 \\
0 & 0 & 0 & 0 & 0 & 1/\mu_{12}
\end{bmatrix}
\begin{bmatrix}
\sigma_{11} \\
\sigma_{22} \\
\sigma_{33} \\
\tau_{23} \\
\tau_{31} \\
\tau_{12}
\end{bmatrix}
\tag{A.82}
$$

Because of symmetry properties there are additional interdependencies:

$$\nu_{21} E_1 = \nu_{12} E_2, \quad \nu_{23} E_3 = \nu_{32} E_2 \quad \text{and} \quad \nu_{31} E_1 = \nu_{13} E_3.$$

(b) Transversal isotropic material behavior

Hereby it is assumed that the material behaves equally (isotropic) in all directions within one plane (x_1, x_2), but with respect to the third coordinate (x_3) other properties hold. This leads to five independent elastic constants. Examples are unidirectional fiber-reinforced composite materials, wood or hexagonal crystals.

$$[C_{\alpha\beta}] = \begin{bmatrix} C_{11} & C_{12} & C_{13} & 0 & 0 & 0 \\ C_{12} & C_{11} & C_{13} & 0 & 0 & 0 \\ C_{13} & C_{13} & C_{33} & 0 & 0 & 0 \\ 0 & 0 & 0 & C_{44} & 0 & 0 \\ 0 & 0 & 0 & 0 & C_{44} & 0 \\ 0 & 0 & 0 & 0 & 0 & \frac{1}{2}(C_{11}-C_{12}) \end{bmatrix} \tag{A.83}$$

(c) Isotropic material behavior

Isotropy is the highest class of symmetry, wherein the elastic behavior is identical in all spatial directions .

$$[C_{\alpha\beta}] = \begin{bmatrix} C_{11} & C_{12} & C_{12} & 0 & 0 & 0 \\ C_{12} & C_{11} & C_{12} & 0 & 0 & 0 \\ C_{12} & C_{12} & C_{11} & 0 & 0 & 0 \\ 0 & 0 & 0 & C_{44} & 0 & 0 \\ 0 & 0 & 0 & 0 & C_{44} & 0 \\ 0 & 0 & 0 & 0 & 0 & C_{44} \end{bmatrix} \quad \text{with} \quad C_{44} = \frac{1}{2}(C_{11}-C_{12}) \tag{A.84}$$

Thus, the elasticity matrix simplifies to two independent elastic constants. Amorphous materials (glass, polymers, et al.) and polycrystalline metallic or ceramic materials behave macroscopically isotropic elastic.

Thermal Strains

It is well known that in many materials a deformation occurs as the result of a temperature change from the initial T_0 to the current temperature T. The shape of these deformations is specified by the *thermal strain tensor* ε_{ij}^{t}. These strains are in first approximation proportional to the temperature difference $\Delta T(x) = T(x) - T_0$, and have anisotropic character that is quantified by the symmetric second-order tensor of *thermal expansion coefficient* α_{ij}. In the special case of isotropy, only thermal strains occur that are equal in all directions. Then the material tensor α_{ij}^{t} is reduced to a diagonal tensor $\alpha_t \delta_{ij}$ and the linear thermal expansion coefficient α_t is left as the only material parameter.

$$\text{anisotropic:} \quad \varepsilon_{ij}^{t}(x) = \alpha_{ij}^{t}\, \Delta T(x)\,, \quad \text{isotropic:} \quad \varepsilon_{ij}^{t} = \alpha_t \delta_{ij}\, \Delta T \tag{A.85}$$

Thermally induced strains appear independent of the stress state and in addition to the elastic strains ε_{ij}^{e}. Therefore, they have to be subtracted from the total strains in Hooke's law.

$$\varepsilon_{ij} = \varepsilon_{ij}^{e} + \varepsilon_{ij}^{t} \quad \Rightarrow \quad \sigma_{ij} = C_{ijkl}\,(\varepsilon_{kl} - \varepsilon_{kl}^{t}) \tag{A.86}$$

The unhindered free heating of a body to a constant temperature does not lead to stresses, whereas a kinematic constraint or inhomogeneous temperature fields may cause large thermal stress or residual stress.

Isotropic Thermoelastic Law

In view of its widespread applications we consider the isotropic elastic material with thermal strains in detail. Since the *modulus of elasticity* E, *Poisson's ratio* $0 \leq \nu \leq 1/2$ and *shear modulus* μ are identical in all directions, the relations (A.82) and (A.85) are written as:

$$\varepsilon_{11} = \frac{1}{E}\left[\sigma_{11} - \nu\left(\sigma_{22} + \sigma_{33}\right)\right] + \alpha_t \Delta T$$

$$\varepsilon_{22} = \frac{1}{E}\left[\sigma_{22} - \nu\left(\sigma_{33} + \sigma_{11}\right)\right] + \alpha_t \Delta T$$

$$\varepsilon_{33} = \frac{1}{E}\left[\sigma_{33} - \nu\left(\sigma_{11} + \sigma_{22}\right)\right] + \alpha_t \Delta T$$

$$\gamma_{12} = \frac{\tau_{12}}{\mu}, \quad \gamma_{23} = \frac{\tau_{23}}{\mu}, \quad \gamma_{31} = \frac{\tau_{31}}{\mu} \quad \text{with} \quad \mu = \frac{E}{2(1+\nu)}. \quad \text{(A.87)}$$

This reads in more general notation

$$\varepsilon_{ij} = \frac{1+\nu}{E}\sigma_{ij} - \frac{\nu}{E}\sigma_{kk}\,\delta_{ij} + \alpha_t\,\Delta T\,\delta_{ij}. \quad \text{(A.88)}$$

Rearranging of (A.87) with respect to the stresses yields:

$$\sigma_{11} = \frac{E}{1+\nu}\left[\varepsilon_{11} + \frac{\nu}{1-2\nu}\left(\varepsilon_{11} + \varepsilon_{22} + \varepsilon_{33}\right)\right] - \frac{E}{1-2\nu}\alpha_t \Delta T$$

$$\sigma_{22} = \frac{E}{1+\nu}\left[\varepsilon_{22} + \frac{\nu}{1-2\nu}\left(\varepsilon_{11} + \varepsilon_{22} + \varepsilon_{33}\right)\right] - \frac{E}{1-2\nu}\alpha_t \Delta T \quad \text{(A.89)}$$

$$\sigma_{33} = \frac{E}{1+\nu}\left[\varepsilon_{33} + \frac{\nu}{1-2\nu}\left(\varepsilon_{11} + \varepsilon_{22} + \varepsilon_{33}\right)\right] - \frac{E}{1-2\nu}\alpha_t \Delta T$$

$$\tau_{12} = \mu\,\gamma_{12}, \quad \tau_{23} = \mu\,\gamma_{23}, \quad \tau_{31} = \mu\,\gamma_{31}.$$

By introducing the isotropic *Hooke's tensor*

$$C_{ijkl} = 2\,\mu\,\delta_{ik}\,\delta_{jl} + \lambda\,\delta_{ij}\,\delta_{kl}, \quad \text{(A.90)}$$

relationship (A.89) is written in a compact form:

$$\sigma_{ij} = C_{ijkl}\left(\varepsilon_{kl} - \varepsilon_{kl}^{t}\right) = 2\,\mu\,\varepsilon_{ij} + \lambda\,\delta_{ij}\,\varepsilon_{kk} - (3\,\lambda + 2\,\mu)\,\alpha_{t}\,\Delta T\,\delta_{ij} \quad \text{(A.91)}$$

Hereby, the Lamé constants μ and λ are introduced, which are related to the other elastic constants E, ν and μ or C_{11}, C_{22} and C_{44} in the following way:

$$\mu = \frac{E}{2\,(1 + \nu)} = C_{44}, \quad \lambda = \frac{E\,\nu}{(1 + \nu)\,(1 - 2\,\nu)} = C_{12}, \quad 2\mu + \lambda = C_{11}. \quad \text{(A.92)}$$

For later constitutive modeling it is appropriate to split Hooke's law into a dilatational part of pure *volume change* and a deviatoric part of *shape change*. Employing the definitions of the spherical tensors σ^{H} and ε^{H} as well as the deviators σ_{ij}^{D} and ε_{ij}^{D}, we can separate (A.91) into the terms:

$$\sigma_{ij}^{D} = 2\,\mu\,\varepsilon_{ij}^{D}, \quad \sigma^{H} = (3\,\lambda + 2\,\mu)(\varepsilon^{H} - \alpha_{t}\Delta T) = 3\,K\,(\varepsilon^{H} - \alpha_{t}\Delta T) \quad \text{(A.93)}$$

$$K = \frac{E}{3\,(1 - 2\,\nu)} = \frac{1}{3}\,(2\,\mu + 3\,\lambda) = \frac{1}{3}\,(C_{11} + 2C_{12}) \quad \text{– modulus of compression.}$$
$$\text{(A.94)}$$

A.4.2 Elastic-Plastic Material Laws

Features of Plastic Deformation

Plastic material behavior is characterized in that the material begins to »flow« after a certain level of stress—the elastic limit—is attained, i.e. inelastic, permanent deformations occur. These plastic strains are a typical feature of most metals and exceed by far the magnitude of elastic deformations. Plastic deformations are irreversible dissipative processes that occur in a (quasi-static) balance between external loading and deformation resistance of the material. Therefore they are not dependent on time or rate of deformation (skleronom). After unloading, the plastic deformations remain existent. The plastic work of deformation is predominantly converted into heat.

In Fig. A.10 the characteristic features of elastic-plastic deformation behavior are exemplified by the uniaxial stress-strain curve. The material behaves elastically until a certain stress value is reached in point (F)-the yield strength σ_{F0}. If the stress exceeds (F), then plastic strains $\varepsilon^{P} > 0$ are formed. In the model of an ideal-plastic material (dotted line in Fig. A.10), unlimited plastic deformation $\varepsilon^{P} \to \infty$ happens now and the load-carrying capacity of the material is exhausted. In real materials the current yield strength σ_{F} increases as a result of plastic deformation, which is

Fig. A.10 Elastic-plastic
material behavior

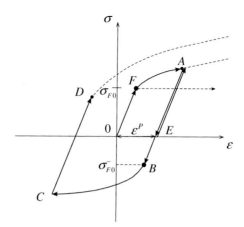

denoted as *hardening* of the material. The course (F)–(A) is called a monotonic yield curve. If the applied stress is reduced to zero (E), the material is relieved by a pure elastic deformation ε^e and only ε^p remains.

A re-loading to the previous value (A) happens again elastically, and only beyond this value does further plastification and hardening resume in the material. If a stress is applied in the opposite direction (tension → compression), then the plastic yielding will usually begin earlier at point (B). This shift of the yield limit at load reversal is known as the Bauschinger effect. If the material is exposed to periodic loading and unloading processes, then alternating plastic deformations $\pm\varepsilon^p$ develop and the stress-strain curve takes the form of a hysteresis loop whose shape may yet change with the cycles.

Basics of Plastic Flow Theory

The theory of plasticity describes the elastic-plastic deformation behavior in the multiaxial stress situation, whereby a number of assumptions are made that will be explained in the following under the restriction of small strains:

- The strains ε_{ij} and their time-rates $d_{ij} \approx \dot{\varepsilon}_{ij}$ (A.31) are composed from an elastic $\dot{\varepsilon}_{ij}^e$ and a plastic part $\dot{\varepsilon}_{ij}^p$.

$$\varepsilon_{ij} = \varepsilon_{ij}^e + \varepsilon_{ij}^p, \quad \dot{\varepsilon}_{ij} = \dot{\varepsilon}_{ij}^e + \dot{\varepsilon}_{ij}^p \quad \text{or} \quad d\varepsilon_{ij} = d\varepsilon_{ij}^e + d\varepsilon_{ij}^p \qquad (A.95)$$

Instead of the above formulation in velocities (»rate form«), Eq. (A.95) is often written in »incremental« form by the change of strain $d\varepsilon_{ij} = \dot{\varepsilon}_{ij}dt$ per time increment. This is allowed since in plasticity, the time has merely the meaning of a loading parameter. Strictly speaking, the additive decomposition is only exact within the frame of infinitesimal strains. But it can be used as a good approximation for large finite strains as well, provided the elastic strain part is small, $\dot{\varepsilon}_{ij}^{e} \ll \dot{\varepsilon}_{ij}^{p}$.

- Plastic flow only begins when the stress state σ_{kl} exceeds a certain limit that is defined by the *yield criterion*:

$$\Phi(\sigma_{kl}, h_\alpha) \begin{cases} < 0 & \text{elastic region} \\ = 0 & \text{plastic region and hardening} \end{cases} \tag{A.96}$$

$$h_\alpha \qquad (\alpha = 1, 2, \ldots, n_H) \quad \text{— hardening variables}$$

This yield condition represents a convex limit surface in stress space (six components of σ_{ij}, or three principal stresses σ_α), which separates the elastic region from the plastified states. It depends on a number n_H of stress-like variables h_α that describe the current state of hardening. As the hardening increases the limit surface $\Phi(\sigma_{kl}, h_\alpha)$ changes and is further-on called a *yield surface*.

- Plastic strains depend on the loading history (loading path in stress space). The current change (increment) of plastic strain $d\varepsilon_{ij}^{p}$ is a direct reaction to the change $d\sigma_{ij}$ in the stress state, but it depends also on the absolute stress state σ_{ij} and the achieved level of hardening h_α. For these reasons, the material law has to be formulated in »incremental form« or »rate form«. From this the name *incremental theory of plasticity* or *flow theory* has arisen.

$$\dot{\varepsilon}_{ij}^{p} = f_{ijkl}(\sigma_{kl}, h_\alpha) \, \dot{\sigma}_{kl} \quad \text{or} \quad d\varepsilon_{ij}^{p} = f_{ijkl}(\sigma_{kl}, h_\alpha) \, d\sigma_{kl} \tag{A.97}$$

- For metallic materials it has been proven experimentally that plastic deformations produce *no change in volume* and that plastic yielding is *not* affected by the hydrostatic stress part σ^{H} (plastic incompressibility). Thus, plastic deformation has a totally shape changing character and can only be described by the deviator

$$\dot{\varepsilon}_{kk}^{p} = 0, \quad \dot{\varepsilon}_{ij}^{p} \equiv \dot{\varepsilon}_{ij}^{pD} \tag{A.98}$$

- Irreversible changes of material conditions during plastic yielding are quantified by so-called *state variables* or *internal variables* z_α $(\alpha = 1, 2, \cdots, n_H)$. For this purpose, commonly strain-like quantities are chosen, which are in terms of thermodynamics work-conjugate with respect to the hardening variables h_α. The quantities z_α specify the hardening state and its changes.

$$h_\alpha = h_\alpha(z_\beta) \tag{A.99}$$

Yield Condition

To formulate the yield condition $\Phi(\sigma_{kl}, h_\alpha)$ for isotropic material behavior it is advantageous to choose the three invariants of the stress tensor. Even better are the invariants of the stress deviator, because due to incompressibility the first invariant $I_1^{\sigma D} = \sigma_{kk}^D = 0$ drops out. Among the variety of existing approaches, we detail the two most common yield conditions.

(a) Yield condition of v. Mises

This yield condition traces back to v. Mises, Huber and Hencky and relies on the *shape change hypothesis*. It is formulated only by means of the 2nd invariant of the stress deviator $I_2^{\sigma D}$.

$$\Phi_{\text{Mises}}(I_2^{\sigma D}) = -I_2^{\sigma D} - \frac{1}{3}\sigma_{F0}^2 = \frac{1}{2}\sigma_{kl}^D \sigma_{kl}^D - \frac{1}{3}\sigma_{F0}^2 = 0 \qquad (\text{A.100})$$

After introducing the v. Mises *equivalent stress* σ_v,

$$\sigma_v = \sqrt{\frac{3}{2}\sigma_{kl}^D \sigma_{kl}^D} = \sqrt{-3\,I_2^{\sigma D}}$$

$$= \sqrt{\frac{1}{2}\left[(\sigma_{11} - \sigma_{22})^2 + (\sigma_{22} - \sigma_{33})^2 + (\sigma_{33} - \sigma_{11})^2\right] + 3(\tau_{12}^2 + \tau_{23}^2 + \tau_{31}^2)}$$

$$= \sqrt{\frac{1}{2}\left[(\sigma_I - \sigma_{II})^2 + (\sigma_{II} - \sigma_{III})^2 + (\sigma_{III} - \sigma_I)^2\right]} \quad \text{in principal stresses,} \qquad (\text{A.101})$$

we obtain

$$\Phi_{\text{Mises}}(\sigma_v) = \sigma_v^2 - \sigma_{F0}^2 = 0 \quad \text{or} \quad \Phi_{\text{Mises}} = \sigma_v - \sigma_{F0} = 0. \qquad (\text{A.102})$$

σ_v is chosen such that in the uniaxial case ($\sigma_I = \sigma$, $\sigma_{II} = \sigma_{III} = 0$) the condition $\sigma_v = \sigma = \sigma_{F0}$ is satisfied. This way the v. Mises equivalent stress compares a multiaxial stress state with an equivalent uniaxial value of the tensile test.

In the special case of plane stress state ($\sigma_{III} = 0$) Eq. (A.101) simplifies to

$$\sigma_v = \sqrt{\sigma_{11}^2 + \sigma_{22}^2 - \sigma_{11}\sigma_{22} + 3\tau_{12}^2} = \sqrt{\sigma_I^2 + \sigma_{II}^2 - \sigma_I \sigma_{II}} \qquad (\text{A.103})$$

and $\Phi_{\text{Mises}} = \sigma_v^2 - \sigma_{F0}^2 = 0$ represents the equation of an ellipse in the coordinate system of the principal stresses (σ_I, σ_{II}) as illustrated in Fig. A.11.

In the general triaxial stress state, the yield criterion is represented in the coordinate system of all three principal stresses as shown in Fig. A.12. The hydrostatic part σ^H of an arbitrary stress state $P \mathrel{\hat{=}} \boldsymbol{\sigma} = \sigma_I \boldsymbol{e}_I + \sigma_{II} \boldsymbol{e}_{II} + \sigma_{III} \boldsymbol{e}_{III}$ is obtained as the projection on the space diagonal $\boldsymbol{e}^H$ (hydrostatic axis). Because the yield criterion is independent of σ^H, it must form a prismatic surface parallel to $\boldsymbol{e}^H$. In these so-called deviatoric π-planes lying perpendicular to $\boldsymbol{e}^H$, the condition $\Phi_{\text{Mises}} = 0$ describes the locus of a circle of radius $R_F = \sqrt{2/3}\,\sigma_{F0}$.

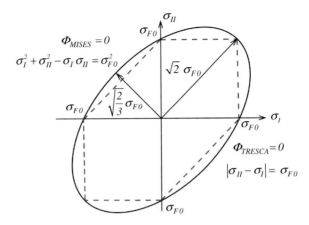

Fig. A.11 Yield conditions of v. Mises and Tresca for plane stress state

b) Yield condition of Tresca

According to this hypothesis, plastic flow occurs if the maximum shear stress (A.64) reaches a limit τ_{F0}.

$$\Phi_{\text{Tresca}}(\sigma_{kl}) = \tau_{\max} - \tau_{F0} = 0$$

$$= \max\{\frac{1}{2}|\sigma_{\text{I}} - \sigma_{\text{II}}|, \frac{1}{2}|\sigma_{\text{II}} - \sigma_{\text{III}}|, \frac{1}{2}|\sigma_{\text{III}} - \sigma_{\text{I}}|\} - \tau_{F0} = 0 \quad \text{(A.104)}$$

For the uniaxial stress state ($\sigma_{\text{I}} = \sigma$, $\sigma_{\text{II}} = \sigma_{\text{III}} = 0$) we get $\tau_{\max} = |\sigma|/2$, providing thus a relation between the yield stresses in shear and tension $\tau_{F0} = \sigma_{F0}/2$. For the plane stress state, Tresca's yield condition is displayed as a hexagon in the (σ_{I}, σ_{II})-plane, see Fig. A.11. In the triaxial stress state $\Phi_{\text{Tresca}} = 0$ corresponds to a cylinder with uniform hexagonal shape in the deviatoric plane as sketched in Fig. A.12.

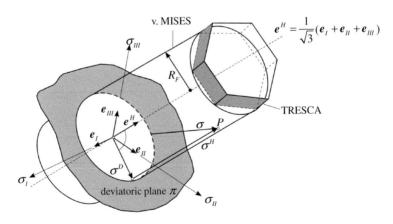

Fig. A.12 Yield conditions of v. Mises and Tresca in 3D space of principal stresses

Flow rule and normality rule

Next we need to know the components of the plastic strain increments $d\varepsilon_{ij}^P$, i.e. the directions and size of plastic flow. They can be determined by the *principle of maximum plastic dissipation* that has been established by Hill and Drucker 1950, see Lubliner [3]. We consider a stress state σ_{ij}^0 on or inside the yield surface and prescribe a plastic strain increment $d\varepsilon_{ij}^P$, cf. Fig. A.13 for this. According to the *principle of maximum plastic dissipation*, among all possible stress states $\tilde{\sigma}_{ij}$ the true stress state σ_{ij} will be realized such that the dissipated energy density (per time dt) $\mathcal{D}^P = \left(\tilde{\sigma}_{ij} d\varepsilon_{ij}^P - h_\alpha dz_\alpha\right)$ attains a maximum. Because this stress state $\tilde{\sigma}_{ij}$ must lie on the yield surface, it requires us to solve an extreme value problem with the yield condition as a constraint. Applying the method of Lagrangian multiplicator $d\Lambda$ we can write

$$\mathcal{D}^P\left(\tilde{\sigma}_{ij}, h_\alpha, d\Lambda\right) = \left[\tilde{\sigma}_{kl} d\varepsilon_{kl}^P - h_\alpha dz_\alpha - \Phi(\tilde{\sigma}_{kl}, h_\alpha) d\Lambda\right] \to \text{max.} \qquad (A.105)$$

Differentiation with respect to $\tilde{\sigma}_{ij}$ gives the *associated yield rule*:

$$d\varepsilon_{ij}^P = d\Lambda \frac{\partial \Phi}{\partial \tilde{\sigma}_{ij}} = d\Lambda \hat{N}_{ij} \text{ at } \tilde{\sigma}_{ij} = \sigma_{ij} \text{ or } \dot{\varepsilon}_{ij}^P = \dot{\Lambda} \frac{\partial \Phi}{\partial \sigma_{ij}} = \dot{\Lambda} \hat{N}_{ij}. \qquad (A.106)$$

This calculation formula is called the *normality rule*, because the direction $\hat{N}_{ij}$ is oriented exactly perpendicular to the yield surface in stress space. For these reasons the function Φ is denoted also as a *plastic dissipation potential*.

The differentiation of (A.105) with respect to the hardening variables h_α yields the law for the temporal evolution of the inner variables z_α:

$$dz_\alpha = -d\Lambda \frac{\partial \Phi}{\partial h_\alpha} \text{ or } \dot{z}_\alpha = -\dot{\Lambda} \frac{\partial \Phi}{\partial h_\alpha}. \qquad (A.107)$$

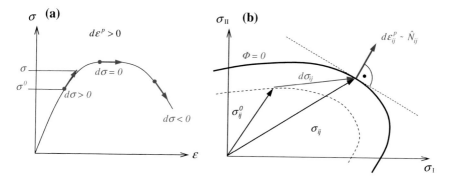

Fig. A.13 Yield condition and normality rule, postulate of Drucker

The non-negative parameter $\dot{\Lambda} = d\Lambda/dt$, which is also called a *plastic multiplicator*, assumes the following values:

$$\dot{\Lambda} \begin{cases} > 0 & \text{at} \quad \Phi = 0 \quad \text{and} \quad \dot{\Phi} = 0 \quad \text{plastic yielding} \\ = 0 & \text{at} \quad \Phi \leq 0 \quad \text{and} \quad \dot{\Phi} < 0 \quad \text{elastic region or unloading} \end{cases} \quad \text{(A.108)}$$

These relations are frequently combined in the *Kuhn-Tucker condition* $\dot{\Lambda}\,\dot{\Phi} = 0$.

In a context with the normality rule and the requirement for convexity, the postulate of Drucker (1950) has to be mentioned that gives an energetic condition for the stability of plastic material behavior. We consider a virtual loading cycle from the initial state σ_{ij}^0 to any final state σ_{ij} and back. Thereby, a strain increment $d\varepsilon_{ij} = d\varepsilon_{ij}^e + d\varepsilon_{ij}^p$ occurs. During this closed loading cycle the stress increment performs an additional plastic work $dU^p = d\sigma_{ij}\,d\varepsilon_{ij}^p$ on the plastic strain increments $d\varepsilon_{ij}^p$ (the elastic part $d\varepsilon_{ij}^e$ is recovered). This is shown in Fig. A.13 for the uniaxial (a) and a multiaxial (b) stress state.

$$\begin{array}{l} \text{1D}: (\sigma - \sigma^0)\,d\varepsilon = d\sigma\,d\varepsilon^p \\ \text{3D}: (\sigma_{ij} - \sigma_{ij}^0)\,d\varepsilon_{ij} = d\sigma_{ij}\,d\varepsilon_{ij}^p \end{array} \begin{cases} \geq 0 & \text{hardening} \quad \Rightarrow \quad \text{stable} \\ = 0 & \text{ideally plastic} \\ \leq 0 & \text{softening} \quad \Rightarrow \quad \text{unstable} \end{cases} \quad \text{(A.109)}$$

In order for the material to remain stable, the scalar product dU^p must not be negative, so $d\varepsilon_{ij}^p$ has to take the normal direction on the yield surface $\Phi = 0$, since $d\sigma_{ij}$ can have any outward direction. For this reason the yield surface must be convex ! This situation can be seen directly from the graph in Fig. A.13 b.

Application of the normality rule (A.106) to the v. Mises yield condition yields:

$$\Phi_{\text{Mises}} = \sqrt{\frac{3}{2}\sigma_{ij}^D\sigma_{ij}^D} - \sigma_{F0} = 0$$

$$\frac{\partial\Phi_{\text{Mises}}}{\partial\sigma_{ij}} = \frac{\partial\Phi_{\text{Mises}}}{\partial\sigma_{kl}^D}\frac{\partial\sigma_{kl}^D}{\partial\sigma_{ij}} = \left(\frac{3\sigma_{kl}^D}{2\sigma_v}\right)\left(\delta_{ki}\delta_{lj} - \frac{1}{3}\delta_{kl}\delta_{ij}\right) = \frac{3}{2}\frac{\sigma_{ij}^D}{\sigma_v} \quad \text{(A.110)}$$

$$\Rightarrow \quad \dot{\varepsilon}_{ij}^p = \frac{3}{2}\frac{\sigma_{ij}^D}{\sigma_v}\dot{\Lambda} \quad \text{(A.111)}$$

This is the isotropic flow law of Prandl-Reuss. The plastic strain rate is proportional to the stress deviator and thus causes a pure distortion (incompressibility).

Types of Hardening

Hardening is specified using the hardening variables h_α that may be scalar or tensor quantities. For their development in the course of loading, evolution laws are set that are proportional to the plastic multiplier. The functions H_α are either assumed empirically or are gained directly via (A.99) and (A.107) from the dissipation potential Φ:

$$\dot{h}_\alpha = H_\alpha(\sigma_{ij}, h_\beta)\, \dot{\Lambda} \quad \text{for example} \quad H_\alpha = -\sum_{\beta=1}^{n_H} \frac{\partial h_\alpha}{\partial z_\beta} \frac{\partial \Phi}{\partial h_\beta}. \qquad (A.112)$$

We explain the two most important types of hardening, which differ especially at load reversal.

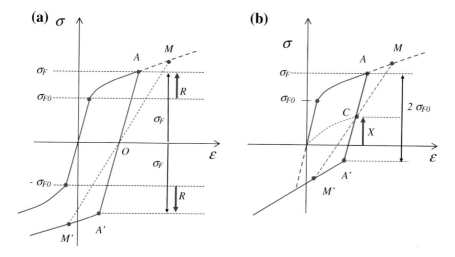

Fig. A.14 **a** Isotropic and **b** kinematic hardening for uniaxial loading

a) Isotropic hardening

After a load reversal, plastic flow will begin again only if the previously attained yield stress $|\sigma_F|$ is exceeded in magnitude anew, i.e. the hardened (elastic) region increases as shown in Fig. A.14a. The stress-strain curve is mirrored with respect to the point O of zero crossing. The isotropic hardening variable $R = R(\varepsilon^P)$ measures how much the yield strength has increased compared to its initial value:

$$\sigma_F(\varepsilon^P) = \sigma_{F0} + R(\varepsilon^P). \qquad (A.113)$$

This is a *scalar hardening variable* $h_1 = R$, expressed as a function of plastic strain. To generalize it to the three-dimensional stress state, the accumulated *equivalent plastic strain* $\varepsilon_v^P \mathrel{\widehat{=}} z_1$ is introduced, which represents the associated internal variable:

$$\varepsilon_v^P = \int_0^t \dot{\varepsilon}_v^P \, dt \,, \qquad \dot{\varepsilon}_v^P = \sqrt{\frac{2}{3} \dot{\varepsilon}_{ij}^P \, \dot{\varepsilon}_{ij}^P} = \dot{\Lambda}. \qquad (A.114)$$

In the special case of uniaxial tension ($\sigma_{11} = \sigma$) it holds that $\dot{\varepsilon}_{11}^P = \dot{\varepsilon}^P$, $\dot{\varepsilon}_{22}^P = \dot{\varepsilon}_{33}^P = -\dot{\varepsilon}^P/2$, and we obtain just $\dot{\varepsilon}_v^P \cong \dot{\varepsilon}^P$. This agreement arises from the *shape change hypothesis*, according to which any multiaxial stress state can be referred to a uniaxial one having an equal dissipation power $\sigma_{ij} \, \dot{\varepsilon}_{ij}^P = \sigma_v \dot{\varepsilon}_v^P$, which is done by means of the v. Mises equivalent stress σ_v and equivalent plastic strain. In this way, each hardening curve $\sigma_F = f(\varepsilon^P)$ measured in a tensile test can be transferred to the three-dimensional case as $\sigma_v = f(\varepsilon_v^P)$.

It can be shown that the equivalent plastic strain rate is identical with the plastic multiplier (A.114) and that the evolution law H_1 associated to R corresponds to the plastic tangent modulus E_p of (A.116).

The relationships for isotropic hardening are summarized as:

Yield condition(v.Mises) : $\Phi(\sigma_{ij}, R) = \sigma_v - \sigma_{F0} - R(\varepsilon_v^P) = 0$

Hardening law: $\sigma_F(\varepsilon_v^P) = \sigma_{F0} + R(\varepsilon_v^P)$ (A.115)

Evolution law: $\dot{R} = H_1 \dot{\Lambda} = \dfrac{dR}{d\varepsilon_v^P} \dot{\varepsilon}_v^P = E_p \dot{\varepsilon}_v^P$ (A.116)

The geometrical representation is the surface of a cylinder with radius $\sigma_F \sqrt{2/3}$, which is enlarging by ε_v^P in all directions (isotropic), see Fig. A.15a. There exist many empirical formulations for the uniaxial isotropic hardening law $R(\varepsilon^P)$, for example:

- linear hardening with a constant plastic tangent modulus $E_p = \frac{dR}{d\varepsilon^P}$:

$$R = E_p \varepsilon^P, \quad \varepsilon^P = \frac{1}{E_p}(\sigma_F - \sigma_{F0}) \qquad (A.117)$$

- power-law hardening of Ramberg-Osgood (exponent $n \geq 1$, parameter α, reference stress σ_0 and reference strain $\varepsilon_0 = \sigma_0/E$):

$$R = \sigma_0 \left(\frac{\varepsilon^P}{\alpha \varepsilon_0}\right)^{1/n} - \sigma_{F0}, \quad \varepsilon^P = \alpha \, \varepsilon_0 \left(\frac{\sigma_F}{\sigma_0}\right)^n \qquad (A.118)$$

- exponential function with saturation R_∞ and slope b:

$$R = R_\infty \left[1 - \exp(-b \, \varepsilon^P)\right] \qquad (A.119)$$

However, the isotropic hardening type does not explain the Bauschinger effect !

b) Kinematic hardening

Kinematic hardening means a shift of the yield condition in the direction of the current loading whereas the size of the elastic regions ($\overline{AA'} = 2\sigma_{F0}$) remains unchanged. As can be seen from Fig. A.14b, this way the Bauschinger effect is captured. The displacement of the reference point of the yield condition is expressed in the 1D case by the *kinematic hardening variable* X that has the dimension of a stress. Various approaches exist for the changing of X as a consequence of plastic deformation.

The simplest evolution equation by (Prager 1959) [3] is a linear shift proportional to the multiplicator $\Lambda = \varepsilon^{\mathrm{P}}$ with the material constant c. Thus the yield condition and the evolution equation read in the 1D case:

$$\Phi(\sigma, X) = |\sigma - X(\varepsilon^{\mathrm{P}})| - \sigma_{F0} = 0$$
$$X = X(\varepsilon^{\mathrm{P}}) = c \, \varepsilon^{\mathrm{P}}. \tag{A.120}$$

In the 3D case X represents a tensor of second order that is called a *back stress tensor*. It is a symmetric deviator, whose components X_{ij} define another six hardening variables, which are work-conjugate with the plastic strain components.

$$X_{ij} \,\widehat{=}\, h_\alpha, \quad \varepsilon_{ij}^{\mathrm{p}} \,\widehat{=}\, z_\alpha \quad (\alpha = 2, 3, \ldots, 7) \tag{A.121}$$

As Fig. A.15b shows, the back stresses describe a shift of the yield surface in the stress space towards the direction of current plastic strains (3D-Bauschinger effect). The yield condition is formulated by the 2nd invariant of $(\sigma_{ij}^{\mathrm{D}} - X_{ij})$, so that the equivalent stress $\bar{\sigma}_{\mathrm{v}} = -3I_2(\sigma_{ij}^{\mathrm{D}} - X_{ij})$ is calculated relatively to the center of the yield surface.

$$\Phi(\sigma_{ij}, X_{ij}) = \underbrace{\sqrt{\frac{3}{2}(\sigma_{ij}^{\mathrm{D}} - X_{ij})(\sigma_{ij}^{\mathrm{D}} - X_{ij})}}_{\bar{\sigma}_{\mathrm{v}}} - \sigma_{F0} = 0$$

$$\text{Evolution law:} \quad \dot{X}_{ij} = c \, \dot{\varepsilon}_{ij}^{\mathrm{p}} = c \, \frac{\partial \Phi}{\partial \sigma_{ij}} \, \dot{\Lambda} \,\widehat{=}\, H_\alpha \, \dot{\Lambda} \tag{A.122}$$

Moreover, *nonlinear* kinematic hardening rules (see e. g. Chaboche [4]) are to be mentioned, which are important to model certain cyclic plastic phenomena (ratcheting, mean stress relaxation). Therefore, the evolution Eq. (A.122) is extended by a second »recall« term $-\gamma X_{ij}\dot{\varepsilon}_{\mathrm{v}}^{\mathrm{p}}$ that causes a saturation of the kinematic hardening (dynamic recovery).

c) Combined hardening

In reality, metallic materials often exhibit a superposition of kinematic and isotropic hardening. This affects both an enlargement and a shift of the yield surface, which is illustrated in Fig. A.15c for the multiaxial case. The equations for a combined isotropic and nonlinear kinematic hardening then have the following form:

yield condition: $\qquad \Phi(\sigma_{ij}, X_{ij}, R) = \bar{\sigma}_v - \sigma_{F0} - R(\varepsilon_v^p) = 0$

evolution equations: $\qquad \dot{R} = E_p \dot{\Lambda}, \quad \dot{X}_{ij} = c\dfrac{\partial \Phi}{\partial \sigma_{ij}}\dot{\Lambda}, \quad \dot{\varepsilon}_v^p = \dot{\Lambda}.$ $\qquad$ (A.123)

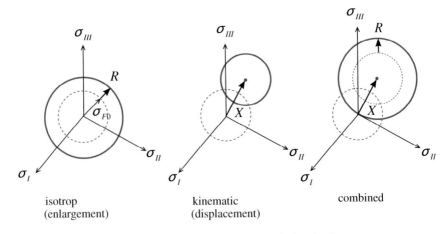

isotrop   kinematic   combined
(enlargement)   (displacement)

Fig. A.15 Representation of various hardening types in the deviatoric plane

Constitutive equations

In the previous section the *flow law* has already been derived, which provides the plastic strain increments as function of the current stress and hardening state. It is valid in the plastic region, i.e. if the yield condition is fulfilled and plastification proceeds ($\Phi = 0$, $\dot{\Lambda} > 0$). During unloading from the plastic region, ($\Phi = 0$, $\dot{\Phi} < 0$, $\dot{\Lambda} = 0$), as well as in the elastic region ($\Phi < 0$, $\dot{\Lambda} = 0$) Hooke's law applies. However, an equation is still missing to determine the plastic multiplier $\dot{\Lambda}$. This is accomplished by means of the so-called *consistency condition*, which says that the yield surface must always maintain a zero value $\Phi = 0$ for further hardening, i.e. its total differential vanishes:

$$\dot{\Phi} = \frac{\partial \Phi}{\partial \sigma_{ij}}\dot{\sigma}_{ij} + \frac{\partial \Phi}{\partial R}\dot{R} + \frac{\partial \Phi}{\partial X_{ij}}\dot{X}_{ij} = 0.$$ $\qquad$ (A.124)

In case of combined isotropic-kinematic hardening the individual terms read:

$$\frac{\partial \Phi}{\partial \sigma_{ij}} = \hat{N}_{ij}\,, \quad \frac{\partial \Phi}{\partial X_{ij}} = -\hat{N}_{ij}\,, \quad \hat{N}_{ij} = \frac{3}{2}\frac{\sigma^{D}_{ij} - X_{ij}}{\bar{\sigma}_{v}}\,, \quad \frac{\partial \Phi}{\partial R} = -1 \quad \text{(A.125)}$$

Together with the evolution laws (A.123) follows:

$$\hat{N}_{ij}\,\dot{\sigma}_{ij} - [\,E_{p} + c\,\underbrace{\hat{N}_{ij}\,\hat{N}_{ij}}_{3/2}\,]\,\dot{\Lambda} = 0 \quad \Rightarrow \quad \dot{\Lambda} = \frac{\hat{N}_{ij}\,\dot{\sigma}_{ij}}{E_{p} + \frac{3}{2}c}\,. \quad \text{(A.126)}$$

The elastic strain rates are obtained from (A.95) by subtracting the plastic part from the total rates, whence the stress rates can be calculated using Hooke's law (A.77):

$$\dot{\sigma}_{ij} = C_{ijkl}\dot{\varepsilon}^{e}_{kl} = C_{ijkl}(\dot{\varepsilon}_{kl} - \dot{\varepsilon}^{p}_{kl}) = C_{ijkl}(\dot{\varepsilon}_{kl} - \dot{\Lambda}\hat{N}_{kl})\,. \quad \text{(A.127)}$$

Inserting of $\dot{\Lambda}$ from (A.126) yields the relationship to the total strain rate $\dot{\varepsilon}_{kl}$

$$\dot{\Lambda} = \frac{\hat{N}_{ij}C_{ijkl}}{\hat{N}_{mn}C_{mnpq}\hat{N}_{pq} + E_{p} + \frac{3}{2}c}\,\dot{\varepsilon}_{kl}\,, \quad \text{(A.128)}$$

from where the sought relation is finally found.

$$\dot{\sigma}_{ij} = \left[C_{ijkl} - \frac{C_{ijmn}\,\hat{N}_{mn}\,\hat{N}_{pq}\,C_{pqkl}}{\hat{N}_{mn}\,C_{mnpq}\,\hat{N}_{pq} + E_{p} + \frac{3}{2}\,c}\right]\dot{\varepsilon}_{kl} = C^{ep}_{ijkl}\,\dot{\varepsilon}_{kl} \quad \text{(A.129)}$$

Thereby the hypoelastic-plastic material law has been gained for an anisotropic-elastic and combined hardening plastic material as a relation between stress rate and total strain rate.

The tensor $C^{ep}_{ijkl}(\sigma_{ij}, X_{ij}, R)$ is denoted as *elastic-plastic continuum tangent*. It depends both on the current stress state and via the hardening variables on the deformation history.

At the end of this section, we write down the important special case, where the elastic material behavior is isotropic, see Sect. A.4.1.

$$C_{ijkl} = 2\,\mu\left[\delta_{ik}\,\delta_{jl} + \frac{\nu}{1 - 2\,\nu}\,\delta_{ij}\,\delta_{kl}\right]\text{see (A.90)}$$

$$C_{ijkl}\,\hat{N}_{kl} = 2\,\mu\,\hat{N}_{ij} \quad \text{since} \quad \hat{N}_{kk} = 0\,, \quad \hat{N}_{mn}\,C_{mnpq}\,\hat{N}_{pq} = 3\,\mu$$

$$C_{ijkl}^{ep} = 2\mu \left\{ \left[\delta_{ik}\delta_{jl} + \frac{\nu}{1-2\nu}\delta_{ij}\delta_{kl} \right] - \beta\frac{3}{2}\frac{(\sigma_{ij}^{D}-X_{ij})(\sigma_{kl}^{D}-X_{kl})}{\bar{\sigma}_{v}^{2}\left[1+\frac{E_{p}}{3\mu}+\frac{c}{2\mu}\right]} \right\}$$

$$\text{(A.130)}$$

$$\beta = \begin{cases} 1 & \text{plastic yielding} \\ 0 & \text{elastic unloading} \end{cases}$$

$$\dot{\sigma}_{ij} = 2\mu\dot{\varepsilon}_{ij} + \lambda\delta_{ij}\dot{\varepsilon}_{kk} - \beta 3\mu\frac{(\sigma_{ij}^{D}-X_{ij})(\sigma_{kl}^{D}-X_{kl})}{\bar{\sigma}_{v}^{2}\left[1+\frac{E_{p}}{3\mu}+\frac{c}{2\mu}\right]}\dot{\varepsilon}_{kl} \qquad \text{(A.131)}$$

Deformation Theory of Plasticity

In contrast to the incremental deformation laws of plastic flow theory presented in the previous section, Hencky (1924) [3] suggested a *finite* deformation law for non-linear material behavior, which is known as the so-called *deformation theory of plasticity* (the name is somewhat misleading). This material model still has importance for fracture mechanics, so we will discuss it in detail. The deformation theory adopts the basic assumptions of isotropic plasticity, inasmuch as the plastic strains are proportional to the stress deviator and thus incompressibility is maintained. Likewise, the shape change hypothesis and v. Mises yield function are used. However, different from the plastic flow theory, instead of the flow rule a proportional relationship is assumed between the total plastic strains and the current stresses.

$$\varepsilon_{ij}^{p} = \Lambda\sigma_{ij}^{D} \qquad \text{(A.132)}$$

The proportionality factor Λ results from transferring the uniaxial hardening curve to the multiaxial case by using v. Mises equivalent stress $\sigma_{v} = \sqrt{\frac{3}{2}\sigma_{ij}^{D}\sigma_{ij}^{D}}$ and plastic equivalent strain $\varepsilon_{v}^{p} = \sqrt{\frac{2}{3}\varepsilon_{ij}^{p}\varepsilon_{ij}^{p}}$, which gives:

$$\sigma_{F} = f(\varepsilon^{p}) \quad \Rightarrow \quad \sigma_{v} = f(\varepsilon_{v}^{p}). \qquad \text{(A.133)}$$

By complementing the elastic strain parts we obtain the finite Hencky material law, decomposed into hydrostatic and deviatoric portions

$$\varepsilon_{ij} = \varepsilon_{ij}^{e} + \varepsilon_{ij}^{p} = \frac{\sigma_{kk}}{3K}\delta_{ij} + \frac{1}{2\mu}\sigma_{ij}^{D} + \frac{3}{2}\frac{\varepsilon_{v}^{p}}{\sigma_{v}}\sigma_{ij}^{D}. \qquad \text{(A.134)}$$

This relation is often used in combination with the Ramberg-Osgood power-law hardening (A.118). The generalization to the multiaxial case is achieved by means of the equivalent quantities σ_v and ε_v^p, so that the following relationship between plastic strains and stresses is obtained:

$$\frac{\varepsilon_v^p}{\varepsilon_0} = \alpha \left(\frac{\sigma_v}{\sigma_0}\right)^n \Rightarrow \left(\frac{\varepsilon_v^p}{\sigma_v}\right) = \frac{\alpha \, \varepsilon_0}{\sigma_0} \left(\frac{\sigma_v}{\sigma_0}\right)^{n-1} \Rightarrow \frac{\varepsilon_{ij}^p}{\varepsilon_0} = \frac{3}{2} \, \alpha \left(\frac{\sigma_v}{\sigma_0}\right)^{n-1} \frac{\sigma_{ij}^D}{\sigma_0}$$

$$(A.135)$$

In deformation theory, this constitutive relation is also denoted as a three-dimensional form of the Ramberg-Osgood law.

As can easily be seen by comparison with the Prandtl-Reuss law (A.111) of incremental plasticity, the deformation theory represents in fact merely a non-linear-elastic (hyperelastic) material law, whose elastic potential is simply calculated by integrating the deformation energy.

$$U(\varepsilon_{ij}) = U^e + U^p = \frac{1}{2} \left[K \, (\varepsilon_{kk}^e)^2 + 2\,\mu\,\varepsilon_{ij}^e \, \varepsilon_{ij}^e \right] + \frac{n}{n+1} \frac{\sigma_0}{(\alpha \, \varepsilon_0)^{1/n}} \, (\varepsilon_v^p)^{\frac{n+1}{n}}$$

$$(A.136)$$

On the other hand, deformation theory has the advantage of being mathematically easier to handle, thereby enabling in some cases even closed solutions of boundary value problems.

Due to the »finite« formulation any influence of loading history is lost. Therefore, the deformation theory is only correct under very restrictive conditions, which one should be aware of:

- The stresses must rise monotonically at each point of the body. A relief would not go on the Hookeian straight line as shown in Fig. A.10, but run along the nonlinear hardening curve in Fig. A.9, which contradicts the true elastic-plastic behavior.
- The stress state must not change qualitatively during the loading process, i.e. the ratios of stress components to each other (principal stress directions) have to remain constant. Such a load path in stress space, running proportionally from zero to a fixed end value σ_{ij}^E, is called »radial«. In this case, all stresses, strains and displacements would increase with the loading parameter $0 \leq t \leq T$.

$$\sigma_{ij}(\boldsymbol{x}, t) = t \, \sigma_{ij}^E(\boldsymbol{x}, T), \quad \varepsilon_{ij}^p(\boldsymbol{x}, t) = t^n \, \varepsilon_{ij}^{pE}(\boldsymbol{x}, t), \quad u_i(\boldsymbol{x}, t) = t^n \, u_i^E(\boldsymbol{x}, t)$$

$$(A.137)$$

At radial loading the finite Hencky law can be derived from the incremental Prandtl-Reuss law directly by integration.

If these conditions are fulfilled in the application, then the deformation theory will provide correct solutions that are identical with those of incremental theory. Otherwise, any stress redistributions or partial reliefs at any point in the structure lead to different results.

A.5 Treatment of Boundary Value Problems

A.5.1 Definition of a Boundary Value Problem

For clarity, the complete set of partial differential equations and relations of the strength of materials is re-compiled (for infinitesimal distortion). In this context, all features of a *boundary value problem* (BVP) or, if time-varying problems are concerned, an *initial boundary value problem* (IBVP) are explained. For this we consider the body represented in Fig. A.16, which extends over the domain V and has the surface A. It contains already a crack whose surface S_c is a part of A. The external loading of the body is divided into body forces $\bar{b}$ and surface loads $\bar{t}$, cf. Sect. A.3. The latter act on the specific part S_t of the surface. On the complementary part of the surface $S_u = A - S_t$, bearing conditions are prescribed in the form of suppressed or imposed displacements that are denoted by $\bar{u}$.

The uniqueness of the solution of a boundary value problem requires that at each part of the surface in each coordinate direction either a displacement or a traction is specified. At time-dependent processes the boundary conditions $\bar{u}(t)$ and $\bar{t}(t)$ as well as the related surface parts $S_u(t)$ and $S_t(t)$ are even themselves functions of time t.

Also, on the crack surface we need to define boundary conditions. These are usually taken as load-free, i.e. on S_c is $\bar{t}_c = 0$. But, for example, stresses could occur due to internal pressure or the contact of crack faces could prevent certain boundary displacements.

According to Sect. A.2, the deformation is represented by a continuous displacement field $u(x)$, from which the strain tensor ε is derived as a symmetric vector gradient by (A.29). In this kinematic relationship ε is a dependent field quantity. If one

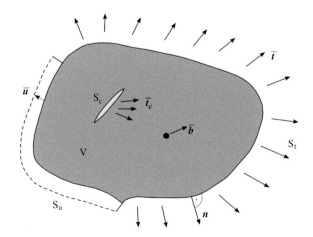

Fig. A.16 Formulation of boundary value problem

wants, however, to formulate the basic equations only with the strains, and to gain the resulting displacement fields by integrating, then this calculation does not lead readily to continuous and unambiguous results. Therefore, some additional conditions have to be imposed on the components of ε, which mathematically ensure the integrability and physically mean that the obtained displacement fields are continuous, i.e. the deformed body is a connected continuum without gaps or overlaps. These necessary relationships are called *compatibility conditions*. In the general three-dimensional case they are expressed by the following six differential equations:

$$\varepsilon_{ij,kl} + \varepsilon_{kl,ij} - \varepsilon_{ik,jl} - \varepsilon_{jl,ik} = 0 \quad \text{with} \quad i, j, k, l = 1, 2, 3. \quad (A.138)$$

Furthermore, there are the static or dynamic equilibrium conditions for the stress tensor σ, see Sect. A.3, which belong to the set of governing equations in the entire domain V as well. Also on the boundary of the domain the equilibrium between the internal stress state and the outer surface loads must be fulfilled according to the Cauchy formula (A.44).

The governing equations are finally completed by the material laws, which establish the connection between stresses and strain or their rates. This is written symbolically with the 4th order material tensor $M_{ijkl}(\varepsilon, \dot{\varepsilon}, h)$ that is representative for all deformation laws described in Sect. A.4.

Overall, the (initial) boundary value problem of strength of materials is defined by the following system of partial differential equations with the appropriate boundary and initial conditions:

Kinematics:	$\varepsilon_{ij} = \dfrac{1}{2}(u_{i,j} + u_{j,i})$	in V
	$u_i = \overline{u}_i$	on S_u
Equilibrium:	$\sigma_{ij,j} + \overline{b}_i = \rho \ddot{u}_i$	in V
	$t_i = \sigma_{ij} n_j = \overline{t}_i$	on S_t
Material law:	$\dot{\sigma}_{ij} = M_{ijkl}(\sigma_{kl}, \varepsilon_{kl}, h_\alpha)\dot{\varepsilon}_{kl}$	in V
	$\dot{h}_\alpha = H_\alpha(\sigma_{ij}, \varepsilon_{ij}, h_\alpha)$	
Initial conditions:	$u_i(x, t = 0) = u_{i0}, \; \dot{u}_i(x, t = 0) = \dot{u}_{i0}$	
	$h_\alpha(t = 0) = h_{\alpha 0}$	(A.139)

The mathematical solution of these IBVP is often quite complicated, especially for finite domains V, three-dimensional structures and non-linear material behavior, so they cannot be solved with analytical calculation methods. In these important practical cases it is therefore essential to rely on numerical calculation methods.

A.5.2 Plane Problems

A general aim in strength of materials and fracture mechanics is to facilitate the computational effort by creating easily manageable models. Thus, in a number of engineering applications, the geometry of the structure, its bearing conditions and loading situation can be simplified in good approximation to two dimensions, so that they can be treated as *plane* boundary value problems. Typical examples are thin-walled structures such as sheets and plates or prismatic components such as shafts, pipes etc.

In this way one can considerably reduce the system of the underlying partial differential equations (PDE) and the number of unknown field variables. Consequently, the mathematical and computational effort for the solution is substantially lower. Often there are also suitable mathematical methods available, which allow an analytical solution of the BVP. For this reason we will, in the following section, consider in detail plane boundary value problems (mainly elasticity theory) and their solution techniques.

In case of two-dimensional problems all field variables σ_{ij}, ε_{ij} and u_i are only functions of (x_1, x_2), and every derivative with respect to $\partial(\cdot)/\partial x_3 = 0$ vanishes. The relation between strains and displacements in the plane reads then:

$$\varepsilon_{11} = \frac{\partial u_1}{\partial x_1}, \quad \varepsilon_{22} = \frac{\partial u_2}{\partial x_2}, \quad \gamma_{12} = \frac{\partial u_1}{\partial x_2} + \frac{\partial u_2}{\partial x_1}. \tag{A.140}$$

The compatibility conditions (A.138) simplify to a single equation between these strain components:

$$\frac{\partial^2 \varepsilon_{11}}{\partial x_2^2} + \frac{\partial^2 \varepsilon_{22}}{\partial x_1^2} - \frac{\partial^2 \gamma_{12}}{\partial x_1 \partial x_2} = 0. \tag{A.141}$$

The external loads act in the same way on all planes $x_3 = $ const. of the component, and cause only stresses in the (x_1, x_2)-plane.

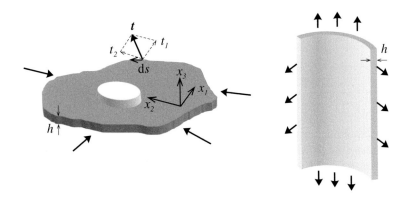

Fig. A.17 Plane stress state in thin sheets (*left*) or shells (*right*)

$$\sigma_{11}(x_1, x_2), \quad \sigma_{22}(x_1, x_2), \quad \tau_{12}(x_1, x_2) \tag{A.142}$$

The equations of motion (A.70) are reduced in this case to:

$$\frac{\partial \sigma_{11}}{\partial x_1} + \frac{\partial \tau_{12}}{\partial x_2} + \bar{b}_1 = \rho \ddot{u}_1, \quad \frac{\partial \tau_{12}}{\partial x_1} + \frac{\partial \sigma_{22}}{\partial x_2} + \bar{b}_2 = \rho \ddot{u}_2. \tag{A.143}$$

Regarding the behavior of the field quantities in the thickness direction x_3, two approximations can be distinguished.

a) Plane stress state

A flat plane surface structure that is only loaded by forces within its plane (x_1, x_2), is called a *sheet*, see Fig. A.17 (left). Likewise in thin-walled containers as shown in Fig. A.17 (right), only stresses σ_{11}, σ_{22} in the membrane plane occur. For such structures with thin walls (thickness $h \ll$ other dimensions) the concept of a *plane stress state* was coined. Since at the top and bottom surface $x_3 = \pm h/2$ no tractions apply, all stresses with a x_3-component have to be zero here. Also inside the sheet it holds with good approximation, that these stress components are negligibly small compared to those in the plane.

$$\sigma_{33}(x_1, x_2) = \tau_{13}(x_1, x_2) = \tau_{23}(x_1, x_2) \equiv 0 \tag{A.144}$$

Thereby Hooke's law (A.87) and (A.89) simplifies to:

$$\varepsilon_{11} = \frac{1}{E} [\sigma_{11} - \nu \sigma_{22}] + \alpha_t \Delta T$$

$$\varepsilon_{22} = \frac{1}{E} [\sigma_{22} - \nu \sigma_{11}] + \alpha_t \Delta T \tag{A.145}$$

$$\gamma_{12} = \frac{\tau_{12}}{\mu} = \frac{2(1+\nu)}{E} \tau_{12}, \quad \gamma_{23} = \gamma_{31} = 0$$

or resolved for the stresses:

$$\sigma_{11} = \frac{E}{1 - \nu^2} [\varepsilon_{11} + \nu \varepsilon_{22} - (1 + \nu) \alpha_t \Delta T]$$

$$\sigma_{22} = \frac{E}{1 - \nu^2} [\varepsilon_{22} + \nu \varepsilon_{11} - (1 + \nu) \alpha_t \Delta T] \tag{A.146}$$

$$\tau_{12} = \frac{E}{2(1+\nu)} \gamma_{12} = \mu \gamma_{12}.$$

The strains ε_{33} are allowed to expand freely in the thickness direction.

$$\varepsilon_{33} = -\frac{1}{1-\nu} [\nu (\varepsilon_{11} + \varepsilon_{22}) - (1 + \nu) \alpha_t \Delta T] \tag{A.147}$$

b) Plane strain state

The conditions of a *plane strain state* apply, if the displacement component u_3 is everywhere zero (or constant). Then all strain components with respect to the x_3-direction vanish.

$$\varepsilon_{33}(x_1, x_2) = \gamma_{13}(x_1, x_2) = \gamma_{23}(x_1, x_2) \equiv 0 \qquad (A.148)$$

The plane strain state applies to prismatic components, if their geometry and loading do not change with the x_3-coordinate and if the u_3-displacement is inhibited by constraints, which is sketched in Fig. A.18.

The stresses are obtained by insertion of (A.148) in the general elasticity law (A.89):

$$\sigma_{11} = \frac{E}{(1 + \nu)(1 - 2\nu)} \left[\varepsilon_{11}(1 - \nu) + \nu \varepsilon_{22} - \alpha_t \Delta T \right]$$

$$\sigma_{22} = \frac{E}{(1 + \nu)(1 - 2\nu)} \left[\varepsilon_{22}(1 - \nu) + \nu \varepsilon_{11} - \alpha_t \Delta T \right] \qquad (A.149)$$

$$\tau_{12} = \mu \gamma_{12} = \frac{E}{2(1 + \nu)} \gamma_{12}, \quad \tau_{23} = \tau_{31} = 0.$$

The conversion for the strains yields:

$$\varepsilon_{11} = \frac{1 - \nu^2}{E} \left[\sigma_{11} - \frac{\nu}{1 - \nu} \sigma_{22} \right] + (1 + \nu) \alpha_t \Delta T$$

$$\varepsilon_{22} = \frac{1 - \nu^2}{E} \left[\sigma_{22} - \frac{\nu}{1 - \nu} \sigma_{11} \right] + (1 + \nu) \alpha_t \Delta T$$

$$\gamma_{12} = \frac{\tau_{12}}{\mu} = \frac{2(1 + \nu)}{E} \tau_{12}. \qquad (A.150)$$

Due to the strain constraint $\varepsilon_{33} = 0$, the axial stress σ_{33} is different from zero, but can be expressed by the stresses σ_{11}, σ_{22} in the plane.

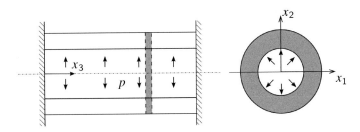

Fig. A.18 Example for a plane strain state in the cross-section of a pipe

$$\sigma_{33} = \nu\,(\sigma_{11} + \sigma_{22}) - E\,\alpha_t\,\Delta T \qquad (A.151)$$

Substituting the elastic constants

$$E \to E' = \frac{E}{1 - \nu^2}, \quad \nu \to \nu' = \frac{\nu}{1 - \nu}, \quad \alpha_t \to \alpha_t' = (1 + \nu)\,\alpha_t, \qquad (A.152)$$

we can write the Eqs. (A.149) and (A.150) for a plane strain state exactly in the same form as it was done for a plane stress state with the relations (A.145) and (A.146). This means, the two approaches differ only in their elastic constants but have an identical mathematical structure.

A.5.3 Method of Complex Stress Functions

For the solution of plane boundary value problems of elasticity, various mathematical methods have been developed. The use of real or complex stress functions belongs to the most important methods that are described in detail in standard textbooks of elasticity theory. In the following these approaches are discussed only in brief form to understand their application in fracture mechanics.

The differential equations of plane elasticity theory can be led back to the determination of a single scalar function. To this end, a stress function $F(x_1, x_2)$ was introduced by Airy which delivers the stresses in such a way that the equilibrium equation (A.143) are satisfied automatically in the static case $\ddot{u} \equiv \overline{b} \equiv 0$:

$$\sigma_{11} = \frac{\partial^2 F}{\partial x_2^2}, \quad \sigma_{22} = \frac{\partial^2 F}{\partial x_1^2}, \quad \tau_{12} = -\frac{\partial^2 F}{\partial x_1 \partial x_2}. \qquad (A.153)$$

After substituting the stresses by the strains using the elasticity law, and inserting them into the compatibility condition (A.138), we get a differential equation of 4th order for the stress function F, which is also known as *bipotential equation*:

$$\frac{\partial^4 F}{\partial x_1^4} + 2\frac{\partial^2 F}{\partial x_1^2}\frac{\partial^2 F}{\partial x_2^2} + \frac{\partial^4 F}{\partial x_2^4} = \Delta\Delta F(x_1, x_2) = -E\alpha_t \Delta T(x_1, x_2). \quad (A.154)$$

Thereby $\Delta(\cdot) = \frac{\partial^2 (\cdot)}{\partial x_1^2} + \frac{\partial^2 (\cdot)}{\partial x_2^2}$ denotes the two-dimensional Laplace-operator in Cartesian coordinates. On the right-hand side of this PDE stands the thermal loading that requires us to find a particular solution. The homogeneous PDE $\Delta\Delta F = 0$ can be satisfied by appropriate Ansatz functions. Their free parameters need to be determined by means of the boundary conditions. Much more elegant and powerful is the usage of the *theory of complex functions*. Here the spatial coordinates (x_1, x_2)

are replaced by the complex variable $z = x_1 + ix_2$ and its conjugate complex quantity $\bar{z} = x_1 - ix_2$. $i = \sqrt{-1}$ denotes the imaginary unit.

$$z = x_1 + ix_2 \,, \quad \bar{z} = x_1 - ix_2 \quad \Rightarrow \quad x_1 = \frac{1}{2}(z + \bar{z})\,, \quad x_2 = \frac{1}{2i}(z - \bar{z}) \quad \text{(A.155)}$$

In complex variables, the homogeneous version of the bipotential equation (A.154) assumes the simple form (A.156). It can be shown (see e.g. Muskhelishvili [5]), that this equation is automatically fulfilled by choosing an ansatz (A.157) with two complex holomorphic functions $\phi(z)$ and $\chi(z)$. (Holomorphic or analytic functions are continuous complex differentiable and obey the Cauchy-Riemann relations.) $\Re(\cdot)$ and $\Im(\cdot)$ mean the real and imaginary part of an expression $(\cdot)$, respectively.

$$4\frac{\partial^4 F(z, \bar{z})}{\partial z^2 \, \partial \bar{z}^2} = 0 \qquad\qquad\qquad \text{(A.156)}$$

$$F(z, \bar{z}) = \Re\left[\bar{z}\phi(z) + \chi(z)\right] \qquad\qquad \text{(A.157)}$$

The relationship to the stress components and displacements in the plane is given by *Kolosov's formulas*

$$\sigma_{11} + \sigma_{22} = 2\left[\phi'(z) + \overline{\phi'(z)}\right] = 4\Re\left[\phi'(z)\right]$$
$$\sigma_{22} - \sigma_{11} + 2i\tau_{12} = 2\left[\bar{z}\phi''(z) + \chi''(z)\right] \qquad \text{(A.158)}$$
$$2\mu(u_1 + iu_2) = \kappa\phi(z) - z\overline{\phi'(z)} - \overline{\chi'(z)}$$

with the elastic constants

$$\kappa = 3 - 4\nu \text{ (plane strain state) or } \kappa = \frac{3 - \nu}{1 + \nu} \text{ (plane stress state). (A.159)}$$

By any choice of complex functions ϕ and χ, all governing equations of a plane elastic BVP are fulfilled in the domain V, i.e. equilibrium conditions, kinematics and Hooke's law. In order to satisfy the prescribed boundary conditions, we need the correlation of ϕ and χ with the required boundary quantities $\boldsymbol{u} = \bar{\boldsymbol{u}} = \bar{u}_1 + i\bar{u}_2$ on S_u and $\boldsymbol{t} = \bar{\boldsymbol{t}} = \bar{t}_1 + i\bar{t}_2$ on S_t. The relationship with $\bar{u}$ is given by the 3rd equation of (A.158). The traction vector at a boundary segment of length ds and normal vector n_j is obtained as $t_i = \sigma_{ij}n_j$ according to the Cauchy formula (A.44), which can be transformed by the complex functions into the expression:

$$\boldsymbol{t} = t_1 + it_2 = -i\frac{d}{ds}\left[\phi(z) + z\overline{\phi'(z)} + \overline{\chi'(z)}\right]. \qquad \text{(A.160)}$$

The *complex method* has the great advantage that proven techniques of complex function theory can be used to solve BVP for two-dimensional structures of finite dimensions. Those techniques are conformal mappings, Cauchy integrals and Laurent series. For example, at this point the conversion of Kolosov's formulas into polar coordinates (r, θ) is to be mentioned. Taking advantage of the Eulerian representation of complex numbers $z = re^{i\theta}$ and $\bar{z} = re^{-i\theta}$, we get the result:

$$\sigma_{rr} + \sigma_{\theta\theta} = 2\left[\phi'(z) + \overline{\phi'(z)}\right]$$

$$\sigma_{\theta\theta} - \sigma_{rr} + 2i\tau_{r\theta} = 2\left[\bar{z}\phi''(z) + \chi''(z)\right]e^{2i\theta} \qquad (A.161)$$

$$2\mu(u_r + iu_\theta) = \left[\kappa\phi(z) - z\overline{\phi'(z)} - \overline{\chi'(z)}\right]e^{-i\theta}.$$

Westergaard [6] introduced another complex stress function $Z(z)$ in fracture mechanics, which is a special case of the above functions of Muskhelishvili [5]. Therefore, the application of $Z(z)$ is confined to certain symmetry properties of the BVP. For symmetry (mode I) there exists the relationship: $\phi' = \frac{1}{2}Z$ and $\chi' = \phi - z\phi'$, whereas for screw-symmetry (mode II) it holds that: $\phi' = \frac{1}{2}Z$ and $\chi' = -\phi - z\phi'$.

A.5.4 Anti-Plane Stress State

The state of *anti-plane* or *longitudinal* shear stresses refers to a pure shear loading that acts perpendicular to the (x_1, x_2)-plane in a prismatic structure, which lies parallel to the x_3-axis, see Fig. A.19. Under the assumption of orthotropy or higher material symmetry regarding the coordinate axes, the deformation states in the plane and perpendicular to it decouple from each other. Therefore, only displacements $u_3(x_1, x_2)$ appear in the x_3-direction. They cause the shear strains γ_{13} and γ_{23} as well as the associated shear stresses τ_{13} and τ_{23}, which all are functions of (x_1, x_2). The loading can be imposed by shear forces or boundary displacements $\bar{u}_3$ as shown in Fig. A.19.

In this very trivial case there remain only the kinematic relations

$$\gamma_{13} = \frac{\partial u_3}{\partial x_1}, \quad \gamma_{23} = \frac{\partial u_3}{\partial x_2}, \quad \frac{\partial \gamma_{13}}{\partial x_2} = \frac{\partial \gamma_{23}}{\partial x_1}, \qquad (A.162)$$

Hooke's law for shear and the equilibrium equations (without body forces)

$$\gamma_{13} = \tau_{13}/\mu, \quad \gamma_{23} = \tau_{23}/\mu, \quad \frac{\partial \tau_{13}}{\partial x_1} + \frac{\partial \tau_{23}}{\partial x_2} = 0. \qquad (A.163)$$

Inserting (A.162) in (A.163) yields a Laplace-equation for the displacement function u_3

Fig. A.19 Anti-plane shear
loading

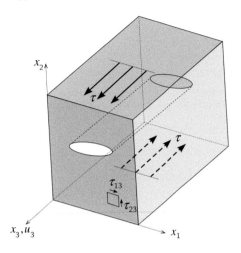

$$\mu\left(\frac{\partial^2 u_3}{\partial x_1^2} + \frac{\partial^2 u_3}{\partial x_2^2}\right) = \mu\Delta u_3(x_1, x_2) = 0. \qquad (A.164)$$

It is well known that both the real and the imaginary part of a holomorphic function
satisfy the Laplace-equation a priori [5, 7].

Therefore, an anti-plane shear problem can easily be solved by assuming the
displacement field to be the real part of a holomorphic function $\Omega(z)$, whose
exact form has to be determined by the boundary conditions. Thereafter, the
shear stresses are calculated from this function by complex differentiation.

$$u_3(x_1, x_2) = \Re\Omega(z)/\mu, \quad \tau_{13} - i\tau_{23} = \Omega'(z) \qquad (A.165)$$

A.5.5 Plates

In order to study cracks in plates later, we recall here briefly Kirchhoff's theory
of thin plates. The reader can find detailed descriptions in any textbook on higher
strength of materials. A *plate* is a planar surface structure (thickness $h \ll$ dimensions
in the plane), whose geometry is defined through the center plane $x_3 = 0$ and the
(x_1, x_2)-coordinates, see Fig. A.20. A plate is loaded perpendicular to its surface by
distributed pressure loads $p(x_1, x_2)$, or by moments imposed on the edges. Again,
the plane stress state exists. The deformation of a plate is primarily described by
the displacement of the center plane in the x_3-direction, which is called deflection

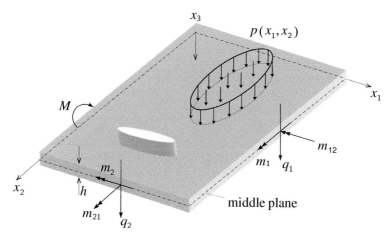

Fig. A.20 Kirchhoff's theory of plates

$u_3 = w(x_1, x_2)$. The first derivatives of w imply the local rotations of the cross sections, and the second derivatives correspond to the curvatures. All field variables of this two-dimensional BVP are just functions of (x_1, x_2).

As it is common in engineering mechanics, the action of stresses in a sectional area is quantified by statically equivalent sectional forces and moments. They are defined along every edge $x_1 = $ const. and $x_2 = $ const. and calculated per arc length, see Fig. A.20.

Shear forces: $\qquad q_1 = \int_{-h/2}^{+h/2} \tau_{13}\,dx_3\,, \qquad\qquad q_2 = \int_{-h/2}^{+h/2} \tau_{23}\,dx_3$

Bending moments: $\quad m_1 = \int_{-h/2}^{+h/2} \sigma_{11}\, x_3\,dx_3\,, \qquad m_2 = \int_{-h/2}^{+h/2} \sigma_{22}\, x_3\,dx_3$

Torsional moments: $\quad m_{12} = m_{21} = \int_{-h/2}^{+h/2} \tau_{12}\, x_3\,dx_3 \qquad\qquad\qquad\text{(A.166)}$

Taking into account the kinematics of deformation and Hooke's law, we get the relationship with the deflection function w:

$$m_1 = -D\left(\frac{\partial^2 w}{\partial x_1^2} + \nu\frac{\partial^2 w}{\partial x_2^2}\right), \quad m_2 = -D\left(\frac{\partial^2 w}{\partial x_2^2} + \nu\frac{\partial^2 w}{\partial x_1^2}\right) \qquad \text{(A.167)}$$

$$m_{12} = m_{21} = -D(1-\nu)\frac{\partial^2 w}{\partial x_1 \partial x_2}\,.$$

The quantity D is called plate stiffness.

$$D = \frac{Eh^3}{12(1-\nu^2)} \tag{A.168}$$

After the equilibrium conditions are deduced between these sectional quantities, we get in the end the well–known equation for the *Kirchhoff plate*:

$$\frac{\partial^4 w}{\partial x_1^4} + 2\frac{\partial^4 w}{\partial x_1^2 \partial x_2^2} + \frac{\partial^4 w}{\partial x_2^4} = \Delta\Delta w = \frac{p(x_1, x_2)}{D}. \tag{A.169}$$

This partial differential equation is a *Bipotential equation* in terms of the deflection function $w(x_1, x_2)$ of the plate, which is to be completed by corresponding boundary conditions either for w, w' or the sectional forces and moments. Thus the entire mathematical algorithm of complex function theory can be used here as well to solve this strength problem. For this purpose, the deflection $w(x_1, x_2)$ is represented by complex variables $z, \bar{z}$ using two complex functions ϕ and χ.

$$w(x_1, x_2) = \Re\left[\bar{z}\phi(z) + \chi(z)\right] \tag{A.170}$$

Finally, we obtain the sectional quantities by the following complex expressions:

$$m_1 + m_2 = -4D(1+\nu)\Re\left[\phi'(z)\right]$$
$$m_2 - m_1 + 2\mathrm{i}\,m_{12} = 2D(1-\nu)\left[\bar{z}\phi''(z) + \chi''(z)\right]$$
$$\frac{\partial w}{\partial x_1} + \mathrm{i}\frac{\partial w}{\partial x_2} = \phi(z) + z\overline{\phi'(z)} + \overline{\chi'(z)} \tag{A.171}$$
$$q_1 - \mathrm{i}q_2 = -4D\,\phi''(z)$$

The similarity with the complex method, employed in Sect. A.5.2 for plane problems in sheets, can clearly be recognized. A corresponding analogy is found in the solution methods.

References

1. Ogden RW (1984) Non-linear elastic deformations. Ellis Horwood and John Wiley, Chichester
2. Haupt P (2000) Continuum mechanics and theory of materials. Springer, Berlin
3. Lubliner J (1990) Plasticity theory. MacMillan, London
4. Lemaitre J, Chaboche JL (1990) Mechanics of solid materials. Cambridge University Press, Cambridge

5. Muskhelishvili NI (1971) Einige Grundaufgaben zur mathematischen Elastizitätstheorie. Fachbuchverlag Leipzig
6. Westergaard HM (1939) Bearing pressures and cracks. J Appl Mech 6:49–53
7. Hahn HG (1976) Bruchmechanik: Einführung in die theoretischen Grundlagen. Mechanik, Stuttgart, Teubner-Studienbücher

Index

A
ansatz functions, 167
assembly, 170
asymptotic near field, 26

B
biaxial parameter, 133
Biaxiality parameter, 292
boundary element method (BEM), 9
Bueckner–singularity, 75

C
cleavage fracture, 17, 14
cohesive elements, 343
cohesive law, 339
cohesive strength, 339
cohesive zone, 338
cohesive zone model, 338
complementary external work, 160
complementary internal work, 160
concept of stress intensity factors, 39
Configurational forces, 263
consistent mass matrix, 189
Constraint-effect, 99
consumption, 80
continuum damage mechanics—CDM, 5
crack arrest, 146
crack arrest concept, 146
crack arrest toughness, 146
Crack at rest, 310
crack closure effect, 120
crack closure integral, 49
    modified, 230, 239
    modified 2D, 231

    modified 3D, 234, 236, 237, 257
    quarter-point elements 2D, 231
    simple, 229
Crack face, 21
Crack front, 21
Crack growth
    J-controlled, 107
crack growth curve, 116
crack growth laws, 116, 131
crack growth rate, 115
crack growth resistance curve, 50, 102, 106
crack growth velocity, 115
crack initiation, 15, 145
Crack opening, 89
crack opening angle, 308
crack opening displacement, 308, 388
crack opening integral, 49
crack opening intensity, 120
Crack opening mode, 22
    mode I, 22, 23, 56, 139, 140
    mode II, 22, 28, 57, 139, 141
    mode III, 22, 30, 57, 137
Crack surface, 21
Crack tip, 21
crack tip elements, 333
    2D hybrid, 214
    3D hybrid, 218, 222
    elastic, 204
    hybrid, 212, 213, 223
    node-distorted, 200
Crack tip field
    anisotrop, 56
    ductile fracture, 108
    electric, 79
    HRR, 93, 95

M. Kuna, *Finite Elements in Fracture Mechanics*, Solid Mechanics and Its
Applications 201, DOI: 10.1007/978-94-007-6680-8,
© Springer Science+Business Media Dordrecht 2013